Buckling and Ultimate Strength of Ship and Ship-like Floating Structures

Buckling and Ultimate Strength of Ship and Ship-like Floating Structures

Tetsuya Yao

Masahiko Fujikubo

AMSTERDAM • BOSTON • HEIDELBERG • LONDON
NEW YORK • OXFORD • PARIS • SAN DIEGO
SAN FRANCISCO • SINGAPORE • SYDNEY • TOKYO

Butterworth-Heinemann is an imprint of Elsevier

ELSEVIER

Butterworth-Heinemann is an imprint of Elsevier
The Boulevard, Langford Lane, Kidlington, Oxford OX5 1GB, United Kingdom
50 Hampshire Street, 5th Floor, Cambridge, MA 02139, United States

Notices
Knowledge and best practice in this field are constantly changing. As new research and experience broaden our
understanding, changes in research methods, professional practices, or medical treatment may become necessary.

Practitioners and researchers must always rely on their own experience and knowledge in evaluating and using
any information, methods, compounds, or experiments described herein. In using such information or methods
they should be mindful of their own safety and the safety of others, including parties for whom they have a
professional responsibility.

To the fullest extent of the law, neither the Publisher nor the authors, contributors, or editors, assume any liability
for any injury and/or damage to persons or property as a matter of products liability, negligence or otherwise, or
from any use or operation of any methods, products, instructions, or ideas contained in the material herein.

Library of Congress Cataloging-in-Publication Data
A catalog record for this book is available from the Library of Congress

British Library Cataloguing-in-Publication Data
A catalogue record for this book is available from the British Library

ISBN: 978-0-12-803849-9

For information on all Butterworth Heinemann publications
visit our website at https://www.elsevier.com/

www.elsevier.com • www.bookaid.org

Publisher: Joe Hayton
Acquisition Editor: Carrie Bolger
Editorial Project Manager: Carrie Bolger
Production Project Manager: Susan Li
Cover Designer: Mark Rogers

Typeset by SPi Global, India

Contents

Preface

It was more than 50 years ago that Timoshenko and Gere published a book titled *Theory of Elastic Stability*. This book fundamentally deals with elastic buckling and postbuckling behavior, and is even at present a good textbook for those who study buckling. On the other hand, problems related to plasticity had been also hot topics in the mid-20th century and many papers were published. However, they were fundamentally based on analytical formulations and were difficult to be applied to practical problems. It was after the 1970s—ie, since the numerical method called finite element method has been developed and performance of computer has been significantly improved—that practical problems related to plasticity have been solved.

The breaking of the structural member in tension was the design criterion for the structure in the 19th century. Then, yielding was introduced as a design criterion, and then in the early 20th century, buckling was also introduced as a criterion. After that, fatigue is considered as one of the design criteria. Now, the ultimate strength is considered as the newest design criterion for ship structures.

On the other hand, although good textbooks have been published relating to "Mathematical Theory of Elasticity," only a few related to buckling/plastic collapse behavior and ultimate strength. From this point of view, we decided to write a new textbook describing in detail what buckling/plastic collapse behavior is and the ultimate strength in ship and ship-like floating structures. As for the external loads acting on ships, description is only given in Chapter 8, where a new integrated motion/collapse analysis system is introduced to simulate the progressive collapse behavior of a ship hull girder in extreme waves. The readers who are interested in the load analyses are recommended to refer to other appropriate textbooks.

This textbook aims at providing better understanding of buckling/plastic collapse behavior of structural members and systems, and derivations of equations are made as concisely as possible. The derivation of some equations is left for readers as exercises, which will be helpful for realizing the essence of the theory.

In Appendix A, a chronological table is given as for research works and events related to buckling. Social events are also indicated in this table. In Appendix B, a brief explanation is made as for the new idealized structural unit method (ISUM) plate element. In Appendix C, structural characteristics and the strength issues to be considered are explained for representative types of ships.

The readers of this textbook are expected to have general knowledge about "strength of materials." In the title, the readers can find "ship and ship-like floating structures." However, the contents up to Chapter 7 are quite general, and are essential not only in the fields of naval architecture and ocean engineering, but also in mechanical engineering and architecture as well as civil engineering. The readers could be graduate students and young engineers who are studying in the field dealing with mainly steel structures.

Tetsuya Yao and Masahiko Fujikubo
May, 2016

Acknowledgments

The contents of this text are mainly from papers published by the authors. The authors are very grateful to the co-authors of the papers, especially to Prof. Yanagihara, who was involved in research works together with the authors. Many Japanese and foreign students are also very much appreciated for their research works under our supervision.

The authors are grateful to Prof. Ueda, who was their supervisor when they started their research carriers. To learn the way of thinking and the attitude for research activity has been very helpful for the authors to carry out research works.

At the end, the authors greatly thank their wives, Mikiko Yao and Keiko Fujikubo, for their patience to let the authors concentrate on research works and for their help in daily life for a long time.

INTRODUCTION

1.1 BUCKLING/PLASTIC COLLAPSE OF SHIP AND SHIP-LIKE FLOATING STRUCTURES

A hull of ship and ship-like floating structures is a box girder structure composed of plates and stiffeners as indicated in Fig. 1.1A and B. The main loads acting on a hull girder are distributed lateral loads such as hull weight and cargo weight as well as buoyancy force and wave force; see Fig. 1.1C and D. Inertia forces also act on a navigating ship in waves. Such distributed loads produce bending moment, torsional moment, and shear force as well as axial force in the cross-section, which are shown in Fig. 1.2.

The distributed loads in the vertical direction may produce bending deformation in a hull girder, as illustrated in Fig. 1.3. Under sagging conditions, the deck plate is subjected to thrust (or in-plane compression) and the bottom plate to tension. Due to this in-plane compression, the deck plate may undergo buckling when extreme bending moment acts. On the other hand, in hogging, the deck plate is in tension and the bottom plate in thrust, and the bottom plate may undergo buckling.

Here, buckling is a phenomenon that a structural member such as a plate, a stiffened plate, a stiffener, a column, etc., which are under thrust load deflect in an out-of-plane direction when the load reaches to a certain critical value. After the buckling, deflection begins to increase in addition to the in-plane (or axial) displacement, which causes a reduction in the in-plane (or axial) stiffness. This is because a deflected structural member shows less resistance against in-plane (or axial) compressive force compared to a flat (or straight) structural member.

One of the structural problems caused by buckling is the reduction in in-plane stiffness mentioned above. Another problem is the earlier occurrence of yielding. This is because bending stress is produced by deflection in addition to in-plane (or axial) stress. The occurrence of yielding further reduces the stiffness.

When a certain structural member undergoes buckling, its load-carrying capacity decreases. This causes redistribution of internal forces in unbuckled structural members and increases the internal forces in these structural members, which may lead to the progressive occurrence of buckling failure of these structural members. If the load increases further, progressive buckling may results in the collapse of a whole structure. This was the reason why occurrence of buckling was not allowed in any members in ship structures in old classification societies' rules.

In a strict sense, buckling is a bifurcation phenomenon that stable deformation changes from in-plane (or axial) deformation to in-plane (or axial) plus out-of-plane deformations. Therefore, to

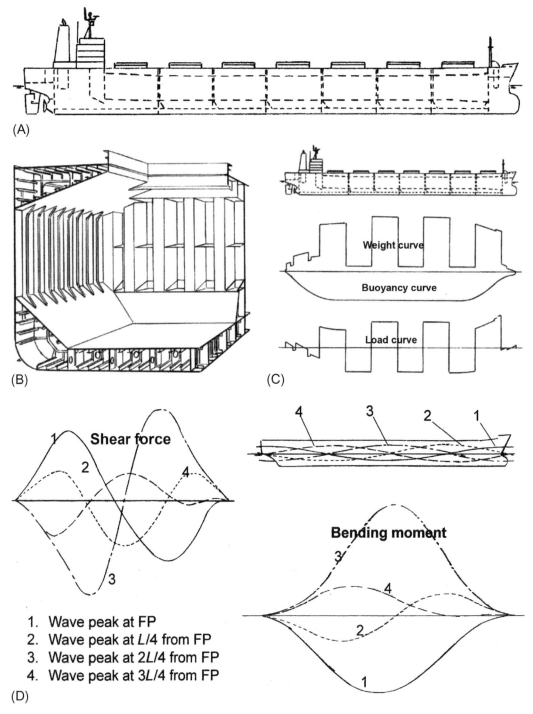

(A)

(B)

(C)

Weight curve

Buoyancy curve

Load curve

Shear force

1
2
4
3

4   3   2   1

Bending moment

3
4
2
1

1. Wave peak at FP
2. Wave peak at L/4 from FP
3. Wave peak at 2L/4 from FP
4. Wave peak at 3L/4 from FP

(D)

FIG. 1.1

Ship's hull girder and loads acting on it. (A) Ship's hull girder (Cape size bulk carrier). (B) Cross-section of hull girder. (C) Distributed loads. (D) Wave loads.

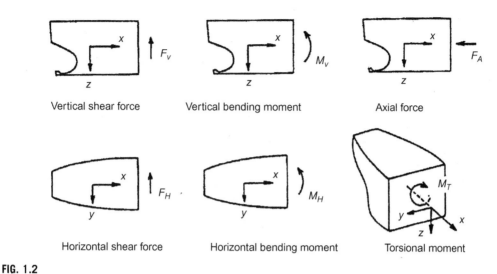

Vertical shear force	Vertical bending moment	Axial force

Horizontal shear force	Horizontal bending moment	Torsional moment

FIG. 1.2

Sectional forces in cross-section.

FIG. 1.3

Hull girder under longitudinal bending. (A) Hogging. (B) Sagging.

have buckling in a strict sense, the structural member has to be completely flat (or straight) before it is loaded; that is, it has to be completely free from initial distortion/deflection.

However, a ship structure is constructed connecting members by welding, and the structural members are accompanied by initial imperfections such as initial distortion/deflection and welding residual stress. This implies that buckling in a strict sense does not occur in actual structures, since they are accompanied by initial distortion or initial deflection.

Here, a straight column member subjected to axial thrust is considered. In this case, deflection increases with no increase in the applied axial load for a while beyond buckling. However, the capacity again starts to increase and a column can sustain further load if its behavior is perfectly elastic. This is called Elastica [1]; see Fig. 1.4. On the other hand, an actual column member undergoes yielding by bending after buckling has occurred, and soon its capacity starts to decrease with an increase in the deflection. In this sense, buckling strength of a column member is the maximum load-carrying capacity and can be regarded as the ultimate strength. Therefore, the occurrence of buckling should not be allowed in column members.

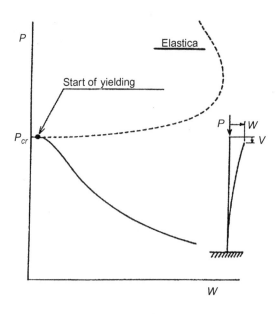

FIG. 1.4

Buckling behavior of column under axial compression.

In the case of a simply supported plate subjected to uniaxial thrust, buckling/plastic collapse behavior is indicated in Fig. 1.5A and B in terms of average stress-central deflection and average stress-average strain relationships, respectively. Behavior of both thin and thick plates is indicated. In the case of a thin plate, lateral deflection starts to develop beyond the buckling point, **A**, when a plate is flat. When a plate is accompanied by small initial deflection, lateral deflection gradually increases from the beginning of loading, although the increasing rate is low. Above the buckling load, deflection starts to increase rapidly as in the case of no initial deflection. Such a phenomenon is also called buckling in a broad sense.

Beyond the buckling, capacity further increases with the increase in buckling deflection, but in-plane stiffness (slope of average stress-average strain curve) decreases to around 0.4 through 0.5 times the Young's modulus, depending on the aspect ratio of the plate. For a while, the in-plane stiffness is almost constant, but again starts to decrease gradually after yielding has started. Finally, the stiffness becomes zero and the ultimate strength is attained. Then, the capacity starts to decrease beyond the ultimate strength.

In the case of a thick plate, yielding starts to take place before the plate undergoes buckling. In this case, the maximum load-carrying capacity—that is the ultimate strength—is nearly equal to the fully plastic strength, and this capacity is kept until buckling takes place if the material does not show remarkable strain hardening. After the buckling has occurred, capacity starts to decrease with the increase in the buckling deflection.

Welding residual stress also affects the buckling strength as well as the ultimate strength. If compressive residual stress exists at the location where buckling deflection develops, buckling strength

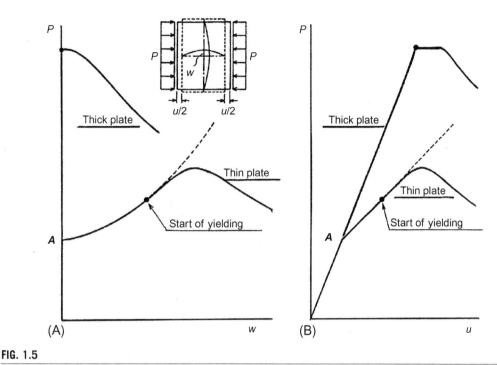

FIG. 1.5

Buckling/plastic collapse behavior of plate under uniaxial thrust. (A) Average stress-central deflection relationships. (B) Average stress-average strain relationships.

is reduced by welding residual stress. On the contrary, welding residual stress increases the buckling strength if tensile residual stress exists in the region where buckling deflection develops.

Buckling/plastic collapse behavior of a stiffened plate is then considered, which is a fundamental structural unit composing a ship's hull girder. This is schematically shown in Fig. 1.6. This is the case of a stiffened plate subjected to thrust (or in-plane compression) in the direction of stiffeners. In actual structure, size of stiffeners are so determined that local panels partitioned by stiffeners buckle before overall buckling of a whole stiffened plate takes place. The figure indicates representative average stress-average strain relationships for such stiffened plates.

When the slenderness ratio of the local panel is high—that is when the local panel is thin—the average stress-average strain relationship follows Curve A. In this case, elastic panel buckling takes place locally at Point 1, and the stiffness decreases hereafter because large deflection in the local panel rapidly develops. At Point 3, yielding starts to take place, and at Point 2, the overall buckling occurs as a stiffened panel. Point 2 stands for the ultimate strength.

When the slenderness ratio of the panel is lower, the average stress-average strain relationship is represented by Curve B. In this case, initial yielding takes place at Point 3, and the ultimate strength is attained at Point 4 by overall buckling as a stiffened plate.

When the panel and the stiffener have a much lower slenderness ratio, the average stress-average strain relationship follows Curve C. In this case, yielding starts at Point 5, and soon the general yielding

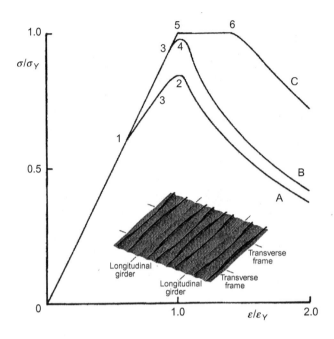

FIG. 1.6

Average stress-average strain relationship of stiffened plate under thrust.

takes place all over the stiffened plates. However, no deflection is produced at this moment. At Point 6, either the panel or the stiffener undergoes buckling, and the capacity decreases hereafter with the increase in deflection in the panel or in the stiffener. After this, plastic deformation concentrates along one line perpendicular to the loading direction and elastic unloading occurs in the rest of the stiffened plate.

In general, ship structures are designed so that buckling collapse does not occur in the primary structural members when a ship is subjected to loads below the design load. However, there could be a mis-loading/unloading of cargoes, or a ship could fail to escape from a storm. In such a case, a ship's hull is exposed to an extreme load, which is above the design load. Even if the working load is below the design load, this could be an extreme load when the ship structures suffer from thickness reduction due to corrosion. In these cases, a ship's hull may break into two as in the case of Prestige, which broke into two in 2002; see Fig. 1.7.

For the safety assessment of ship and ship-like floating structures, it is very important to know the extreme loads acting on them and what shall happen when the structures are exposed to extreme loads as mentioned above. For this, it is necessary to understand wave loads as well as buckling/plastic collapse behavior of structural members and systems of ship and ship-like floating structures including the capacity of members beyond their ultimate strength. In the present textbook, however, attention is focused on the latter strength issue.

As for elastic buckling and postbuckling behavior of columns and plates in general, a comprehensive textbook was written by Timoshenko and Gere [1] more than 50 years ago. Since that time, there has been significant development in the method of analysis of nonlinear behavior and the ultimate strength

FIG. 1.7

Prestige broken in two.

of structures including both material and geometrical nonlinearities, and plenty of new knowledge has been obtained in relation to the above issues.

On the basis of such new findings, in this textbook, fundamentals are explained as for buckling/plastic collapse behavior and the ultimate strength of plates and stiffened plates in general, and also for those of hull girders of ship and ship-like floating structures including the assessment of existing rule formulas to evaluate the ultimate strength.

1.2 **SHORT HISTORICAL REVIEW ON RESEARCH WORKS**

To evaluate the ultimate strength of members and systems of ship and ship-like floating structures, it is necessary to perform progressive collapse analysis, taking into account of the influences of yielding and buckling. However, in past times, it was not possible to perform such analysis. At that time, tensile strength of the material was considered as a parameter which controls the capacity. In other words, the breaking strength of the material in tension was the criterion to prevent structural failure. For example, when Sir Isambard Kingdom Brunel designed a huge iron ship, "Great Eastern," in the middle of the 19th century, he determined the thickness of deck and bottom plating on the condition that they do not break in tension under the extreme loads [2]. He applied *Beam Theory* to evaluate the working stress in the stead of performing progressive collapse analysis to evaluate the ultimate strength, which was not possible to perform at that time.

It was Bryan [3] who first considered buckling as a criterion to determine the thickness of plating in ship structures. He solved the buckling problem of panels theoretically, and derived formulas to evaluate the buckling load of a rectangular plate subjected to thrust.

The first attempt to evaluate the ultimate strength of a ship structure was made by Caldwell [4]. He applied *Rigid Plastic Mechanism Analysis* to evaluate the ultimate hull girder strength under longitudinal bending. He modeled a cross-section of a ship's hull composed of stiffened plates as that of a box girder composed of plates with equivalent thicknesses. Then, fully plastic bending moment was calculated, which was considered as the ultimate hull girder strength. The influence of buckling was taken into account by reducing the yield stress of the material of plating which locates in the compression side of longitudinal bending.

In 1956, Turner et al. presented a paper entitled "Stiffness and deflection analysis of complex structures" in the *Journal of Aeronautical Science* [5], which was a debut paper of the finite element method (FEM). Soon after this, the FEM was introduced into the analysis of ship structures modeling a plated structure as a frame structure. In the 1960s, it had become possible to model ship structures with plate elements. In 1971, MSC Nastran developed by NASA was released as the first commercial code for practical use of the FEM in structural analysis.

At the beginning, the FEM was applied only to the analyses of elastic behavior of structural members and systems. Then, from the early 1970s, it became possible to perform collapse analysis applying the FEM. Papers by Bergan [6] and Ohtsubo [7] are examples of pioneer papers at that time. Collapse analysis usually employs incremental calculation assuming linear behavior within a small increment. This is fundamentally different from a linear elastic analysis.

In early times, collapse analyses could be performed only on structural members. However, with the developments of computer performance and computational environments, it became possible to apply the FEM to the collapse analysis of structural systems.

A similar method was also proposed, called the finite strip method (FSM) [8]. This method was more analytical than the FEM and considered larger structural unit as an element. The FSM could be applied to elastoplastic large deflection analysis [9]. However, the applicability was limited when compared with the FEM.

On the other hand, some simplified methods were proposed for some special problems. For example, to perform progressive collapse analysis on a ship's hull girder in longitudinal bending, Smith proposed a simple but efficient method [10].

An alternative method that can be applied to simulate collapse behavior of various structures may be the idealized structural unit method (ISUM), which was proposed by Ueda and Rashed [11] more than 40 years ago. This method uses a larger structural unit as an element and the yielding condition is considered in terms of sectional forces, although its formulation is in a framework of the FEM.

Applying these analysis methods as tools together with experimental results, it has now become possible to understand buckling/plastic collapse behavior and the ultimate strength of structural members and systems in ship and ship-like floating structures.

1.3 **CONTENTS OF THE TEXT**

In the present text, the attention is focused firstly on the buckling/plastic collapse behavior and the ultimate strength of plates and stiffened plates as well as girders in ship and ship-like floating structures, and then those of ship and ship-like floating structures as systems mainly focusing on hull girder. Buckling/plastic collapse behavior and strength of double bottom, transverse bulkhead, triangular corner bracket, and hatch cover are also briefly explained introducing literatures. For the structural members, in-plane uniaxial and/or biaxial thrust loads in combination with lateral pressure load are mainly considered, but bending and shearing loads are also considered in some cases.

In Chapter 2, it is explained how and what initial imperfections are produced by welding such as welding residual stress and initial deflection in plating and stiffeners on the basis of the measured results in ship structures. Simple formulas are shown to estimate welding residual stress and initial deflection in panels and stiffeners.

In Chapter 3, at the beginning, fundamentals of buckling/plastic collapse behavior of plates and stiffened plates are explained briefly, showing representative deformations. The fundamental theories and methods to simulate such collapse behavior and to evaluate the ultimate strength are then explained. That is, fundamental ideas and theories for buckling strength analysis, elastic large deflection analysis, and elastoplastic large deflection analysis are briefly explained.

In Chapter 4, buckling strength, postbuckling behavior, secondary buckling, and the ultimate strength of rectangular plates are explained. Applied load is fundamentally uniaxial thrust. Most of them are briefly explained in Section 4.1, and then in detail in the following sections. Buckling/plastic collapse behavior under extreme cyclic loading is also explained at the end of this chapter.

In Chapter 5, buckling/plastic collapse behavior and the ultimate strength of rectangular plates subjected to combined loads are explained. Combined loads are longitudinal/transverse thrust and lateral pressure. Combined thrust and bending and combined thrust and shear are also considered.

Chapter 6 deals with stiffened plates subjected to longitudinal thrust, transverse thrust, lateral pressure load, and their combinations. The buckling/plastic collapse behavior and the ultimate strength are explained on the basis of the results of nonlinear FEM. Then, a simple method is introduced to evaluate the ultimate strength under the above-mentioned single and combined loads.

In Chapter 7, plate girders subjected to pure bending, shear load, and combined bending and shear loads are considered. Firstly, girders with web panel free from perforation and stiffeners are considered, and then those with perforated and stiffened web panel. The latter is girder and floor in double bottom structure. After explaining the buckling/plastic collapse behavior of plate girders, simple methods are introduced that enable us to evaluate their ultimate strength under bending, shearing, and combined bending and shear loads.

In Chapter 8, a short historical review is firstly made regarding the research works on the ultimate hull girder strength. Then, as a simple but efficient method of analysis, Smith's method is explained and the results of analysis applying Smith's method are introduced. After this, showing the calculated results applying explicit and implicit nonlinear FEM, progressive collapse behavior of the hull girder in longitudinal bending is explained. Then, results of collapse tests on large-scale hull girder models are introduced. At the end, as a new method, a total system and its applications are introduced combining load/pressure analysis and progressive collapse analysis to simulate actual collapse behavior of hull girders in extreme sea conditions.

Chapter 9 deals with assessment of rule formulas. Formulas to evaluate the ultimate strength of plates and stiffened plates in CSR-B and panel ultimate limit state (PULS) are introduced. The formulas are applied to evaluate the ultimate strength of stiffened plates in Chapter 6, and the calculated results are compared with those by the FEM. The average stress-average strain relationships specified by H-CSR are also assessed through comparison with FEM results.

Chapter 10 deals with the collapse behavior and the ultimate strength of structural members and systems in ship and ship-like floating structures. In Section 10.2, those of double bottom structures of bulk carriers are explained on the basis of the calculated results applying the nonlinear FEM and the ISUM.

Section 10.3 deals with the progressive collapse behavior and the ultimate strength of watertight transverse bulkheads of bulk carriers under the flooding condition on the basis of the calculated results by the nonlinear FEM analysis. A simple method is introduced to evaluate the collapse load with high accuracy considering the influence of local buckling of the flange plate. The influences of shedder and gusset plates are also considered.

Section 10.4 deals with the triangular corner bracket with an arbitrary shape. Firstly, collapse behavior and the ultimate strength of triangular corner brackets are explained on the basis of the results of collapse tests and nonlinear FEM analysis. Then, a simple method is introduced to estimate the optimum thickness of the triangular corner bracket from the condition that the beam with brackets at its both ends and the bracket collapse at the same time.

In Section 10.5, progressive collapse behavior and the ultimate strength of hatch covers of bulk carriers are explained on the basis of the results of nonlinear FEM analysis. Hatch covers of a folding type used in Handy size bulk carrier as well as of side-sliding type used in Panamax size and Cape size bulk carriers are considered. For each type, two hatch covers are considered which are designed by the old ICLL (International Convention on Load Lines) rule and the new IACS (International Association of Classification Societies) rule. The influence of corrosion margin is also considered. A simple method is introduced also for this case to evaluate the collapse strength of the hatch cover under uniformly distributed lateral load.

In Appendix A, a chronological table is given for the research works and events in buckling/plastic collapse of structures, and in Appendix B, fundamentals in ISUM. In Appendix C, structural characteristics of representative ship and ship-like floating structures are indicated.

EXERCISES

1.1 What was the strength criterion in the 1800s?
1.2 What are the sectional forces produced in the cross-section of a ship's hull girder?

REFERENCES

[1] Timoshenko S, Gere J. Theory of elastic stability. McGraw-Hill Kogakusha, Ltd; 1961.
[2] Rutherford S, Caldwell J. Ultimate longitudinal strength of ships: a case study. Trans SNAME 1990;98: 441–471.
[3] Bryan G. On the stability of a plane plate under thrust in its own plane with application to the buckling of the side of a ship. Proc Lond Math Soc 1881;22:54–67.
[4] Caldwell J. Ultimate longitudinal strength. Trans RINA 1965;107:411–430.
[5] Turner M, Clough R, Martin H, Topp L. Stiffness and deflection analysis of complex structures. J Aeronaut Sci 1956;23-9:805–823.
[6] Bergan G. Non-linear analysis of plates considering geometric and material effects. Structural Engineering Lab., Report No. UCSESM 71-7; 1971.
[7] Ohtsubo H. A generalized method of analysis of large-deformed elastic-plastic plate problems. Ultimate strength of compressive plates with initial deflection. J Soc Naval Arch Jpn 1971;130:173–182 [in Japanese].

[8] Cheung Y. Finite strip method analysis of elastic slabs. J Eng Mech ASCE 1968;EM6:1365–1378.

[9] Ueda Y, Matsuishi M, Yamauchi Y, Tanaka M. Non-linear analysis of plates using the finite strip method. J Kansai Soc Naval Arch Jpn 1974;154:83–92 [in Japanese].

[10] Smith C. Influence of local compressive failure on ultimate longitudinal strength of a ship's hull. In: Proc int symp on practical design in shipbuilding, Tokyo, Japan; 1977. p. 73–79.

[11] Ueda Y, Rashed S. An ultimate transverse strength analysis of ship structure. J Soc Naval Arch Jpn 1974;136:309–324 [in Japanese].

INITIAL IMPERFECTIONS DUE TO WELDING

2.1 INITIAL IMPERFECTIONS DUE TO WELDING

Ship structure is a box girder composed of stiffened plates such as deck plating, side shell plating, and bottom plating. Ship-like floating structure also has the same structure as ship structure. The deck plating is stiffened by longitudinal stiffeners, girders, deck beams, and frames, and can be regarded as an orthogonally stiffened panel. Side shell plating and bottom plating also have a similar structure.

The longitudinal and transverse stiffeners are fitted on the plate by fillet welding, which produces initial deflection and welding residual stresses in the plate. Such initial imperfections are also produced by butt welding of the plates. Fig. 2.1 schematically shows initial deflection and welding residual stress produced in plating by fillet weld and butt weld.

Buckling may take place in plates and stiffened plates as indicated in Fig. 2.1 when excess external loads act on ship or ship-like floating structures. Initial deflection and welding residual stress often reduce the buckling/ultimate strength and stiffness of plate and stiffened plate elements. From this point of view, it is important to know what the magnitude and shape of initial deflection and welding residual stress are. In this chapter, general characteristics of welding residual stress and initial deflection are briefly explained on the basis of the measured results on deck and bottom plating as examples.

2.2 WELDING RESIDUAL STRESS
2.2.1 WELDING RESIDUAL STRESS IN PANELS

When welding is performed, the weld metal melts together with mother plates and then solidifies again. During the solidification process, the solidified part shrinks, which is constrained by the surrounding unmelted part. Because of this, tensile stress is produced in the solidified part and compressive stress in the neighboring region so that equilibrium condition is satisfied as a whole. In general, residual stress in the welding direction is in tension near the weld line, whereas it is in compression in the area adjacent to the tensile residual stress field.

The dashed line in Fig. 2.2A shows the welding residual stress due to fillet welding of stiffeners to plating, and that in Fig. 2.2B is due to butt welding between two plates. Such welding residual stress can be approximated by solid lines of a rectangular shape in Fig. 2.2A and B.

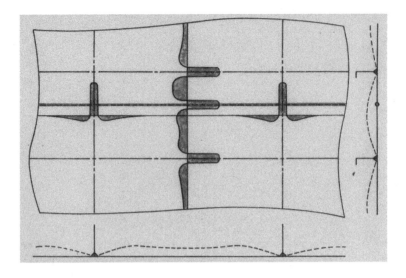

FIG. 2.1

Initial imperfections in stiffened plating produced by welding.

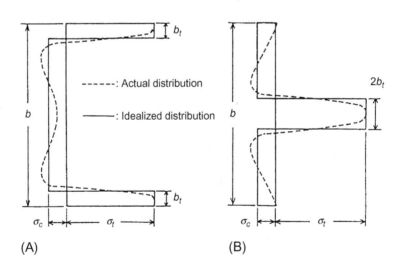

FIG. 2.2

Welding residual stress in plating. (A) Fillet weld. (B) Butt weld.

Welding residual stress is self-equilibrating, since it is produced with no action of external loads. For the cases shown in Fig. 2.2, the self-equilibrium condition of welding residual stress gives the following equation:

$$2b_t\sigma_t = (b - 2b_t)\sigma_c \tag{2.1}$$

where b_t represents the breadth of the area where tensile residual stress is produced; see Fig. 2.2. This breadth is determined from the welding condition—that is, weld heat input, ΔQ_{max} (Cal/cm), and is given for fillet weld as [1]:

$$b_t = t_w/2 + 0.26\Delta Q_{max}/(t_w + 2t_p) \quad \text{(in mm)} \tag{2.2}$$

and for butt weld as [2]:

$$b_t = 0.13\Delta Q_{max}/t_p \quad \text{(in mm)} \tag{2.3}$$

where t_p (mm) and t_w (mm) represent thicknesses of panel and stiffener web, respectively.

When multipass welding is performed, ΔQ_{max} is taken as the maximum heat input during one pass among multipasses. When CO_2 welding is performed, ΔQ_{max} can be expressed by empirical formulas as [3]:

$$\Delta Q_{max} = 78.7 \times \ell^2 \tag{2.4}$$

where ℓ represents the leg length of fillet weld, and can be estimated as:

$$\ell = \begin{cases} 0.7 \times t_w \text{ (mm)} & \text{(when } 0.7 \times t_w < 7.0 \text{ mm)} \\ 7.0 \text{ (mm)} & \text{(when } 0.7 \times t_w \geq 7.0 \text{ mm)} \end{cases} \tag{2.5}$$

On the other hand, the magnitude of tensile residual stress depends on the material and is expressed as follows [4]:

SM41: $\sigma_t = \sigma_Y$
SM50: $\sigma_t = \sigma_Y$
SM53: $\sigma_t = \sigma_Y$
SM58: $\sigma_t = 0.8\sigma_Y$
HT80: $\sigma_t = 0.55\sigma_Y$

where σ_Y is the yield stress of the material. It is known that the tensile residual stress is equal to the yield stress of the material in the case of ordinary steel. However, for the steel with higher yield stress, the tensile residual stress, σ_t, does not reach the yield stress of the material. This is because of the volume expansion associated with transformation of crystalline structure at melted part during its cooling process.

When σ_t and b_t are known, the compressive residual stress, σ_c, is determined with self-equilibrium condition expressed by Eq. (2.1) as:

$$\sigma_c = \frac{2b_t}{b - 2b_t}\sigma_t \tag{2.6}$$

2.2.2 WELDING RESIDUAL STRESS IN STIFFENED PANELS

When a stiffened plate is considered, the welding residual stress can be approximated as illustrated in Fig. 2.3. In this case, the compressive residual stress in panel and stiffener can be represented as follows:

$$\sigma_{cp} = \frac{2b_t}{b - 2b_t}\sigma_{tp}, \quad \sigma_{cs} = \frac{b_{ts}}{b - b_{ts}}\sigma_{ts} \tag{2.7}$$

or

$$\sigma_{cp} = \sigma_{cs} = \frac{2b_t t_p \sigma_{tp} + b_{ts} t_w \sigma_{ts}}{(b - 2b_t)t_p + A_s - b_{ts}t_w} \tag{2.8}$$

where

$$b_{ts} = \frac{t_w}{t_p} \times 0.26\Delta Q_{max} \tag{2.9}$$

where A_s is a cross-sectional area of the stiffener, and σ_{tp} and σ_{ts} are the tensile residual stress in the panel and the stiffener, respectively. Eq. (2.7) is the case when self-equilibrium condition of residual stress is independently considered in the panel and in the stiffener. On the other hand, to derive Eq. (2.8), self-equilibrium condition is considered as a whole assuming that compressive residual stress is the same in the panel and the stiffener.

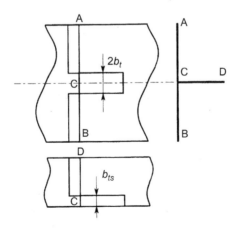

FIG. 2.3

Welding residual stress in stiffened plating.

2.3 INITIAL DISTORTION/DEFLECTION

2.3.1 MECHANISM PRODUCING INITIAL DISTORTION/DEFLECTION AND ITS MEASUREMENT

In Fig. 2.1, initial deflection only in the local panel was indicated. In general, in a continuous stiffened plate, initial deflection is produced in both panels and stiffeners due to fillet welding. Firstly, a stiffener deflects as illustrated in Fig. 2.4A due to thermal shrinkage of the weld metal. At the same time, a panel also deflects; this deflection is divided into three modes as indicated in Fig. 2.4B.

Mode I is due to the angular distortion along four sides of the local panel. Modes II and III are due to the thermal shrinkage of the weld metal in the transverse directions along the transverse frame and in the longitudinal direction along the longitudinal stiffeners, respectively. As a result, initial deflection in a local panel is represented as:

$$w_0 = w_{0P} + w_{0T} + w_{0L} \tag{2.10}$$

where

$$w_{0P} = \sum_{i=1}^{m_p} \sum_{j=1}^{n_p} A_{ij} \sin \frac{i\pi x}{a} \sin \frac{j\pi y}{b} \tag{2.11}$$

$$w_{0T} = \sum_{k=1}^{n_s} \left(\frac{x}{a}B + \frac{a-x}{a}C \right) \sin \frac{k\pi y}{b} \tag{2.12}$$

$$w_{0L} = \sum_{\ell=1}^{m_s} \left(\frac{y}{b}D + \frac{b-y}{b}E \right) \sin \frac{\ell\pi x}{a} \tag{2.13}$$

Deflection components, w_{0P}, w_{0T}, and w_{0L} correspond to Modes I, II, and III in Fig. 2.4B. In a stiffener, initial deflections in a vertical direction and a horizontal direction are produced. The former is represented by Eq. (2.13) substituting $y = y_i$ and the latter as follows:

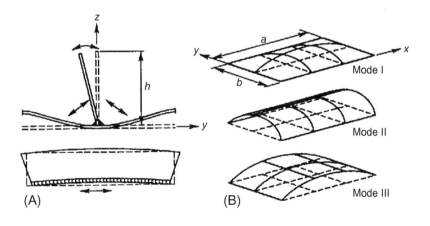

FIG. 2.4

Initial deflection in stiffened plate [4]. (A) Stiffener. (B) Local panel.

$$v_{0L} = \frac{z}{h} \sum_{m=1}^{m_h} F_m \sin \frac{m \pi x}{a} \qquad (2.14)$$

A series of measurements of initial deflection was conducted on inner bottom plates of an existing Handy-size bulk carrier [4,5]; see Fig. 2.5. The local panel is partitioned by longitudinal stiffeners and floors.

As indicated in Fig. 2.6, initial deflection in local panel, $w_{0P} + w_{0T}$, was measured at five points toward its width direction and every 100 mm toward the length direction. Initial deflections of longitudinal stiffeners in vertical and horizontal directions, w_{0L} and v_{0L}, were also measured at five points within each span. The measurements were carried out on 45 panels and 30 spans. A typical measured mode is shown in Fig. 2.7.

It is seen that the initial deflection in panels is in a so-called thin-horse mode, which looks like a breast of thin-horse. Longitudinal stiffeners are also deflected in vertical and horizontal directions.

2.3.2 SHAPE OF INITIAL DEFLECTION IN PANELS

As is known from Fig. 2.7, the shape of initial deflection is complex. However, in general, initial deflection in actual plating can be classified into two types. One is the case of as-weld condition, which is illustrated in Fig. 2.8A. Another is shown in Fig. 2.8B. In this case, heat treatment was carried out to remove excess initial deflection after welding. If Modes II and III in Fig. 2.4 are removed, initial deflection in a panel is expressed by Eq. (2.11), which can be rewritten as follows:

$$w_0 = \sum_{i=1}^{m} \sum_{j=1}^{n} A_{0ij} \sin \frac{i \pi x}{a} \sin \frac{j \pi y}{b} \qquad (2.15)$$

When compressive load acts in x-direction, the deflection components in y-direction decrease except the first term with one half-wave as the applied load increases as far as the secondary buckling does not occur. In this case, only the first term ($n = 1$) may play a dominant role as for the deflection in y-direction, and a simpler form of initial deflection can be used for analysis as follows:

$$w_0 = \sum_{i=1}^{m} A_{0i} \sin \frac{i \pi x}{a} \sin \frac{\pi y}{b} \qquad (2.16)$$

On the other hand, when load acts in y-direction, the dominant deflection component is one half-wave both in x- and y-directions. In this case, initial deflection of one half-wave mode ($m = n = 1$) dominates the behavior of the compressed plate. However, in all cases, every deflection components play important roles when the compressive stress exceeds the buckling stress. This will be explained later in Chapters 4 and 5.

On the basis of the measured initial deflection at an upper deck of a bulk carrier and a strength deck of a pure car carrier, deflection components are calculated for the deflection along the center line of the local plate applying the *Method of Least Square* with Eq. (2.16). Calculated results are summarized in Table 2.1 [6,7] and Table 2.2 [8]. Typical two patterns of measured initial deflection are shown in Fig. 2.9A and B, which correspond to those in Fig. 2.8A and B, respectively. The as-weld initial

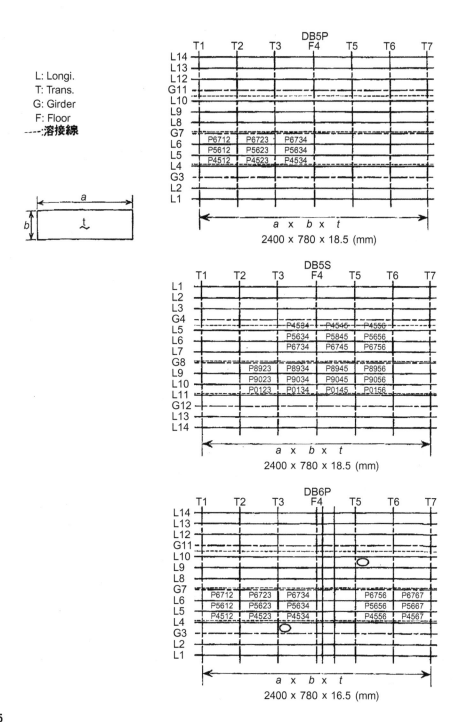

L: Longi.
T: Trans.
G: Girder
F: Floor
----:溶接線

FIG. 2.5

Inner bottom plating of bulk carrier on which initial deflection was measured.

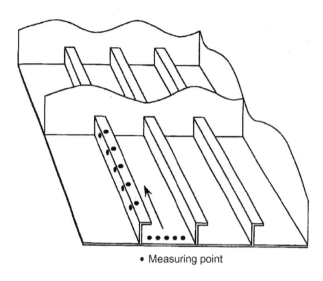

• Measuring point

FIG. 2.6

Location where initial deflection was measured on local panel and longitudinal stiffeners.

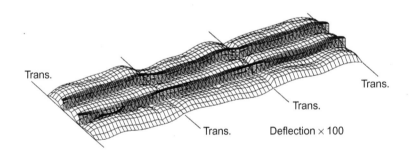

FIG. 2.7

Example of measured initial deflection.

deflection is of a thin-horse mode, whereas initial deflection after heat treatment is of a complex mode which is not possible to observe a specific characteristic.

Initial deflection along the centerline of local panel measured in Ueda and Yao [6,7] are illustrated in Fig. 2.10.

Typical deflection components of initial deflection measured on six panels each from a bulk carrier and a pure car carrier are plotted against the wave number in Fig. 2.11A and B, respectively. The deflection components in Fig. 2.11A corresponds to the initial deflection shown in Fig. 2.8A or 2.9A, which is of a thin-horse mode. These components have characteristics as:

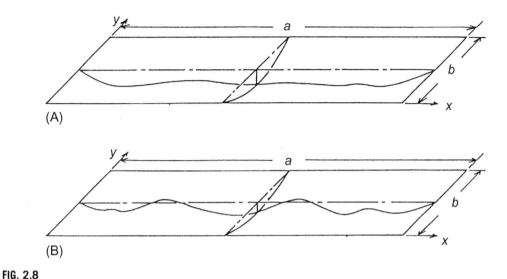

FIG. 2.8

Typical modes of initial deflection in plating. (A) As-weld mode. (B) After heat treatment.

(1) Components of odd half-waves are larger than those of even half-waves.

(2) One half-wave component is the largest, and then three, five, seven half-wave components, and so on.

On the other hand, deflection components in Fig. 2.11B shows almost no regularity. This may be because deflection mode changes according to the method of heat treatment and heat input to remove excess initial deflection. The only regularity is that the magnitude of higher order component decreases with the increase of wave number.

In Ueda and Yao [6,7], the idealized mode of initial deflection is proposed, which is composed of quarters of a sine wave at both ends and flat part at the middle part of the plate. This mode is composed of only odd terms, of which magnitudes are indicated by solid circles in Fig. 2.10A and B, and in Table 2.3. It may be said that the components of the idealized initial deflection show good correlations with the measured results for the initial deflection in as-weld condition.

An alternative mode of idealized initial deflection which includes even deflection components was also proposed [8]. Although there is no strict background, magnitude of the $2n$th component is defined as 1/5 of that of the $(2n - 1)$th component, and the sign is alternatively changed. This comes from the observation that even components are smaller than odd components. Magnitudes of even half-wave components are given in Table 2.3, and the deflection modes in Fig. 2.12.

2.3.3 MAGNITUDE OF INITIAL DEFLECTION IN PANELS

For all the panels in Tables 2.1 and 2.2, the probabilistic distributions of the maximum magnitude of initial deflection are shown in Fig. 2.13A-C.

Table 2.1 Measured Components of Initial Deflection (1/2) [6,7]

No.	$a \times b \times t$ (in mm)	w_{0max}	A_{01}	A_{02}	A_{03}	A_{04}	A_{05}	A_{06}	A_{07}	A_{08}	A_{09}	A_{010}	A_{011}
Bulk carrier (in mm)													
1	2400 × 800 × 34.5	−1.320	−1.346	−0.090	−0.224	−0.036	−0.087	−0.082	0.102	−0.036	−0.014	0.026	−0.028
2	2400 × 800 × 34.5	−0.770	−0.746	−0.026	−0.124	0.001	−0.066	−0.023	0.112	0.016	0.016	0.065	−0.008
3	2400 × 800 × 34.5	−1.550	−1.235	−0.339	−0.344	0.129	−0.061	−0.015	−0.031	−0.031	−0.033	0.075	−0.018
4	2400 × 800 × 34.5	−1.260	−1.247	0.178	−0.090	−0.002	−0.004	0.034	0.023	0.018	0.029	0.009	0.023
5	2400 × 800 × 34.5	−1.040	−0.968	−0.176	−0.133	−0.052	0.043	−0.020	0.030	0.007	−0.000	−0.000	0.030
6	2400 × 800 × 34.5	−1.090	−1.136	−0.107	−0.102	−0.007	0.012	−0.014	0.015	−0.011	−0.039	−0.004	−0.006
7	2400 × 800 × 34.5	−0.590	−0.563	−0.021	−0.179	−0.044	−0.014	−0.039	−0.009	0.011	0.020	−0.043	0.042
8	2400 × 800 × 34.5	−1.330	−1.101	−0.481	−0.102	−0.032	−0.020	−0.043	0.032	0.025	0.036	−0.014	0.019
9	2400 × 800 × 34.5	−1.680	−1.239	0.483	−0.268	0.050	0.029	−0.029	−0.046	0.110	0.056	0.004	−0.008
10	2400 × 800 × 34.5	−1.420	−1.339	−0.094	−0.090	−0.024	−0.176	−0.040	−0.049	−0.027	−0.050	0.009	−0.031
11	2400 × 800 × 34.5	−1.210	−1.158	−0.142	−0.091	−0.059	−0.057	−0.062	−0.055	0.038	−0.015	−0.022	−0.035
12	2400 × 800 × 34.5	−1.390	−1.481	0.024	−0.283	−0.028	−0.062	−0.013	−0.047	0.003	−0.032	−0.019	−0.022
13	2800 × 800 × 15	−2.390	−2.668	0.286	−0.686	0.097	−0.534	−0.008	−0.234	0.058	−0.105	0.044	−0.073
14	2800 × 800 × 15	−4.150	−4.535	−0.183	−1.419	0.387	−0.656	0.138	−0.341	0.043	−0.155	0.092	−0.135
15	2800 × 800 × 15	−3.740	−3.955	−0.218	−1.204	−0.095	−0.359	−0.025	0.023	−0.080	0.137	0.016	0.015
16	2800 × 800 × 19	−3.180	−3.339	−0.188	−0.297	0.059	−0.194	0.128	−0.143	0.030	−0.070	0.014	−0.042
17	2800 × 800 × 19	−3.680	−4.187	0.441	−0.867	0.150	−0.388	−0.110	−0.181	−0.045	−0.035	−0.037	−0.074
18	2800 × 800 × 19	−3.810	−4.161	0.434	−0.921	−0.013	−0.559	−0.015	−0.243	0.004	−0.083	−0.022	−0.093
19	2800 × 800 × 19	−3.010	−3.092	0.336	−0.362	0.056	0.017	0.076	−0.010	0.058	0.060	0.030	−0.013
20	2800 × 800 × 19	−3.360	−3.687	−0.049	−0.749	−0.055	−0.382	−0.016	−0.142	0.048	0.041	−0.013	−0.092
21	2800 × 800 × 19	−3.110	−3.634	−0.035	−0.937	0.043	−0.336	0.069	−0.078	0.119	0.069	0.027	−0.046
Car carrier (in mm)													
1	3440 × 780 × 11	−5.290	−5.949	−0.223	−1.936	−0.070	−0.598	−0.148	−0.455	0.031	0.020	0.022	0.008
2	3440 × 780 × 11	−5.360	−5.611	0.735	−1.948	−0.075	−0.793	0.075	−0.547	0.116	−0.030	0.013	−0.030
3	3440 × 780 × 11	−4.620	−4.643	0.702	−1.456	−0.147	−1.065	−0.073	−0.362	0.146	0.128	0.057	0.083
4	3440 × 780 × 11	−5.870	−3.425	−1.500	−2.965	0.525	−0.966	−0.380	−0.332	0.124	−0.125	−0.026	0.005
5	3440 × 780 × 11	−5.470	−5.125	0.586	−1.832	0.318	−0.945	0.107	−0.629	−0.013	−0.082	−0.064	−0.057
6	3440 × 780 × 11	−5.650	−5.647	0.133	−2.167	0.147	−1.201	0.209	−0.270	0.114	−0.006	0.051	−0.003
7	3440 × 780 × 8	−3.970	−2.635	1.455	−0.871	0.328	0.349	0.194	0.154	0.027	0.114	0.074	0.108
8	3440 × 780 × 8	1.300	0.605	−0.508	0.037	−0.308	−0.349	0.082	0.058	−0.058	−0.107	0.002	0.014
9	3440 × 780 × 8	2.330	0.269	−0.198	−1.195	0.621	0.144	−0.350	−0.085	0.029	−0.066	0.006	−0.080
10	3440 × 780 × 8	1.510	−0.142	0.720	0.482	0.012	0.261	0.073	0.158	0.086	0.120	0.006	−0.014
11	3440 × 780 × 8	−1.350	−0.497	−0.338	0.339	−0.159	0.200	−0.252	0.154	−0.032	0.031	−0.120	0.017
12	3440 × 780 × 8	−2.960	0.460	1.023	−0.981	0.259	−0.862	0.100	−0.156	0.096	0.032	0.023	0.076

Table 2.2 Measured Components of Initial Deflection (2/2) (in mm) [8]

a	b	t	A_{01}	A_{02}	A_{03}	A_{04}	A_{05}	A_{06}	A_{07}	A_{08}	A_{09}	A_{010}	A_{011}
4200	835	22.0	0.3543	0.0972	−0.0051	−0.0433	−0.0316	−0.0169	−0.0938	−0.0183	−0.0485	−0.0324	−0.0232
4200	835	22.0	0.1491	−0.2048	0.0273	−0.0339	−0.0707	−0.0011	−0.0862	0.0400	−0.0629	−0.0079	−0.0242
4200	835	22.0	0.0888	−0.0546	−0.0299	−0.0270	−0.0671	−0.0066	−0.1129	−0.0073	−0.0519	0.0016	−0.0377
4200	835	22.0	−0.3520	0.1582	−0.0950	−0.0065	−0.0973	0.0135	−0.0713	0.0122	−0.0555	−0.0081	−0.0281
4200	835	22.0	−0.0118	−0.0583	−0.1383	−0.1626	−0.0891	0.0326	0.0133	0.0550	−0.0160	0.0015	−0.0031
4200	835	22.0	0.0024	−0.0443	−0.1856	−0.0412	0.0018	−0.0345	−0.0558	−0.0095	−0.0599	−0.0178	−0.0658
4200	835	22.0	−0.4787	0.0400	−0.1955	−0.0585	−0.1073	0.0214	−0.1294	−0.0260	−0.0985	−0.0285	−0.0897
4200	835	22.0	0.2234	−0.0346	0.0316	0.0539	−0.1029	−0.0406	−0.1019	0.0065	−0.0553	0.0305	−0.0480
4200	835	22.0	0.4608	0.1789	0.1286	0.2052	−0.1289	−0.3564	−0.0787	−0.1372	−0.0006	0.0273	−0.2124
4200	835	22.0	0.1393	−0.2752	−0.0038	−0.3023	−0.3681	0.2372	0.0024	0.0728	−0.0912	0.1732	0.0403
4200	835	22.0	0.5030	−0.2348	0.1042	−0.2198	−0.1956	−0.0137	−0.0844	0.1779	−0.1007	0.0658	−0.0779
4200	835	22.0	0.0640	0.4891	−0.0157	0.4574	−0.0981	−0.1975	−0.1717	−0.0344	−0.0518	0.1162	−0.1282
4200	835	18.5	−0.2987	0.3663	−0.1887	0.0900	−0.1411	0.0602	−0.1343	0.0120	−0.1004	0.0130	−0.0416
4200	835	18.5	−0.7076	0.0366	−0.3604	−0.0564	−0.2804	−0.0501	−0.1594	−0.0134	−0.0710	−0.0215	−0.0439
4200	835	18.5	−0.1522	0.4035	−0.2150	0.0370	−0.1285	0.0392	−0.1166	−0.0035	−0.0910	−0.0068	−0.0410
4200	835	18.5	−0.9636	−0.0354	−0.3891	−0.0209	−0.2351	−0.0209	−0.1189	−0.0015	−0.0624	−0.0092	−0.0479
4200	835	18.5	−0.0010	0.1572	−0.1135	0.0611	−0.0993	0.0233	−0.0709	−0.0029	−0.0488	0.0079	−0.0493
4200	835	18.5	−0.2528	−0.0950	−0.1960	0.0559	−0.0592	0.0214	−0.0723	−0.0090	−0.0947	0.0123	−0.0447
4200	835	18.5	−0.4102	−0.1521	−0.2231	0.0077	−0.1858	−0.0148	−0.1628	−0.0191	−0.0880	0.0136	−0.0638
4200	835	18.5	0.0308	0.0812	−0.0204	−0.0281	−0.1990	−0.0297	−0.1560	−0.0129	−0.1024	0.0304	−0.0341
4200	835	18.5	−0.8188	0.1770	−0.3146	0.0428	−0.1725	0.0287	−0.0934	−0.0135	−0.0572	−0.0296	−0.0447
4200	835	18.5	−0.8505	−0.2816	−0.2026	−0.0429	−0.1707	0.0113	−0.1120	0.0175	−0.0963	−0.0126	−0.0433
4200	835	18.5	−0.0788	−0.1869	−0.1329	−0.0449	−0.1251	−0.0353	−0.0628	−0.0355	−0.0423	−0.0327	−0.0510

Continued

Table 2.2 Measured Components of Initial Deflection (2/2) (in mm) [8]—cont'd

a	b	t	A_{01}	A_{02}	A_{03}	A_{04}	A_{05}	A_{06}	A_{07}	A_{08}	A_{09}	A_{010}	A_{011}
4200	835	18.5	−0.5282	−0.0469	−0.2116	−0.0908	−0.1235	0.0360	−0.0749	−0.0144	−0.0697	0.0132	−0.0462
4200	835	13.0	−4.1787	0.4700	−0.5195	0.3016	−0.5868	−0.1786	−0.1462	−0.2333	−0.1641	−0.0767	−0.0397
4200	835	13.0	−2.9667	0.2718	−1.5489	−0.3930	−0.9848	−0.0842	−0.2113	0.0018	−0.0797	0.0288	−0.0634
4200	835	13.0	−3.4087	0.4793	−1.1008	0.0849	−0.6974	−0.0126	−0.2636	−0.0801	−0.1396	−0.0824	−0.0064
4200	835	13.0	−3.8272	1.2715	−1.0577	−0.1182	−0.4843	−0.2805	−0.3610	−0.0706	−0.0927	−0.0031	0.0250
4200	835	13.0	−3.0192	0.4541	−0.9384	−0.1844	−0.6087	0.1203	−0.2253	−0.1455	−0.1181	−0.1153	−0.0936
4200	835	13.0	−2.7932	0.7274	−0.9647	−0.1594	−0.7269	0.1866	−0.1623	−0.3599	−0.2548	−0.1927	−0.0336
4200	835	13.0	−4.5272	0.0463	−1.1956	−0.0386	−0.6717	−0.0967	−0.2407	−0.0345	−0.0554	0.0024	0.0092
4200	835	13.0	−3.5429	−0.1604	−1.5268	−0.0882	−0.7295	−0.1600	−0.3718	−0.1658	−0.0716	−0.0383	−0.0391
4200	835	13.0	−1.8620	−0.3439	−1.0984	−0.3603	−0.5491	−0.2338	−0.1545	−0.0026	−0.1640	0.0081	−0.0597
4200	835	13.0	−3.9369	−0.4384	−1.2481	0.0893	−0.4174	0.1101	−0.1261	−0.0799	−0.1247	−0.0494	−0.0031
4200	645	18.5	−0.6498	−0.1087	−0.2214	−0.0023	−0.1043	−0.0216	−0.0849	−0.0213	−0.0601	−0.0638	−0.0509
4200	645	18.5	0.4333	−0.3425	−0.3048	0.4072	−0.2201	−0.1943	0.1573	−0.0112	−0.2052	0.1487	0.0015
4200	645	18.5	−0.4370	0.0121	−0.2586	0.0340	−0.1808	0.0423	−0.0920	0.0573	−0.0505	0.0343	−0.0866
4200	645	18.5	−0.3803	0.1116	−0.0997	−0.0479	−0.0798	0.0360	−0.0706	0.0357	−0.0183	0.0300	−0.0259
4200	645	22.0	0.1412	−0.3820	0.0933	−0.2244	0.2123	0.0415	−0.0602	0.0253	0.0636	−0.0889	−0.0429
4200	645	22.0	0.5526	0.2027	−0.2360	0.0891	0.1569	−0.1236	−0.2258	0.1067	−0.0996	−0.1475	0.0488
4200	645	22.0	1.6133	−0.1954	0.2959	0.0390	0.0883	−0.2888	0.1389	−0.1461	0.1248	−0.1042	0.0111
4200	645	22.0	0.2199	−0.2865	0.0536	−0.0651	−0.1622	0.1722	0.0080	0.0468	−0.0139	0.1544	0.0035

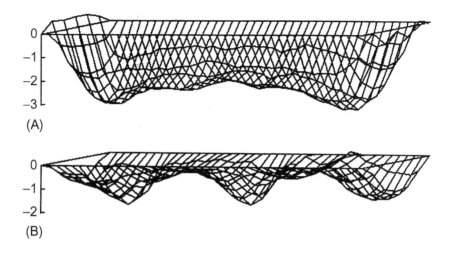

FIG. 2.9

Examples of measured initial deflection in upper deck plating. (A) As-weld condition. (B) With heat treatment.

Under a certain combination of welding condition and plate dimensions, the magnitude of initial deflection becomes too large. In this case, heat treatment is conducted to reduce excess initial deflection below the critical value specified by rules and/or standards.

For example, Japan Shipbuilding Quality Standards (JSQS) [9] specify the maximum magnitude of initial deflection at strength deck plating as follows:

Location	Standard (mm)	Allowable (mm)
Midship part	4	6
Fore and aft	6	9
Nonnaked part	7	9

On the other hand, Smith et al. [10] proposed a measure to represent magnitude of initial deflection as follows:

$$w_{0\,\max} = \alpha \left(b/t\sqrt{\sigma_Y/E}\right)^2 \cdot t_p \tag{2.17}$$

where α is specified according to the magnitude of initial deflection as:

$\alpha = 0.025$: slight
$\alpha = 0.1$: average
$\alpha = 0.3$: severe

Fig. 2.14 shows the comparison of the maximum initial deflection specified by Eq. (2.17) and the measured results given in Tables 2.1 and 2.2. It is seen that measured magnitudes are between average and slight lines. It is also known that the maximum magnitude is below the allowable limit specified by JSQS [9].

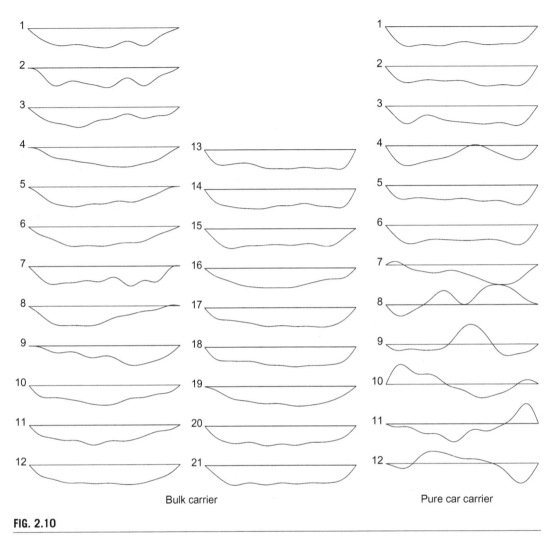

FIG. 2.10

Measured mode of initial deflection in a local panel in Table 2.1.

2.3.4 INITIAL DEFLECTION IN LONGITUDINAL STIFFENERS

Fig. 2.7 indicates that longitudinal stiffeners are accompanied by initial deflection in both vertical and horizontal directions. Fig. 2.15 shows measured results on initial deflection in longitudinal stiffeners in horizontal and vertical directions. On the other hand, Fig. 2.13 shows the probabilistic distribution of the measured maximum initial deflection in longitudinal stiffeners. Data are all from measurement along the top line of the stiffener; see Fig. 2.6.

In Fig. 2.15, solid and dashed lines represent initial deflection in horizontal and vertical directions, respectively. In accordance with the mechanism to produce initial deflection in Fig. 2.4, all initial deflections in vertical direction have to be upwards. However, as is known from Fig. 2.15, vertical

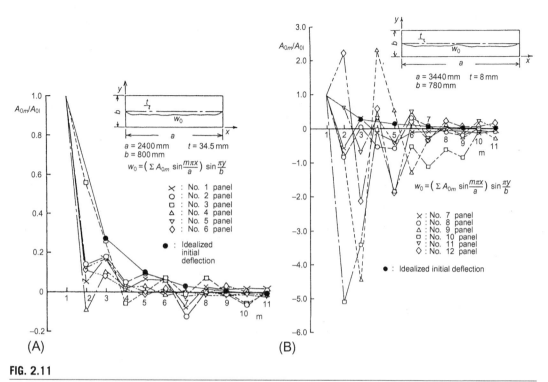

FIG. 2.11

Deflection components of measured initial deflection [6]. (A) Bulk carrier
($a \times b \times t_p = 2400 \times 800 \times 34.5$ mm). (B) Pure car carrier ($a \times b \times t_p = 3440 \times 780 \times 8$ mm).

deflections are in the opposite direction at seven span ends among 20 measured points. On the other hand, as for initial deflection in horizontal direction, it is at four span ends among 20 measured points that initial deflection is in an opposite direction in the adjacent spans.

Maximum, minimum, and average values of measured initial deflection in vertical and horizontal directions are summarized in Tables 2.4 and 2.5, respectively. The maximum value in the vertical direction is −0.0007 to 0.0006 times the span length, whereas that in the horizontal direction is −0.00125 to 0.000135 times the span length. The span length in this case is $a = 2400$ mm.

2.4 SETTING OF INITIAL IMPERFECTIONS DUE TO WELDING IN BUCKLING/PLASTIC COLLAPSE ANALYSIS

In this section, it is described how the initial imperfections have to be set when buckling/plastic collapse analysis is performed on panels and stiffened panels in deck or bottom of ship and ship-like floating structures applying the finite element method (FEM).

In the buckling/plastic collapse analysis applying the FEM, it is necessary to give initial imperfection produced by welding. Especially, it is essential to give initial deflection when a plate under

Table 2.3 Components of Idealized Initial Deflection

a/b	A_{01}/t	A_{02}/t	A_{03}/t	A_{04}/t	A_{05}/t	A_{06}/t	A_{07}/t	A_{08}/t	A_{09}/t	$A_{0\,10}/t$	$A_{0\,11}/t$
$1 < a/b < \sqrt{2}$	1.1158	−0.0276	0.1377	0.0025	−0.0123	−0.0009	−0.0043	0.0008	0.0039	−0.0002	−0.0011
$\sqrt{2} \leq a/b < \sqrt{6}$	1.1421	−0.0457	0.2284	0.0065	0.0326	−0.0022	−0.0109	0.0010	−0.0049	−0.0005	0.0027
$\sqrt{6} \leq a/b < \sqrt{12}$	1.1458	−0.0616	0.3079	0.0229	0.1146	−0.0065	0.0327	0.0000	0.0000	−0.0015	−0.0074
$\sqrt{12} \leq a/b < \sqrt{20}$	1.1439	−0.0677	0.3385	0.0316	0.1579	−0.0149	0.0743	0.0059	0.0293	−0.0012	0.0062
$\sqrt{20} \leq a/b < \sqrt{30}$	1.1271	−0.0697	0.3483	0.0375	0.1787	−0.0199	0.0995	0.0107	0.0537	−0.0051	0.0256

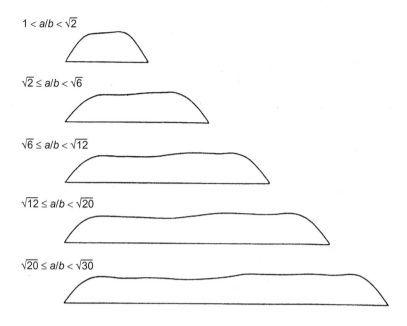

FIG. 2.12

Modes of idealized initial deflection.

in-plane load is analyzed. If the plate is completely flat, only in-plane displacements are produced by in-plane loads, and buckling never occurs. This is because buckling is a bifurcation phenomenon and no buckling occurs without any disturbance or trigger to cause bifurcation.

The simplest way is to give initial deflection of a buckling mode. Then, buckling/plastic collapse behavior can be simulated. However, initial deflection of a buckling mode is not realistic when its magnitude is large, because initial deflection in actual plate members are in many cases of a thin-horse mode. Although deflection components of a buckling mode is included in a thin-horse mode, its magnitude is in general relatively small if it is compared with the maximum magnitude of initial deflection.

On this basis, it is recommended to assume initial deflection of an idealized mode shown in Fig. 2.12. So, the recommended initial deflection in the plate is

$$w_{0pl} = \left| \sum_{i=1}^{m} A_{0i} \sin \frac{i\pi x}{a} \sin \frac{\pi y}{b} \right| + A_{0sw} \sin \frac{\pi x}{a} \sin \frac{\pi y}{B} \tag{2.18}$$

using the deflection components given in Table 2.3 as A_{0i}. a and b are the length and the breadth of a local panel, and B is the breadth of a stiffened panel. The first and the second terms in Eq. (2.18) are initial deflections in a local panel and a whole stiffened panel, respectively. As for the magnitude of the initial deflection in a local panel, on the basis of Fig. 2.14, use of the smaller one of 6.0 mm specified by JSQS and 0.1 $\beta^2 t$ by Smith is recommended.

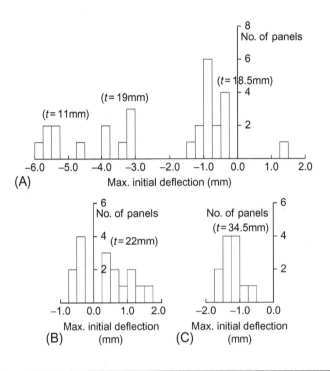

FIG. 2.13

Probabilistic distributions of measured maximum initial deflection. (A) $t = 11, 18.5$, and 19 mm.
(B) $t = 22$ mm. (C) $t = 34.5$ mm.

When symmetry condition can be introduced at the mid-span plane, only odd deflection components in Table 2.4 have to be used imposing symmetry condition at the mid-span plane. As for A_{0ws}, Eq. (2.18), which will appear later, is used.

Here, what has to be noticed is that the initial deflection in local panels is of a thin-horse mode and deflection is in the same direction. Also when a continuous panel without stiffeners is analyzed setting initial deflection of an overall mode as zero ($A_{0ws} = 0$), no bifurcation takes place because each panel constrains each other and prevents the occurrence of buckling. To avoid this, magnitude of initial deflection of a thin-horse mode has to be varied among the adjacent panels by several percent.

As for the ith longitudinal stiffener located at $y = y_i$, initial deflection in the vertical direction can be approximated as:

$$w_{0si} = A_{0sw} \sin \frac{\pi x}{a} \sin \frac{\pi y_i}{B} \tag{2.19}$$

On the basis of the measured results in Table 2.4, magnitude of initial deflection in a vertical direction, A_{0sw}, can be defined as follows:

$$A_{0sw} = 0.001 \times a \tag{2.20}$$

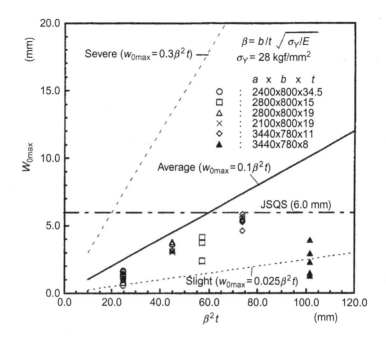

FIG. 2.14

Comparison of measured and predicted initial deflection in plating.

The initial deflection in a horizontal direction can be defined as:

$$v_{0si} = A_{0sv} \sin \frac{\pi x}{a} \tag{2.21}$$

where

$$A_{0sv} = 0.001 \times a \tag{2.22}$$

Both A_{0sw} and A_{0sv} are determined from the consideration on measured results in Tables 2.4 and 2.5.

On the other hand, as for welding residual stress, the breadth of the region where tensile residual stress exists is defined in accordance with the previous equations, Eqs. (2.2), (2.8) for local panel and stiffener web, respectively, which were defined as:

$$b_t = t_w/2 + 0.26 \Delta Q_{max}/(t_w + 2t_p) \quad \text{(in mm)} \tag{2.2}$$

$$b_{ts} = \frac{t_w}{t_p} \times 0.26 \Delta Q_{max} \quad \text{(in mm)} \tag{2.8}$$

where t_p and t_w are thickness of panel and stiffener web, respectively, and

$$\Delta Q_{max} = 78.7 \times \ell^2 \tag{2.4}$$

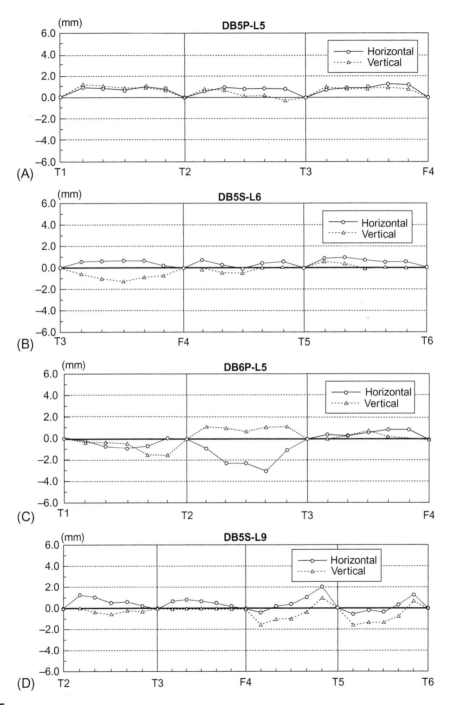

FIG. 2.15

Measured initial deflection in longitudinal stiffeners [5]. (A) DB5P. (B) DB5S. (C) DB6P. (D) DB5S-L9.

Table 2.4 Maximum Initial Deflection of Stiffener in Vertical Direction [5]

Block	Max (mm)	Min (mm)	μ (mm)	$\mu+$ (mm)	$\mu-$ (mm)	σ^2
DB5P	1.20	0.16	0.81	0.81	–	0.171
DB5S	1.42	−1.76	−0.29	0.68	−1.02	−1.030
DB6P	1.27	−1.55	0.40	0.83	−1.25	0.888

Notes: *Max, maximum initial deflection in vertical direction; Min, minimum initial deflection in vertical direction; μ, mean of maximum initial deflection; $\mu+$, mean of maximum initial deflection in positive side; $\mu-$, mean of maximum initial deflection in negative side; σ^2, COV of maximum initial deflection.*

Table 2.5 Maximum Initial Deflection of Stiffener in Horizontal Direction [5]

Block	Max (mm)	Min (mm)	μ (mm)	$\mu+$ (mm)	$\mu-$ (mm)	σ^2
DB5P	3.23	0.98	1.57	1.57	–	0.763
DB5S	2.03	−0.58	1.08	1.21	−0.58	0.383
DB6P	0.98	−3.00	−0.91	0.79	−1.34	1.288

Notes: *Max, maximum initial deflection in vertical direction; Min, minimum initial deflection in vertical direction; μ, mean of maximum initial deflection; $\mu+$, mean of maximum initial deflection in positive side; $\mu-$, mean of maximum initial deflection in negative side; σ^2, COV of maximum initial deflection.*

$$\ell = \begin{cases} 0.7 \times t_w \ (\text{mm}) & (\text{when } 0.7 \times t_w < 7.0 \text{ mm}) \\ 7.0 \ (\text{mm}) & (\text{when } 0.7 \times t_w \geq 7.0 \text{ mm}) \end{cases} \tag{2.5}$$

In the FEM analysis, element boundary has to be created in the local panel at the location of b_t apart, and in the stiffener web at the location of b_{st} apart, respectively, from the panel-stiffener web intersection line.

In the following chapters, buckling/plastic collapse behavior and the ultimate strength will be explained considering the influences of welding residual stress and initial deflection explained in this chapter.

EXERCISES

2.1 Explain how initial deflection is produced.

2.2 Explain how welding residual stress is produced.

2.3 Explain the characteristics of initial deflection in a thin-horse mode.

2.4 Explain how to define the initial imperfections due to welding when nonlinear FEM analysis is performed to simulate buckling/plastic collapse behavior of stiffened plate.

2.5 Explain what idealized initial deflection is.

REFERENCES

[1] Yao T. Compressive ultimate strength of structural members in ship structure [Doctoral thesis]. Osaka University; 1980 [in Japanese].

[2] Sato K, Terasaki T. Effects of welding conditions on welding residual stresses and welding distortions of structural members. J Jpn Welding Soc 1976;45(1):42–50 [in Japanese].

[3] Matsuoka K, Yoshii T. Weld residual stress in corner boxing joints. J Soc Naval Arch Jpn 1997;180:753–761 [in Japanese].

[4] Yao T, Fujikubo M, Yanagihara D, Varghese B, Niho O. Influences of welding imperfections on buckling/ultimate strength of ship bottom plating subjected to combined bi-axial thrust and lateral pressure. In: Proc int symp thin-walled structures, Singapore; 1998. p. 425–432.

[5] Hirakawa S, Hayashida K. Measurement of initial deflection in stiffened panels and simulation of their buckling/plastic collapse behaviour. Graduation Thesis. Hiroshima University; 1998 [in Japanese].

[6] Ueda Y, Yao T, Nakacho K, Tanaka Y, Handa K. Compressive ultimate strength of rectangular plates with initial imperfections due to welding (3rd rep). Prediction method of compressive ultimate strength. J Soc Naval Arch Jpn 1983;154:345–355 [in Japanese].

[7] Ueda Y, Yao T. The influence of complex initial deflection modes on the behaviour and ultimate strength of rectangular plate in compression. J Constr Steel Res 1985;5:265–302.

[8] Yao T, Nikolov PI, Miyagawa Y. Influence of welding imperfections on stiffness of rectangular plate under thrust. In: Karlsson L, Lindgren LE, Jonsson M, editors. Mechanical effects of welding. Springer-Verlag; 1992. p. 261–268.

[9] Society of Naval Architects. Research Committee on Steel Ship Construction: Japan Shipbuilding Quality Standards (JSQS). Society of Naval Architects, Japan; 1979, 28 pp.

[10] Smith C, Davidson P, Chapman J, Dowling P. Strength and stiffness of ships plating under in-plane compression and tension. Trans RINA 1987;W.6:277–296.

FUNDAMENTAL THEORY AND METHODS OF ANALYSIS TO SIMULATE BUCKLING/PLASTIC COLLAPSE BEHAVIOR

3.1 DEFLECTION MODE OF PLATES AND STIFFENED PLATES IN BUCKLING/PLASTIC COLLAPSE BEHAVIOR

In general, ship and ship-like floating structures are constructed with thin plates. To maintain strength and stiffness of thin plates effectively, stiffeners are provided as indicated in Fig. 3.1. It can be said that the minimum structural units composing structures is plates and stiffened plates.

To simulate buckling/plastic collapse behavior of such thin-plated structures, both analytical and numerical methods can be applied. With the numerical method such as finite element method (FEM), it is possible to simulate structural behaviors considering both material and geometrical nonlinearities. Reasonable accuracy can be expected with proper meshing using proper elements under proper boundary condition and loading condition. However, results are obtained only numerically, and consideration is necessary on physical meaning of the calculated results. On the other hand, in the analytical method, deflection modes produced by buckling are assumed with proper functions. Although the application is limited to relatively simple problems such as elastic behaviors under fundamental load and boundary conditions, the solutions having physical meaning can be obtained explicitly. The accurate solution, however, cannot be obtained if the assumed deflection functions are not proper. In this sense, it is necessary to consider carefully the loading and boundary conditions when deflection functions are assumed.

In this section, it is explained what deflection functions have to be assumed in accordance with the buckling behavior to be simulated. Such deflection functions are derived from intuition together with observations of the results of numerical calculations explained in the continuing chapters.

3.1.1 BUCKLING COLLAPSE BEHAVIOR OF PLATES

The minimum structural unit in the structure is a plate in stiffened plate partitioned by longitudinal and transverse stiffeners. This plate is in general rectangular and is considered to be simply supported along its four sides. Actually, when this plate buckles and lateral deflection develops, the stiffeners along four sides resist against the rotation of the plates. So, exactly saying, such rectangular plate is elastically supported along its four sides. However, the constraint by the stiffeners is considered weak in general, and the rectangular plate is assumed to be simply supported along its four sides.

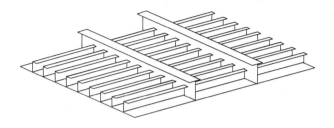

FIG. 3.1

Ship structure composed of stiffened plates.

When the in-plane load acting on the plate exceeds a certain critical value, the plate undergoes buckling. Here, rectangular plate of which aspect ratio (ratio of length to breadth) is two is considered, that is $a/b = 2.0$, as shown in Fig. 3.2. When the longitudinal thrust load acts, the plate buckles as indicated in Fig. 3.2A with the deflection mode:

$$w_{\text{long}} = A_{m1} \sin \frac{m\pi x}{a} \sin \frac{\pi y}{b} \tag{3.1}$$

When $a/b = 2.0$, buckling is in two half-waves in the loading direction and $m = 2$.

On the other hand, when the transverse thrust load acts, the plate buckles in one half-wave mode as indicated in Fig. 3.2B. In this case, buckling mode can be expressed as follows:

$$w_{\text{trans}} = A_{11} \sin \frac{\pi x}{a} \sin \frac{\pi y}{b} \tag{3.2}$$

When in-plane bending moment acts along the longer side of the plate, the plate buckles as indicated in Fig. 3.2C. The buckling mode in this case cannot be expressed by one representative term, but is expressed in the form:

$$w_{\text{bend}} = \left(\sum_{i=1}^{m} A_{i1} \sin \frac{i\pi x}{a} \right) \sin \frac{\pi y}{b} \tag{3.3}$$

The deflection is in one half-wave mode in the transverse (y-) direction, but sum of several half-waves in the longitudinal direction. Taking the number of deflection components as infinite, exact solution is obtained. However, an accurate solution can be obtained with around 10 components.

In the case when in-plane shear load acts along four sides of the plate, buckling mode is more complex as indicated in Fig. 3.2D. In this case more deflection components in both longitudinal and transverse directions are necessary to represent the buckling mode, that is

$$w_{\text{shear}} = \sum_{i=1}^{m} \sum_{j=1}^{n} A_{ij} \sin \frac{i\pi x}{a} \sin \frac{j\pi y}{b} \tag{3.4}$$

Here, Eqs. (3.1) through (3.4) give accurate solution of the buckling strength. However, as will be explained in Sections 4.2 and 4.5, deflection components other than buckling mode also start to develop beyond the buckling. Other deflection components are, at the beginning, very small, and the postbuckling behavior can be approximately simulated with only the buckling mode until the load

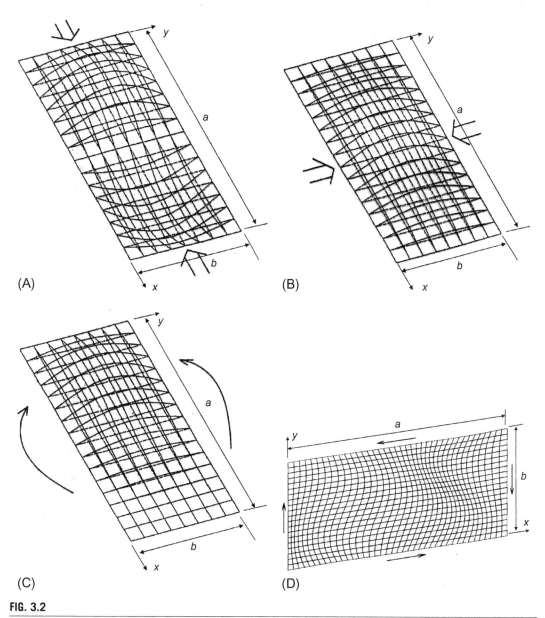

FIG. 3.2

Buckling modes of simply supported rectangular plate. (A) Longitudinal thrust. (B) Transverse thrust.
(C) Bending moment. (D) Shear force.

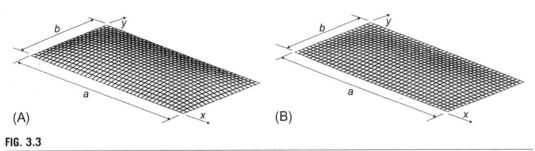

(A)   (B)

FIG. 3.3

Assumed deflection modes under lateral pressure load. (A) Simply supported condition. (B) Clamped condition.

approaches to a certain level. Then, the influence of other deflection components becomes not negligible (see Figs. 4.6 and 4.33 in Chapter 4).

It should be noticed that calculated deflection is only in the range of assumed deflection. So, if it is enough to simulate the postbuckling behavior, for example, up to two times the buckling stress, it is enough if only deflection of a buckling mode is considered. However, if analysis is necessary to simulate secondary buckling behavior, as will be explained in Section 4.4, deflection of a general form, Eq. (3.4), has to be used.

Here, lateral loads are considered acting in the perpendicular direction to the plate surface as pressure load acting on a ship's surface. When an isolated plate which is simply supported is considered, the plate may deflect as indicated in Fig. 3.3A, and the deflection can be approximately expressed by Eq. (3.3) with different deflection coefficients compared to the plate in Fig. 3.2B.

However, in the actual structure, isolated plate does not exist, but a rectangular plate is a part of continuous plating as indicated in Fig. 3.4A. In this case, all the plates deflect in the same direction as indicated in Fig. 3.4B. Deflection produced by lateral pressure load (Fig. 3.3B) of a clamped mode can be expressed as

$$w_{pr2} = \frac{1}{2} \sum_{i=1}^{m} A_{i1} f_i(x) \left(1 - \cos \frac{\pi y}{2b}\right) \tag{3.5}$$

When analytical method is applied to simulate buckling and postbuckling behavior of a rectangular plate, deflection mode expressed by Eqs. (3.1) through (3.5) or their combinations can be assumed depending on the boundary and loading conditions.

3.1.2 BUCKLING COLLAPSE BEHAVIOR OF STIFFENED PLATES

Here, stiffened plate is considered on which stiffeners of the same size are provided with equal distances. Thrust load is acting in the direction of the stiffeners. If the stiffeners have enough flexural stiffness and little deflect under the action of thrust load, the plate locally buckles and the stiffeners do not buckle as indicated in Fig. 3.5A. In this case, local buckling mode of the plate is expressed as:

$$w_{long} = A_{mn} \sin \frac{m\pi x}{a} \sin \frac{n\pi y}{B} \tag{3.6}$$

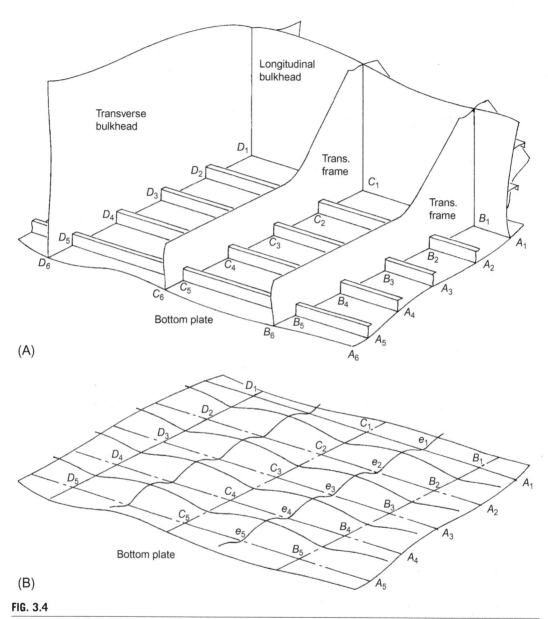

FIG. 3.4

Deflection under lateral pressure load. (A) Stiffened plate structure. (B) Deflection produced by lateral pressure load.

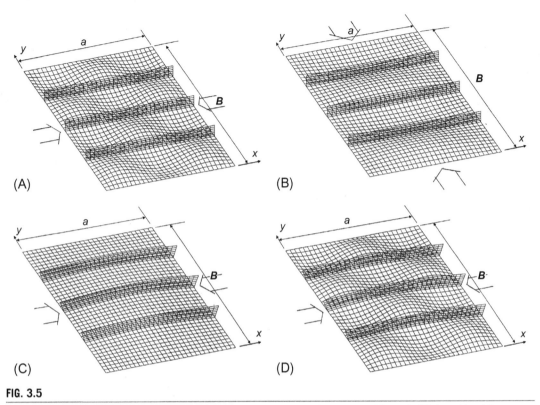

FIG. 3.5

Buckling mode of stiffened plate under thrust. (A) Local buckling (longitudinal thrust). (B) Local buckling (transverse thrust). (C) Overall buckling. (D) Overall buckling after local buckling.

where m and n are the numbers of half wave of the local buckling mode in longitudinal and transverse directions, respectively. $m = 3$ and $n = 4$ when local buckling takes place as indicated in Fig. 3.5A. It is known that $n = N + 1$, where N is the number of stiffeners.

On the other hand, when thrust load acts on the same plate but in transverse direction, local buckling of a different mode occurs as indicated in Fig. 3.5B when the stiffeners are stiff enough. The local buckling mode in this case is one half-wave mode between stiffeners, and is expressed as follows:

$$w_{\text{trans}} = A_{1n} \sin \frac{\pi x}{a} \sin \frac{n\pi y}{B} \tag{3.7}$$

Contrary to these, when the stiffeners are not stiff enough, overall buckling takes place as indicated in Fig. 3.5C. In this case, buckling mode is in one half-wave mode regardless of the direction of thrust load whether it is in longitudinal direction or in transverse direction. The overall buckling mode is expressed as:

$$w_{\text{overall}} = A_{11} \sin \frac{\pi x}{a} \sin \frac{\pi y}{B} \tag{3.8}$$

In addition to the above, overall buckling often takes place as the secondary buckling after the local buckling has occurred as the primary buckling in the case of a stiffened plate; see Fig. 3.5D. The buckling mode is expressed as the sum of Eq. (3.6) (or Eq. 3.7) and Eq. (3.8).

3.1.3 BUCKLING COLLAPSE BEHAVIOR OF STIFFENERS

There are several possible buckling modes of stiffeners predominantly under thrust, as shown in Fig. 3.6. The lateral torsional buckling, Fig. 3.6A, is a rotation of stiffeners about a plate-stiffener connection line. For more slender stiffeners such as a flat-bar stiffener, more localized buckling, called a tripping, as shown in Fig. 3.6B, likely takes place and often interacts with the local panel buckling. For the combination of a slender web with a relatively stiff flange, a local web buckling as shown in Fig. 3.6C may take place also accompanied by the interaction with plate. It is a common design philosophy for stiffeners that the local panel buckling must be preceded by these stiffener buckling.

When overall buckling as shown in Fig. 3.5C takes place, deflection of a stiffener located at $y = y_i$ is expressed as:

$$w_{si} = A_{11} \sin \frac{\pi x}{a} \sin \frac{\pi y_i}{B} \tag{3.9}$$

In the following, theories and methods for buckling analysis as well as elastic and elastoplastic large deflection analyses will be briefly introduced.

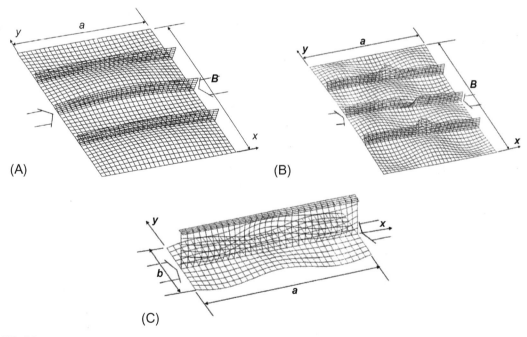

(A)

(B)

(C)

FIG. 3.6

Buckling mode of stiffener in stiffened plate. (A) Lateral torsional buckling. (B) Tripping. (C) Web buckling.

3.2 BUCKLING STRENGTH ANALYSIS

3.2.1 GENERAL THEORY FOR A RECTANGULAR PLATE

As the most representative and important case, a simply supported rectangular plate with welding residual stresses under combined in-plane stresses is considered, as shown in Fig. 3.7. The plate is accompanied by the residual stresses, σ_{xr}, σ_{yr}, and τ_{xyr} (initial stresses), and the stresses σ_x, σ_y, and τ_{xy} are produced by a unit intensity of load ($\lambda=1.0$). Then, it is assumed that buckling occurs when the applied load intensity reaches to $\lambda^{cr}(\neq 1.0)$. At the onset of buckling, very small deflection is produced under the prescribed boundary conditions for deflection. Such small deflection will not produce stretching nor shortening of middle plane of the plate. However, small in-plane displacements shall be produced due to lateral deflection.

The buckling strength is determined from the condition that the work done by the applied forces with in-plane displacements produced by lateral deflection fully changes to the strain energy of bending produced by buckling deflection [1]. The same condition can be derived from the *Principle of Minimum Potential Energy* considering the strain energy due to the buckling deflection and the work done by the internal forces during the buckling deformation [1].

Strain energy by bending stored in the elastic and plastic regions of the plate are represented as [2]:
Elastic region:

$$U_e = \int_{S_e} \frac{D}{2} \left[\left(\frac{\partial^2 w}{\partial x^2} \right)^2 + 2(1 - v) \left(\frac{\partial^2 w}{\partial x \partial y} \right)^2 + 2v \left(\frac{\partial^2 w}{\partial x^2} \right) \left(\frac{\partial^2 w}{\partial y^2} \right) \right.$$
$$\left. + \left(\frac{\partial^2 w}{\partial y^2} \right)^2 \right] dx \, dy \tag{3.10}$$

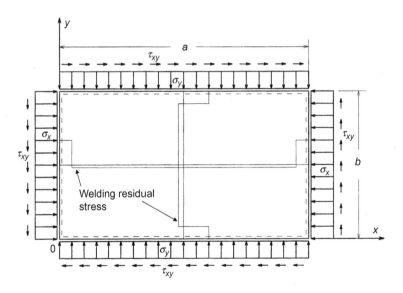

FIG. 3.7

Rectangular plate with welding residual stresses under combined loads.

Plastic region:

$$U_p = \int_{S_p} \frac{D_s}{2} \left[C_1 \left(\frac{\partial^2 w}{\partial x^2} \right)^2 - C_2 \left(\frac{\partial^2 w}{\partial x^2} \right) \left(\frac{\partial^2 w}{\partial x \partial y} \right) + C_{31} \left(\frac{\partial^2 w}{\partial x \partial y} \right)^2 \right.$$

$$\left. + C_{32} \left(\frac{\partial^2 w}{\partial x^2} \right) \left(\frac{\partial^2 w}{\partial y^2} \right) - C_4 \left(\frac{\partial^2 w}{\partial y^2} \right) \left(\frac{\partial^2 w}{\partial x \partial y} \right) + C_5 \left(\frac{\partial^2 w}{\partial y^2} \right)^2 \right] dx \, dy \tag{3.11}$$

where

$$C_1 = 1 - (1 - v_p^2)\kappa(\lambda\sigma_x + \sigma_{xr})^2/\sigma_i^2$$

$$C_2 = 4(1 - v_p^2)\kappa(\lambda\sigma_x + \sigma_{xr})\lambda\tau_{xy} + \tau_{xyr}/\sigma_i^2$$

$$C_{31} = 2[(1 - v_p) - (1 - v_p^2)\kappa(\lambda\tau_{xy} + \tau_{xyr})/\sigma_i^2]$$

$$C_{32} = 2[v_p - (1 - v_p^2)\kappa(\lambda\sigma_x + \sigma_{xr})(\lambda\sigma_y + \sigma_{yr})/\sigma_i^2]$$

$$C_4 = 4(1 - v_p^2)\kappa(\lambda\sigma_y + \sigma_{yr})(\lambda\tau_{xy} + \tau_{xyr})/\sigma_i^2$$

$$C_5 = 1 - (1 - v_p^2)\kappa(\lambda\sigma_y + \sigma_{yr})/\sigma_i^2$$

$$\kappa = 1 - E_t/E_s$$

$$\sigma_i = \sqrt{(\lambda\sigma_x + \sigma_{xr})^2 - (\lambda\sigma_x + \sigma_{xr})(\lambda\sigma_y + \sigma_{yr}) + (\lambda\sigma_y + \sigma_{yr})^2 + 3(\lambda\tau_{xy} + \tau_{xyr})^2}$$

$$D = Et^3/12(1 - v^2), \quad D_s = E_s t^3/12(1 - v_p^2)$$

S_e and S_p represent the elastic and plastic region, respectively. t is a plate thickness. D is a bending stiffness of the elastic region, and E and v are the elastic modulus and Poisson's ratio, respectively. E_t and E_s are the tangent modulus and secant modulus of the material in the plastic region when the buckling takes place. Eq. (3.11) is derived applying the *Deformation Theory of Plasticity*, in which an elastic unloading is not considered in the plastic region. The details of the plasticity theory can be found in Ueda [2].

The work done by internal forces (in-plane stress components) with in-plane displacements (in-plane strain components) due to the buckling deformation can be expressed as [1]:

$$T = \int_0^a \int_0^b \frac{t}{2} \left[(\lambda\sigma_x + \sigma_{xr}) \left(\frac{\partial w}{\partial x} \right)^2 + 2(\lambda\tau_{xy} + \tau_{xyr}) \left(\frac{\partial w}{\partial x} \right) \left(\frac{\partial w}{\partial y} \right) \right.$$

$$\left. + (\lambda\sigma_y + \sigma_{yr}) \left(\frac{\partial w}{\partial y} \right)^2 \right] dx \, dy \tag{3.12}$$

Total potential energy is then expressed as:

$$\Pi = U_e + U_p - T \tag{3.13}$$

Applying the *Principle of Minimum Potential Energy*, the following expression is obtained.

$$\delta\Pi = \delta(U_e + U_p - T) = 0 \tag{3.14}$$

The buckling strength, that is λ when buckling occurs, can be obtained from Eq. (3.14).

In the analytical method, the buckling deflection is approximated by an analytical function that satisfies the boundary conditions for buckling deflection. For example, for a rectangular plate of the length a and width b, simply supported along its four sides, that is, the deflection is constrained while rotation about the side is free, the following double trigonometric series can be used as a general form of buckling deflection.

$$w = \sum_{i=1}^{m}\sum_{j=1}^{n} A_{ij} \sin\frac{i\pi x}{a} \sin\frac{j\pi y}{b} \tag{3.15}$$

Substituting this deflection function into Eq. (3.14), the following expression is obtained:

$$\frac{\partial}{\partial A_{ij}}(U_e + U_p - T)\delta A_{ij} = 0 \tag{3.16}$$

Since U_e, U_p, and T are the quadratic functions of the deflection amplitudes A_{ij}, Eq. (3.16) results in linear simultaneous equations with respect to A_{ij} of which number is $(m \times n)$. From the condition that Eq. (3.16) holds regardless of δA_{ij}, the following equation is derived:

$$([G_1] - \lambda[G_2] - [G_3])\{A_{ij}\} = 0 \tag{3.17}$$

where the matrix $[G_1]$ is obtained from Eqs. (3.10), (3.11), and $[G_2]$ and $[G_3]$ from Eq. (3.12). $[G_2]$ is a function of the stress components, σ_x, σ_y, and τ_{xy}, for $\lambda = 1$ and $[G_3]$ is that of the residual stress components, σ_{xr}, σ_{yr}, and τ_{xyr}. For the occurrence of buckling, nonzero $\{A_{ij}\}$s have to appear as the solutions of coefficient matrix in Eq. (3.17). This corresponds to the condition that the determinant of Eq. (3.17) has to be zero, that is

$$| [G_1 + G_3] - \lambda_{cr}[G_2] | = 0 \tag{3.18}$$

λ_{cr} that satisfies Eq. (3.18) gives the buckling load.

3.2.2 ELASTIC BUCKLING STRENGTH OF A RECTANGULAR PLATE UNDER UNI-AXIAL THRUST

Here, a simple example of a rectangular plate under uni-axial compression is considered as shown in Fig. 3.8, where welding residual stress and buckling mode are illustrated. As explained in Section 2.2, the tensile residual stresses equal to the yield strength σ_Y are assumed near the longitudinal welded lines and the compressive residual stresses, σ_c, in the central part of the plate. These stresses satisfy the self-equilibrium condition, Eq. (2.1).

The plate is assumed to be simply supported and its buckling mode is represented with one deflection component with m half-waves in x-direction and one half-wave in y-direction as:

$$\sigma_c = 2b_t\sigma_Y/(b - 2b_t)$$

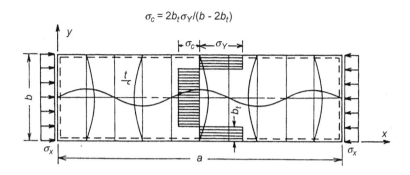

FIG. 3.8

Rectangular plate with welding residual stresses under thrust.

$$w = A_m \sin \frac{m\pi x}{a} \sin \frac{\pi y}{b} \tag{3.19}$$

Three kinds of buckling behavior are possible to occur depending on a slenderness of the plate. When the plate is thin, buckling takes place before the central region has been yielded in compression. This is called elastic buckling because all over the plate is elastic when buckling takes place. If the plate becomes thick, buckling takes place with the stress state that the central region is yielded in compression while the edges region is elastic. This is called elastoplastic buckling. If the plate becomes very thick, all over the plate is yielded in compression when buckling takes place, including the edge parts where residual stress was in tension. This is called plastic buckling. For each case, using Eq. (3.10) for the elastic region and Eq. (3.11) for the plastic region, Eq. (3.18) gives the buckling strength as follows:

Elastic buckling:

$$\sigma_{cr} = \sigma_{cr0} - \frac{\sigma_Y}{\pi(1-\mu)} \sin\mu\pi \tag{3.20}$$

$$\varepsilon_{cr} = \sigma_{cr}/E \tag{3.21}$$

where

$$\sigma_{cr0} = \frac{\pi^2 E}{12(1-v^2)} \left(\frac{t}{b}\right)^2 \left(\frac{mb}{a} + \frac{a}{mb}\right)^2 \tag{3.22}$$

$$\mu = 2b_t/b = \sigma_c/(\sigma_Y + \sigma_c) \tag{3.23}$$

Elastoplastic buckling:

$$\sigma_{cr} = \sigma_Y - \sigma_c - \sigma_1 \tag{3.24}$$

$$\varepsilon_{cr} = (\sigma_Y - \sigma_c - \sigma_1/\mu)/E \tag{3.25}$$

where σ_1 is derived by solving the following equation:

$$
\frac{\pi^2 E t^2}{12}\left(\frac{a}{m}\right)^2\left[\frac{\sigma_Y}{(1-v_p^2)(\sigma_Y+\sigma_1/\mu)}\left\{\left(v_p\frac{m^2}{a^2}+\frac{1}{b^2}\right)^2\left(1-\mu+\frac{1}{\pi}\sin\mu\pi\right)\right.\right.
$$

$$
\left.+2(1-v_p)\frac{m^2}{a^2 b^2}\left(1-\mu-\frac{1}{\pi}\sin\mu\pi\right)\right\}+\frac{1}{(1-v^2)}\left\{\left(\frac{m^4}{a^4}+2v\frac{m^2}{a^2 b^2}+\frac{1}{b^4}\right)\left(\mu-\frac{1}{\pi}\sin\mu\pi\right)\right.
$$

$$
\left.\left.+2(1-v)\frac{m^2}{a^2 b^2}\left(\mu+\frac{1}{\pi}\sin\mu\pi\right)\right\}\right]
$$

$$
-\left[\sigma_Y\left(1-\mu+\frac{1}{\pi}\sin\mu\pi\right)+\left(\frac{\sigma_1}{\mu}-\sigma_c\right)\left(\mu-\frac{1}{\pi}\sin\mu\pi\right)\right]=0 \tag{3.26}
$$

Plastic buckling:

$$
\sigma_{cr}=\sigma_Y \tag{3.27}
$$

$$
\varepsilon_{cr}=2\sigma_Y/E+\varepsilon_1 \tag{3.28}
$$

where ε_1 is obtained by solving the following equation:

$$
\frac{\pi^2 t^2}{12(1-v_p^2)}\left(\frac{a}{m}\right)^2\left[\frac{1}{2\varepsilon_Y+\varepsilon_1}\left\{\left(v_p\frac{m^2}{a^2}+\frac{1}{b^2}\right)^2\left(1-\mu+\frac{1}{\pi}\sin\mu\pi\right)\right.\right.
$$

$$
\left.+2(1-v_p)\frac{m^2}{a^2 b^2}\left(1-\mu-\frac{1}{\pi}\sin\mu\pi\right)\right\}+\frac{1}{\varepsilon_Y+\varepsilon_1}\left\{\left(v_p\frac{m^2}{a^2}+\frac{1}{b^2}\right)^2\left(\mu-\frac{1}{\pi}\sin\mu\pi\right)\right.
$$

$$
\left.\left.+2(1-v_p)\frac{m^2}{a^2 b^2}\left(\mu+\frac{1}{\pi}\sin\mu\pi\right)\right\}\right]-1=0 \tag{3.29}
$$

The characteristics of the buckling strength obtained by these equations are explained in Chapter 4. In the above example, the deflection was represented by double trigonometric series. This expression is effective for the plate with all edges simply supported. For more general edges conditions including the clamped edges, the biharmonic function given in Masaoka et al. [3] is effective.

3.3 ELASTIC LARGE DEFLECTION ANALYSIS OF RECTANGULAR PLATE SUBJECTED TO COMBINED LOADS

Fundamental formulation for the elastic large deflection analysis of a rectangular plate in analytical manner is explained. To simulate the secondary buckling behavior explained in Section 4.4, the formulation is performed assuming general expression of deflection as indicated by Eq. (3.4). A simply supported rectangular plate of uniform thickness with initial deflection and welding residual stress, Fig. 3.7, is considered. After showing the general formulation for elastic large deflection analysis, the results for a specific case shall be presented.

3.3.1 ASSUMED DEFLECTION MODE

The following double trigonometric deflection function that satisfy the simply supported condition is assumed:

$$w = \sum_{i=1}^{m}\sum_{j=1}^{n} A_{ij}\sin\frac{i\pi x}{a}\sin\frac{j\pi y}{b} \tag{3.30}$$

It is also assumed that initial deflection has the same deflection components as total deflection as

$$w_0 = \sum_{i=1}^{m}\sum_{j=1}^{n} A_{0ij}\sin\frac{i\pi x}{a}\sin\frac{j\pi y}{b} \tag{3.31}$$

In Eqs. (3.30), (3.31), A_{ij} and A_{0ij} are the magnitudes of deflection components of total deflection and initial deflection with i half-waves in x-direction and j half-waves in y-direction, respectively.

3.3.2 IN-PLANE STRESS AND STRAIN

Denoting the displacements at the neutral plane in x-, y-, and z-directions as $u(x,y)$, $v(x,y)$, and $w(x,y)$, the in-plane strain-displacement relationship is expressed as follows:

$$\epsilon_{xp} = \frac{\partial u}{\partial x} + \frac{1}{2}\left\{\left(\frac{\partial w}{\partial x}\right)^2 - \left(\frac{\partial w_0}{\partial x}\right)^2\right\} \tag{3.32}$$

$$\epsilon_{yp} = \frac{\partial v}{\partial y} + \frac{1}{2}\left\{\left(\frac{\partial w}{\partial y}\right)^2 - \left(\frac{\partial w_0}{\partial y}\right)^2\right\} \tag{3.33}$$

$$\gamma_{xyp} = \frac{\partial u}{\partial y} + \frac{\partial v}{\partial x} + \left(\frac{\partial w}{\partial x}\right)\left(\frac{\partial w}{\partial y}\right) - \left(\frac{\partial w_0}{\partial x}\right)\left(\frac{\partial w_0}{\partial y}\right) \tag{3.34}$$

The nonlinear components as a function of the total deflection w and the initial deflection w_0 represent the membrane strains due to large deflection. By eliminating the displacements u and v, the compatibility condition is obtained as

$$\frac{\partial^2 \epsilon_{xp}}{\partial y^2} + \frac{\partial^2 \epsilon_{yp}}{\partial x^2} - \frac{\partial^2 \gamma_{xyp}}{\partial x\partial y} = \left(\frac{\partial^2 w}{\partial x\partial y}\right)^2 - \left(\frac{\partial^2 w}{\partial x^2}\right)\left(\frac{\partial^2 w}{\partial y^2}\right)$$
$$- \left(\frac{\partial^2 w_0}{\partial x\partial y}\right)^2 + \left(\frac{\partial^2 w_0}{\partial x^2}\right)\left(\frac{\partial^2 w_0}{\partial y^2}\right) \tag{3.35}$$

According to the definition of Airy's stress function, the in-plane stress components, σ_{xp}, σ_{yp}, and τ_{xyp}, are expressed as follows:

$$\sigma_{xp} = \frac{\partial^2 F}{\partial y^2} \tag{3.36}$$

$$\sigma_{yp} = \frac{\partial^2 F}{\partial x^2} \tag{3.37}$$

$$\tau_{xyp} = -\frac{\partial^2 F}{\partial x \partial y} \tag{3.38}$$

These stresses automatically satisfy the following equilibrium condition in the $x - y$ plane:

$$\frac{\partial \sigma_{xp}}{\partial x} + \frac{\partial \tau_{xyp}}{\partial y} = 0 \tag{3.39}$$

$$\frac{\partial \tau_{xyp}}{\partial x} + \frac{\partial \sigma_{yp}}{\partial y} = 0 \tag{3.40}$$

Assuming plane stress condition, the in-plane strain components are expressed in the following forms:

$$\epsilon_{xp} = \frac{1}{E}(\sigma_{xp} - \nu\sigma_{yp}) = \frac{1}{E}\left(\frac{\partial^2 F}{\partial y^2} - \nu\frac{\partial^2 F}{\partial x^2}\right) \tag{3.41}$$

$$\epsilon_{yp} = \frac{1}{E}(\sigma_{yp} - \nu\sigma_{xp}) = \frac{1}{E}\left(\frac{\partial^2 F}{\partial x^2} - \nu\frac{\partial^2 F}{\partial y^2}\right) \tag{3.42}$$

$$\gamma_{xyp} = \frac{2(1+\nu)}{E}\tau_{xyp} = -\frac{2(1+\nu)}{E}\frac{\partial^2 F}{\partial x \partial y} \tag{3.43}$$

Substituting Eqs. (3.41)–(3.43) to Eq. (3.35), the compatibility condition is given in the form

$$\frac{\partial^4 F}{\partial x^4} + 2\frac{\partial^4 F}{\partial x^2 \partial y^2} + \frac{\partial^4 F}{\partial y^4} = E\left[\left(\frac{\partial^2 w}{\partial x \partial y}\right)^2 - \left(\frac{\partial^2 w}{\partial x^2}\right)\left(\frac{\partial^2 w}{\partial y^2}\right)\right.$$
$$\left. - \left(\frac{\partial^2 w_0}{\partial x \partial y}\right)^2 + \left(\frac{\partial^2 w_0}{\partial x^2}\right)\left(\frac{\partial^2 w_0}{\partial y^2}\right)\right] \tag{3.44}$$

Substituting the deflections, Eqs. (3.30) and (3.31), into Eq. (3.44), and considering the loading condition and welding residual stress, Airy's stress function, F, can be derived in the form

$$F = F(A_{0ij}, A_{0kl}, A_{ij}, A_{kl}, i, j, k, l, a, b, \alpha, k(x), f(y), \overline{\sigma}_x, \overline{\sigma}_y, \overline{\tau}_{xy}, E, x, y) \tag{3.45}$$

where $\overline{\sigma}_x$ and $\overline{\sigma}_y$ are the average normal stresses in longitudinal and transverse directions, respectively, and $\overline{\tau}_{xy}$ is the average shear stress. $\alpha = a/b$ is the aspect ratio of the plate. $k(x)$ and $f(y)$ are the functions giving the welding residual stress in y- and x-directions, respectively, and residual stress, $\sigma_{yr}(x)$ and $\sigma_{xr}(y)$, are given by

$$\sigma_{yr}(x) = \frac{\partial^2 k(x)}{\partial x^2}, \quad \sigma_{xr}(y) = \frac{\partial^2 f(y)}{\partial y^2} \tag{3.46}$$

Airy's stress function of Eq. (3.45) is given in Section 3.5.1 in Section 3.5.

3.3.3 BENDING STRESS AND STRAIN

According to the definition of plate bending [1], bending strain components are given as

$$\epsilon_{xb} = -z\frac{\partial^2}{\partial x^2}(w - w_0) \tag{3.47}$$

$$\epsilon_{yb} = -z\frac{\partial^2}{\partial y^2}(w - w_0) \tag{3.48}$$

$$\gamma_{xyb} = -2z\frac{\partial^2}{\partial x \partial y}(w - w_0) \tag{3.49}$$

and under the plane stress condition, the bending stress components are given as follows:

$$\sigma_{xb} = \frac{E}{1 - v^2}(\epsilon_{xb} + v\epsilon_{yb}) = -\frac{E}{1 - v^2}z\left(\frac{\partial^2}{\partial x^2} + v\frac{\partial^2}{\partial y^2}\right)(w - w_0) \tag{3.50}$$

$$\sigma_{yb} = \frac{E}{1 - v^2}(\epsilon_{yb} + v\epsilon_{xb}) = -\frac{E}{1 - v^2}z\left(\frac{\partial^2}{\partial y^2} + v\frac{\partial^2}{\partial x^2}\right)(w - w_0) \tag{3.51}$$

$$\tau_{xyb} = \frac{E}{2(1 + v)}\gamma_{xyp} = -\frac{E}{(1 + v)}z\frac{\partial^2}{\partial x \partial y}(w - w_0) \tag{3.52}$$

Substituting Eqs. (3.30) and (3.31) into Eqs. (3.47)–(3.52), bending strain and stress components are derived; see Section 3.5.3 in Section 3.5.

3.3.4 OVERALL SHRINKAGE AND SHEAR DEFORMATION CONSIDERING LARGE DEFLECTION EFFECTS

To evaluate the work done by the applied in-plane loads, it is necessary to know the overall shrinkage and shear deformation of the rectangular plate considering large deflection effects. The average shrinkages, $\bar{u}_t$, and $\bar{v}_t$ in x- and y-directions are derived as:

$$\bar{u}_t = \frac{1}{b}\int_0^a\int_0^b\frac{\partial u}{\partial x}dx\, dy$$
$$= -\frac{a}{E}(\bar{\sigma}_x - v\bar{\sigma}_y) - \frac{\pi^2}{8a}\sum_{i=1}^m\sum_{j=1}^n i^2\left(A_{ij}^2 - A_{0ij}^2\right) \tag{3.53}$$

$$\bar{v}_t = \frac{1}{a}\int_0^a\int_0^b\frac{\partial v}{\partial y}dx\, dy$$
$$= -\frac{b}{E}(\bar{\sigma}_y - v\bar{\sigma}_x) - \frac{\pi^2}{8b}\sum_{i=1}^m\sum_{j=1}^n j^2\left(A_{ij}^2 - A_{0ij}^2\right) \tag{3.54}$$

On the other hand, denoting the average shear displacements in x- and y-directions as $\bar{u}_s$ and $\bar{v}_s$, the following equation with regard to the shear deformation is obtained:

$$
a\bar{u}_s + b\bar{v}_s = \int_0^a \int_0^b \left(\frac{\partial u}{\partial y} + \frac{\partial v}{\partial x} \right) dx\, dy
$$

$$
= \sum_{i=1}^{m} \sum_{j=1}^{n} \sum_{k=1}^{o} \sum_{l=1}^{p} T(i,j,k,l)(A_{ij}A_{kl} - A_{0ij}A_{0kl}) + ab\frac{2(1+v)}{E}\bar{\tau}_{xy}
$$

$$
- \sum_{i=1}^{m} \sum_{j=1}^{n} \sum_{k=1}^{o} \sum_{l=1}^{p} S(i,j,k,l)(A_{ij}A_{kl} - A_{0ij}A_{0kl}) \tag{3.55}
$$

where $T(i,j,k,l)$ and $S(i,j,k,l)$ are the functions of $a, b, t, E, v, i, j, k,$ and l, and are given in Section 3.5.4 in Section 3.5.

The average compressive strains, $\bar{\epsilon}_x$ in x-direction and $\bar{\epsilon}_y$ in y-direction, and the average shear strain, $\bar{\gamma}_{xy}$, can be obtained as follows:

$$
\bar{\epsilon}_x = \frac{\bar{u}_t}{a} \tag{3.56}
$$

$$
\bar{\epsilon}_y = \frac{\bar{v}_t}{b} \tag{3.57}
$$

$$
\bar{\gamma}_{xy} = \frac{1}{ab}(a\bar{u}_s + b\bar{v}_s) = \frac{\bar{u}_s}{b} + \frac{\bar{v}_s}{a} \tag{3.58}
$$

3.3.5 ELASTIC LARGE DEFLECTION BEHAVIOR

The remaining condition to be considered for the formulation of the elastic large deflection behavior is an equilibrium condition of forces in out-of-plane direction (z-direction). This condition can be given by applying the *Principle of Virtual Work* considering all stress and strain components and external load actions.

The general form of the *Principle of Virtual Work* is expresses as

$$
\int_V \{\delta\epsilon\}^T \{\sigma\} dV = \{\delta\bar{u}\}^T \{F\} \tag{3.59}
$$

where δ represents virtual values, and $\{\epsilon\}, \{\sigma\}, \{\bar{u}\}$, and $\{F\}$ are a strain vector, a stress vector, a displacement vector, and an external load vector, respectively, which are given as

$$
\{\epsilon\} = \left\{ \begin{array}{c} \epsilon_{xp} + \epsilon_{xb} \\ \epsilon_{yp} + \epsilon_{yb} \\ \gamma_{xyp} + \gamma_{xyb} \end{array} \right\}, \quad \{\sigma\} = \left\{ \begin{array}{c} \sigma_{xp} + \sigma_{xb} \\ \sigma_{yp} + \sigma_{yb} \\ \tau_{xyp} + \tau_{xyb} \end{array} \right\}
$$

$$
\{\bar{u}\} = \left\{ \begin{array}{c} -\bar{u}_t \\ -\bar{v}_t \\ \bar{u}_s \\ \bar{v}_s \end{array} \right\}, \quad \{F\} = \left\{ \begin{array}{c} bt\bar{\sigma}_x \\ at\bar{\sigma}_y \\ at\bar{\tau}_{xy} \\ bt\bar{\tau}_{xy} \end{array} \right\}
$$

Note that $\bar{u}_t$ and $\bar{v}_t$ are positive for the shortening, and thus the negative sign is added in $\{\bar{u}\}$.

Substitution of the total and initial deflections, Eqs. (3.30) and (3.31), into Eq. (3.59) yields the third-order simultaneous equations with respect to the $(m \times n)$ deflection components, A_{ij}. Hence, for the practical application, it is convenient to reformulate the equations into an incremental form. In addition, in order to simulate the dynamic buckling behavior, such as a dynamic snap-through behavior due to the secondary buckling, as will be explained in Section 4.4, the inertia and damping forces need to be included.

The *Principle of Virtual Work*, Eq. (3.59), with consideration of the dynamic load effects is given by

$$\int_V \{\delta\epsilon\}^T\{\sigma\}dV = \{\delta\bar{u}\}^T\{F\} + \int_V \{\delta u\}^T\{-\rho\ddot{u}\}dV - \int_V \{\delta\epsilon\}^T\{\sigma_{sd}\}dV \tag{3.60}$$

and its incremental form by

$$\int_V \{\delta\Delta\epsilon\}^T\{\sigma + \Delta\sigma\}dV = \{\delta\Delta\bar{u}\}^T\{F + \Delta F\} + \int_V \{\delta\Delta u\}^T\{-\rho(\ddot{u} + \Delta\ddot{u})\}dV$$
$$- \int_V \{\delta\Delta\epsilon\}^T\{\sigma_{sd} + \Delta\sigma_{sd}\}dV \tag{3.61}$$

where ρ is a density of material, $\{\sigma_{sd}\}$ is a structural damping force vector, and $\{\ddot{u}\}$ and $\{\dot{u}\}$ are the acceleration and velocity vector, respectively. In the following, the structural damping force is assumed to be proportional to the strain rate as

$$\{\sigma_{sd}\} = C_{sd}[D^e]\{\dot{\epsilon}\} = C_{sd}[D^e][D^e]^{-1}\{\dot{\sigma}\} = C_{sd}\{\dot{\sigma}\} \tag{3.62}$$

where C_{sd} is structural damping coefficient, which is here evaluated as

$$C_{sd} = \frac{2\zeta}{\omega} \tag{3.63}$$

and ζ and ω are the specified damping ratio and eigen frequency, respectively.

Ignoring the higher-order terms of deflection increments and retaining the linear increment term in Eq. (3.61), the final equation for the increments of the deflection components are given in the form

$$t\sum_{g=1}^{s}\sum_{h=1}^{t}(\lceil M_{gh}\rfloor\{\Delta\ddot{A}\} + \lceil C_{gh}\rfloor\{\Delta\dot{A}\} + \lceil K_{gh}\rfloor\{\Delta A\} - \lceil R_{gh}\rfloor\{\Delta\bar{\sigma}\} - Q_{gh})\tilde{\Delta}A_{gh} = 0 \tag{3.64}$$

where

$$\lceil M_{gh}\rfloor = \lceil M_{gh-11} \quad M_{gh-12} \quad M_{gh-13} \quad \cdots \rfloor$$
$$\lceil C_{gh}\rfloor = \lceil C_{gh-11} \quad C_{gh-12} \quad C_{gh-13} \quad \cdots \rfloor$$
$$\lceil K_{gh}\rfloor = \lceil K_{gh-11} \quad K_{gh-12} \quad K_{gh-13} \quad \cdots \rfloor$$
$$\{\Delta A\} = \lceil \Delta A_{11} \quad \Delta A_{12} \quad \Delta A_{13} \quad \cdots \rfloor^T$$
$$\lceil R_{gh}\rfloor = \lceil R_{x-gh} \quad R_{y-gh} \quad R_{xy-gh}\rfloor$$
$$\{\Delta\bar{\sigma}\} = \lceil \Delta\bar{\sigma}_x \quad \Delta\bar{\sigma}_y \quad \Delta\bar{\tau}_{xy}\rfloor^T$$

Eq. (3.64) represents the relationship among increments of deflection coefficients, $\{\Delta A\}$, their velocity, $\{\Delta\dot{A}\}$, their acceleration, $\{\Delta\ddot{A}\}$ and increment of average stress, $\{\Delta\bar{\sigma}\}$, and is written in the matrix form as follows:

$$[M]\{\Delta\ddot{A}\} + [C]\{\Delta\dot{A}\} + [K]\{\Delta A\} = [R]\{\Delta\bar{\sigma}\}\sigma + \{Q\} \tag{3.65}$$

where

$$[M] = [\ \lceil M_{11}\rfloor^T \quad \lceil M_{12}\rfloor^T \quad \lceil M_{13}\rfloor^T \quad \cdots\ ]^T$$
$$[C] = [\ \lceil C_{11}\rfloor^T \quad \lceil C_{12}\rfloor^T \quad \lceil C_{13}\rfloor^T \quad \cdots\ ]^T$$
$$[K] = [\ \lceil K_{11}\rfloor^T \quad \lceil K_{12}\rfloor^T \quad \lceil K_{13}\rfloor^T \quad \cdots\ ]^T$$
$$[R] = [\ \lceil R_{11}\rfloor^T \quad \lceil R_{12}\rfloor^T \quad \lceil R_{13}\rfloor^T \quad \cdots\ ]^T$$
$$\{Q\} = \lceil Q_{11}\ Q_{12}\ Q_{13}\ \cdots\ \rfloor^T$$

$\{Q\}$ represents the load correction term which compensates the unbalancing force caused by linear approximation of nonlinear behavior during a small load increment. The load correction term is equal to zero, that is $\{Q\} = \{0\}$, when the plate is in completely equilibrium state.

In the case when $\{Q\} = \{0\}$, Eq. (3.64) reduces to the following equation:

$$t\sum_{i=1}^{m}\sum_{j=1}^{n}\rho\frac{ab}{4}\ddot{A}_{ij}\tilde{\Delta}A_{kl} + C_{sd}\,t\sum_{i=1}^{m}\sum_{j=1}^{n}\sum_{k=1}^{o}\sum_{l=1}^{p}\sum_{e=1}^{q}\sum_{f=1}^{r}\sum_{g=1}^{s}\sum_{h=1}^{t}H(i,j,k,l,e,f,g,h)$$

$$\times\left[(\dot{A}_{ij}A_{kl} + A_{ij}\dot{A}_{kl})(A_{ef}\tilde{\Delta}A_{gh} + A_{gh}\tilde{\Delta}A_{ef})\right]$$

$$+ C_{sd}\,t\sum_{e=1}^{q}\sum_{f=1}^{r}\sum_{g=1}^{s}\sum_{h=1}^{t}T(e,f,g,h)(\dot{\bar{\tau}}_{xy} + \ddot{\bar{\tau}}_{xy}\Delta t)(A_{ef}\tilde{\Delta}A_{gh} + A_{gh}\tilde{\Delta}A_{ef})$$

$$+ C_{sd}\,\frac{\pi^4 Dab}{4}\sum_{i=1}^{m}\sum_{j=1}^{n}\left(\frac{i^2}{a^2} + \frac{j^2}{b^2}\right)^2\dot{A}_{ij}\tilde{\Delta}A_{ij}$$

$$+ t\sum_{i=1}^{m}\sum_{j=1}^{n}\sum_{k=1}^{o}\sum_{l=1}^{p}\sum_{e=1}^{q}\sum_{f=1}^{r}\sum_{g=1}^{s}\sum_{h=1}^{t}H(i,j,k,l,e,f,g,h)\left[(A_{ij}A_{kl} - A_{0ij}A_{0kl})\right.$$

$$\times\left(A_{ef}\tilde{\Delta}A_{gh} + A_{gh}\tilde{\Delta}A_{ef}\right)\bigg]$$

$$- t\sum_{e=1}^{q}\sum_{f=1}^{r}\sum_{g=1}^{s}\sum_{h=1}^{t}\frac{\pi^2\alpha^2}{4}J(e,f,g,h)(A_{ef}\tilde{\Delta}A_{gh} + A_{gh}\tilde{\Delta}A_{ef})$$

$$+ \frac{\pi^4 Dab}{4}\sum_{i=1}^{m}\sum_{j=1}^{n}\left(\frac{i^2}{a^2} + \frac{j^2}{b^2}\right)^2(A_{ij} - A_{0ij})\tilde{\Delta}A_{ij}$$

$$- \frac{\pi^2 bt}{4a}\sum_{i=1}^{m}\sum_{j=1}^{n}i^2\bar{\sigma}_x A_{ij}\tilde{\Delta}A_{ij} - \frac{\pi^2 at}{4b}\sum_{i=1}^{m}\sum_{j=1}^{n}j^2\bar{\sigma}_y A_{ij}\tilde{\Delta}A_{ij}$$

$$+ t\sum_{e=1}^{q}\sum_{f=1}^{r}\sum_{g=1}^{s}\sum_{h=1}^{t}S(e,f,g,h)\bar{\tau}_{xy}(A_{ef}\tilde{\Delta}A_{gh} + A_{gh}\tilde{\Delta}A_{ef}) = 0 \tag{3.66}$$

For the static analysis, the equilibrium equation is obtained by setting $\rho = 0$ and $C_{sd} = 0$ in Eq. (3.66), and the following equation is obtained:

$$[\mathbf{K'}]\{\Delta\mathbf{A}\} - [\mathbf{R}]\{\Delta\bar{\sigma}\} = \{\mathbf{Q'}\} \tag{3.67}$$

Eq. (3.67) is equivalent to the static equilibrium equation derived by applying *Principle of Virtual Work*:

$$
\begin{aligned}
&\frac{\pi^2}{4}abt\left[\sum_{i=1}^{m}\sum_{j=1}^{n}\left(\frac{i^2}{a^2}\bar{\sigma}_x + \frac{j^2}{b^2}\bar{\sigma}_y\right)\Delta A_{ij}\right.\\
&\quad + \frac{4}{\pi^2 ab}\sum_{e=1}^{o}\sum_{f=1}^{p}\sum_{g=1}^{s}\sum_{h=1}^{t}\{T(e,f,g,h) - S(e,f,g,h)\}\bar{\tau}_{xy}\Delta A_{ef}\\
&\quad + \sum_{e=1}^{q}\sum_{f=1}^{r}\frac{e^2}{a^2}A_{ef}\Delta\bar{\sigma}_x + \sum_{e=1}^{q}\sum_{f=1}^{r}\frac{f^2}{b^2}A_{ef}\Delta\bar{\sigma}_y\\
&\quad + \frac{4}{\pi^2 ab}\sum_{e=1}^{o}\sum_{f=1}^{p}\sum_{g=1}^{s}\sum_{h=1}^{t}\{T(e,f,g,h) - S(e,f,g,h)\}A_{gh}\Delta\bar{\tau}_{xy}\\
&\quad \left. + \sum_{i=1}^{m}\sum_{j=1}^{n}\left(\frac{i^2}{a^2}\bar{\sigma}_x + \frac{j^2}{b^2}\bar{\sigma}_y\right)A_{ij}\right] \times \delta\Delta A_{ij}\\
&= \frac{\pi^2}{4}abt\left[\frac{\pi^2\alpha^2 E}{16a^2b^2}\sum_{i=1}^{m}\sum_{j=1}^{n}\sum_{k=1}^{o}\sum_{l=1}^{p}\sum_{e=1}^{q}\sum_{f=1}^{r}\sum_{g=1}^{s}\sum_{h=1}^{t}H(i,j,k,l,e,f,g,h)\{A_{ij}A_{ef}\Delta A_{kl}\right.\\
&\quad + A_{ef}A_{kl}\Delta A_{ij} + (A_{ij}A_{kl} - A_{0ij}A_{0kl})\Delta A_{ef} + (A_{ij}A_{kl} - A_{0ij}A_{0kl})A_{ef}\}\\
&\quad + \frac{\alpha}{ab}\sum_{e=1}^{q}\sum_{f=1}^{r}\sum_{g=1}^{s}\sum_{h=1}^{t}\{(Q_1(e,f,g,h)\bar{\sigma}_x + Q_2(e,f,g,h)\bar{\sigma}_y - Q_3(e,f,g,h)\bar{\tau}_{xy} - RS)\Delta A_{ef}\\
&\quad + Q_1(e,f,g,h)A_{ef}\Delta\bar{\sigma}_x + Q_2(e,f,g,h)A_{ef}\Delta\bar{\sigma}_y - Q_3(e,f,g,h)A_{ef}\Delta\bar{\tau}_{xy}\\
&\quad + (Q_1(e,f,g,h)\bar{\sigma}_x + Q_2(e,f,g,h)\bar{\sigma}_y - Q_3(e,f,g,h)\bar{\tau}_{xy}\\
&\quad \left.\left. - RS(e,f,g,h))A_{ef}\right]\}\right] \times \delta\Delta A_{gh}\\
&\quad + \frac{\pi^2}{4}abt\left[\frac{\pi^2\alpha^2 E}{16a^2b^2}\sum_{i=1}^{m}\sum_{j=1}^{n}\sum_{k=1}^{o}\sum_{l=1}^{p}\sum_{e=1}^{q}\sum_{f=1}^{r}\sum_{g=1}^{s}\sum_{h=1}^{t}H(i,j,k,l,e,f,g,h)\{A_{ij}A_{gh}\Delta A_{kl}\right.\\
&\quad + A_{gh}A_{kl}\Delta A_{ij} + (A_{ij}A_{kl} - A_{0ij}A_{0kl})\Delta A_{gh} + (A_{ij}A_{kl} - A_{0ij}A_{0kl})A_{gh}\}\\
&\quad + \frac{\alpha}{ab}\sum_{e=1}^{q}\sum_{f=1}^{r}\sum_{g=1}^{s}\sum_{h=1}^{t}\{(Q_1(e,f,g,h)\bar{\sigma}_x + Q_2(e,f,g,h)\bar{\sigma}_y - Q_3(e,f,g,h)\bar{\tau}_{xy} - RS)\Delta A_{ef}\\
&\quad + Q_1(e,f,g,h)A_{gh}\Delta\bar{\sigma}_x + Q_2(e,f,g,h)A_{gh}\Delta\bar{\sigma}_y - Q_3(e,f,g,h)A_{gh}\Delta\bar{\tau}_{xy}\\
&\quad + < Q_1(e,f,g,h)\bar{\sigma}_x + Q_2(e,f,g,h)\bar{\sigma}_y - Q_3(e,f,g,h)\bar{\tau}_{xy}\\
&\quad \left. - RS(e,f,g,h) > A_{gh}\}\right] \times \delta\Delta A_{ef}
\end{aligned}
$$

$$+ \frac{\pi^2}{4}abt \left[\frac{\pi^2 t^2 E}{12(1-v^2)} \sum_{i=1}^{m}\sum_{j=1}^{n} \left(\frac{i^2}{a^2} + \frac{j^2}{b^2} \right)^2 (\Delta A_{ij} + A_{ij} - A_{0ij}) \right] \times \delta \Delta A_{ij} \qquad (3.68)$$

Coefficient functions, $H(i,j,k,l,e,f,g,h)$, $T(e,f,g,h)$, $S(e,f,g,h)$, $J(e,f,g,h)$, $Q_1(e,f,g,h)$, $Q_2(e,f,g,h)$, $Q_3(e,f,g,h)$, and $RS(e,f,g,h)$, are the functions of parameters representing plate dimensions and material properties as well as integer parameters, i,j,k,l,e,f,g, and h and are given in Section 3.5.4 in Section 3.5.

Coefficient of the cross term of $\delta \Delta A_{kl}$ and ΔA_{ij} in Eq. (3.68) comes at (kl, ij) in $[K]$ in Eq. (3.67), that is at the klth row and the ijth column in the matrix.

3.3.6 ELASTIC LARGE DEFLECTION ANALYSIS OF RECTANGULAR PLATE SUBJECTED TO UNI-AXIAL THRUST

As the simplest case, a rectangular plate subjected to longitudinal thrust load, as was shown in Fig. 3.8, is considered. The total deflection and initial deflection are expressed by one term of the elastic buckling mode as

$$w = A_m \sin \frac{m\pi x}{a} \sin \frac{\pi y}{b} \qquad (3.69)$$

$$w_0 = A_{0m} \sin \frac{m\pi x}{a} \sin \frac{\pi y}{b} \qquad (3.70)$$

The equilibrium equation corresponding to Eq. (3.68) becomes

$$\frac{m^2\pi^2 a^2 E}{16} \left(\frac{1}{a^4} + \frac{1}{b^4} \right)(A_m^2 - A_{0m}^2)A_m + \frac{\pi^2 E}{12(1-v^2)} \left(\frac{t}{b} \right)^2 \left(\frac{\alpha}{m} + \frac{m}{\alpha} \right)^2 (A_m - A_{0m}) + (\zeta - \sigma)A_m = 0 \quad (3.71)$$

where

$$\zeta = -\frac{1}{b} \int_0^b g(y) \cos \frac{2\pi y}{b} dy = \frac{\sigma_Y}{\pi(1-\mu)} \sin \mu\pi \qquad (3.72)$$

μ represents the breadth where tensile residual stress is produced, and is defined by Eq. (3.23).

Dividing the average shortening expressed by Eq. (3.53) with the length, a, average stress-average strain relationship is derived as:

$$\bar{\varepsilon}_x = -\frac{1}{E}\bar{\sigma}_x - \frac{m^2\pi^2}{8a^2}(A_m^2 - A_{0m}^2) \qquad (3.73)$$

3.4 ELASTOPLASTIC LARGE DEFLECTION ANALYSIS
3.4.1 FUNDAMENTALS

In the elastic large deflection analysis, the stress-strain relationship of the material is elastic and the same all over the whole plate. Therefore, accurate solution can be obtained in an analytical form if the assumed deflection mode is accurate. On the other hand, in the elastoplastic large deflection analysis, plastic stress-strain relationship has to be used in the plastic region, and elastic stress-strain relationship in the elastic region. This makes it almost impossible to perform volume integration analytically over the whole plate when *Energy Theorem* is applied. This is the reason why elastoplastic large deflection analysis is in general performed applying numerical methods such as the FEM dividing the plate into small elements.

In the following, fundamental formulation for elastoplastic large deflection analysis is explained applying the FEM. The element that will be introduced here is an isoparametric shell element with four nodes [4] which is characterized as a bi-linear degenerated shell element with reduced integration. Both material and geometrical nonlinearities are considered. This shell element has following characteristics:

(1) The virtual work equation is expressed in terms of updated Green's strain components, and the calculated Kirchhoff's stress increments which are transformed into Jaumann's stress increments [5] at each incremental step. This enables to analyze ultra-large deflection problems.
(2) In this transformation, normal strain in the thickness direction is determined so that a plane stress condition is satisfied.
(3) *Plastic Flow Theory* is applied considering von Mises's yield condition as a plastic potential.
(4) The virtual stiffness against in-plane rotation [6] is incorporated in the element stiffness.
(5) The virtual stiffness to control hourglass mode instability [7] is implemented in the element stiffness.

Fig. 3.9 shows the coordinate system of the isoparametric shell element. There exist three coordinate systems, which are

(a) x-, y-, z-coordinates: reference coordinate system which defines spatial location of the element.
(b) x'-, y'-, z'-coordinates: element coordinate system which defines direction of stress and strain components in the element.
(c) ξ-, η-coordinates: local coordinate system which defines spatial location in the element.

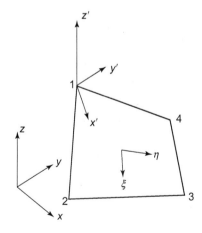

FIG. 3.9

Bilinear quadrilateral shell element.

The origin of the x'-, y'-, z'-coordinate system locates at nodal point 1, and z'-axis is so defined that it is perpendicular to both diagonals, $\overline{13}$ and $\overline{24}$. Then, x'-axis is defined in the plane containing z'-axis and the side $\overline{12}$ with the condition that it is perpendicular to z'-axis. y'-axis is defined so as to be perpendicular to both x'- and z'-axes.

3.4.2 DISPLACEMENT FIELD IN THE ELEMENT

Displacement field in an element is defined by the nodal point displacements, which are three translations and three rotations. The nodal point displacement vectors at nodal point i with respect to reference coordinate system and element coordinate system are defined as:

$$\begin{aligned}
\{h_i\}^T &= \lfloor u_i \; v_i \; w_i \; \theta_{xi} \; \theta_{yi} \; \theta_{zi} \rfloor \\
\{h_i'\}^T &= \lfloor u_i' \; v_i' \; w_i' \; \theta_{xi}' \; \theta_{yi}' \; \theta_{zi}' \rfloor
\end{aligned} \right\} \tag{3.74}$$

The displacement vector, $\{h_i\}$, with respect to the reference coordinate system can be transformed to the displacement vector, $\{h_i'\}$, with respect to the element coordinate system through coordinate transformation matrix as:

$$\{h_i'\} = [L_i]\{h_i\} \tag{3.75}$$

where

$$[L_i] = \begin{bmatrix} [T_i] & 0 \\ 0 & [T_i] \end{bmatrix} \tag{3.76}$$

$$[T_i] = \begin{bmatrix} \lambda_{x'x} & \lambda_{x'y} & \lambda_{x'z} \\ \lambda_{y'x} & \lambda_{y'y} & \lambda_{y'z} \\ \lambda_{z'x} & \lambda_{z'y} & \lambda_{z'z} \end{bmatrix} \tag{3.77}$$

$\{h_i\}$ and $\{h_i'\}$ are defined as the displacements at nodal point i on the middle plane of the plate. λ_{ij} in Eq. (3.77) is the direction cosine between the axes, "i" and "j."

The displacements (u_m', v_m', w_m') at point $P_m(\xi, \eta)$ are defined as:

$$\left\{ \begin{array}{c} u_m' \\ v_m' \\ w_m' \end{array} \right\} = \sum_{i=1}^{4} N_i(\xi, \eta) \left\{ \begin{array}{c} u_i' \\ v_i' \\ w_i' \end{array} \right\} \tag{3.78}$$

where $N_i (i = 1 \sim 4)$ is bi-linear interpolation function defined as:

$$\begin{aligned}
N_1 &= (1 - \xi)(1 - \eta)/4 \\
N_2 &= (1 + \xi)(1 - \eta)/4 \\
N_3 &= (1 + \xi)(1 + \eta)/4 \\
N_4 &= (1 - \xi)(1 + \eta)/4
\end{aligned} \right\} \tag{3.79}$$

Then, at point P which locates at z' toward the thickness direction from point P_m on the middle plane, displacements can be expressed as follows:

$$\left\{ \begin{array}{c} u' \\ v' \\ w' \end{array} \right\} = \sum_{i=1}^{4} N_i(\xi, \eta) \left\{ \begin{array}{c} u_i' \\ v_i' \\ w_i' \end{array} \right\} + \sum_{i=1}^{4} z' N_i(\xi, \eta) \left\{ \begin{array}{c} \theta_{yi}' \\ -\theta_{xi}' \\ 0 \end{array} \right\} \tag{3.80}$$

3.4.3 STRAIN-DISPLACEMENT RELATIONSHIPS

Five updated Green's strain components are defined except the normal strain in the thickness direction in an incremental form as:

$$
\left\{
\begin{array}{c}
\Delta\varepsilon_x{}^* \\
\Delta\varepsilon_y{}^* \\
\Delta\gamma_{xy}{}^* \\
\Delta\gamma_{yz}{}^* \\
\Delta\gamma_{zx}{}^*
\end{array}
\right\}
=
\left\{
\begin{array}{c}
\frac{\partial \Delta u'}{\partial x'} \\[4pt]
\frac{\partial \Delta v'}{\partial y'} \\[4pt]
\frac{\partial \Delta u'}{\partial y'} + \frac{\partial \Delta v'}{\partial x'} \\[4pt]
\frac{\partial \Delta v'}{\partial z'} + \frac{\partial \Delta w'}{\partial y'} \\[4pt]
\frac{\partial \Delta w'}{\partial x'} + \frac{\partial \Delta u'}{\partial z'}
\end{array}
\right\}
+
\left\{
\begin{array}{c}
\frac{1}{2}\left(\frac{\partial \Delta w'}{\partial x'}\right)^2 \\[4pt]
\frac{1}{2}\left(\frac{\partial \Delta w'}{\partial y'}\right)^2 \\[4pt]
\frac{\partial \Delta w'}{\partial x'} \cdot \frac{\partial \Delta w'}{\partial y'} \\[4pt]
0 \\[4pt]
0
\end{array}
\right\}
$$

$$
=
\left\{
\begin{array}{c}
\Delta\varepsilon_x \\
\Delta\varepsilon_y \\
\Delta\gamma_{xy} \\
\Delta\gamma_{yz} \\
\Delta\gamma_{zx}
\end{array}
\right\}
+
\left\{
\begin{array}{c}
\Delta\overline{\varepsilon}_x \\
\Delta\overline{\varepsilon}_y \\
\Delta\overline{\gamma}_{xy} \\
0 \\
0
\end{array}
\right\}
= \{\Delta\varepsilon\} + \{\Delta\overline{\varepsilon}\}
\tag{3.81}
$$

$\{\Delta\varepsilon\}$ and $\{\Delta\overline{\varepsilon}\}$ represent linear and nonlinear strain components, respectively. The linear strain components are expressed as:

$$
\left.
\begin{array}{l}
\Delta\varepsilon_x = \sum_{i=1}^{4}\left\{\Delta u_i'\left(\frac{\partial N_i}{\partial x'}\right) + z'\left(\Delta\theta_{yi}'\right)\left(\frac{\partial N_i}{\partial x'}\right)\right\} \\[6pt]
\Delta\varepsilon_y = \sum_{i=1}^{4}\left\{\Delta v_i'\left(\frac{\partial N_i}{\partial y'}\right) - z'\left(\Delta\theta_{xi}'\right)\left(\frac{\partial N_i}{\partial y'}\right)\right\} \\[6pt]
\Delta\gamma_{xy} = \sum_{i=1}^{4}\left\{\Delta u_i'\left(\frac{\partial N_i}{\partial y'}\right) + z'\left(\Delta\theta_{yi}'\right)\left(\frac{\partial N_i}{\partial y'}\right)\right. \\[6pt]
\qquad \left. + \Delta v_i'\left(\frac{\partial N_i}{\partial x'}\right) - z'\left(\Delta\theta_{xi}'\right)\left(\frac{\partial N_i}{\partial x'}\right)\right\} \\[6pt]
\Delta\gamma_{xz} = \sum_{i=1}^{4}\left\{N_i\left(\Delta\theta_{yi}'\right) + \Delta w_i'\left(\frac{\partial N_i}{\partial x'}\right)\right\} \\[6pt]
\Delta\gamma_{yz} = \sum_{i=1}^{4}\left\{-N_i\left(\Delta\theta_{xi}'\right) + \Delta w_i'\left(\frac{\partial N_i}{\partial y'}\right)\right\}
\end{array}
\right\}
\tag{3.82}
$$

Here, performing reduced integration at point $\xi = \eta = 0$, the following expressions are obtained:

$$
\left.
\begin{array}{l}
N_1 = N_2 = N_3 = N_4 = \frac{1}{4} \\[8pt]
\frac{\partial N_1}{\partial x'} = \frac{1}{2S}\left(y_2' - y_4'\right), \quad \frac{\partial N_2}{\partial x'} = \frac{1}{2S}\left(y_3' - y_1'\right) \\[8pt]
\frac{\partial N_3}{\partial x'} = \frac{1}{2S}\left(y_4' - y_2'\right), \quad \frac{\partial N_4}{\partial x'} = \frac{1}{2S}\left(y_1' - y_3'\right) \\[8pt]
\frac{\partial N_1}{\partial y'} = \frac{1}{2S}\left(x_4' - x_2'\right), \quad \frac{\partial N_2}{\partial y'} = \frac{1}{2S}\left(x_1' - x_3'\right) \\[8pt]
\frac{\partial N_3}{\partial y'} = \frac{1}{2S}\left(x_2' - x_4'\right), \quad \frac{\partial N_4}{\partial y'} = \frac{1}{2S}\left(x_3' - x_1'\right)
\end{array}
\right\}
\tag{3.83}
$$

where S is the area of the element and is expressed as:

$$S = \frac{1}{2}\left\{(x_3' - x_1')(y_4' - y_2') + (x_2' - x_4')(y_3' - y_1')\right\} \tag{3.84}$$

The derivation of Eq. (3.83) is shown in Section 3.6.

Substituting Eq. (3.83) into Eq. (3.82), the following relationship is obtained between increments of strain components and nodal displacement components.

$$\{\Delta\varepsilon\} = [\mathbf{B}]\{\Delta h'\} \tag{3.85}$$

$$[B] = [[B^{(1)}] \quad [B^{(2)}] \quad [B^{(3)}] \quad [B^{(4)}]] \tag{3.86}$$

$$[B^{(i)}] = \frac{1}{2S}
\begin{bmatrix}
y(i) & 0 & 0 & 0 & z'y(i) & 0 \\
0 & x(i) & 0 & -z'x(i) & 0 & 0 \\
x(i) & y(i) & 0 & -z'y(i) & z'x(i) & 0 \\
0 & 0 & y(i) & 0 & S/2 & 0 \\
0 & 0 & x(i) & -S/2 & 0 & 0
\end{bmatrix} \tag{3.87}$$

where

$$\begin{aligned}
x(1) &= y_2' - y_4', & y(1) &= x_4' - x_2' \\
x(2) &= y_3' - y_1', & y(2) &= x_1' - x_3' \\
x(3) &= y_4' - y_2', & y(3) &= x_2' - x_4' \\
x(4) &= y_1' - y_3', & y(4) &= x_3' - x_1'
\end{aligned} \tag{3.88}$$

On the other hand, nonlinear strain component can be expressed in the following form:

$$\{\Delta\bar{\varepsilon}\} = \frac{1}{2}[C][B_1]\{\Delta h'\} \tag{3.89}$$

where

$$[C] =
\begin{bmatrix}
\frac{\partial w'}{\partial x'} & 0 \\
0 & \frac{\partial w'}{\partial y'} \\
\frac{\partial w'}{\partial y'} & \frac{\partial w'}{\partial x'}
\end{bmatrix} \tag{3.90}$$

and $[B_1]$ is defined as:

$$\left\{ \begin{matrix} \frac{\partial \Delta w'}{\partial x'} \\ \frac{\partial \Delta w'}{\partial y'} \end{matrix} \right\} = [B_1]\{\Delta h'\} \tag{3.91}$$

where $[B_1]$ is expressed as follows:

$$[B_1] = [[B_1^{(1)}] \quad [B_1^{(2)}] \quad [B_1^{(3)}] \quad [B_1^{(4)}]] \tag{3.92}$$

$$[B_1^{(i)}] =
\begin{bmatrix}
0 & 0 & y(i) & 0 & 0 & 0 \\
0 & 0 & x(i) & 0 & 0 & 0
\end{bmatrix} \tag{3.93}$$

3.4.4 STRESS-STRAIN RELATIONSHIPS

Updated Kirchhoff stress increments, $\{\Delta\sigma^*\}$, is related to updated Green strain increments, $\{\Delta\varepsilon^*\}$, through stress-strain matrix, $[D]$, as:

$$\{\Delta\sigma^*\} = [D]\{\Delta\varepsilon^*\} \tag{3.94}$$

where

$$\{\Delta\sigma^*\}^T = \begin{bmatrix} \Delta\sigma_x^* & \Delta\sigma_y^* & \Delta\sigma_z^* & \Delta\tau_{xy}^* & \Delta\tau_{yz}^* & \Delta\tau_{zx}^* \end{bmatrix} \tag{3.95}$$

$$\{\Delta\varepsilon^*\}^T = \begin{bmatrix} \Delta\varepsilon_x^* & \Delta\varepsilon_y^* & \Delta\sigma_z^* & \Delta\gamma_{xy}^* & \Delta\gamma_{yz}^* & \Delta\gamma_{zx}^* \end{bmatrix} \tag{3.96}$$

In the elastic range, the stress-strain relationship is expressed as follows:

$$[D] = [D^e] = \frac{E}{1+v} \begin{bmatrix} \frac{1-v}{1-2v} & \frac{v}{1-2v} & \frac{v}{1-2v} & 0 & 0 & 0 \\ \frac{v}{1-2v} & \frac{1-v}{1-2v} & \frac{v}{1-2v} & 0 & 0 & 0 \\ \frac{v}{1-2v} & \frac{v}{1-2v} & \frac{1-v}{1-2v} & 0 & 0 & 0 \\ 0 & 0 & 0 & \frac{1}{2} & 0 & 0 \\ 0 & 0 & 0 & 0 & \frac{\kappa}{2} & 0 \\ 0 & 0 & 0 & 0 & 0 & \frac{\kappa}{2} \end{bmatrix} \tag{3.97}$$

where κ is an effective shear coefficient, which is equal to 5/6 for isotropic material.

On the other hand, in the plastic region, it is assumed that the material follows the combined kinematic and isotropic hardening law. In this case, Mises's yielding condition is given as:

$$f^2 = \left[(\sigma_x - \alpha_x)^2 - (\sigma_x - \alpha_x)(\sigma_y - \alpha_y) + (\sigma_y - \alpha_y)^2 + 3(\tau_{xy} - \alpha_{xy})^2 \right.$$
$$\left. + 3(\tau_{yz} - \alpha_{yz})^2 + 3(\tau_{zx} - \alpha_{zx})^2 \right] - \sigma_Y^2 = 0 \tag{3.98}$$

where α_x, α_y, and α_{xy} are parameters to represent the shift of the center of the yield surface. Considering f as a plastic potential, stress-strain matrix in the plastic region is derived as follows [5]:

$$[D] = [D^p] = [D^e] - \frac{[D^e]\{\partial f/\partial\sigma\}\{\partial f/\partial\sigma\}^T[D^e]}{H' + H_k' + \{\partial f/\partial\sigma\}^T[D^e]\{\partial f/\partial\sigma\}}$$

$$= \frac{9G^2}{S_0^2} \begin{bmatrix} \sigma_x^a\sigma_x^a & \sigma_x^a\sigma_y^a & \sigma_x^a\sigma_z^a & \sigma_x^a\tau_{xy}^a & \sigma_x^a\tau_{yz}^a & \sigma_x^a\tau_{zx}^a \\ \sigma_y^a\sigma_x^a & \sigma_y^a\sigma_y^a & \sigma_y^a\sigma_z^a & \sigma_y^a\tau_{xy}^a & \sigma_y^a\tau_{yz}^a & \sigma_y^a\tau_{zx}^a \\ \sigma_z^a\sigma_x^a & \sigma_z^a\sigma_y^a & \sigma_z^a\sigma_z^a & \sigma_z^a\tau_{xy}^a & \sigma_z^a\tau_{yz}^a & \sigma_z^a\tau_{zx}^a \\ \tau_{xy}^a\sigma_x^a & \tau_{xy}^a\sigma_y^a & \tau_{xy}^a\sigma_z^a & \tau_{xy}^a\tau_{xy}^a & \tau_{xy}^a\tau_{yz}^a & \tau_{xy}^a\tau_{zx}^a \\ \tau_{yz}^a\sigma_x^a & \tau_{yz}^a\sigma_y^a & \tau_{yz}^a\sigma_z^a & \tau_{yz}^a\tau_{xy}^a & \tau_{yz}^a\tau_{yz}^a & \tau_{xy}^a\tau_{zx}^a \\ \tau_{zx}^a\sigma_x^a & \tau_{zx}^a\sigma_y^a & \tau_{zx}^a\sigma_z^a & \tau_{zx}^a\tau_{xy}^a & \tau_{zx}^a\tau_{yz}^a & \tau_{zx}^a\tau_{zx}^a \end{bmatrix} \tag{3.99}$$

where

$$S_0 = (H' + 3G)\{(\sigma_x - \alpha_x)^2 - (\sigma_x - \alpha_x)(\sigma_y - \alpha_y) + (\sigma_y - \alpha_y)^2$$
$$+ 3(\tau_{xy} - \alpha_{xy})^2 + 3(\tau_{yz} - \alpha_{yz})^2 + 3(\tau_{zx} - \alpha_{zx})^2\}$$

$$G = \frac{E}{2(1+v)}$$

$$H' = d\bar{\sigma}/d\bar{\varepsilon}_p \quad \text{(isotropic strain hardening rate)}$$

$$\sigma_x^a = \sigma_x' - \alpha_x', \quad \sigma_y^a = \sigma_y' - \alpha_y', \quad \sigma_z^a = \sigma_z' - \alpha_z'$$

$$\sigma_x' = \frac{1}{3}(2\sigma_x - \sigma_y - \sigma_z), \quad \alpha_x' = \frac{1}{3}(2\alpha_x - \alpha_y - \alpha_z)$$

$$\sigma_y' = \frac{1}{3}(2\sigma_y - \sigma_z - \sigma_x), \quad \alpha_y' = \frac{1}{3}(2\alpha_y - \alpha_z - \alpha_x)$$

$$\sigma_z' = \frac{1}{3}(2\sigma_z - \sigma_x - \sigma_y), \quad \alpha_z' = \frac{1}{3}(2\alpha_z - \alpha_x - \alpha_y)$$

$$\tau_{xy}^a = \tau_{xy} - \alpha_{xy}$$

$$\tau_{yz}^a = \tau_{yz} - \alpha_{yz}$$

$$\tau_{zx}^a = \tau_{zx} - \alpha_{zx}$$

In the above expression, H' and H_k' represent isotropic hardening rate and kinematic hardening rate, respectively.

Regarding the kinematic hardening law, there are Prager's law and Ziegler's law [5]. According to Ziegler's law, the increment of the parameter, $\{\alpha\}$, is given as:

$$\{d\alpha\} = \begin{Bmatrix} d\alpha_x \\ d\alpha_y \\ d\alpha_z \\ d\alpha_{xy} \\ d\alpha_{yz} \\ d\alpha_{zx} \end{Bmatrix} = \frac{H_k' d\bar{\varepsilon}^p}{S_1} \begin{Bmatrix} \sigma_x - \alpha_x \\ \sigma_y - \alpha_y \\ \sigma_z - \alpha_z \\ \sigma_{xy} - \alpha_{xy} \\ \sigma_{yz} - \alpha_{yz} \\ \sigma_{zx} - \alpha_{zx} \end{Bmatrix} \tag{3.100}$$

where

$$S_1 = \frac{3}{2\bar{\sigma}}\{(\sigma_x - \alpha_x)(\sigma_x' - \alpha_x') + (\sigma_y - \alpha_y)(\sigma_y' - \alpha_x') + (\sigma_z - \alpha_z)(\sigma_z' - \alpha_z')$$
$$+ (\tau_{xy} - \alpha_{xy})^2 + (\tau_{yz} - \alpha_{yz})^2 + (\tau_{zx} - \alpha_{zx})^2\}$$
$$\bar{\sigma}^2 = (\sigma_x - \alpha_x)^2 - (\sigma_x - \alpha_x)(\sigma_y - \alpha_y) + (\sigma_y - \alpha_y)^2$$
$$+ 3(\tau_{xy} - \alpha_{xy})^2 + 3(\tau_{yz} - \alpha_{yz})^2 + 3(\tau_{zx} - \alpha_{zx})^2$$

$$d\bar{\varepsilon}^p = \frac{1}{\bar{\sigma}}\{\sigma\}^T \left\{\frac{\partial f}{\partial \sigma}\right\} \frac{\left\{\frac{\partial f}{\partial \sigma}\right\}^T [D^e]\{d\varepsilon\}}{\frac{1}{\bar{\sigma}}\{\sigma\}^T \left\{\frac{\partial f}{\partial \sigma}\right\} H' + \left\{\frac{\partial f}{\partial \sigma}\right\}^T [D^e]\left\{\frac{\partial f}{\partial \sigma}\right\}}$$

To simulate ultra-large deflection behavior, it is necessary to transform updated Kirchhoff's stress increments to Jaumann's stress increments, assuming small strain increments. This transformation can be written as follows [5]:

$$\{\Delta\sigma^J\} = \{\Delta\sigma^*\} + [D']\{\Delta\varepsilon^*\} \tag{3.101}$$

where

$$\{\Delta\sigma^J\}^T = \begin{bmatrix} \Delta\sigma_x^J & \Delta\sigma_y^J & \Delta\sigma_z^J & \Delta\tau_{xy}^J & \Delta\tau_{yz}^J & \Delta\tau_{zx}^J \end{bmatrix} \tag{3.102}$$

$$\{\Delta\sigma^*\}^T = \begin{bmatrix} \Delta\sigma_x^* & \Delta\sigma_y^* & \Delta\sigma_z^* & \Delta\tau_{xy}^* & \Delta\tau_{yz}^* & \Delta\tau_{zx}^* \end{bmatrix} \tag{3.103}$$

and

$$[D'] = \begin{bmatrix}
\sigma_x & -\sigma_x & -\sigma_x & \tau_{xy} & \tau_{zx} & 0 \\
-\sigma_y & \sigma_y & -\sigma_y & \tau_{xy} & 0 & \tau_{yz} \\
-\sigma_z & -\sigma_z & \sigma_z & 0 & \tau_{zx} & \tau_{yz} \\
0 & 0 & -\tau_{xy} & \frac{\sigma_x+\sigma_y}{2} & \frac{\tau_{yz}}{2} & \frac{\tau_{zx}}{2} \\
-\tau_{xy} & 0 & 0 & \frac{\tau_{yz}}{2} & \frac{\sigma_x+\sigma_y}{2} & \frac{\tau_{zx}}{2} \\
0 & -\tau_{zx} & 0 & \frac{\tau_{yz}}{2} & \frac{\tau_{zx}}{2} & \frac{\sigma_x+\sigma_y}{2}
\end{bmatrix} \tag{3.104}$$

Thus, the stress components, $\{\sigma\}^{(n+1)}$ at the $(n+1)$-th step is represented with the stress components, $\{\sigma\}^n$ at the nth step and Jaumann's stress increments, $\{\Delta\sigma^J\}$, as:

$$\{\sigma\}^{(n+1)} = \{\sigma\}^{(n)} + \left\{\Delta\sigma^J\right\} \tag{3.105}$$

To maintain plane stress condition at each incremental step, the following condition has to be satisfied:

$$\Delta\sigma_z^J = 0 \tag{3.106}$$

Eliminating $\Delta\varepsilon_z^*$ using this condition, the following transformation equation is obtained:

$$\left\{\Delta\varepsilon^*\right\} = [A]\left\{\Delta\varepsilon'\right\} \tag{3.107}$$

where

$$\left\{\Delta\varepsilon'\right\} = \left\{\Delta\varepsilon_x^* \;\; \Delta\varepsilon_y^* \;\; \Delta\gamma_{xy}^* \;\; \Delta\gamma_{yz}^* \;\; \Delta\gamma_{zx}^*\right\}^T \tag{3.108}$$

$$[A] = -\frac{1}{d_{33}}\begin{bmatrix}
-d_{33} & 0 & 0 & 0 & 0 \\
0 & -d_{33} & 0 & 0 & 0 \\
d_{31} & d_{32} & d_{34} & d_{36}+\tau_{yz} & d_{35}+\tau_{xz} \\
0 & 0 & 0 & 0 & -d_{33} \\
0 & 0 & 0 & -d_{33} & 0 \\
0 & 0 & 0 & 0 & -d_{33}
\end{bmatrix} \tag{3.109}$$

where d_{ij} is the element located at the ith line and the jth column of the stress-strain matrix (Eq. 3.94) between updated Kirchhoff's stress increments and updated Green's strain increments.

3.4.5 INTRODUCTION OF VIRTUAL STIFFNESS

In the ordinary small displacement analysis, the lack of stiffness against rotation with respect to z-axis causes no problem. However, in the ultra-large deflection analysis, the lack of this stiffness against this rotation results in singular solution when all the elements surrounding a certain nodal point come in the same plane. To prevent this problem, penalty energy, W_T, is considered which is produced by virtual stress, R_θ, and virtual strain, r_θ. That is [6]

$$W_T = \frac{1}{2}R_\theta r_\theta St \tag{3.110}$$

$$R_\theta = 2\kappa_T G r_\theta \tag{3.111}$$

$$r_\theta = \theta'_z - \frac{1}{2}\left(\frac{\partial v'}{\partial x'} - \frac{\partial u'}{\partial y'}\right) = \{T_\theta\}^T \{h'\} \tag{3.112}$$

where S and t are the area and the thickness of the element, respectively. κ_T is a parameter to define the magnitude of penalty energy, and Kanok-Nukulchai [6] proposed to use $\kappa_T > 0.1$.

On the other hand, reduced integration at one point sometimes causes the occurrence of hourglass mode deformation, which results in unrealistic deformation. To prevent the occurrence of hourglass mode instability, Flanagon and Belytschko [7] proposed to introduce virtual elastic stiffness. They defined the hourglass mode vector, $\{r\}$, as:

$$r_i = H_i - \left[(x'_1 - x'_2 + x'_3 - x'_4)\frac{\partial N_i}{\partial x'} + (y'_1 - y'_2 + y'_3 - y'_4)\frac{\partial N_i}{\partial y'}\right] \quad (i = 1 \sim 4) \tag{3.113}$$

where

$$\{H\}^T = \lfloor 1 \quad -1 \quad 1 \quad -1 \rfloor \tag{3.114}$$

Then, virtual strain components corresponding to hourglass mode vector are defined as follows:

$$\left.\begin{array}{lll} q_1 = \sum_i r_i u'_i, & q_2 = \sum_i r_i v'_i, & q_3 = \sum_i r_i w'_i \\ q_4 = \sum_i r_i \theta'_{xi}, & q_5 = \sum_i r_i \theta'_{yi}, & q_6 = \sum_i r_i \theta'_{zi} \end{array}\right\} \tag{3.115}$$

or in the matrix form as:

$$\{q\} = [B_H]\{\delta'\} \tag{3.116}$$

where

$$\{q\}^T = \lfloor q_1 \ q_2 \ q_3 \ q_4 \ q_5 \ q_6 \rfloor \tag{3.117}$$

$$[B_H] = \begin{bmatrix} r_1 & 0 & 0 & 0 & 0 & 0 & r_2 & 0 & 0 & 0 & 0 & 0 \\ 0 & r_1 & 0 & 0 & 0 & 0 & 0 & r_2 & 0 & 0 & 0 & 0 \\ 0 & 0 & r_1 & 0 & 0 & 0 & 0 & 0 & r_2 & 0 & 0 & 0 \\ 0 & 0 & 0 & r_1 & 0 & 0 & 0 & 0 & 0 & r_2 & 0 & 0 \\ 0 & 0 & 0 & 0 & r_1 & 0 & 0 & 0 & 0 & 0 & r_2 & 0 \\ 0 & 0 & 0 & 0 & 0 & r_1 & 0 & 0 & 0 & 0 & 0 & r_2 \\ r_3 & 0 & 0 & 0 & 0 & 0 & r_4 & 0 & 0 & 0 & 0 & 0 \\ 0 & r_3 & 0 & 0 & 0 & 0 & 0 & r_4 & 0 & 0 & 0 & 0 \\ 0 & 0 & r_3 & 0 & 0 & 0 & 0 & 0 & r_4 & 0 & 0 & 0 \\ 0 & 0 & 0 & r_3 & 0 & 0 & 0 & 0 & 0 & r_4 & 0 & 0 \\ 0 & 0 & 0 & 0 & r_3 & 0 & 0 & 0 & 0 & 0 & r_4 & 0 \\ 0 & 0 & 0 & 0 & 0 & r_3 & 0 & 0 & 0 & 0 & 0 & r_4 \end{bmatrix} \tag{3.118}$$

Here, virtual stress components against virtual strain components are defined as:

$$\left.\begin{array}{lll} Q_1 = C_1 q_1, & Q_2 = C_2 q_2, & Q_3 = C_3 q_3, \\ Q_4 = C_4 q_4, & Q_5 = C_5 q_5, & Q_6 = C_6 q_6 \end{array}\right\} \tag{3.119}$$

or in the matrix form as:

$$\{Q\} = [C_H]\{q\}$$

$$\{Q\}^T = \lfloor Q_1 \ Q_2 \ Q_3 \ Q_4 \ Q_5 \ Q_6 \rfloor \tag{3.120}$$

where

$$[C_H] = \begin{bmatrix} C_1 & 0 & 0 & 0 & 0 & 0 \\ 0 & C_2 & 0 & 0 & 0 & 0 \\ 0 & 0 & C_3 & 0 & 0 & 0 \\ 0 & 0 & 0 & C_4 & 0 & 0 \\ 0 & 0 & 0 & 0 & C_5 & 0 \\ 0 & 0 & 0 & 0 & 0 & C_6 \end{bmatrix} \tag{3.121}$$

$$\left. \begin{aligned} C_1 &= \tfrac{r_m}{8} EtS \sum_i \left\{ \left(\tfrac{\partial N_i}{\partial x'} \right)^2 + \left(\tfrac{\partial N_i}{\partial y'} \right)^2 \right\} \\ C_2 &= C_1 \\ C_3 &= \tfrac{r_w}{12} \kappa G t^3 \sum_i \left\{ \left(\tfrac{\partial N_i}{\partial x'} \right)^2 + \left(\tfrac{\partial N_i}{\partial y'} \right)^2 \right\} \\ C_4 &= \tfrac{r_t}{192} E t^3 S \sum_i \left\{ \left(\tfrac{\partial N_i}{\partial x'} \right)^2 + \left(\tfrac{\partial N_i}{\partial y'} \right)^2 \right\} \\ C_5 &= C_4 \\ C_6 &= r_a GtS \sum_i \left\{ \left(\tfrac{\partial N_i}{\partial x'} \right)^2 + \left(\tfrac{\partial N_i}{\partial y'} \right)^2 \right\} \end{aligned} \right\} \tag{3.122}$$

where r_m, r_w, r_t, r_a are parameters related to the magnitude of virtual stiffness against hourglass mode instability, and are recommended to use the value between 0.01 and 0.05 by Flanagon and Belytschko [6]. According to them, Toi et al. [4] used 0.03 in the elastic region. They suggested that, in the elastoplastic range, 0.03 should be multiplied by the ratio of the average of E or H' in the layer toward the thickness direction to E. They used 0.1 as for κ_T.

3.4.6 DERIVATION OF THE STIFFNESS MATRIX

Here, the stiffness matrix is derived on the basis of the updated Lagrangian approach. That is, the incremental form of virtual work equation for the $(n + 1)$-th step is obtained using updated Green's strain components and updated Kirchhoff's stress components as follows:

$$S \int \{\delta \Delta \varepsilon\}^T \{\Delta \sigma\} dz' + S \int \{\delta \Delta \bar{\varepsilon}\}^T \{\sigma\} dz' + \{\delta \Delta q\}^T \{\Delta Q\} + \{\delta \Delta r_\theta\}^T \{\Delta R_\theta\}$$
$$= \{\delta \Delta h\}^T \{\Delta f\} + \{\delta \Delta h\}^T \{f_r\} \tag{3.123}$$

Hence,

$$[L_0]([K_0] + [K_G] + [K_H] + [K_T])[L_0]\{\Delta h\} = \{\Delta f\} + \{f_r\} \tag{3.124}$$

where

$[K_0]$: stiffness come from virtual work made by linear stress and strain components
$[K_G]$: stiffness come from geometrically nonlinear strain and total stress components; being called initial stress stiffness matrix
$[K_H]$: stiffness to control hourglass mode instability
$[K_T]$: stiffness to prevent inplane rotation

which are represented as:

$$[K_0] = \int [B]^T [A]^T [D][A][B] dz' \tag{3.125}$$

$$[K_G] = \int [B_1]^T [\sigma_0][B_1] dz' \tag{3.126}$$

$$[K_H] = [B_H]^T [C][B_H] \tag{3.127}$$

$$[K_T] = \{T_\theta\} \alpha \{T_\theta\}^T \tag{3.128}$$

where

$$\alpha = 2\kappa_T G t S \tag{3.129}$$

$$[\sigma_0] = \begin{bmatrix} \sigma_x & \tau_{xy} \\ \tau_{xy} & \sigma_y \end{bmatrix} \tag{3.130}$$

$\{f_r\}$ in the right-hand side of Eq. (3.124) represents the unbalanced force vector produced by cumulated differences between external and internal nodal forces caused by linear approximation during each small increment, and is expressed as follows:

$$\{f_r\} = \{f\} - \int [L_0]^T [B]^T \{\sigma\} dz' - [L_0]^T [B_H]^T \{q\} - [L_0]^T [T_\theta] R_\theta \tag{3.131}$$

where $\{f\}$ represents total external force vector.

Derivation of the initial stress stiffness matrix, $[K_G]$, is shown in Section 3.7.

According to the formulation explained above, FEM code, *ULSAS* is developed to perform elastic and elastoplastic ultra-large deflection analysis. The accuracy of the calculated results are confirmed through many benchmark calculation in the Technical Committee III.1 in the International Ship and Offshore Structures Congress (ISSC) [8–12]. Most of the calculations in this text are performed using *ULSAS*.

EXERCISES

3.1 What type of deflection function is appropriate for the buckling analysis of a rectangular plate with four sides simply supported, and why?

3.2 What is the appropriate deflection function for a local plate part of a continuous stiffened plate under lateral pressure load?

3.3 Explain the four possible types of buckling modes with respect to stiffeners under longitudinal thrust.

3.4 Derive the elastic strain energy of a plate, Eq. (3.10), with reference to the textbook, for example, Timoshenko and Gere, assuming that a whole plate is elastic [1].

3.5 Obtain the elastic buckling strength of Eq. (3.20) assuming the buckling mode of Eq. (3.19).

3.6 Obtain the matrices $[G_1]$ and $[G_2]$ assuming the following two deflection components and ignoring welding residual stresses:

$$w = A_{mn} \sin \frac{m\pi x}{a} \sin \frac{n\pi y}{b} + A_{pq} \sin \frac{p\pi x}{a} \sin \frac{q\pi y}{b} \quad (m \neq p, n \neq q)$$

3.7 Derive the compatibility condition of Eq. (3.35) using Eqs. (3.32)–(3.34), and then derive Eq. (3.44) further using Eqs. (3.41)–(3.43).

3.8 Derive the overall shrinkage, Eqs. (3.53), (3.54), using Eqs. (3.32)–(3.34) and assuming the deflections of Eqs. (3.30), (3.31).

3.9 Obtain the deflection-stress relationship of Eq. (3.71) by substituting Eqs. (3.69), (3.70) to the *Principle of Virtual Work* of Eq. (3.59).

3.10 Explain how many degrees of freedom are considered in the shell element of Fig. 3.9, and how the displacement in x, y, and z directions are interpolated in the element region.

3.11 Derive the initial stress stiffness matrix $[K_G]$ following the formulation in Section 3.7.

3.12 Explain what kind of nonlinear behaviors must be taken into account in the buckling/plastic collapse analysis, and in the elastic large deflection analysis.

3.5 APPENDIX: FUNDAMENTAL EQUATIONS FOR ELASTIC LARGE DEFLECTION ANALYSIS ASSUMING GENERAL DEFLECTION MODE

3.5.1 AIRY'S STRESS FUNCTION

Airy's stress function expressed by Eq. (3.45) is expressed as:

$$
\begin{aligned}
F = \frac{E\alpha^2}{4} \sum_{i=1}^{m}\sum_{j=1}^{n}\sum_{k=1}^{o}\sum_{l=1}^{p} & \left(A_{ij}A_{kl} - A_{0ij}A_{0kl} \right) \\
\times & \left[\alpha_1 \cos\frac{(i+k)\pi x}{a}\cos\frac{(j+l)\pi y}{b} + \alpha_2 \cos\frac{(i+k)\pi x}{a}\cos\frac{(j-l)\pi y}{b} \right. \\
& \left. + \alpha_3 \cos\frac{(i-k)\pi x}{a}\cos\frac{(j+l)\pi y}{b} + \alpha_4 \cos\frac{(i-k)\pi x}{a}\cos\frac{(j-l)\pi y}{b} \right] \\
& - \frac{y^2}{2}\sigma_x - \frac{x^2}{2}\sigma_x - xy\overline{\tau}_{xy}; k(x) + f(y)
\end{aligned}
$$

(3.132)

where $\alpha = a/b$ and

$$
\alpha_1 = \frac{il(jk - il)}{\left\{(i+k)^2 + \alpha^2(j+l)^2\right\}^2}, \quad \alpha_2 = \frac{il(jk + il)}{\left\{(i+k)^2 + \alpha^2(j-l)^2\right\}^2}
$$

$$
\alpha_3 = \frac{il(jk + il)}{\left\{(i-k)^2 + \alpha^2(j+l)^2\right\}^2}, \quad \alpha_4 = \frac{il(jk - il)}{\left\{(i-k)^2 + \alpha^2(j-l)^2\right\}^2}
$$

3.5.2 IN-PLANE STRESS AND STRAIN COMPONENTS

Substituting Eq. (3.45) into Eqs. (3.36)–(3.38), (3.41)–(3.43), the in-plane stress and strain components are derived as follows:

$$\sigma_{xp} = -\frac{\pi^2 \alpha^2 E}{4b^2} \sum_{i=1}^{m} \sum_{j=1}^{n} \sum_{k=1}^{o} \sum_{l=1}^{p} \left(A_{ij}A_{kl} - A_{0ij}A_{0kl} \right)$$

$$= \left[\alpha_1 (j+l)^2 \cos \frac{(i+k)\pi x}{a} \cos \frac{(j+l)\pi y}{b} \right.$$

$$+ \alpha_2 (j-l)^2 \cos \frac{(i+k)\pi x}{a} \cos \frac{(j-l)\pi y}{b}$$

$$+ \alpha_3 (j+l)^2 \cos \frac{(i-k)\pi x}{a} \cos \frac{(j+l)\pi y}{b}$$

$$\left. + \alpha_4 (j-l)^2 \cos \frac{(i-k)\pi x}{a} \cos \frac{(j-l)\pi y}{b} \right] - \overline{\sigma}_x + \sigma_{xr}(y) \qquad (3.133)$$

$$\sigma_{yp} = -\frac{\pi^2 \alpha^2 E}{4a^2} \sum_{i=1}^{m} \sum_{j=1}^{n} \sum_{k=1}^{o} \sum_{l=1}^{p} \left(A_{ij}A_{kl} - A_{0ij}A_{0kl} \right)$$

$$= \left[\alpha_1 (i+k)^2 \cos \frac{(i+k)\pi x}{a} \cos \frac{(j+l)\pi y}{b} \right.$$

$$+ \alpha_2 (i+k)^2 \cos \frac{(i+k)\pi x}{a} \cos \frac{(j-l)\pi y}{b}$$

$$+ \alpha_3 (i-k)^2 \cos \frac{(i-k)\pi x}{a} \cos \frac{(j+l)\pi y}{b}$$

$$\left. + \alpha_4 (i-k)^2 \cos \frac{(i-k)\pi x}{a} \cos \frac{(j-l)\pi y}{b} \right] - \overline{\sigma}_y + \sigma_{yr}(x) \qquad (3.134)$$

$$\tau_{xyp} = -\frac{\pi^2 \alpha^2 E}{4ab} \sum_{i=1}^{m} \sum_{j=1}^{n} \sum_{k=1}^{o} \sum_{l=1}^{p} \left(A_{ij}A_{kl} - A_{0ij}A_{0kl} \right)$$

$$= \left[\alpha_1 (i+k)(j+l) \sin \frac{(i+k)\pi x}{a} \sin \frac{(j+l)\pi y}{b} \right.$$

$$+ \alpha_2 (i+k)(j-l) \sin \frac{(i+k)\pi x}{a} \sin \frac{(j-l)\pi y}{b}$$

$$+ \alpha_3 (i-k)(j+l) \sin \frac{(i-k)\pi x}{a} \sin \frac{(j+l)\pi y}{b}$$

$$\left. + \alpha_4 (i-k)(j-l) \sin \frac{(i-k)\pi x}{a} \sin \frac{(j-l)\pi y}{b} \right] + \overline{\tau}_{xy} \qquad (3.135)$$

$$\epsilon_{xp} = -\frac{\pi^2 \alpha^2}{4} \sum_{i=1}^{m} \sum_{j=1}^{n} \sum_{k=1}^{o} \sum_{l=1}^{p} \left(A_{ij}A_{kl} - A_{0ij}A_{0kl} \right)$$

$$= \left[\alpha_1 \left(\frac{(j+l)^2}{b^2} - \frac{v(i+k)^2}{a^2} \right) \cos \frac{(i+k)\pi x}{a} \cos \frac{(j+l)\pi y}{b} \right.$$

$$+ \alpha_2 \left(\frac{(j-l)^2}{b^2} - \frac{v(i+k)^2}{a^2} \right) \cos \frac{(i+k)\pi x}{a} \cos \frac{(j-l)\pi y}{b}$$

$$+ \alpha_3 \left(\frac{(j+l)^2}{b^2} - \frac{v(i-k)^2}{a^2} \right) \cos \frac{(i-k)\pi x}{a} \cos \frac{(j+l)\pi y}{b}$$

$$
+\alpha_4 \left(\frac{(j-l)^2}{b^2} - \frac{v(i-k)^2}{a^2} \right) \cos \frac{(i-k)\pi x}{a} \cos \frac{(j-l)\pi y}{b} \Bigg]
$$

$$
- \frac{1}{E}(\bar{\sigma}_x - v\bar{\sigma}_y) + \frac{1}{E}\left(\sigma_{xr}(y) - v\sigma_{yr}(x) \right) \tag{3.136}
$$

$$
\epsilon_{yp} = - \frac{\pi^2 \alpha^2}{4} \sum_{i=1}^{m} \sum_{j=1}^{n} \sum_{k=1}^{o} \sum_{l=1}^{p} \left(A_{ij}A_{kl} - A_{0ij}A_{0kl} \right)
$$

$$
= \Bigg[\alpha_1 \left(\frac{(i+k)^2}{a^2} - \frac{v(j+l)^2}{b^2} \right) \cos \frac{(i+k)\pi x}{a} \cos \frac{(j+l)\pi y}{b}
$$

$$
+ \alpha_2 \left(\frac{(i+k)^2}{a^2} - \frac{v(j-l)^2}{b^2} \right) \cos \frac{(i+k)\pi x}{a} \cos \frac{(j-l)\pi y}{b}
$$

$$
+ \alpha_3 \left(\frac{(i-k)^2}{a^2} - \frac{v(j+l)^2}{b^2} \right) \cos \frac{(i-k)\pi x}{a} \cos \frac{(j+l)\pi y}{b}
$$

$$
+ \alpha_4 \left(\frac{(i-k)^2}{a^2} - \frac{v(j-l)^2}{b^2} \right) \cos \frac{(i-k)\pi x}{a} \cos \frac{(j-l)\pi y}{b} \Bigg]
$$

$$
- \frac{1}{E}(\bar{\sigma}_y - v\bar{\sigma}_x) + \frac{1}{E}\left(\sigma_{yr}(x) - v\sigma_{xr}(y) \right) \tag{3.137}
$$

$$
\gamma_{xyp} = - \frac{\pi^2 \alpha^2 (1+v)}{2ab} \sum_{i=1}^{m} \sum_{j=1}^{n} \sum_{k=1}^{o} \sum_{l=1}^{p} \left(A_{ij}A_{kl} - A_{0ij}A_{0kl} \right)
$$

$$
= \Bigg[\alpha_1 (i+k)(j+l) \sin \frac{(i+k)\pi x}{a} \sin \frac{(j+l)\pi y}{b}
$$

$$
+ \alpha_2 (i+k)(j-l) \sin \frac{(i+k)\pi x}{a} \sin \frac{(j-l)\pi y}{b}
$$

$$
+ \alpha_3 (i-k)(j+l) \sin \frac{(i-k)\pi x}{a} \sin \frac{(j+l)\pi y}{b}
$$

$$
+ \alpha_4 (i-k)(j-l) \sin \frac{(i-k)\pi x}{a} \sin \frac{(j-l)\pi y}{b} \Bigg] + \frac{2(1+v)}{E}\bar{\tau}_{xy} \tag{3.138}
$$

where $\sigma_{yr}(x)$ and $\sigma_{xr}(y)$ represent welding residual stress and are obtained from Eq. (3.46), or can be newly defined on the basis of the assumed distribution of welding residual stress.

3.5.3 BENDING STRAIN AND STRESS COMPONENTS

$$
\epsilon_{xb} = \frac{\pi^2 z}{a^2} \sum_{i=1}^{m} \sum_{j=1}^{n} i^2 \left(A_{ij} - A_{0ij} \right) \sin \frac{i\pi x}{a} \sin \frac{j\pi y}{b} \tag{3.139}
$$

$$
\epsilon_{yb} = \frac{\pi^2 z}{b^2} \sum_{i=1}^{m} \sum_{j=1}^{n} j^2 \left(A_{ij} - A_{0ij} \right) \sin \frac{i\pi x}{a} \sin \frac{j\pi y}{b} \tag{3.140}
$$

$$\gamma_{xyb} = -\frac{2\pi^2 z}{ab} \sum_{i=1}^{m} \sum_{j=1}^{n} ij \left(A_{ij} - A_{0ij}\right) \cos \frac{i\pi x}{a} \cos \frac{j\pi y}{b} \tag{3.141}$$

$$\sigma_{xb} = \frac{\pi^2 zE}{1-v^2} \sum_{i=1}^{m} \sum_{j=1}^{n} \left(\frac{i^2}{a^2} + \frac{vj^2}{b^2}\right) \left(A_{ij} - A_{0ij}\right) \sin \frac{i\pi x}{a} \sin \frac{j\pi y}{b} \tag{3.142}$$

$$\sigma_{yb} = \frac{\pi^2 zE}{1-v^2} \sum_{i=1}^{m} \sum_{j=1}^{n} \left(\frac{vi^2}{a^2} + \frac{j^2}{b^2}\right) \left(A_{ij} - A_{0ij}\right) \sin \frac{i\pi x}{a} \sin \frac{j\pi y}{b} \tag{3.143}$$

$$\tau_{xyb} = -\frac{\pi^2 zE}{ab(1+v)} \sum_{i=1}^{m} \sum_{j=1}^{n} ij \left(A_{ij} - A_{0ij}\right) \cos \frac{i\pi x}{a} \cos \frac{j\pi y}{b} \tag{3.144}$$

3.5.4 COEFFICIENTS IN FUNDAMENTAL EQUATIONS

$$
\begin{aligned}
H(i,j,k,l,e,f,g,h) =& [\alpha^2 \{a_1(d_1A_1B_1 + d_2A_1B_2 + d_3A_2B_1 + d_4A_2B_2) \\
& + a_2(d_1A_1B_3 + d_2A_1B_4 + d_3A_2B_3 + d_4A_2B_4) \\
& + a_3(d_1A_3B_1 + d_2A_3B_2 + d_3A_4B_1 + d_4A_4B_2) \\
& + a_4(d_1A_3B_3 + d_2A_3B_4 + d_3A_4B_3 + d_4A_4B_4)\} \\
& + \{b_1(e_1A_1B_1 + e_2A_1B_2 + e_3A_2B_1 + e_4A_2B_2) \\
& + b_2(e_1A_1B_3 + e_2A_1B_4 + e_3A_2B_3 + e_4A_2B_4) \\
& + b_3(e_1A_3B_1 + e_2A_3B_2 + e_3A_4B_1 + e_4A_4B_2) \\
& + b_4(e_1A_3B_3 + e_2A_3B_4 + e_3A_4B_3 + e_4A_4B_4)\} \\
& + 2\alpha^2(1+v)\{c_1(f_1C_1D_1 + f_2C_1D_2 + f_3C_2D_1 + f_4C_2D_2) \\
& + c_2(f_1C_1D_3 + f_2C_1D_4 + f_3C_2D_3 + f_4C_2D_4) \\
& + c_3(f_1C_3D_1 + f_2C_3D_2 + f_3C_4D_1 + f_4C_4D_2) \\
& + c_4(f_1C_3D_3 + f_2C_3D_4 + f_3C_4D_3 + f_4C_4D_4\}
\end{aligned}
$$

$$
\begin{aligned}
T(i,j,k,l) =& -\frac{\pi^2\alpha^2(1+v)}{2} \{f_1(i,j,k,l)E_2(i,k)F_2(j,l) \\
& + f_2(i,j,k,l)E_2(i,k)F_3(j,l) + f_3(i,j,k,l)E_3(i,k)F_2(j,l) \\
& + f_4(i,j,k,l)F_3(i,k)F_3(j,l)\}
\end{aligned}
$$

$$S(i,j,k,l) = \frac{\pi^2 il}{4} E_4(i,k)F_4(j,l) \tag{3.145}$$

$$
\begin{aligned}
RS(e,f,g,h) =& d_3(e,f,g,h)E_1(e,g)RSX_1(f,h) + d_4(e,f,g,h)E_1(e,g)RSC_2(f,h) \\
& + e_2(e,f,g,h)F_1(e,g)RSY_1(f,h) + e_4(e,f,g,h)F_1(e,g)RSY_2(f,h)
\end{aligned}
$$

$$
\begin{cases}
Q_1(e,f,g,h) = & d_4(e,f,g,h)E_1(e,g)F_1(f,h) \\
Q_2(e,f,g,h) = & e_4(e,f,g,h)E_1(e,g)F_1(f,h) \\
Q_3(e,f,g,h) = & 2\alpha(1+v)(f_1(e,f,g,h)E_2(e,g)F_2(f,h) \\
& + f_2(e,f,g,h)E_2(e,g)F_3(f,h) + f_3(e,f,g,h)E_3(e,g)F_2(f,h) \\
& + f_4(e,f,g,h)E_3(e,g)F_3(f,h))
\end{cases}
$$

where

$$a_1(i,j,k,l) = \alpha_1(i,j,k,l)(j+l)^2, \qquad a_2(i,j,k,l) = \alpha_2(i,j,k,l)(j-l)^2$$

$$a_3(i,j,k,l) = \alpha_3(i,j,k,l)(j+l)^2, \qquad a_4(i,j,k,l) = \alpha_4(i,j,k,l)(j-l)^2$$

$$b_1(i,j,k,l) = \alpha_1(i,j,k,l)(i+k)^2, \qquad b_2(i,j,k,l) = \alpha_2(i,j,k,l)(i+k)^2$$

$$b_3(i,j,k,l) = \alpha_3(i,j,k,l)(i-k)^2, \qquad b_4(i,j,k,l) = \alpha_4(i,j,k,l)(i-k)^2$$

$$c_1(i,j,k,l) = \alpha_1(i,j,k,l)(i+k)(j+l), \qquad c_2(i,j,k,l) = \alpha_2(i,j,k,l)(i+k)(j-l)$$

$$c_3(i,j,k,l) = \alpha_3(i,j,k,l)(i-k)(j+l), \qquad c_4(i,j,k,l) = \alpha_4(i,j,k,l)(i-k)(j-l)$$

$$d_1(e,f,g,h) = \alpha_1(e,f,g,h)\{(f+h)\alpha^2 - \nu(e+g)^2\}$$

$$d_2(e,f,g,h) = \alpha_2(e,f,g,h)\{(f-h)\alpha^2 - \nu(e+g)^2\}$$

$$d_3(e,f,g,h) = \alpha_2(e,f,g,h)\{(f+h)\alpha^3 - \nu(e-g)^2\}$$

$$d_4(e,f,g,h) = \alpha_2(e,f,g,h)\{(q-s)\alpha^4 - \nu(e-g)^2\}$$

$$e_1(e,f,g,h) = \alpha_1(e,f,g,h)\{(e+g)\alpha^1 - \nu(f+h)^2\}$$

$$e_2(e,f,g,h) = \alpha_1(e,f,g,h)\{(e+g)\alpha^2 - \nu(f-h)^2\}$$

$$e_3(e,f,g,h) = \alpha_1(e,f,g,h)\{(e-g)\alpha^3 - \nu(f+h)^2\}$$

$$e_4(e,f,g,h) = \alpha_1(e,f,g,h)\{(e-g)\alpha^4 - \nu(f-h)^2\}$$

$$f_1(e,f,g,h) = \alpha_1(e,f,g,h)(e+g)(f+h), \qquad f_2(e,f,g,h) = \alpha_2(e,f,g,h)(e+g)(f-h)$$

$$f_3(e,f,g,h) = \alpha_3(e,f,g,h)(e-g)(f+h), \qquad f_4(e,f,g,h) = \alpha_4(e,f,g,h)(e-g)(f-h)$$

$$A_1(i,k,e,g) = \int_0^a \cos\frac{(i+k)\pi x}{a} \cos\frac{(e+g)\pi x}{a} dx, \qquad A_2(i,k,e,g) = \int_0^a \cos\frac{(i+k)\pi x}{a} \cos\frac{(e-g)\pi x}{a} dx$$

$$A_3(i,k,e,g) = \int_0^a \cos\frac{(i-k)\pi x}{a} \cos\frac{(e+g)\pi x}{a} dx, \qquad A_4(i,k,e,g) = \int_0^a \cos\frac{(i-k)\pi x}{a} \cos\frac{(e-g)\pi x}{a} dx$$

$$B_1(j,l,f,h) = \int_0^b \cos\frac{(j+l)\pi y}{b} \cos\frac{(f+h)\pi y}{b} dy, \qquad B_2(j,l,f,h) = \int_0^b \cos\frac{(j+l)\pi y}{b} \cos\frac{(f-h)\pi y}{b} dy$$

$$B_3(j,l,f,h) = \int_0^b \cos\frac{(j-l)\pi y}{b} \cos\frac{(f+h)\pi y}{b} dy, \qquad B_4(j,l,f,h) = \int_0^b \cos\frac{(j-l)\pi y}{b} \cos\frac{(f-h)\pi y}{b} dy$$

$$C_1(i,k,e,g) = \int_0^a \sin\frac{(i+k)\pi x}{a} \sin\frac{(e+g)\pi x}{a} dx, \qquad C_2(i,k,e,g) = \int_0^a \sin\frac{(i+k)\pi x}{a} \sin\frac{(e-g)\pi x}{a} dx$$

$$C_3(i,k,e,g) = \int_0^a \sin\frac{(i-k)\pi x}{a} \sin\frac{(e+g)\pi x}{a} dx, \qquad C_4(i,k,e,g) = \int_0^a \sin\frac{(i-k)\pi x}{a} \sin\frac{(e-g)\pi x}{a} dx$$

$$D_1(j,l,f,h) = \int_0^b \sin\frac{(j+l)\pi y}{b} \sin\frac{(f+h)\pi y}{b} dy, \qquad D_2(j,l,f,h) = \int_0^b \sin\frac{(j+l)\pi y}{b} \sin\frac{(f-h)\pi y}{b} dy$$

$$D_3(j,l,f,h) = \int_0^b \sin\frac{(j-l)\pi y}{b} \sin\frac{(f+h)\pi y}{b} dy, \qquad B_4(j,l,f,h) = \int_0^b \sin\frac{(j-l)\pi y}{b} \sin\frac{(f-h)\pi y}{b} dy$$

$$E_1(e,g) = \int_0^a \cos\frac{(e-g)\pi x}{a} dx, \qquad E_2(e,g) = \int_0^a \sin\frac{(e+g)\pi x}{a} dx, \qquad E_3(e,g) = \int_0^a \sin\frac{(e-g)\pi x}{a} dx$$

$$F_1(f,h) = \int_0^b \cos\frac{(f-h)\pi y}{b} dy, \qquad F_2(f,h) = \int_0^b \sin\frac{(f+h)\pi y}{b} dy, \qquad E_3(f,h) = \int_0^b \sin\frac{(f-h)\pi y}{b} dy$$

$$RSX_1(f,h) = \int_0^b \cos\frac{(f+h)\pi y}{b}\sigma_{xr}(y)dy, \quad RSX_2(f,h) = \int_0^b \cos\frac{(f-h)\pi y}{b}\sigma_{xr}(y)dy$$

$$RSY_1(e,g) = \int_0^a \cos\frac{(e+g)\pi x}{a}\sigma_{yr}(x)dx, \quad RSY_2(e,g) = \int_0^a \cos\frac{(e-g)\pi x}{a}\sigma_{yr}(x)dx$$

Here, the signe, $\pm$, in CN_{ij} and SN_{ij} is the same with that in the coefficients, a_i, aa_j, b_i, bb_j, c_i, and cc_j.

$$J(i,j,k,l) = J_1 + J_2$$

where

$$J_1(i,j,k,l) = \begin{cases} 0 & \text{(when } i-k \neq 0) \\ -\frac{\pi k^2 l}{4\alpha}\frac{2\sigma_Y}{1-\mu_b}\frac{1}{(j+l)^2}\sin\frac{(j+l)\mu_b\pi}{2} & \text{(when } i-k=0 \text{ and } j-l=0) \\ -\frac{\pi k^2 l}{4\alpha}\frac{2\sigma_Y}{1-\mu_b}\left\{\frac{1}{(j+l)^2}\sin\frac{(j+l)\mu_b\pi}{2}+\frac{1}{(j-l)^2}\sin\frac{(j-l)\mu_b\pi}{2}\right\} \\ \qquad \text{(when } i-k=0 \text{ and } j-l \neq 0 \text{ and } j \pm l \text{ is an even number)} \\ 0 \qquad \text{(when } i-k=0 \text{ and } j-l \neq 0 \text{ and } j \pm l \text{ is an odd number)} \end{cases}$$

$$J_2(i,j,k,l) = \begin{cases} = 0 & \text{(when } j-l \neq 0) \\ -\frac{\pi \alpha i j^2}{4}\frac{2\sigma_Y}{1-\mu_a}\frac{1}{(i+k)^2}\sin\frac{(i+k)\mu_a\pi}{2} & \text{(when } i-k=j-l=0) \\ -\frac{\pi \alpha i j^2}{4}\frac{2\sigma_Y}{1-\mu_a}\left\{\frac{1}{(i+k)^2}\sin\frac{(i+k)\mu_a\pi}{2}\right. \\ \qquad \left. -\frac{1}{(i-k)^2}\sin\frac{(i-k)\mu_a\pi}{2}\right\} \\ \qquad \text{(when } j-l=0 \text{ and } i-k \neq 0 \text{ and } i \pm k \text{ is an even number)} \\ 0 \qquad \text{(when } n-l=0 \text{ and } m-k \neq 0 \text{ and } m \pm k \text{ is an odd number)} \end{cases}$$

where

$$\alpha = a/b, \quad \mu_b = 2b_t/b, \quad \mu_a = 2a_t/a$$

3.6 APPENDIX: DERIVATION OF EQ. 3.83 FOR STRAIN-DISPLACEMENT RELATIONSHIP

N_i and its partial derivatives are represented as:

$$\left.\begin{array}{lll} N_1 = \frac{1}{4}(1-\xi)(1-\eta) & \frac{\partial N_1}{\partial \xi} = -\frac{1}{4}(1-\eta) & \frac{\partial N_1}{\partial \eta} = -\frac{1}{4}(1-\xi) \\ N_2 = \frac{1}{4}(1+\xi)(1-\eta) & \frac{\partial N_2}{\partial \xi} = \frac{1}{4}(1-\eta) & \frac{\partial N_2}{\partial \eta} = -\frac{1}{4}(1+\xi) \\ N_3 = \frac{1}{4}(1+\xi)(1+\eta) & \frac{\partial N_3}{\partial \xi} = \frac{1}{4}(1+\eta) & \frac{\partial N_3}{\partial \eta} = \frac{1}{4}(1+\xi) \\ N_4 = \frac{1}{4}(1-\xi)(1+\eta) & \frac{\partial N_4}{\partial \xi} = -\frac{1}{4}(1+\eta) & \frac{\partial N_4}{\partial \eta} = \frac{1}{4}(1-\xi) \end{array}\right\} \qquad (3.146)$$

Considering that ξ and η are functions of x' and y', the partial derivatives can be expressed as:

$$\frac{\partial N_i}{\partial \xi} = \frac{\partial N_i}{\partial x'}\frac{\partial x'}{\partial \xi} + \frac{\partial N_i}{\partial y'}\frac{\partial y'}{\partial \xi} \tag{3.147}$$

$$\frac{\partial N_i}{\partial \eta} = \frac{\partial N_i}{\partial x'}\frac{\partial x'}{\partial \eta} + \frac{\partial N_i}{\partial y'}\frac{\partial y'}{\partial \eta} \tag{3.148}$$

or in the matrix form as:

$$\begin{bmatrix} \frac{\partial x'}{\partial \xi} & \frac{\partial y'}{\partial \xi} \\ \frac{\partial x'}{\partial \eta} & \frac{\partial y'}{\partial \eta} \end{bmatrix} \left\{ \begin{matrix} \frac{\partial N_i}{\partial x'} \\ \frac{\partial N_i}{\partial y'} \end{matrix} \right\} = \left\{ \begin{matrix} \frac{\partial N_i}{\partial \xi} \\ \frac{\partial N_i}{\partial \eta} \end{matrix} \right\} \tag{3.149}$$

Solving above equation with respect to $\partial N_i/\partial x'$ and $\partial N_i/\partial y'$, the following expression is derived:

$$\left\{ \begin{matrix} \frac{\partial N_i}{\partial x} \\ \frac{\partial N_i}{\partial y} \end{matrix} \right\} = \frac{1}{\frac{\partial x'}{\partial \xi}\frac{\partial y'}{\partial \eta} - \frac{\partial y'}{\partial \xi}\frac{\partial x'}{\partial \eta}} \begin{bmatrix} \frac{\partial y'}{\partial \eta} & -\frac{\partial y'}{\partial \xi} \\ -\frac{\partial x'}{\partial \eta} & \frac{\partial x'}{\partial \xi} \end{bmatrix} \left\{ \begin{matrix} \frac{\partial N_i}{\partial \xi} \\ \frac{\partial N_i}{\partial \eta} \end{matrix} \right\} \tag{3.150}$$

Here,

$$\left. \begin{matrix} x' = N_1 x_1' + N_2 x_2' + N_3 x_3' + N_4 x_4' \\ y' = N_1 y_1' + N_2 y_2' + N_3 y_3' + N_4 y_4' \end{matrix} \right\} \tag{3.151}$$

The partial derivatives of x' can be written as:

$$\frac{\partial x'}{\partial \xi} = \frac{\partial N_1}{\partial \xi}x_1' + \frac{\partial N_2}{\partial \xi}x_2' + \frac{\partial N_3}{\partial \xi}x_3' + \frac{\partial N_4}{\partial \xi}x_4'$$
$$= -\frac{1}{4}(1-\eta)x_1' + \frac{1}{4}(1-\eta)x_2' + \frac{1}{4}(1+\eta)x_3' - \frac{1}{4}(1+\eta)x_4' \tag{3.152}$$

$$\frac{\partial x'}{\partial \eta} = \frac{\partial N_1}{\partial \eta}x_1' + \frac{\partial N_2}{\partial \eta}x_2' + \frac{\partial N_3}{\partial \eta}x_3' + \frac{\partial N_4}{\partial \eta}x_4'$$
$$= -\frac{1}{4}(1-\xi)x_1' - \frac{1}{4}(1+\xi)x_2' + \frac{1}{4}(1+\xi)x_3' + \frac{1}{4}(1-\xi)x_4' \tag{3.153}$$

At the center of geometry, $\eta = \xi = 0$, Eqs. (3.152), (3.153) result in

$$\frac{\partial x'}{\partial \xi} = \frac{1}{4}(-x_1' + x_2' + x_3' - x_4') \tag{3.154}$$

$$\frac{\partial x'}{\partial \eta} = \frac{1}{4}(-x_1' - x_2' + x_3' + x_4') \tag{3.155}$$

In the same manner, partial derivatives of y' with respect to ξ and η are derived as follows:

$$\frac{\partial y'}{\partial \xi} = \frac{1}{4}(-y_1' + y_2' + y_3' - y_4') \tag{3.156}$$

$$\frac{\partial y'}{\partial \eta} = \frac{1}{4}(-y_1' - y_2' + y_3' + y_4') \tag{3.157}$$

and hence,

$$\frac{\partial x'}{\partial \xi}\frac{\partial y'}{\partial \eta} - \frac{\partial y'}{\partial \xi}\frac{\partial x'}{\partial \eta} = \frac{1}{8}\{(x'_2 - x'_4)(y'_3 - y'_1) + (x'_3 - x'_1)(y'_4 - y'_2)\} = \frac{S}{4} \tag{3.158}$$

From the above equations, Eq. (3.83) can be derived. For example,

$$
\begin{aligned}
\frac{\partial N_1}{\partial x'} &= \frac{4}{S}\left(\frac{\partial y'}{\partial \eta}\frac{\partial N_1}{\partial \xi} - \frac{\partial y'}{\partial \xi}\frac{\partial N_1}{\partial \eta}\right)\\
&= \frac{4}{S}\left\{\frac{1}{4}(-y'_1 - y'_2 + y'_3 + y'_4)\left(-\frac{1}{4}\right) + \frac{1}{4}(y'_1 - y'_2 - y'_3 + y'_4)\left(-\frac{1}{4}\right)\right\}\\
&= \frac{1}{2S}(y'_2 - y'_4) \tag{3.159}
\end{aligned}
$$

Other partial derivatives in Eq. (3.83) can be obtained in the same manner.

3.7 APPENDIX: DERIVATION OF INITIAL STRESS STIFFNESS MATRIX

The initial stress stiffness matrix is written as:

$$[K_G] = \int [B_1]^T [\sigma_0][B_1]\,dz' \tag{3.160}$$

To derive the initial stress stiffness matrix, $[K_G]$, the following relationship is considered:

$$\left\{\begin{array}{c}\frac{\partial \Delta w'}{\partial x'}\\\frac{\partial \Delta w'}{\partial y'}\end{array}\right\} = [B_1]\{\Delta h'\} \tag{3.161}$$

Eq. (3.126) is derived from the virtual work done by nonlinear virtual strain increments and stress components, that is

$$\overline{W}_i^{(\text{nonlinear})} = \int \{\delta \Delta \overline{\varepsilon}\}^T\{\sigma\}dz' \tag{3.162}$$

Here, nonlinear strain vector is expressed as:

$$\{\Delta \overline{\varepsilon}\} = \left\{\begin{array}{c}\Delta \overline{\varepsilon}_x\\\Delta \overline{\varepsilon}_y\\\Delta \overline{\gamma}_{xy}\\0\\0\end{array}\right\} = \left\{\begin{array}{c}\frac{1}{2}\left(\frac{\partial \Delta w'}{\partial x'}\right)^2\\\frac{1}{2}\left(\frac{\partial \Delta w'}{\partial y'}\right)^2\\\frac{\partial \Delta w'}{\partial x'}\cdot\frac{\partial \Delta w'}{\partial y'}\\0\\0\end{array}\right\} \tag{3.163}$$

Hence, the virtual nonlinear strain components are expressed as follows:

$$\{\delta\Delta\bar{\varepsilon}\} = \left\{ \begin{array}{c} \frac{\partial\delta\Delta w'}{\partial x'}\frac{\partial\Delta w'}{\partial x'} \\[2mm] \frac{\partial\delta\Delta w'}{\partial y'}\frac{\partial\Delta w'}{\partial y'} \\[2mm] \frac{\partial\delta\Delta w'}{\partial x'}\cdot\frac{\partial\Delta w'}{\partial y'}+\frac{\partial\delta\Delta w'}{\partial y'}\cdot\frac{\partial\Delta w'}{\partial x'} \\[2mm] 0 \\[2mm] 0 \end{array} \right\} \tag{3.164}$$

Then, the term included in the integration of Eq. (3.162) becomes as follows:

$$\{\delta\Delta\bar{\varepsilon}\}^{T}\{\sigma\} = \left\{ \frac{\partial\delta\Delta w'}{\partial x'}\frac{\partial\Delta w'}{\partial x'} \quad \frac{\partial\delta\Delta w'}{\partial y'}\frac{\partial\Delta w'}{\partial y'} \quad \frac{\partial\delta\Delta w'}{\partial x'}\cdot\frac{\partial\Delta w'}{\partial y'}+\frac{\partial\delta\Delta w'}{\partial y'}\cdot\frac{\partial\Delta w'}{\partial x'} \quad 0 \; 0 \right\} \left\{ \begin{array}{c} \sigma_x \\ \sigma_y \\ \tau_{xy} \\ \tau_{yz} \\ \tau_{xz} \end{array} \right\}$$

$$= \frac{\partial\delta\Delta w'}{\partial x'}\frac{\partial\Delta w'}{\partial x'}\sigma_x + \frac{\partial\delta\Delta w'}{\partial y'}\frac{\partial\Delta w'}{\partial y'}\sigma_y + \frac{\partial\delta\Delta w'}{\partial x'}\frac{\partial\Delta w'}{\partial y'}\tau_{xy} + \frac{\partial\delta\Delta w'}{\partial y'}\frac{\partial\Delta w'}{\partial x'}\tau_{xy}$$

$$= \left(\frac{\partial\delta\Delta w'}{\partial x'}\sigma_x + \frac{\partial\delta\Delta w'}{\partial y'}\tau_{xy}\right)\frac{\partial\Delta w'}{\partial x'} + \left(\frac{\partial\delta\Delta w'}{\partial x'}\tau_{xy} + \frac{\partial\delta\Delta w'}{\partial y'}\sigma_y\right)\frac{\partial\Delta w'}{\partial y'}$$

$$= \left\{ \frac{\partial\delta\Delta w'}{\partial x'}\sigma_x + \frac{\partial\delta\Delta w'}{\partial y'}\tau_{xy} \quad \frac{\partial\delta\Delta w'}{\partial x'}\tau_{xy} + \frac{\partial\delta\Delta w'}{\partial y'}\sigma_y \right\} \left\{ \begin{array}{c} \frac{\partial\Delta w'}{\partial x'} \\[2mm] \frac{\partial\Delta w'}{\partial y'} \end{array} \right\}$$

$$= \left\{ \frac{\partial\delta\Delta w'}{\partial x'} \quad \frac{\partial\delta\Delta w'}{\partial y'} \right\} \left[\begin{array}{cc} \sigma_x & \tau_{xy} \\ \tau_{xy} & \sigma_y \end{array} \right] \left\{ \begin{array}{c} \frac{\partial\Delta w'}{\partial x'} \\[2mm] \frac{\partial\Delta w'}{\partial y'} \end{array} \right\}$$

$$= \{\delta\Delta h'\}^{T}[B_1]^{T}[\sigma_0][B_1]\{\Delta h'\} \tag{3.165}$$

Thus, Eq. (3.126) has been derived.

REFERENCES

[1] Timoshenko S, Gere J. Theory of elastic stability. McGraw-Hill Kogakusha Ltd.; 1961.

[2] Ueda Y. Elastic, elastic-plastic and plastic buckling of plates with residual stresses. Lehigh University. PhD Dissertation; 1962.

[3] Masaoka K, Rashed S, Ueda Y. An efficient method of buckling analysis of rectangular plates. J Soc Naval Arch Jpn 1992;172:409–416 [in Japanese].

[4] Toi Y, Yuge K, Kawai T. Basic studies on the crashworthyness of structural elements; part 1. Crush analysis by the finite element method. J Soc Naval Arch Jpn 1986;159:248–257 [in Japanese].

[5] Yamada Y. Plasticity and visco-elasticity. Baifukan; 1980 [in Japanese].

[6] Kanok-Nukulchai W. A simple and efficient finite element for general shell analysis. Int J Numer Methods Eng 1979;14:179.

[7] Flanagan D, Belytschko T. A uniform strain hexahedron and quadrilateral with orthogonal hourglass control. Int J Numer Methods Eng 1981;18:679.

[8] Dow R, Amdahl J, Camisetti C, Chen T, Estefen S, Kmiecik M, et al. Ductile collapse. In: Report of committee III.1, the 11th international ship and offshore structures congress (ISSC), Wuxi, China; vol. I; 1991. p. 319–403.

[9] Jensen J, Amdahl J, Caridis P, Chen T, Cho S, Damonte R, et al. Ductile collapse. In: Report of committee III.1, the 12th international ship and offshore structures congress (ISSC), St. John's, Canada; vol. I; 1994. p. 299–387.

[10] Jensen J, Caridis P, Cho S, Damonte R, Dow R, Gordo J, et al. Ultimate strength. In: Report of committee III.1, the 13th international ship and offshore structures congress (ISSC), Trondheim, Norway; vol. I; 1997. p. 233–283.

[11] Yao T, Astrup O, Caridis P, Chen YN, Cho SR, Dow R, et al. Ultimate hull girder strength. In: Proc 14th international ship and offshore structures congress (ISSC); vol. 2; 2000. p. 321–391.

[12] Yao T, E B, Cho SR, Choo Y, Czujko J, Estefen S, et al. Ductile collapse. In: Report of committee III.1. Proc 16th international ship and offshore structures congress; vol. 1; 2006. p. 353–437.

BUCKLING/PLASTIC COLLAPSE BEHAVIOR AND STRENGTH OF RECTANGULAR PLATE SUBJECTED TO UNI-AXIAL THRUST

4.1 POSSIBLE BUCKLING MODES/BEHAVIOR

4.1.1 BUCKLING

A plate is one of the fundamental and the simplest structural members in ship and ship-like floating structures. What have to be considered for its buckling collapse are primary buckling, postbuckling behavior, secondary buckling, yielding, the ultimate strength and postultimate strength behavior.

As for buckling, it may be said that buckling collapse of a rectangular plate subjected to in-plane uni-axial thrust is the simplest buckling collapse phenomenon. In the case of a thin plate shown in Fig. 4.1, buckling occurs as the applied thrust load reaches a certain critical value. When a flat plate free from initial stress is simply supported along its four sides, the elastic buckling stress is expressed as [1]:

$$\sigma_{cr} = \frac{\pi^2 k E}{12(1 - \nu^2)} \left(\frac{t}{b}\right)^2 \tag{4.1}$$

where

$$k = \left(\frac{mb}{a} + \frac{a}{mb}\right)^2 \tag{4.2}$$

a, b, and t are the length, breadth, and thickness of the plate and m is a number of buckling half-waves in the loading direction. k is called buckling coefficient. In this case, the buckling mode is expressed as follows (Fig. 4.2):

$$w = A_m \sin \frac{m\pi x}{a} \sin \frac{\pi y}{b} \tag{4.3}$$

The buckling coefficient is calculated and plotted in Fig. 4.2. Buckling mode changes at the aspect ratio, $\sqrt{m(m+1)}$, from m half-waves mode to $(m+1)$ half-waves mode.

Buckling and Ultimate Strength of Ship and Ship-like Floating Structures. http://dx.doi.org/10.1016/B978-0-12-803849-9.00004-6

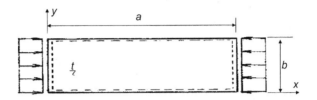

FIG. 4.1

Simply supported rectangular plate subjected to uni-axial longitudinal thrust.

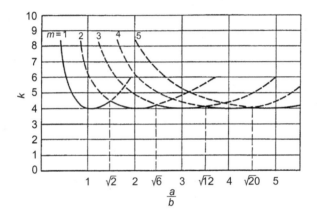

FIG. 4.2

Buckling coefficient for rectangular plate under longitudinal thrust.

4.1.2 ELASTIC POSTBUCKLING BEHAVIOR OF SHORT PLATE

Behavior of a plate beyond the buckling point is called postbuckling behavior. Postbuckling behavior of a rectangular plate subjected to uni-axial thrust in Fig. 4.1 can be approximately simulated if the deflection of a buckling mode, that is Eq. (4.3) is assumed to develop. Here, initial deflection same with the buckling mode is assumed as:

$$w_0 = A_{0m} \sin \frac{m\pi x}{a} \sin \frac{\pi y}{b} \tag{4.4}$$

In this case, according to the theory explained in Section 3.4, elastic postbuckling behavior can be simulated by Eqs. (3.71), (3.73), which can be rewritten when there is no welding residual stress as follows:

$$\alpha_1(A_m^2 - A_{0m}^2)A_m + \sigma_{cr0}(A_m - A_{0m}) - \sigma A_m = 0 \tag{4.5}$$

$$\varepsilon = \frac{1}{E}\sigma + \frac{\pi^2 m^2}{8a^2}(A_m^2 - A_{0m}^2) \tag{4.6}$$

where

$$\alpha_1 = \frac{m^2 \pi^2 a^2 E}{16}\left(\frac{1}{a^4} + \frac{1}{m^4 b^4}\right)$$

$$\sigma_{cr0} = \frac{\pi^2 E}{12(1-\nu^2)}\left(\frac{t}{b}\right)^2 \left(\frac{a}{mb} + \frac{mb}{a}\right)^2$$

In the above equations, compressive stress, σ, and compressive strain, ε, are taken positive.

When the plate is not accompanied by initial deflection, relationships among average stress, average strain, and deflection coefficient are written as follows:

$$\alpha_1 A_m^2 + \sigma_{cr0} - \sigma = 0 \tag{4.7}$$

$$\varepsilon = \frac{1}{E}\sigma + \frac{\pi^2 m^2}{8a^2}A_m^2 \tag{4.8}$$

Eliminating A_m from Eqs. (4.7), (4.8), the average stress-average strain relationship is derived as follows:

$$\sigma = E\frac{1 + (a/mb)^4}{3 + (a/mb)^4}\varepsilon + \frac{2}{3 + (a/mb)^4}\sigma_{cr} \quad (\varepsilon \geq \sigma_{cr}/E) \tag{4.9}$$

Fig. 4.3A and B shows examples of calculated average stress-deflection and average stress-average strain relationships, respectively. Plate dimensions are $a \times b \times t = 800 \times 1000 \times 10$ mm, and the number of buckling half-waves is one, that is $m = 1$. The magnitude of initial deflection is varied as $A_0/t = 0.0, 0.01, 0.1, 0.25, 0.5, 0.75$, and 1.0.

When initial deflection is zero, the deformed shape changes from a flat state to a deflected state. This is a bifurcation phenomenon. After the buckling has taken place, the average stress-deflection coefficient relationship is of a parabola of the second order as is known from Eq. (4.7). On the other hand, average stress-average strain relationship is linear. From Eq. (4.9), the tangential stiffness above the buckling stress is obtained as:

$$D_T = \frac{d\sigma}{d\varepsilon} = \frac{1 + (a/mb)^4}{3 + (a/mb)^4}E \tag{4.10}$$

Eq. (4.10) indicates that tangent stiffness beyond buckling is a function of the aspect ratio, a/b, of the plate. When $m = 1$, depending on a/b, it changes as:

$a/mb = 0.0$ $D_T/E = 0.3333$
$a/mb = 0.2$ $D_T/E = 0.3337$
$a/mb = 0.4$ $D_T/E = 0.3362$
$a/mb = 0.6$ $D_T/E = 0.3609$
$a/mb = 0.8$ $D_T/E = 0.4134$
$a/mb = 1.0$ $D_T/E = 0.5000$
$a/mb = 1.2$ $D_T/E = 0.6058$
$a/mb = 1.4$ $D_T/E = 0.7077$

which are plotted against aspect ratio in Fig. 4.4.

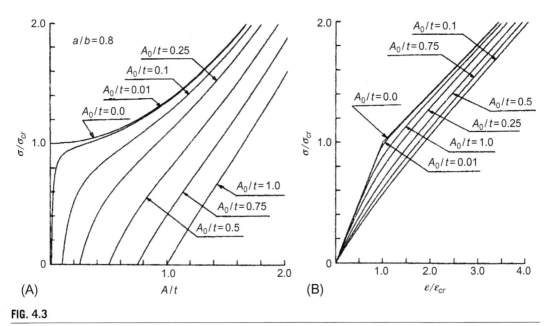

FIG. 4.3

Elastic large deflection behavior of rectangular plate subjected to uni-axial thrust. (A) Average stress-deflection coefficient relationships. (B) Average stress-average strain relationships.

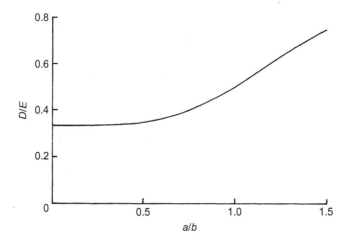

FIG. 4.4

In-plane stiffness beyond buckling.

On the other hand, when the plate is accompanied by initial deflection, average stress-deflection coefficient relationship is a cubic function of deflection coefficient. In this case, deflection increases from the beginning of loading. However, when the magnitude of initial deflection is small, deflection does not increase largely until the average stress increases up to near the buckling stress. Above the buckling stress, deflection rapidly increases as the case of no initial deflection.

When the magnitude of initial deflection is large, for example the case of $A_0/t = 1.0$, deflection increases almost monotonously from the beginning of loading. In this case, the average stress-average strain relationship is almost linear, whereas it is nearly bi-linear when the magnitude of initial deflection is very small. In all cases of different magnitude of initial deflection, the tangential stiffness of the plate, which is represented as the slope of the average stress-average strain relationship, seems not to be lower than the tangential stiffness, D_H, defined by Eq. (4.10).

Here, it was said that relationships indicated in Fig. 4.3A and B are the approximation of the real phenomenon on the basis of simple deflections expressed by Eqs. (4.3), (4.4). To examine the accuracy of this approximation, elastic large deflection analysis is performed on a rectangular plate of $a \times b \times t = 800 \times 1000 \times 12$ mm applying finite element method (FEM). Initial deflection expressed by Eq. (4.4) is assumed taking $m = 1$ and $A_{0m}/t = 0.01$. Calculated results are compared in Fig. 4.5A and B with analytical solution obtained by Eqs. (4.5), (4.6).

Some differences are observed between the two results, which may indicate that deflection above the buckling stress can no more be expressed by a simple form such as that of Eq. (4.3). General form of deflection in the case of a simply supported rectangular plate can be expressed as follows:

$$w = \sum_{i=1}^{m} \sum_{j=1}^{n} A_{ij} \sin \frac{i\pi x}{a} \sin \frac{j\pi y}{b} \tag{4.11}$$

Applying the *Method of Least Squares* (see Section 4.8) with nodal point displacements obtained by the FEM analysis, deflection components, A_{ij}, are calculated taking $m = n = 9$. The magnitudes of deflection components divided by that of one half-wave component, A_{11}, are plotted against the average compressive strain in Fig. 4.6. The average strain is divided by the yield strain of the material ($\varepsilon_Y = \sigma_Y/E = 313.6/205{,}800$), and the buckling point on the horizontal axis is $\varepsilon_{cr}/\varepsilon_Y = \sigma_{cr}/\sigma_Y = 0.3589$.

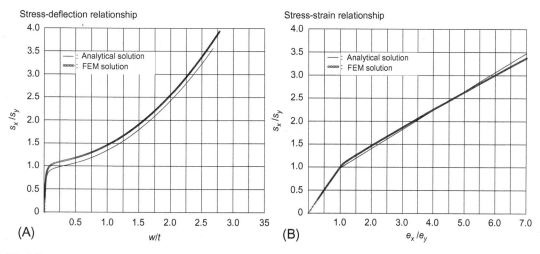

(A)

(B)

FIG. 4.5

Rectangular plate subjected to uni-axial thrust. (A) Average stress-deflection coefficient relationships. (B) Average stress-average strain relationships.

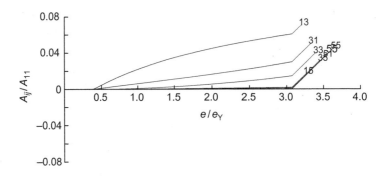

FIG. 4.6

Average strain-deflection component relationships.

It is seen that deflection components other than one half-wave component start to develop from the buckling point as the average strain increases. Although the magnitudes of these components are not so large, deflection mode gradually changes from a sinusoidal mode to a different mode.

4.1.3 ELASTIC POSTBUCKLING BEHAVIOR OF LONG PLATE

Postbuckling behavior of a rectangular plate subjected to longitudinal thrust is a little complicated compared to a short rectangular plate. Here, results of elastic large deflection analysis on strength deck panel of an existing pure car carrier are shown in Fig. 4.7A-C. The size of the panel is $a \times b \times t = 4200 \times 835 \times 13$ mm. Measured initial deflection is given to the plate, which is represented as:

$$w_0 = \sum_{i=1}^{11} A_{0i} \sin \frac{i\pi x}{a} \sin \frac{\pi y}{b} \quad (4.12)$$

where

$$
\begin{array}{llll}
A_{01}/t = -0.3482 & A_{02}/t = 0.0031 & A_{03}/t = -0.0920 & A_{04}/t = -0.0030 \\
A_{05}/t = -0.0517 & A_{06}/t = -0.0074 & A_{07}/t = -0.0185 & A_{08}/t = -0.0027 \\
A_{09}/t = -0.0426 & A_{0\,10}/t = 0.0002 & A_{0\,11}/t = 0.0007 &
\end{array}
$$

The magnitudes of deflection components on the horizontal axis in Fig. 4.7A are the values of deflection components composing the initial deflection. Until the average stress approaches the primary buckling stress, all the components increases with the increase in the applied average stress. Above the buckling stress, deflection component of five half-waves continues to increase whereas other components decrease as indicated in Fig. 4.7A. Consequently, deflection mode changes from the solid line to the dashed line, and then to the chain line as illustrated in Fig. 4.7A. The five half-waves mode correspond to the buckling mode of this plate, since the aspect ratio of this plate is 5.03.

The dashed lines in Fig. 4.7A-C are the results of analysis on the plate with initial deflection of a buckling mode. For this analysis, the magnitude of initial deflection of the buckling mode is taken as $A_{05}/t = -0.0517$, and all other components are set as zero. Fig. 4.7A indicates that the five half-waves mode develops more easily in this case compared to the case with initial deflection of a thin-horse mode.

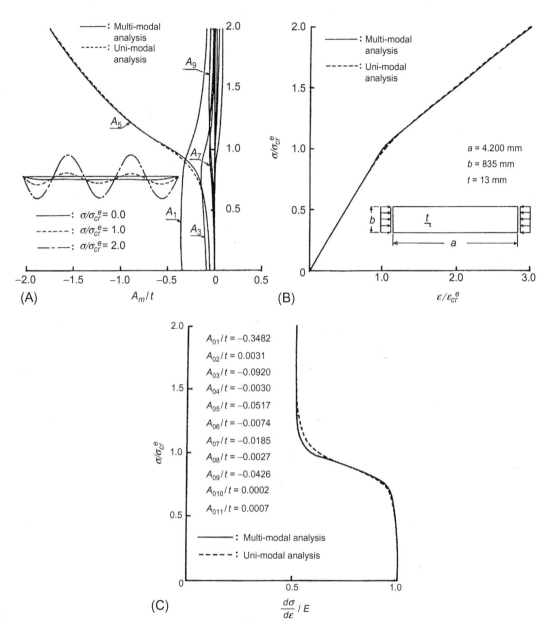

FIG. 4.7

Elastic postbuckling behavior of long rectangular plate subjected to uni-axial thrust. (A) Average stress-deflection coefficient relationships. (B) Average stress-average strain relationship. (C) Average stress-inplane stiffness relationship.

This is because other deflection components constrain the development of buckling mode deflection when they exist.

The difference observed in the deflection components in the postbuckling range reflects on the average stress-average strain and average stress-in-plane stiffness relationships as indicated in Fig. 4.7B and C. It can be said that the postbuckling behavior of both cases is almost the same. This is very important when the ultimate compressive strength of a rectangular plate is considered as shall be explained later. That is, it is not the maximum magnitude of initial deflection but the magnitude of the buckling mode component that dominates the postbuckling behavior and the ultimate compressive strength.

Here, it seems there exist some contradiction in the facts indicated in Figs. 4.6 and 4.7A. In Fig. 4.6, deflection components other than the buckling mode start to develop above the buckling stress, whereas they start to decrease above the buckling stress in Fig. 4.7A. This is because of the difference in half-wave numbers of buckling mode. Deflection components A_{13}, A_{31}, A_{33}, etc., in Fig. 4.6 corresponds to $A_{5\ 15}, A_{15\ 5}, A_{15\ 15}$, etc., in Fig. 4.6, which are not indicated in the figure.

4.2 BUCKLING STRENGTH

4.2.1 INFLUENCE OF BOUNDARY CONDITION ON BUCKLING STRENGTH

The elastic buckling strength of a plate subjected to various loads can be expressed as follows [1]:

$$\sigma_{cr} = \frac{\pi^2 kE}{12(1 - \nu^2)} \left(\frac{t}{b}\right)^2 \tag{4.13}$$

where t and b are the thickness and the breadth of the plate, respectively.

On the other hand, k is called "buckling coefficient," which is determined depending on loading and boundary conditions.

Here, buckling strength of a rectangular plate with various boundary conditions is considered under uni-axial thrust. Four cases are considered which are

Case (A): all sides simply supported
Case (B): loaded sides clamped and nonloaded sides simply supported
Case (C): loaded sides simply supported and nonloaded sides clamped
Case (D): all sides clamped

The buckling coefficients in four cases are plotted against aspect ratio, a/b, of the plate in Fig. 4.8. For Case (A), the buckling coefficient, k, is expressed as:

$$k = \left(\frac{a}{mb} + \frac{mb}{a}\right)^2 \tag{4.14}$$

where a and b are the length and the breadth of the plate (see Fig. 4.1) and m corresponds to the number of half-waves of a buckling mode expressed as:

$$w = A_m \sin\frac{m\pi x}{a} \sin\frac{\pi y}{b} \tag{4.15}$$

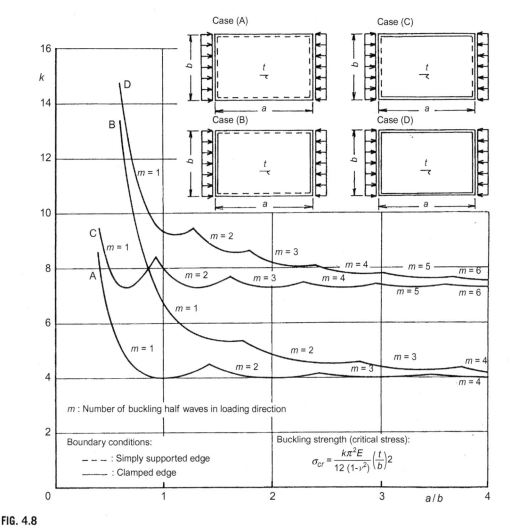

FIG. 4.8

Buckling strength of rectangular plates with various boundary conditions under uni-axial thrust.

It is seen that Case (A) gives the lowest buckling strength. That is, the buckling strength of a rectangular plate under uni-axial thrust is the lowest when its four sides are simply supported. It is also known that the lowest buckling strength with all sides simply supported condition is attained when the aspect ratio is an integer, that is when $a/mb = 1.0$, as is known from Eq. (4.14).

Comparing Case (A) and Case (B), buckling strength is different when the aspect ratio of the plate is low, but it gradually become the same as the aspect ratio increases, and may coincide when it increases to infinity. As is known from Eq. (4.15), the buckling mode of Case (A) is periodical in the loading direction (x-direction), whereas that of Case (B) is not periodical, and may be expressed as:

$$w = A_m f(x) \cdot \sin \frac{\pi y}{b} \tag{4.16}$$

However, when the aspect ratio of the plate increases, the buckling mode of Case (B) becomes almost the same with that of Case (A) except near the clamped both ends where thrust load is applied. This is the reason why buckling strength of Case (B) converges to that of Case (A) in the range of higher aspect ratio of the plate.

The buckling mode of Case (C) is also periodical to the loading direction as Case (A), and can be expressed as follows:

$$w = \frac{1}{2} A_m \sin \frac{m\pi x}{a} \left(1 - \cos \frac{2\pi y}{b} \right) \tag{4.17}$$

So, the variation of the buckling coefficient of Case (C) is the same as that of Case (A) although the absolute value is different.

On the other hand, buckling mode for Case (D) is expressed as:

$$w = \frac{1}{2} A_m g(x) \left(1 - \cos \frac{2\pi y}{b} \right) \tag{4.18}$$

With the same reason regarding relationship between Cases (A) and (B), buckling strength of Case (D) converges to that of Case (C) in the range of higher aspect ratio of the plate.

In design, it is usually considered that the buckling coefficient for a rectangular plate subjected to uni-axial thrust is 4.0. The meaning of "4.0" can be seen in Fig. 4.1, that is the most conservative value of buckling coefficient for this loading condition under various boundary conditions.

4.2.2 INFLUENCE OF WELDING RESIDUAL STRESS ON BUCKLING STRENGTH

Here, influence of welding residual stress on buckling strength of a rectangular plate subjected to uni-axial thrust load is explained. It is assumed that the plate is simply supported and is accompanied by welding residual stress as illustrated in the figure in Fig. 4.9. The buckling mode is represented by Eq. (4.15). Elastic buckling strength, elastoplastic buckling strength, and plastic buckling strength are plotted against slenderness ratio, $(b/t)\sqrt{\sigma_Y/E}$ [2].

It is known that reduction due to welding residual stress depends on the magnitude of compressive residual stress in the center of the plate. When the plate becomes thin, the buckling strength curve intersects the horizontal axis. Above this slenderness ratio, negative buckling strength is calculated. This physically implies that the plate has buckled due to compressive residual stress with no action of external thrust.

It is indicated in Fig. 4.9 that welding residual stress reduce the buckling strength of plates. However, this is the case that welding residual stress is in compression at the location where buckling deflection develops. Here, two representative cases are considered shown in Fig. 4.10A and B. In Fig. 4.10A, compressive residual stress is produced at the center of the plate, whereas in Fig. 4.10B, tensile residual stress is produced at the center of the plate.

Assuming a simple buckling mode indicated by Eq. (4.15), the buckling strength is derived as follows:

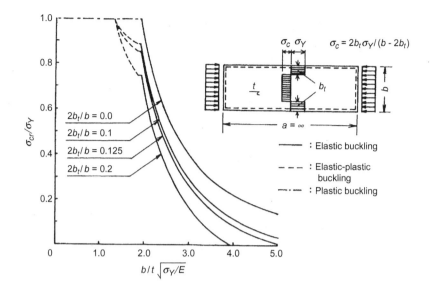

FIG. 4.9

Influence of welding residual stress on compressive buckling strength of rectangular plate [2].

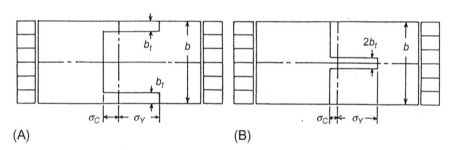

(A)   (B)

FIG. 4.10

Schematic representation of typical welding residual stress. (A) Due to fillet weld. (B) Due to butt weld.

Fillet weld F:

$$\sigma_{cr} = \frac{\pi^2 E}{12(1-v^2)} \left(\frac{t}{b}\right)^2 \left(\frac{\alpha}{m} + \frac{m}{\alpha}\right)^2 - \frac{\sigma_Y}{\pi(1-\mu)} \sin \mu\pi \qquad (4.19)$$

Butt weld F:

$$\sigma_{cr} = \frac{\pi^2 E}{12(1-v^2)} \left(\frac{t}{b}\right)^2 \left(\frac{\alpha}{m} + \frac{m}{\alpha}\right)^2 + \frac{\sigma_Y}{\pi(1-\mu)} \sin(1-\mu)\pi \qquad (4.20)$$

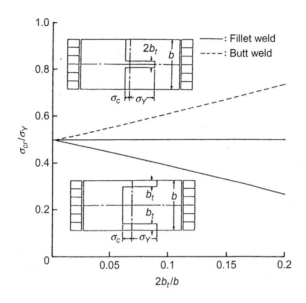

FIG. 4.11

Influence of welding residual stress by fillet and butt weld on compressive buckling strength of rectangular plate.

where $\alpha = a/b$ is the aspect ratio of the plate. Buckling strength calculated by Eqs. (4.19), (4.20) is plotted in Fig. 4.11. It should be noticed that tensile residual stress in the center of the plate, where buckling deflection develops, increases the buckling strength.

Another example related to this issue can be seen in Masubuchi [3]. It is shown that residual stress increases or decreases the buckling strength depending on its distribution and the slenderness ratio of the plate.

Elastic, elastoplastic, and plastic buckling strength of rectangular plates subjected to pure bending as well as combined bending and thrust is calculated in Ueda and Yamakawa [4]. Buckling strength under pure shear load is also calculated in Terazawa and Ueda [5]. In both calculation, influence of welding residual stress is examined.

4.3 LOCAL BUCKLING STRENGTH OF STIFFENED PLATE CONSIDERING WEB-PLATE INTERACTIONS

4.3.1 INTERACTION BETWEEN PLATE AND STIFFENER

Stiffened plates are so designed that local panel buckling takes place before overall buckling occurs. For the design of such a stiffened plate, the buckling coefficient, k, is usually taken as 4.0 as described in Section 4.1. However, local buckling strength of a stiffened plate is affected by the interaction between plate and stiffener web, and is expected to increase. To account for the influence of the

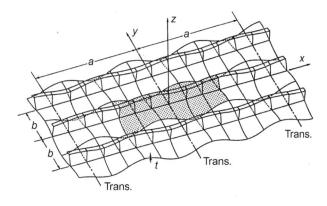

FIG. 4.12

Continuous stiffened plate.

interaction between the plate and the stiffener web accurately, the stiffener web has to be treated as a plate.

Here, continuous stiffened plate illustrated in Fig. 4.12 is considered. On the plate, stiffeners of the same size are fitted with equal distances. The interaction between the plate and the stiffener web can be simulated by assuming the following deflection modes as buckling modes [6].

For plate:

$$w = \sin\frac{m\pi x}{a}\left\{W_1\sin\frac{\pi y}{b} + \frac{1}{2}W_2\left(1 - \cos\frac{2\pi y}{b}\right)\right\} \tag{4.21}$$

For stiffener web:

$$v_{\text{web}} = \sin\frac{m\pi x}{a}\left\{V_1\frac{z}{h} + V_2\left(1 - \cos\frac{\pi z}{2h}\right) + V_3\sin\frac{\pi z}{2h}\right\} \tag{4.22}$$

In Eq. (4.21), the first term represents the buckling mode of a simply supported plate, and the second term that of a clamped plate. The latter term produces bending moment which is transferred to the stiffener web. Three deflection components of a stiffener web in Eq. (4.22) are illustrated in Fig. 4.13. The first term represents a rigid-body rotation about the line of attachment of stiffener web to plate. The second term produces the bending moment transferred to the panel and the third term that to the stiffener flange from the stiffener web.

4.3.2 DERIVATION OF INTERACTIVE BUCKLING STRENGTH

In the following, a method of analysis is explained to evaluate the local panel buckling strength considering the interaction between the plate and the stiffener web as well as that between stiffener web and flange. In the case of a continuous stiffened plating, symmetry condition or periodically continuous condition can be imposed along the centerlines in both longitudinal and transverse directions. So, it is enough if the shaded region partitioned by four centerlines is analyzed instead of analyzing the whole stiffened plating; see Fig. 4.12.

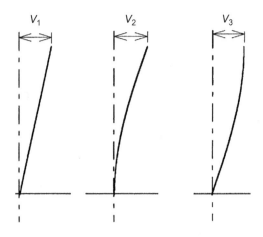

FIG. 4.13

Assumed deflection components in stiffener web.

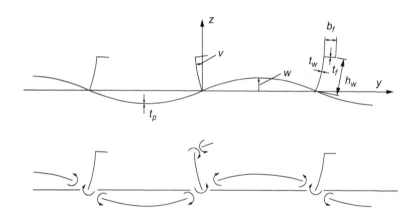

FIG. 4.14

Equilibrium condition and deflection mode.

The following boundary conditions are considered in the formulation.

(1) Continuity condition for rotation angle along panel/web intersection:

$$\left.\frac{\partial w}{\partial y}\right|_{y=0} = \left.\frac{\partial v_{\text{web}}}{\partial z}\right|_{z=0} \tag{4.23}$$

(2) Equilibrium condition for bending/torsional moment along panel/web intersection (see Fig. 4.14):

$$D_w \left(\frac{\partial^2 v_{web}}{\partial z^2} + v \frac{\partial^2 v_{web}}{\partial x^2} \right)\bigg|_{z=0} + 2D_p \left(\frac{\partial^2 w}{\partial y^2} + v \frac{\partial^2 w}{\partial x^2} \right)\bigg|_{y=0} = 0 \tag{4.24}$$

where,

$$D_w = \frac{Et_w^3}{12(1 - v^2)}, \quad D_p = \frac{Et_p^3}{12(1 - v^2)}$$

and t_p and t_w are the thicknesses of panel and stiffener web.

(3) Equilibrium condition for bending/torsional moment along web/flange intersection of stiffener considering continuity of rotation (see Fig. 4.14):

$$GJ_f \frac{\partial^3 v_{web}}{\partial x^2 \partial z}\bigg|_{z=h} - D_w \left(\frac{\partial^2 v_{web}}{\partial z^2} + v \frac{\partial^2 v_{web}}{\partial x^2} \right)\bigg|_{z=h} = 0 \tag{4.25}$$

where GJ_f is a torsional stiffness of the flange of a stiffener

Applying the *Principle of Minimum Potential Energy*, the elastic buckling interaction equation is finally derived in the following form:

$$\kappa_1 \sigma_x^2 + \kappa_2 \sigma_x \sigma_y + \kappa_3 \sigma_y^2 - \kappa_4 \sigma_x - \kappa_5 \sigma_y + \kappa_6 = 0 \tag{4.26}$$

Detail of the derivation and the coefficients in Eq. (4.26) are given in Fujikubo and Yao [6].

4.3.3 INFLUENCE OF PLATE-STIFFENER WEB INTERACTION ON LOCAL BUCKLING STRENGTH

Here, calculated results of local buckling strength of continuous stiffened plate are shown in Table 4.1. The local panel is fixed as $a \times b \times t = 2400 \times 800 \times 10$ mm. As a stiffener, different sizes of flat-bar, angle-bar, and tee-bar are considered. Typical buckling modes are shown in Fig. 4.15A-C.

σ_1/σ_0 ratio in Table 4.1 indicates the ratio of the local buckling strength to buckling strength of a simply supported plate. It is seen that local buckling strength is increased by as high as 45% to 60% when angle-bar or tee-bar stiffener is fitted. Contrary to this, low and thin flat-bar stiffener is not so effective to increase local buckling strength. However, thicker flat-bar stiffener is effective as angle-bar and tee-bar stiffeners.

On the other hand, σ_2/σ_1 ratio in Table 4.1 indicates the ratio of FEM result to analytical solution, which represents the accuracy of Eq. (4.26). It is seen that local buckling strength calculated by Eq. (4.26) is between -1.5% and 4% of the FEM results.

In the case of the flat-bar stiffener of 100×10 mm, σ_2/σ_1 ratio is 1.0865. This is because the buckling was in overall mode, which is not considered in the present formulation. Also in the case of flat-bar stiffener with 10 mm thickness but different height, it does not contributes to increase the local buckling strength although the assumed buckling mode is rational. This may be because deflection of V_1 mode in Fig. 5.13 is dominant and deflection of V_2 mode which produces bending moment is small. This can be seen in Fig. 4.15A. However, it is known that thicker flat-bar stiffener resists more against the rotation of the plate edge along the plate/web intersection line and increases the local buckling strength.

Table 4.1 Buckling Strength of Stiffened Plating Under Thrust

Type	h	t_w	b_f	t_w	σ_1/σ_0	σ_2/σ_1	Type	h	t_w	b_f	t_f	σ_1/σ_0	σ_2/σ_1
Flat	100.0	10.0	0.0	0.0	0.9850	1.0856	Tee	250.0	12.0	90.0	16.0	1.5388	1.0104
Flat	100.0	15.0	0.0	0.0	1.1479	0.9893	Angle	250.0	12.0	120.0	16.0	1.5537	1.0148
Flat	100.0	20.0	0.0	0.0	1.4345	0.9908	Angle	250.0	12.0	150.0	16.0	1.5597	1.0214
Flat	195.0	10.0	0.0	0.0	1.0900	0.9880	Angle	400.0	12.0	90.0	16.0	1.4461	1.0214
Flat	195.0	15.0	0.0	0.0	1.3183	0.9886	Angle	400.0	12.0	150.0	16.0	1.4580	1.0403
Flat	195.0	20.0	0.0	0.0	1.5835	0.9870	Tee	150.0	12.0	90.0	12.0	1.5912	0.9848
Flat	310.0	10.0	0.0	0.0	0.9965	0.9920	Tee	150.0	12.0	150.0	12.0	1.5904	1.0167
Flat	310.0	15.0	0.0	0.0	1.3066	0.9922	Tee	250.0	12.0	90.0	16.0	1.5118	0.9965
Flat	310.0	20.0	0.0	0.0	1.6300	0.9926	Tee	250.0	12.0	150.0	16.0	1.5578	1.0129
Flat	400.0	10.0	0.0	0.0	0.8297	1.0071	Tee	400.0	12.0	90.0	12.0	1.4229	1.0015
Flat	400.0	20.0	0.0	0.0	1.6194	1.0006	Tee	400.0	12.0	120.0	12.0	1.4387	1.0068
Angle	150.0	12.0	90.0	12.0	1.5763	1.0197	Tee	400.0	12.0	90.0	16.0	1.4506	1.0247
Angle	150.0	12.0	150.0	12.0	1.5906	1.0439	Tee	400.0	12.0	150.0	16.0	1.4581	1.0355
Angle	250.0	12.0	90.0	16.0	1.5388	1.0104	Tee	400.0	12.0	150.0	12.0	1.4458	1.0111

σ_0: buckling stress of simply supported panel

σ_1: buckling stress by proposed method

σ_2: buckling stress by FEM

panel: $a \times b \times t = 2400 \times 800 \times 10$mm

h, t_w, b_f, t_f: in mm

flat-bar   angle-bar   tee-bar

Comparing Fig. 4.15B and C, it is found that flange is more effective to prevent the horizontal displacement of the stiffener top in case of angle-bar stiffener compared to that of tee-bar stiffener.

Fig. 4.16A and B shows the buckling strength of local panel calculated both by Eq. (4.26) and FEM changing the panel thickness but keeping the panel size as $a \times b = 2400 \times 800$ mm [6]. Firstly, good correlations are observed between buckling strength calculated by Eq. (4.26) and FEM. It is seen that the increase of buckling strength becomes low as the panel thickness increases.

Under the longitudinal thrust, the buckling mode is three half-waves in the loading direction when the local plate is assumed to be simply supported. However, it changes to four half-waves when the panel becomes thin. This is because the rotational constraint from the stiffener web along the panel/web intersection becomes larger as the panel thickness decreases.

Under the transverse thrust, the local buckling strength rapidly increases as the panel thickness decreases. It is also observed that the local buckling strength increases as the stiffener size increases. However, 300IA and 350IA, where "IA" stands for "Invert Angle," have almost the same effect on the increase of buckling strength.

On the other hand, under the longitudinal thrust, 300IA gives the highest buckling strength and buckling strength of 350IA is lower even than that of 250IA. That is, with an increase in stiffener size,

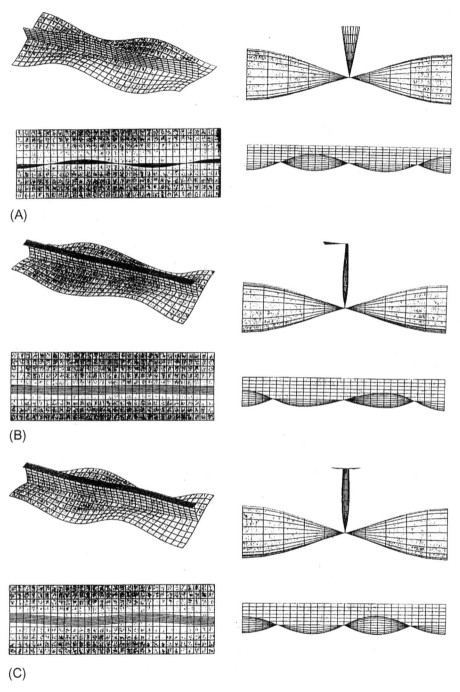

FIG. 4.15

Some typical buckling modes of stiffened plate under thrust. (A) Flat-bar ($h \times t_w = 195 \times 20$ mm).
(B) Angle-bar ($250 \times 90 \times 12/16$ mm). (C) Tee-bar ($250 \times 90 \times 12/16$ mm).

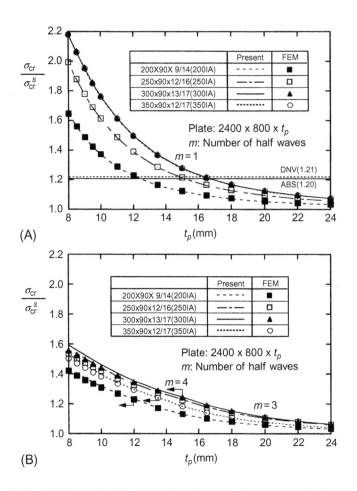

FIG. 4.16

Influence of stiffener on local buckling strength of continuous stiffened plates under thrust. (A) Transverse thrust. (B) Longitudinal thrust.

the effect of the stiffeners reaches the maximum at a certain size and then starts to decrease depending on the web slenderness. This implies that an optimum stiffener size exists which gives the highest elastic local buckling strength. The flexibility of the stiffener web may depend not only on so-called "size" but also on its height to thickness ratio.

The elastic local buckling strength of a continuous stiffened plate under bi-axial thrust is calculated by Eq. (4.26) and the FEM considering the influence of welding residual stress. Welding residual stress shown in Fig. 2.3 is assumed of which magnitude in compression is given by Eq. (2.8) in Chapter 2.

Calculation is performed on the bottom plating of existing bulk carrier and VLCC. The results of calculation are plotted in Fig. 4.17A and B, respectively [6]. Three cases are plotted, which are stiffened plates with and without welding residual stress and a simply supported plate without welding residual stress.

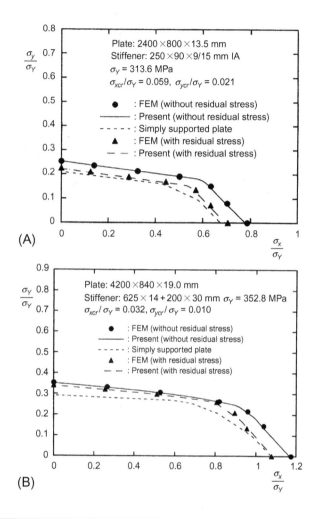

FIG. 4.17

Combined influence of panel/web interaction and welding residual stress on local buckling strength of continuous stiffened plates under thrust. (A) Bulk carrier. (B) Tanker (VLCC).

It is seen that local buckling strength calculated by Eq. (4.26) show good correlations with those calculated by the FEM. It is interesting that increased local buckling strength owing to the stiffener is reduced by welding residual stress, and consequently for the BC panel, the buckling strength considering both stiffener effect and welding residual stress is very close to that of a simply supported plate free from welding residual stress.

The same result can be observed for the VLCC panel when longitudinal thrust is dominant. However, the reduction of the local buckling strength due to welding residual stress is very small when transverse thrust is dominant. This is partly because the compressive residual stress in the transverse direction is smaller and partly because high tensile stress exists near the web/flange intersection of the stiffener where lateral deflection of the web and the flange is the largest.

Here, a series of local buckling strength estimation is performed applying Eq. (4.26) choosing 10 stiffened plates from the deck and the bottom plating of existing bulk carriers, container ships, and tankers. The calculated results are summarized in Table 4.2 [6].

In most cases, the increase in the local buckling strength due to stiffeners is only slightly larger than the decrease due to welding residual stresses, and the resulting buckling strength is close to that of a simply supported plate free from welding residual stress for both longitudinal and transverse thrust. From these results, it can be concluded that as far as stiffened plates in the ordinary ships are concerned, traditional calculation method of elastic buckling strength assuming simply supported sides is valid from the viewpoint of practical application. In other words, it can be said that the reduction in buckling strength due to welding residual stress is compensated for by the increase in buckling strength owing to the stiffeners.

4.4 SECONDARY BUCKLING IN RECTANGULAR PLATE SUBJECTED TO UNI-AXIAL THRUST

4.4.1 ELASTIC SECONDARY BUCKLING OF SIMPLY SUPPORTED RECTANGULAR PLATE

A rectangular plate subjected to uni-axial thrust undergoes the secondary buckling after the primary buckling has taken place. To simulate secondary buckling behavior in an analytical manner, deflection of a general form has to be assumed, which is

$$w = \sum_{i=1}^{m} \sum_{j=1}^{n} \sin \frac{i\pi x}{a} \sin \frac{i\pi y}{b} \tag{4.27}$$

Inclusion of deflection components in the transverse direction other than one half-wave mode is essential for the simulation of the secondary buckling behavior.

4.4.2 STATIC EQUILIBRIUM PATH IN SECONDARY BUCKLING OF SHORT PLATE UNDER UNI-AXIAL THRUST

Here, elastic large deflection analysis is performed firstly on a square plate under uni-axial thrust. Static analysis is performed applying the analytical method. The length and the breadth are taken as 1000 mm, and the thickness as 10 mm. The plate is assumed to be simply supported along its four sides and is flat—that is with no initial deflection.

In the analysis, the following deflection is assumed.

$$w = \sum_{m=1}^{9} \sum_{n=1}^{5} A_{mn} \sin \frac{m\pi x}{a} \sin \frac{n\pi y}{b} \tag{4.28}$$

m and n are only the odd numbers, and thus the number of deflection components is in total 15.

Table 4.2 Local Buckling Strength of Continuous Stiffened Plates Considering Influences of Stiffeners and Welding Residual Stresses

Type and No.	Panel ($a \times b \times t_p$ mm³)	Stiffener (mm)	σ_Y(MPa)	Longitudinal Thrust			Transverse Thrust		
				σ_0(MPa)	σ_1/σ_0	σ_2/σ_0	σ_0(MPa)	σ_1/σ_0	σ_2/σ_0
BC1	3200 × 850 × 18.5	250 × 25	313.6	353.7	1.226	1.159	101.0	1.124	1.059
BC2	2500 × 820 × 16.5	300 × 90 × 10/16	352.8	301.3	1.101	1.022	92.4	1.140	1.048
BC3	2400 × 800 × 14.5	250 × 90 × 10/16	313.6	244.4	1.162	1.064	75.4	1.192	1.078
CS1	4000 × 800 × 20.5	300 × 90 × 13/17	352.8	488.6	1.103	1.063	132.1	1.034	1.000
CS2	3100 × 854 × 18.0	350 × 100 × 11/17	313.6	333.7	1.076	1.025	95.7	1.117	1.058
CS3	3270 × 880 × 15.5	350 × 100 × 12/17	313.6	232.1	1.161	1.080	66.4	1.182	1.093
TK1	5630 × 955 × 20.5	400 × 11 + 150 × 19	313.6	342.9	1.059	0.988	90.7	1.028	1.004
TK2	5630 × 955 × 22.0	700 × 14 + 200 × 25	313.6	395.0	0.889	0.950	104.5	1.093	1.072
TK3	5095 × 850 × 18.0	650 × 13.5 + 150 × 25	313.6	333.7	0.918	0.996	88.1	1.111	1.087
TK4	4000 × 795 × 22.0	330 × 25	235.2	569.8	1.023	0.996	153.9	1.063	1.040

Notes: BC, bulk carrier; CS, container ship; TK, tanker; σ_0, buckling strength of simply supported plate; σ_1, buckling strength considering stiffeners; σ_2, buckling strength considering stiffeners and welding residual stresses.

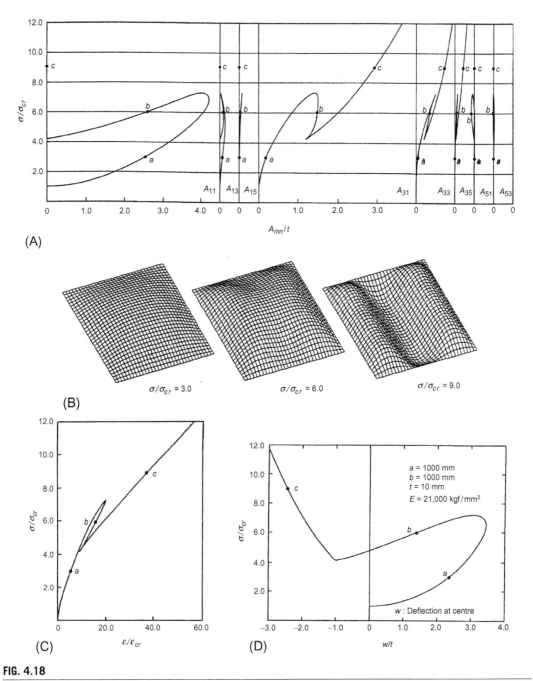

FIG. 4.18

Secondary buckling behavior of simply supported square plate under thrust [7]. (A) Average stress-deflection coefficients relationships. (B) Change in deflection mode. (C) Average stress-average strain relationship. (D) Average stress-deflection relationship.

The relationships between the average stress and deflection components are shown in Fig. 4.18A, and the mode changes in Fig. 4.18B [7]. The average stress-average strain and the average stress-central deflection relationships are also shown in Fig. 4.18C and D, respectively [7]. The deflection components which are not indicated in Fig. 4.18A are very small compared to those in the same figure.

The primary buckling mode of a simply supported square plate is, as well known, one half-wave of sine mode with deflection coefficient of A_{11}. However, the components other than one half-wave component starts to develop with the increase in the applied stress as indicated in Fig. 4.18A. At point a ($\sigma/\sigma_{cr} = 3.0$) on the curves, deflection component of one half-wave, which is the buckling mode, is dominant, and the deflection mode is as illustrated at the left-hand side in Fig. 4.18B.

With a further increase in the applied stress, deflection components reach their maximum values, and then start to decrease. During this process, the applied stress attains its local maximum and then start to decrease. The deflection mode at point b is shown in the middle of Fig. 4.18B. It is seen that deflection mode is changing from one half-wave mode to other mode. Beyond point b, deflection components of A_{11}, A_{13}, A_{15}, A_{51}, and A_{55} decrease, and the applied stress attains its local minimum.

After this, the above-mentioned deflection components do not increase with the increase in the applied stress and reaches point c moving along the vertical axes. On the other hand, components of A_{31}, A_{33}, and A_{35} increase after they attain their local minimum. Consequently, deflection mode changes to three half-waves mode, which is illustrated at the right-hand side in Fig. 4.18B.

It is known from Fig. 4.18C that beyond the point at which applied stress attains its local maximum, both average stress and average strain decrease until they attain their local minimum. This unloading path is an unstable equilibrium path along which the potential energy of this system shows its local maximum. Beyond the local minimum point, both average stress and average strain start to increase and the path is again a stable equilibrium path. This phenomenon is called secondary buckling.

In the actual behavior, load is applied through forced displacement or directly by force. That is, force or displacement continues to increase monotonously. In this case, the unloading path does not exist and jumping takes place from the local maximum point of the load to the continued stable equilibrium path. That is, so-called snap-through takes place, which will be explained later.

4.4.3 STATIC EQUILIBRIUM PATH IN SECONDARY BUCKLING OF LONG PLATE UNDER UNI-AXIAL THRUST

As explained in the previous section, postbuckling behavior of a long rectangular plate is a little complicated when it is compared with that of a short rectangular plate. Here, postbuckling behavior of a long plate is explained including the secondary buckling behavior after the primary buckling.

Considering a rectangular plate with initial deflection, of which aspect ratio is $a/b = 5.0$ as an example, the secondary buckling behavior is explained in the following. For this plate, the numbers of deflection components is taken as 21 and 3 in the longitudinal and transverse directions, respectively. That is

$$w = \sum_{i=1}^{21} \sum_{j=1}^{3} A_{ij} \sin \frac{i\pi x}{a} \sin \frac{j\pi y}{b} \tag{4.29}$$

In the above expression, only odd terms are included.

Initial deflection is expressed with the same deflection components as:

$$w_0 = \sum_{i=1}^{21} \sum_{j=1}^{3} A_{0ij} \sin\frac{i\pi x}{a} \sin\frac{j\pi y}{b} \tag{4.30}$$

where

$$A_{0\,11} = 1.1271, \quad A_{0\,31} = 0.3483, \quad A_{0\,51} = 0.1787$$
$$A_{0\,71} = 0.0995, \quad A_{0\,91} = 0.0537, \quad A_{0\,111} = 0.0256$$

All other coefficients are set as zero.

Fig. 4.19 shows the relationships between average stress and representative deflection coefficients. As the average stress increases, all the deflection components increase at the beginning, but start to decrease as the average stress comes near the primary buckling stress except the five half-waves component. Then, the average stress attains its local maximum at point B and bifurcation takes place. After bifurcation, some deflection components start to increase whereas component of five half-waves starts to decrease. After a while, all the components gradually decreases until the average stress reaches its local minimum at point D. Beyond point D, only the deflection component of seven half-waves starts to increase. This path coincides with a loading path represented by the dashed line which is for the case when the primary buckling takes place in a seven half-waves mode.

In the above-mentioned behavior, the equilibrium path from point B to point D is an unstable path along which the total potential energy takes its local maximum, and physically does not exist. The change in deflection mode from point A to point E is illustrated in Fig. 4.20. It is know how the deflection mode changes from five half-waves to seven half-waves.

The mode of the primary buckling for this plate is of a five half-waves mode. As mentioned above, however, the stable deflection mode is of a seven half-waves mode after the secondary buckling. This

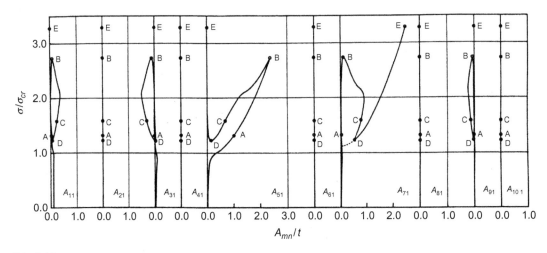

FIG. 4.19

Influence of initial deflection on secondary buckling behavior.

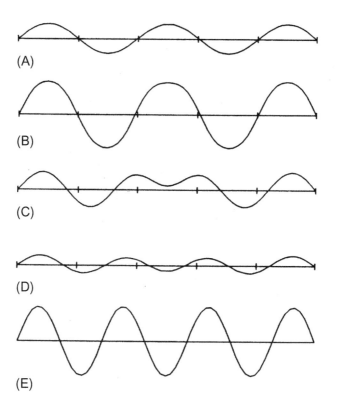

(A)

(B)

(C)

(D)

(E)

FIG. 4.20

Mode changes in secondary buckling.

may indicate that the energy level for a seven half-waves mode is lower than that for a five half-waves mode as the deflection becomes larger.

4.4.4 SECONDARY BUCKLING STRENGTH OF RECTANGULAR PLATE UNDER UNI-AXIAL THRUST

A series of static analysis is performed changing the aspect ratio of the plate. The breadth and the thickness are taken as $b = 1000$ mm and $t = 10$ mm, respectively, and the length is varied. The calculated secondary buckling strength is plotted against the aspect ratio of the plate in Fig. 4.21 together with the primary buckling strength.

$p \rightarrow q$ in the figure indicates the secondary buckling from p half-waves mode to q half-waves mode. There is an abrupt change in the secondary buckling strength at the aspect ratio where primary buckling mode changes. In each range of an aspect ratio, secondary buckling strength decreases with the increase in the aspect ratio.

Here, physical meaning of the secondary buckling is considered. Fig. 4.22A shows schematically the average stress-average stress relationships when the plate buckles in m half-waves mode and $m + 2$ half-waves mode, respectively. The strain energy stored in the plate is at the beginning the case of m

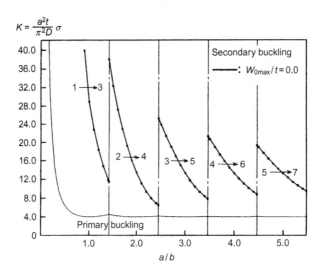

FIG. 4.21

Secondary bucking strength of flat rectangular plate.

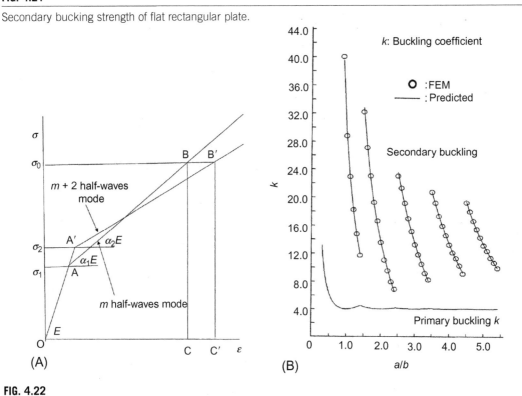

FIG. 4.22

Simple estimation of secondary bucking strength. (A) Average stress-average strain relationships.
(B) Secondary bucking strength.

Table 4.3 Coefficients to Estimate Secondary Buckling Stress

a/b	p	q	r
$a/b \leq \sqrt{2}$	-1.5781	3.0009	0.5074
$\sqrt{2} \leq a/b \leq \sqrt{6}$	-1.0896	3.4933	-0.9392
$\sqrt{6} \leq a/b \leq \sqrt{12}$	-0.8335	3.9999	-2.3695
$\sqrt{12} \leq a/b \leq \sqrt{20}$	-0.3294	1.6265	1.1598
$\sqrt{20} \leq a/b \leq \sqrt{30}$	-0.0318	-0.6786	0.7513

half-waves mode buckling is lower than the case of $m+2$ half-waves mode buckling. However, the strain energy of the latter case becomes larger than that of the former case as the average strain increases. Both strain energy becomes the same when the average stress reaches σ_0 which is expressed as:

$$\sigma_0 = \sqrt{\left(\frac{\alpha_2 - 1}{\alpha_2} \sigma_2^2 - \frac{\alpha_1 - 1}{\alpha_1} \sigma_1^2 \right) \bigg/ \left(\frac{1}{\alpha_1} - \frac{1}{\alpha_2} \right)} \qquad (4.31)$$

where

$$\sigma_1 = \frac{\pi^2 kE}{1281 - \nu^2} \left(\frac{a}{mb} + \frac{mb}{a} \right)^2, \quad \sigma_2 = \frac{\pi^2 kE}{1281 - \nu^2} \left\{ \frac{a}{(m+2)b} + \frac{(m+2)b}{a} \right\}^2 \qquad (4.32)$$

$$\alpha_1 = \frac{1 + (a/mb)^4}{3 + +(a/mb)^4}, \quad \alpha_2 = \frac{1 + \{a/(m+2)b\}^4}{3 + \{a/(m+2)b\}^4} \qquad (4.33)$$

Actual secondary buckling stress shown in Fig. 4.21 is roughly 0.9 through 3.0 times σ_0 calculated by Eq. (4.31). The secondary buckling strength can then be estimated as:

$$\sigma_{cr2} = \sigma_0 \{ p(a/b)^2 + q(a/b) + r \} \qquad (4.34)$$

The coefficients, p, q, and r, are given in Table 4.3. Estimated results are compared with FEM results in Fig. 4.22B. Good correlations are observed between two results. Although it is not clear what could be a trigger for occurrence of secondary buckling, the secondary buckling behavior could be a jumping from the higher energy level to the lower energy level.

4.4.5 INFLUENCE OF OUT-OF-PLANE BOUNDARY CONDITIONS ON SECONDARY BUCKLING BEHAVIOR

To investigate into the influence of boundary condition regarding the out-of-plane deflection on the secondary buckling behavior, a square plate is analyzed considering four boundary conditions as follows:

Case A: all sides simply supported
Case B: loaded sides simply supported and unloaded sides clamped
Case C: loaded sides clamped and unloaded sides simply supported
Case D: all sides clamped

Initial deflection of the primary buckling mode is assumed of which magnitude is 10% of the plate thickness. The unloaded sides are assumed to move freely in the in-plane direction keeping a straight line. The FEM is applied in combination with the Arc Length Method. The average stress-central deflection relationships and the average stress-average strain relationships are summarized in Fig. 4.23A and B, respectively [6].

The primary buckling strength is the highest when the all sides are clamped whereas it is the lowest when the all sides are simply supported. Contrary to this, the secondary buckling strength is the highest when all sides are simply supported and is the lowest for all sides clamped condition. Comparing Cases C and D, Case C shows higher primary buckling strength but lower secondary buckling strength. However, the shape of the average stress-central deflection relationships are similar and bifurcation is observed at the secondary buckling point. In Case B, the primary buckling mode is of a two half-waves mode in the loading direction, and the initial deflection of a two half-waves mode is assumed. However, secondary buckling is not observed within the stress range analyzed here. The in-plane stiffness beyond the secondary buckling point is the lowest in Case A and is the highest in Case D.

It should be noticed that there exists no simply supported plate in actual ship and ship-like floating structures. A plate is a part of continuous plating and the rotation along its four sides are elastically constrained. If the plate is simply supported along its four sides, the secondary buckling strength is pretty high more than two times the primary buckling strength. However, when its four sides are elastically constrained, the secondary buckling strength is much reduced in accordance with the magnitude of constraint. This should be kept in mind.

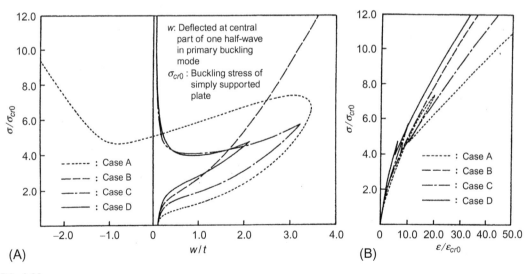

FIG. 4.23

Influence of out-of-plane boundary conditions on secondary buckling behavior of square plate under uni-axial thrust [6]. (A) Average stress-central deflection relationships. (B) Average stress-average strain.

4.4.6 DYNAMIC PHENOMENA IN SECONDARY BUCKLING BEHAVIOR

Taking the plate of which aspect ratio is 1.4 as an example, snap through phenomenon in secondary buckling is investigated applying the analytical method for dynamic elastic large deflection analysis [8]. Initial deflection is expressed as:

$$w_0 = \left(A_{01} \sin \frac{\pi x}{a} + A_{03} \sin \frac{3\pi x}{a} \right) \sin \frac{\pi y}{b} \tag{4.35}$$

is imposed to the plate assuming $A_{01}/t = A_{03}/t = 0.01$. This mode is the superposition of the primary and the secondary buckling modes which are shown in Fig. 4.24. The external load is applied with average stress ratio of 10 kgf/mm^2 per s.

4.4.6.1 Loading path

Fig. 4.25 shows the fundamental secondary buckling behavior of a short rectangular plate subjected to uni-axial thrust [7,8]. Fig. 4.25A shows the average stress-average strain relationship and Fig. 4.25B shows the average stress-central deflection relationship. In each figure, the dotted line is the static equilibrium path obtained by static analysis applying the Arc Length Method. On the other hand, solid lines are obtained by dynamic analysis. For this analysis, load is applied with the stress rate of 10 kgf/mm^2 per s. With this stress rate, primary buckling takes place almost statically, but secondary buckling is rationally simulated as a dynamic phenomenon.

As explained in Section 4.4.1, deflection of a primary buckling mode increases above the primary buckling strength and in-plane stiffness is reduced. With further increase in the applied stress, deflection mode starts to change and the plate reaches the secondary buckling strength at point a. Beyond this point, the static equilibrium path during unloading is an unstable path, which physically does not exist. The physically possible path beyond point a is a snap-through path which is obtained by dynamic analysis. This path is indicated by the solid line. Some oscillation is observed around point b for a while after the snap-through. When a forced displacement is applied dynamically, snap-through takes place from point a to point c.

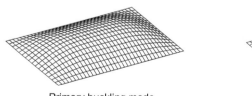
Primary buckling mode

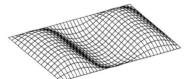
Secondary buckling mode

FIG. 4.24

Primary and secondary buckling modes.

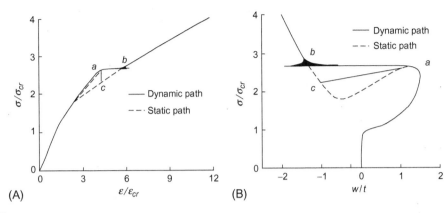

FIG. 4.25

Secondary buckling behavior of short rectangular plate subjected to uni-axial thrust [8]. (A) Average stress-average strain relationship. (B) Average stress-central deflection relationship.

4.4.6.2 Unloading path

Fig. 4.26A-C shows unloading path as well as loading path when secondary buckling takes place. As mentioned above, snap-through takes place from Point A to Point C in the loading path. On the other hand, snap-through takes place from Point B to Point D during unloading. This implies that statically unstable path between points A and B is skipped because this path is not physically possible to exist and cannot be traced.

Fig. 4.26C shows how the deflection components dynamically vary during loading and unloading. After the snap-through in the loading path, deflection component, A_{11}, vanishes after some oscillation, while the component, A_{31}, re-starts to increase. On the other hand, after the snap-through in the unloading path, deflection component, A_{11}, abruptly increases at point B while component, A_{13}, continues to decrease.

4.5 POSTBUCKLING BEHAVIOR AND ULTIMATE STRENGTH

4.5.1 SECONDARY BUCKLING AND BUCKLING/PLASTIC COLLAPSE BEHAVIOR

Up to here, in this chapter, buckling behavior, postbuckling behavior, and secondary buckling behavior have been explained as fundamental buckling collapse of a rectangular plate subjected to uni-axial thrust in the longitudinal direction. It should be noticed that no yielding has occurred in the above behavior. The solid line in Fig. 4.27 shows the above-mentioned elastic large deflection behavior schematically in the case of a rectangular plate subjected to uni-axial thrust. The plate undergoes primary buckling at point a, and buckling deflection of m half-waves mode develops beyond point a. Then, at point b, secondary buckling takes place, and deflection mode changes from m half-waves mode to $(m + 2)$ half-waves mode beyond point b. Similar behavior takes place at point c, beyond

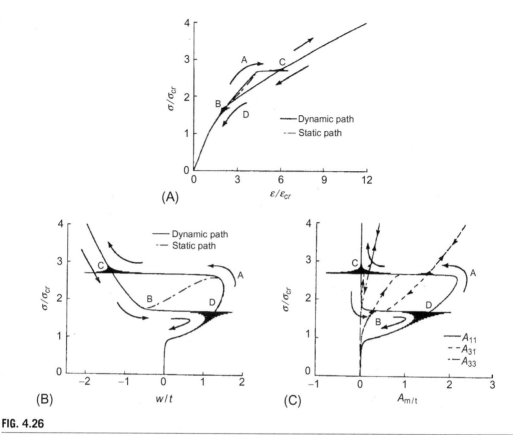

FIG. 4.26

Secondary buckling behavior of short rectangular plate subjected to uni-axial thrust during loading and unloading [8]. (A) Average stress-average strain relationship. (B) Average stress-central deflection relationship. (C) Average stress-deflection component relationship.

which deflection mode changes from $(m + 2)$ half-waves mode to $(m + 4)$ half-waves mode. Similarly, beyond point d, deflection mode changes from $(m + 4)$ to $(m + 6)$ half-waves mode. This is the case when the plate is completely elastic.

According to Fig. 4.6, deflection components other than buckling component also start to develop after the primary buckling has taken place. Deflection components of $(1, 3)$, $(3, 1)$, $(3, 3)$, etc., in Fig. 4.6 correspond to those of $(1, 3m)$, $(3m, 1)$, $(3m, 3m)$, etc., respectively, for the case indicated in Fig. 4.27. On the other hand, Fig. 4.27 indicates that only the buckling mode develops above the buckling load. The buckling mode for the case in Fig. 4.27 is m half-waves mode, and other deflection components except those of $(1, 3m)$, $(3m, 1)$, $(3m, 3m)$, etc., decrease above the bucking load.

In the actual structure, yielding takes place on a certain point along the average stress-average strain curve indicated in Fig. 4.27 depending on the slenderness ratio of the plate, $\beta = b/t \cdot \sqrt{\sigma_Y/E}$, where b and t are the breadth and the thickness of the plate, and σ_Y and E are the yield stress and Young's modulus of the material.

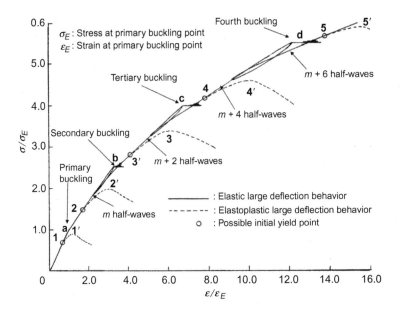

FIG. 4.27

Schematic representation of buckling collapse behavior of rectangular plate under longitudinal thrust.

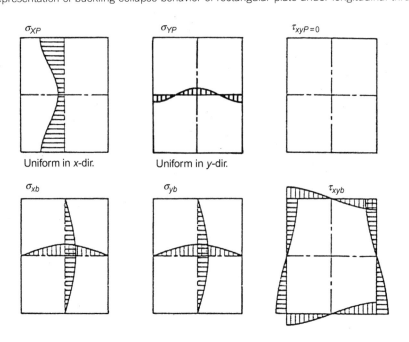

FIG. 4.28

Stress distributions in rectangular plate under longitudinal thrust.

Points 1 through 5 are the possible initial yielding point. Beyond initial yielding, yielded region spreads and the stiffness gradually decreases and reaches its ultimate strength. Beyond the ultimate strength, load carrying capacity starts to drop. Points 1′ through 5′ represent the ultimate strength for individual cases.

The secondary buckling strength is in general very high compared to the primary buckling strength. So, the possible initial yielding points are 1 and 2 in plate members in ship and ship-like floating structures. However, in some cases, elastoplastic secondary buckling can take place after the initial yielding takes place at point 2.

In this section, in the following, buckling/plastic collapse behavior shall be explained for rectangular plates subjected to uni-axial thrust.

4.5.2 BUCKLING/PLASTIC COLLAPSE BEHAVIOR: SHORT PLATES

4.5.2.1 Fundamentals in buckling/plastic collapse behavior

In the case of a short plate subjected to uni-axial thrust, fundamental equations are given in Section 4.1.2, where is assumed as for initial and total deflections as:

$$w_0 = A_0 \sin \frac{\pi x}{a} \sin \frac{\pi y}{b} \tag{4.36}$$

$$w = A \sin \frac{\pi x}{a} \sin \frac{\pi y}{b} \tag{4.37}$$

From Eqs. (3.133)–(3.137), in-plane stress components are derived as follows:

$$\begin{cases} \sigma_{px} = -\sigma - \frac{\pi^2 E}{32b^2}(A^2 - A_0^2)\cos \frac{2\pi y}{b} \\ \sigma_{py} = \frac{\pi^2 E}{32a^2}(A^2 - A_0^2)\cos \frac{2\pi x}{a} \\ \tau_{pxy} = 0 \end{cases} \tag{4.38}$$

On the other hand, bending stress components are derived from Eqs. (3.142)–(3.144) as:

$$\begin{cases} \sigma_{bx} = -\frac{\pi^2 E}{(1-v^2)}\left(\frac{1}{a^2} + \frac{v}{b^2}\right)(A - A_0)\sin \frac{\pi x}{a} \sin \frac{\pi y}{b} \\ \sigma_{by} = -\frac{\pi^2 E}{(1-v^2)}\left(\frac{v}{a^2} + \frac{1}{b^2}\right)(A - A_0)\sin \frac{\pi x}{a} \sin \frac{\pi y}{b} \\ \tau_{bxy} = \frac{\pi^2 E}{2(1+v)ab}(A - A_0)\cos \frac{\pi x}{a} \cos \frac{\pi y}{b} \end{cases} \tag{4.39}$$

The stress distributions obtained by Eqs. (4.38), (4.39) are illustrated in Fig. 4.28. Due to the effect of large deflection, in-plane normal stress in tension is produced in the middle of the plate. This is the reason why in-plane compressive stress, σ_{px}, is low in the center of the plate. The in-plane stress produced by large deflection is called membrane stress and is self-balancing in the cross-section. This can be observed in the distribution of σ_{py} under the condition that no thrust load is applied in y-direction. That is, to cancel the tensile force produced by tensile stress at the middle part, compressive stress is produced at the edge parts. On the other hand, according to the assumed deflection mode, the normal bending stress is maximum at the center of the plate, whereas shear bending stress is maximum at the four corners.

The equivalent stress can be expressed as:

$$\bar{\sigma} = \sqrt{(\sigma_{px} + \sigma_{bx})^2 - (\sigma_{px} + \sigma_{bx})(\sigma_{py} + \sigma_{bx}) + (\sigma_{py} + \sigma_{bx})^2 + 3(\tau_{pxy} + \tau_{bxy})^2} \qquad (4.40)$$

The equivalent stress could be the highest at surface of four corners or the center of the plate. These point could be the candidate where initial yielding takes place.

To explain the fundamentals in buckling/plastic collapse behavior of a rectangular plate subjected to thrust, a plate of which length and the breadth are $a \times b = 800 \times 1000$ mm is analyzed. The in-plane displacement of the unloaded sides in the direction perpendicular to the loading direction is assumed to be free but uniform. The thickness is chosen as 10 and 20 mm, which represents typical thin and thick plates, respectively. Average stress-central deflection relationships and average stress-average strain relationships are plotted in Fig. 4.29A and B, respectively. The distributions of equivalent stress and strain as well as stress components toward the thickness direction in an upper right-hand side quarter of the plate are illustrated in Fig. 4.30A and B for thin and thick plate, respectively. The changes in deflection mode and spread of yielded region are also shown in Fig. 4.31, in which A and B are for thin and thick plates, respectively.

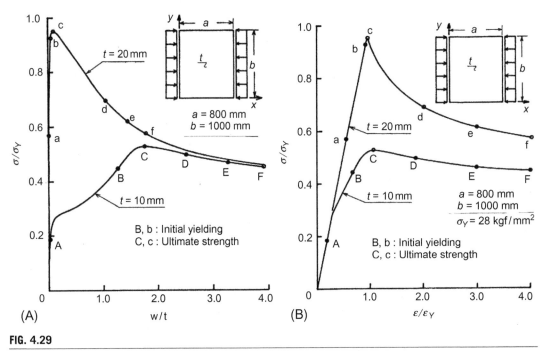

FIG. 4.29

Short rectangular plate subjected to uni-axial thrust. (A) Average stress-central deflection relationships. (B) Average stress-average strain relationships.

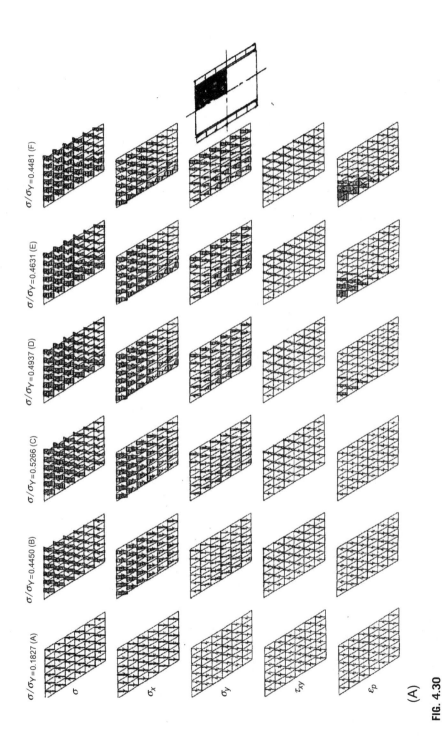

FIG. 4.30

Change in stress distributions during buckling/plastic collapse. (A) Thin plate.

(Continued)

FIG. 4.30, CONT'D

(B) Thick plate.

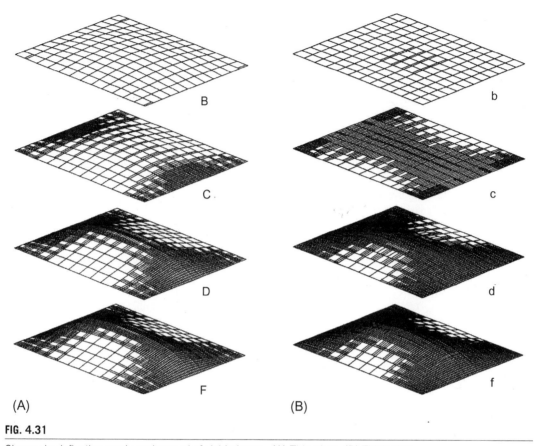

FIG. 4.31

Change in deflection mode and spread of yielded zone. (A) Thin plate. (B) Thick plate.

In Fig. 4.30, stress and strain calculated at 10 layers at the center of each element are illustrated. On the other hand, in Fig. 4.31, horizontal lines in each element indicate the yielding from the back side of the element and the vertical lines from the face side. The number of lines shows the number of yielded layers from the surface and the back of the plate.

In the following, collapse behavior is explained according to the letters on the average stress-average strain and average stress-central deflection curves in Fig. 4.29. Firstly in the case of a thin plate, at Point A before buckling takes place, deflection is still small and stress in the loading direction, σ_x, is almost uniform whereas other stress components are almost zero; see Fig. 4.30A. After the average stress goes over the buckling stress, deflection starts to increase rapidly as can be seen in Fig. 4.29A. Because of this, in-plane stiffness decreases to 41% of that before buckling (see Eq. 4.10) beyond the buckling as indicated in Fig. 4.29B. With further increase in the applied load, yielding starts at Point B. At this stress level, lateral deflection is large, and large bending stress is produced in the plate as can be seen in Fig. 4.30A.

As indicated in Fig. 4.31A, yielding starts at four corners of the plate where high in-plane normal stresses and bending shear stress are produced by large deflection; see Fig. 4.28. With further increase of the applied load, yielding spreads along the sides and toward the center of the plate. This yielding gradually reduces the in-plane stiffness, which becomes zero at Point C. The average stress at this point is called the ultimate strength. At this point, edge parts of the mid-length cross-section is fully yielded as is seen in Fig. 4.31A. This is evident also from Fig. 4.30A. The deflection is approximately of a sinusoidal mode until the ultimate strength has been attained. However, after the ultimate strength, deflection mode gradually changes to a roof mode as indicated in Fig. 4.31A. During this process, plastic strain is cumulated in the mid-length cross-section and the edge parts as is seen in Fig. 4.30A.

In the case of a thick plate, yielding starts at Point b on the average stress-central deflection and average stress-average strain curves before buckling as shown in Fig. 4.29. The location where yielding starts is the center of the plate where normal bending stress is the maximum. At this stage, lateral deflection is small and the membrane stress which reduces the compressive stress at the center of the plate is low. However, normal bending stress is a little high because of the thicker thickness. This is the reason why yielding starts at the center of the plate.

Soon after the initial yielding, the plate attains its ultimate strength at Point c. At this stage, deflection is still small and so the bending stress components as illustrated in Fig. 4.30B. It is seen in Fig. 4.31B that yielding is only from the compression side (back side) of plate bending except the corner parts. Beyond the ultimate strength, capacity rapidly decreases with the increase of lateral deflection. In this process, deflection mode changes from a sinusoidal mode to a roof mode as in the case of a thin plate. At Point d on the average stress-average strain and average stress-deflection curves, all the edges are fully yielded.

4.5.2.2 Influence of yielding on development of deflection components

Here, it is examined how deflection components change during buckling/plastic collapse, and how their changes are influenced by yielding. A series of nonlinear FEM analysis is performed on short rectangular plates subjected to uni-axial thrust. Keeping the breadth of the plate as $b = 1000$ mm, the aspect ratio of the plate, a/b, is changed as 0.6, 0.8, and 1.0, and the thickness, t, as 12 and 24 mm. For all cases, initial deflection expressed by Eq. (4.36) is given taking A_0/t as 0.01.

Fig. 4.32A and B shows average stress-average strain relationships and average stress-central deflection relationships for the plates of $t = 12$ mm, respectively. On the other hand, Fig. 4.33A-C shows the relationships between average strain and the coefficients of deflection components for plates of which aspect ratios, a/b, are 0.6, 0.8, and 1.0.

Also for the plates of $t = 24$ mm, average stress-average strain and average stress-central deflection relationships are given in Fig. 4.34A and B, respectively, and the relationships between average strain and coefficients of deflection components in Fig. 4.35A-C, which are for a/b being 0.6, 0.8, and 1.0.

In Figs. 4.33 and 4.35, the dashed lines are the results of elastic large deflection analyses and the solid lines those of elastoplastic large deflection analyses. The dashed and solid lines coincide until the initial yielding takes place at an open circle on the curves.

Firstly, the case of thin plates is considered. Fig. 4.32 indicates that the buckling strength is the minimum when the aspect ratio, a/b, is 1.0, whereas the ultimate strength is the maximum in this case. This will be discussed later in Section 4.5.2.7.

Regarding the deflection components, the most significant differences between the elastic and elastoplastic behavior can be seen in the component, A_{31}, which is three half-waves mode in the loading direction and a half-wave mode in the direction perpendicular to the loading direction. This component increases above the buckling stress in the same direction as one half-wave component, A_{11}. It continues

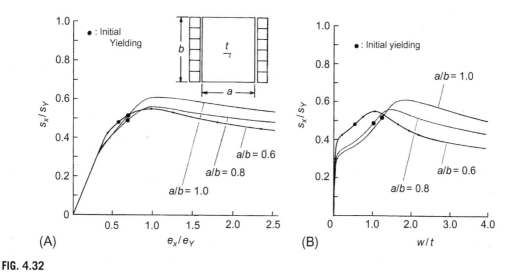

FIG. 4.32

Collapse behavior of rectangular plates with different aspect ratio under uni-axial thrust (thin plate; $t = 12$ mm). (A) Average stress-average strain relationships. (B) Average stress-central deflection relationships.

to increase in the same direction with A_{11} when the behavior is elastic. However, in the elastoplastic behavior, after yielding takes place, A_{31} starts to develop to an opposite direction, and its value turns from positive to negative.

Fig. 4.36 schematically shows the resulting deflection modes in elastic and elastoplastic behavior. When yielding does not takes place, both A_{31} and A_{11} are positive and the deflection shape is flattened in the loading direction. This finally leads to the mode change from one half-wave mode to three half-waves mode, which is known as the secondary buckling explained in Section 4.4. On the other hand, the elastoplastic deflection mode is sharpened after yielding takes place since A_{31} becomes negative whereas A_{11} keeps to be positive. This leads to the formation of a plastic mechanism collapse mode called as a roof mode.

In the elastoplastic behavior, A_{33} also changes its value from positive to negative after yielding has taken place. Other components such as A_{13}, A_{15}, and A_{51} keep to be positive after the yielding has taken place, but their magnitudes are larger when they are compared with those obtained by elastic large deflection analysis.

The increase in A_{31} results in the flattening or sharpening of the deflection mode in the loading direction. This can be seen in Fig. 4.37, in which elastoplastic and elastic deflections are illustrated when the average strain is 3.75 times the yielding strain.

4.5.2.3 Relationship between initial yielding strength and ultimate strength

The ultimate strength of a thin plate is largely different from that of a thick plate. The behavior of both plates is also different until the ultimate strength is attained. However, the average stress-central deflection curves of thin and thick plates come near as the deflection increases; see Fig. 4.29.

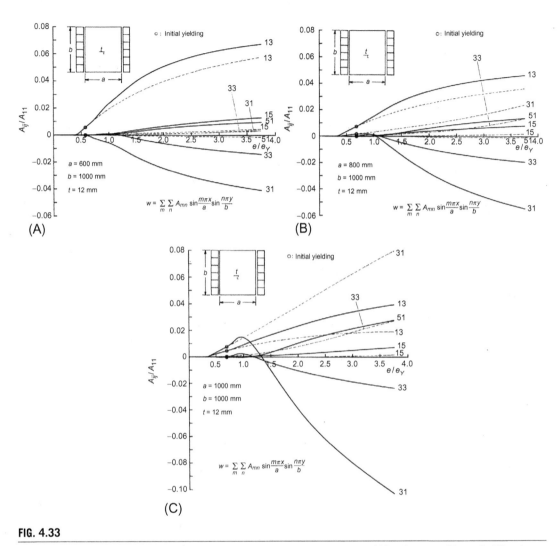

FIG. 4.33

Comparison of deflection components in elastic and elastoplastic behavior (thin plate; $t = 12$ mm). (A) $a/b = 0.6$. (B) $a/b = 0.8$. (C) $a/b = 1.0$.

Here, varying the slenderness ratio of the plate, a series of elastoplastic large deflection analysis is performed on the same short plate analyzed in Fig. 4.29. The initial deflection is of a sinusoidal shape expressed by Eq. (4.36) with the amplitude of $A_0/t = 0.01$. The average stress-central deflection and average stress-average strain relationships are summarized in Fig. 4.38A and B, respectively.

In these figures, the average compressive stress is divided by yielding stress. It is seen in Fig. 4.38A that the average stress-central deflection curves converges to one curve as the deflection increases regardless of the slenderness ratio of the plate. This may correspond to the fact that deflection mode far beyond the ultimate strength becomes a roof mode regardless of the slenderness ratio of the plate as indicated in Fig. 4.31.

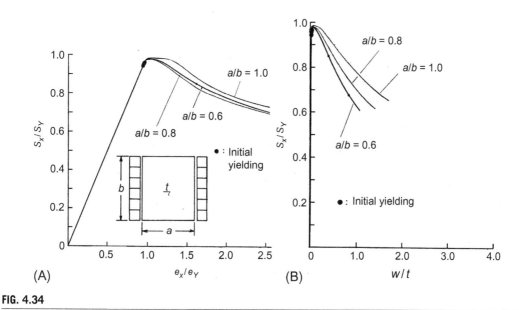

FIG. 4.34

Collapse behavior of rectangular plates with different aspect ratio under uni-axial thrust (thick plate; $t = 24$ mm). (A) Average stress-average strain relationships. (B) Average stress-central deflection relationships.

In Fig. 4.38B, yielding takes place before buckling occurs when the plate thickness is equal to or above 24 mm. In the case of the plate of which thickness is 27 mm, the plate undergoes plastic buckling when the average strain reaches around 1.5 times the yield strain. In the case of the 30 mm thickness plate, buckling does not occur within the analyzed range of the average strain.

Here, open circles on the curves in Fig. 4.38 represent the initial yielding point. The initial yielding strength is far above the buckling strength when the plate is thin. However, the initial yielding strength and the ultimate strength become close as the thickness increases. In Fig. 4.39, initial yielding strength is plotted against the slenderness ratio of the plate together with the ultimate strength. It is seen that the reserve strength beyond the initial yielding is the smallest when the slenderness ratio is around 2.2. It decreases as the slenderness ratio decreases below 1.6.

In Fig. 4.39, the dashed line is elastic buckling strength, σ_E, which is equal to σ_{cr} expressed by Eq. (4.1). On the other hand, the solid line represents the buckling strength with Johnson's plasticity correction, which is expressed as:

$$\sigma_{cr} = \left(1 - \frac{\sigma_Y}{4\sigma_E}\right)\sigma_Y \tag{4.41}$$

Eq. (4.41) is applicable when σ_E is greater than $0.5 \times \sigma_Y$.

4.5.2.4 Influence of in-plane boundary conditions on buckling/plastic collapse behavior

When a rectangular plate is subjected to uni-axial thrust in longitudinal direction, the plate expands in a transverse direction by Poisson's effect. However, after buckling has taken place, membrane stress is produced due to large deflection in the central part of the plate as indicated in Fig. 4.28. This tensile

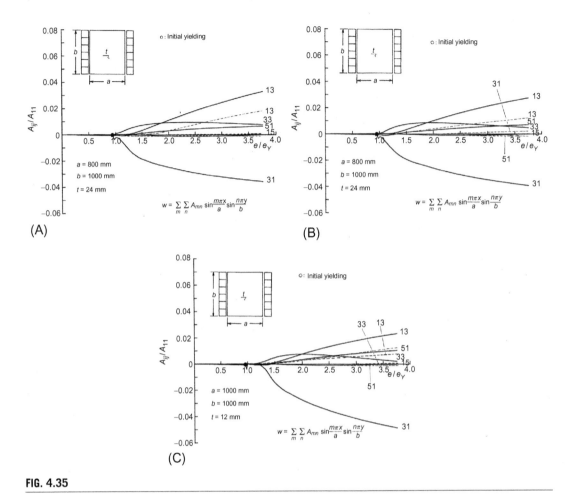

FIG. 4.35

Comparison of deflection components in elastic and elastoplastic behavior (thick plate; $t = 24$ mm). (A) $a/b = 0.6$. (B) $a/b = 0.8$. (C) $a/b = 1.0$.

stress produces pull-in effect of the nonloaded sides. In the analysis performed in Section 4.5.2.1 of this section, the nonloaded sides are assumed to move freely keeping a straight line. Here, including this case, three cases are considered to examine the influence of in-plane boundary condition along the nonloaded sides on the buckling/plastic collapse behavior of a plate subjected to uni-axial thrust.

A square plate is considered of which side length is 500 mm and the thickness is varied as 9 and 18 mm. The assumed three boundary conditions are illustrated in Fig. 4.40, which are

- Case (A): completely free (nonloaded side can deform freely)
- Case (B): uniform displacements (nonloaded side is kept straight)
- Case (C): no displacement in transverse direction

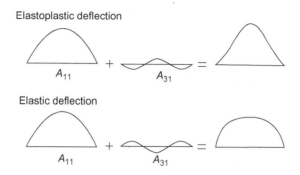

FIG. 4.36

Comparison between elastic and elastoplastic deflection modes.

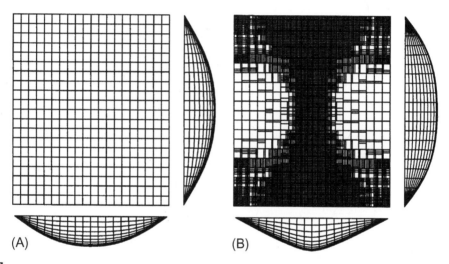

(A)

(B)

FIG. 4.37

Comparison of elastic and elastoplastic deflection modes. (A) Elastoplastic deflection. (B) Elastic deflection.

For the out-of-plane deflection, all sides are assumed to be simply supported. Initial deflection represented by Eq. (4.36) is assumed with the amplitude of $A_0/t = 0.01$. Average stress-central deflection and average stress-transverse displacement relationships are plotted in Fig. 4.41A and B, respectively. The transverse displacement is measured at the center of the nonloaded sides.

In Cases (A) and (B), the plate expands in the transverse direction until buckling takes place as indicated in Fig. 4.41B. However, as lateral deflection starts to increase, the nonloaded sides are pulled in due to the membrane stress produced by large deflection. In the case of a thin plate, the transverse displacement by pull-in effect beyond the buckling is different between Cases (A) and (B) as shown in Fig. 4.41B although the average stress-central deflection relationships of the two cases are almost the same for a while beyond the buckling. Then, with a further increase in the lateral deflection, difference

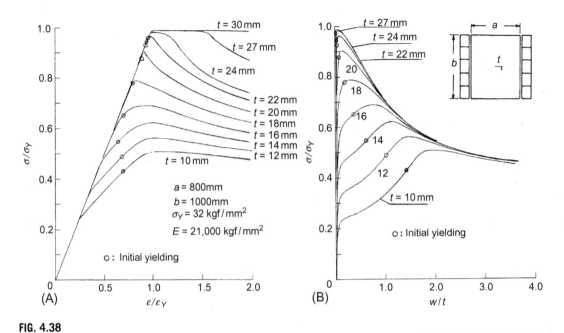

FIG. 4.38

Buckling/plastic collapse behavior of short rectangular plate subjected to uni-axial thrust. (A) Average stress-central deflection relationships. (B) Average stress-average strain relationships.

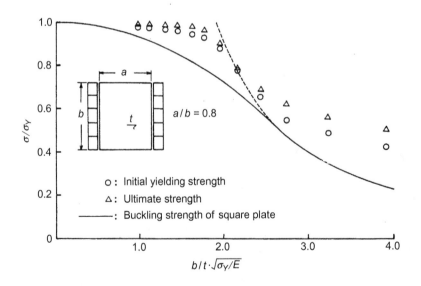

FIG. 4.39

Initial yielding and ultimate strength of short rectangular plate subjected to uni-axial thrust.

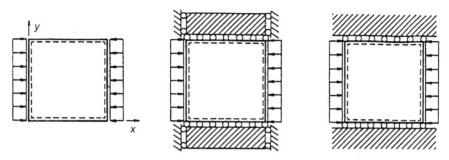

FIG. 4.40

Assumed boundary conditions for in-plane displacement along sides.

appears also in average stress-central deflection relationships of the two cases, and Case (B) shows the higher ultimate strength.

In Case (C), in-plane expansion in the transverse direction is constrained from the beginning of loading. Because of this, the plate is subjected to bi-axial thrust of which stress ratio is $1:\nu$, where ν is Poisson's ratio of the material. Because of this, Case (C) shows the lowest buckling strength. However, beyond the buckling, pull-in effect of the unloaded sides is completely constrained and the capacity of Case (C) exceeds those of Cases (A) and (B), which results in the highest ultimate strength of Case (C).

On the other hand, when the plate is thick, the ultimate strength is attained before lateral deflection develops. Because of this, Cases (A) and (B) show almost the same collapse behavior. However, due to the bi-axial compression, Case (C) shows the lowest ultimate strength as well as lowest buckling strength.

The experiment on a single panel may correspond to Case (A). The most possible boundary condition for panel members in the actual thin plated structures is Case (B). Case (C) may corresponds to a certain special case.

4.5.2.5 Influence of initial deflection on buckling/plastic collapse behavior
A series of elastoplastic large deflection analysis is performed on short plates of which length and the breadth are $a = 800$ mm and $b = 1000$. Yield stress is set as 32 kgf/mm^2.

The results of calculation for two cases with thicknesses of 10 and 20 mm, respectively, are shown in Fig. 4.42, where (A) shows the average stress-average strain relationships and (B) the average stress-central deflection relationships. Initial deflection expressed by Eq. (4.36) is given varying its magnitude as 0.01, 0.25, 0.5, and 1.0 times the plate thickness. The stress is nondimensionalized by the yield stress of the material, and the deflection by plate thickness.

It is known that initial deflection reduces the ultimate strength in accordance with the magnitude of initial deflection. Much reduction is observed in the case of thick plate than the case of thin plate.

Fig. 4.42A shows that average stress-central deflection relationship converges to a certain curve regardless of the magnitude of initial deflection when the plate is thin. Also in the case of a thick plate,

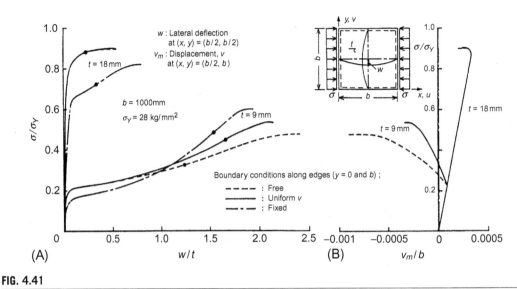

FIG. 4.41

Influence of boundary conditions for in-plane displacement on buckling/plastic collapse behavior. (A) Average stress-central deflection relationships. (B) Average stress-transverse displacement relationships.

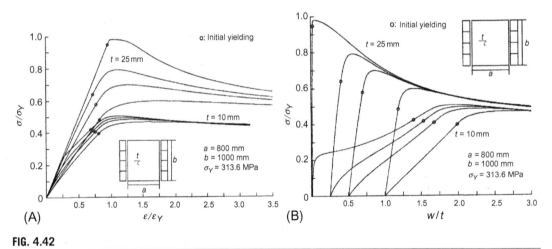

FIG. 4.42

Influence of initial deflection on buckling/plastic collapse behavior of short rectangular plate under uni-axial thrust. (A) Average stress-average strain relationships. (B) Average stress-deflection relationships.

the same tendency is observed, although complete convergence is not seen in the range shown in this figure. As can be expected from Fig. 4.38, two curves for $t = 10$ mm and $t = 25$ mm shall converge to a unique curve.

Average stress-central deflection relationships also converge to a unique curve regardless of the slenderness ratio and the magnitude of initial deflection after the ultimate strength has been attained.

4.5.2.6 Influence of welding residual stress on buckling/plastic collapse behavior

Firstly, a flat plate is considered which is not accompanied by welding residual stress. The material is assumed to be elastic-perfectly plastic. If this plate is subjected to uni-axial tensile or compressive load, the average stress-average strain relationship follows that of the material, which is indicated by dashed line in Fig. 4.43A.

Then, the influence of welding residual stress is firstly considered for the most fundamental case when buckling does not take place. The welding residual stress of a rectangular distribution is assumed as indicated in Fig. 4.43A. The tensile residual stress is equal to the yield stress of the material. Such welding residual stress is produced in the panel partitioned by stiffeners attached to plating by fillet welding. In this case, the average stress-average strain relationship follows the solid line under uni-axial tensile or compressive load.

When tensile load is applied on this plate, the average stress-average strain relationship follows the path, OABCI. In this case, in-plane stiffness along the path, OA, is given as:

$$D_t = (1 - \mu)E \qquad (4.42)$$

where $\mu = 2b_t/b = \sigma_c/(\sigma_Y + \sigma_c)$. The reduction in the in-plane stiffness is because the edge parts which is already yielded in tension cannot sustain the tensile load any more. Beyond point A, all part of the plate is yielded in tension.

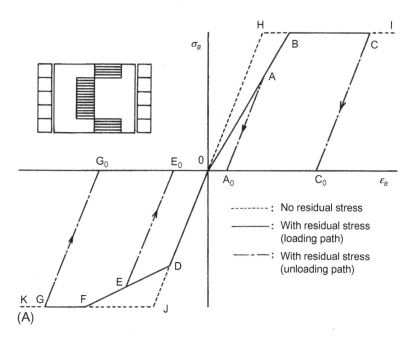

FIG. 4.43

Influence of welding residual stress on elastoplastic behavior of plate under uni-axial load sides. (A) Assumed residual stress and average stress-average strain relationship.

(Continued)

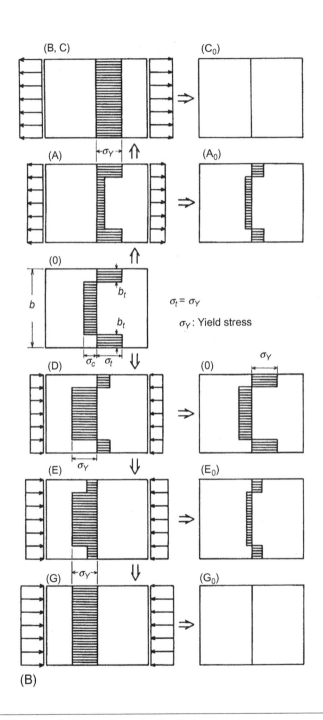

FIG. 4.43, CONT'D

(B) Change in stress distribution during loading and unloading.

On the other hand, under the compressive load, the average stress-average strain relationship is represented by a solid line, ODEFGK. At point D, the stress in the central part of the plate reaches the yield stress of the material at the average stress of $\sigma_Y - \sigma_c$. The in-plane stiffness along the path, DEF, is expressed as follows:

$$D = \mu E \qquad (4.43)$$

The reduction in the in-plane stiffness is because the central part which is yielded in compression cannot sustain any more compressive load. Beyond point F, all part of the plate is yielded in compression.

Along the unloading paths from points A, C, E, and G, the in-plane stiffness is equal to Young's modulus, E, since all the plate is in an elastic state along these paths.

The stress distributions at each point is illustrated in Fig. 4.43B. It should be noticed that at points A_0 and E_0, residual stress is released and become lower owing to unloading. This is because the plate behaves elastically in all region during the unloading although it is partly plastic just before the unloading. On the other hand, at points C_0 and G_0, the residual stress is completely removed. This is because all over the plate is plastic before unloading. It should also be noticed that there exist nonuniform residual strain although residual stress is zero at points C_0 and G_0.

When a plate undergoes buckling, the collapse behavior in compression is different from what is shown in Fig. 4.43 because of lateral deflection produced by buckling. Here, the results of analysis on a short plate is shown in Fig. 4.44 [9]. The size of the plate is $a \times b \times t = 800 \times 1000 \times 10$ mm. The material is assumed to be elastic-perfectly plastic with the yield stress of 28 kgf/mm². Two cases are considered as welding residual stress. One is due to fillet welding along two sides of the plate and the

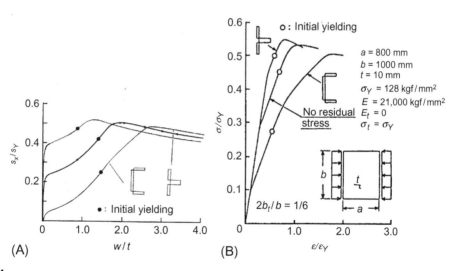

FIG. 4.44

Influence of welding residual stress on buckling/plastic collapse behavior of plate under uni-axial load.
(A) Average stress-average strain relationship. (B) Average stress-central deflection relationship.

other due to butt welding along the centerline of the plate. The breadth where tensile residual stress exists is set as $2b_t/b = 1/6$ in both cases.

It is seen that the butt welding increases the buckling strength, whereas fillet welding reduces the buckling strength. Regarding the in-plane stiffness beyond the buckling, it is the same in all cases regardless of the welding residual stress until yielding starts. This is because elastic behavior continues until yielding starts at circle marks.

In the case of a plate with welding residual stress by fillet welding, reserve strength after the occurrence of buckling until yielding starts is high. This is partly because the average stress at buckling is low and partly because high tensile residual stress exists along the nonloaded sides where initial yielding occurs when there is no residual stress. The yielding spreads toward the breadth direction from the edge to the center within the region where compressive residual stress exists.

When the butt welding is performed, the reserve strength after the occurrence of buckling is low. This is partly because high average stress is attained at buckling and partly because compressive residual stress exists where initial yielding is supposed to occur. With the same reasons, the ultimate strength is attained soon after the start of yielding.

A series of elastoplastic large deflection analysis is performed on the plate of which length is $a = 800$ mm and the breadth is $b = 1000$ mm. The thickness, t, is taken as 12 and 24 mm. The average stress-average strain relationships and the average stress-central deflection relationships are summarized in Fig. 4.45A and B for $t = 12$ mm, and in Fig. 4.46A and B for $t = 24$ mm, respectively.

Firstly, a thin plate is considered. With the increase in the welding residual stress, the buckling strength decreases. When $2b_t/b$ is 0.3, buckling strength is almost zero. It becomes negative when $2b_t/b$ is 0.4. In the latter case, the plate buckles due to high compressive residual stress at the central

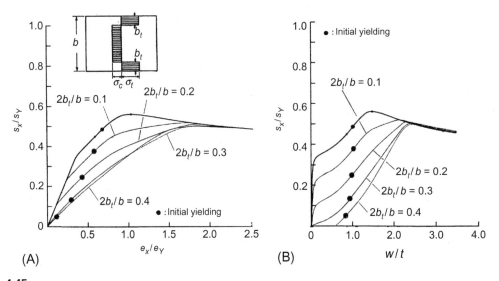

FIG. 4.45

Influence of welding residual stress on buckling/plastic collapse behavior of plate under uni-axial load ($t = 12$ mm). (A) Average stress-average strain relationships. (B) Average stress-central deflection relationships.

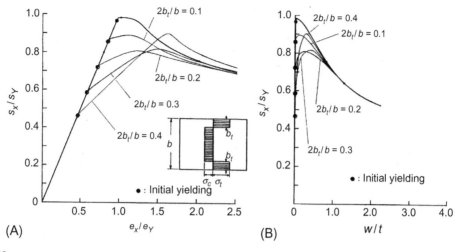

FIG. 4.46

Influence of welding residual stress on buckling/plastic collapse behavior of plate under uni-axial load ($t = 24$ mm). (A) Average stress-average strain relationships. (B) Average stress-central deflection relationships.

part of the plate before compressive load is applied. This is the reason why central deflection is about 60% of the plate thickness before applying the compressive load. The ultimate strength decreases with the increase in the residual stress.

When the plate is thick, yielding takes place before lateral deflection starts to develop. Because of this, the tangential stiffness of the plate after yielding is approximately expressed by Eq. (4.43). In this case, the ultimate strength is the minimum when $2b_t/b$ is 0.2 among the four cases. Here, the compressive residual stress in the central part of the plate reduce the buckling and yielding strength. On the other hand, the tensile residual stress along the sides increase the initial yielding strength. It is considered that the ultimate strength becomes minimum when $2b_t/b$ is 0.2 because of the opposite influences of tensile and compressive residual stresses under the condition that lateral deflection is small.

Even in the case of thin plate accompanied by large deflection, the reduction in the ultimate strength becomes to saturate when $2b_t/b$ becomes near to its limiting value, 0.5; see Fig. 4.45.

4.5.2.7 Dependence of ultimate strength on aspect ratio of short rectangular plates

Varying the aspect ratio of the plate between 0.25 and 1.5, a series of elastoplastic large deflection analysis is performed [10]. The breadth of the plate is fixed as 1000 mm, and the length is varied. The thickness is changed as 12, 18, and 24 mm. Initial deflection of the following form is imposed.

$$w_0 = A_{01} \sin \frac{\pi x}{a} \sin \frac{\pi y}{b} + A_{02} \sin \frac{2\pi x}{a} \sin \frac{\pi y}{b} \qquad (4.44)$$

Up to the aspect ratio of $a/b = 1.2$, $A_{01}/t = 0.01$, and $A_{02}/t = 0.0$, whereas in the range of aspect ratio above $a/b \geq 1.3$, initial deflection is set as $A_{01}/t = A_{02}/t = 0.01$.

The ultimate strength obtained for three thicknesses are plotted in Fig. 4.47A-C [11]. Firstly, case of a thin plate ($t = 12$ mm) is considered. As mentioned in Section 4.1.1, the minimum buckling strength

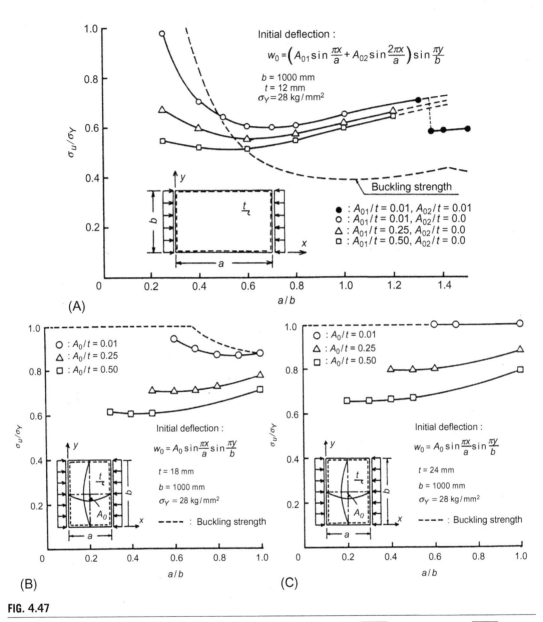

FIG. 4.47

Ultimate strength of rectangular plate under uni-axial thrust [10]. (A) $b/t\sqrt{\sigma_Y/E} = 3.04$. (B) $b/t\sqrt{\sigma_Y/E} = 2.03$. (C) $b/t\sqrt{\sigma_Y/E} = 1.52$.

is attained when the aspect ratio is 1.0. However, the aspect ratio which gives the minimum ultimate strength is lower than 1.0 and depends on the magnitude of initial deflection. The aspect ratio which gives the minimum ultimate strength is around 0.7 when the magnitude of initial deflection is 0.01 times the plate thickness. As the aspect ratio decreases, the curvature in the loading direction increases for the same magnitude of deflection, which results in the higher bending stress. This reduces the yielding strength. On the other hand, the buckling strength increases as the aspect ratio decreases from 1.0. Because of these opposite influences, the minimum ultimate strength may be attained when the aspect ratio is about 0.7. This aspect ratio decreases as the magnitude of initial deflection increases.

When the plate thickness is 18 mm, the buckling stress is 25.84 kgf/mm^2, which is 92% of the yield stress of the material. In this case, the aspect ratio which gives the minimum ultimate strength is about 0.9 when the magnitude of initial deflection is 0.01 times the plate thickness.

In the case of the plate of $t = 24$ mm, plastic buckling takes place and the ultimate strength is equal to the yield stress of the material regardless of the aspect ratio when the magnitude of initial deflection is small. It is known that the aspect ratio which gives the minimum buckling strain is between 0.7 and 0.8 when plastic buckling stakes place [2]. When the magnitude of initial deflection is 0.25 and 0.5 times the plate thickness, the minimum ultimate strength is attained at the aspect ratio of 0.5 and 0.25, respectively.

4.5.3 BUCKLING/PLASTIC COLLAPSE BEHAVIOR: LONG PLATES

To clarify how the ultimate strength of a uni-axially compressed plate varies with the aspect ratio, a series of nonlinear FEM analysis is performed [12,13]. For the analysis, initial deflection given by Eq. (4.44) is assumed as in the case of the plate in Fig. 4.47 but changing the combination of coefficients of deflection components as indicated in the table in Fig. 4.48. Here, Eq. (4.44) is rewritten below.

$$w_0 = A_{01} \sin \frac{\pi x}{a} \sin \frac{\pi y}{b} + A_{02} \sin \frac{2\pi x}{a} \sin \frac{\pi y}{b} \qquad (4.45)$$

Three series are considered regarding the initial deflection, which are

Case (1): Initial deflection of one half-wave mode.
Case (2): Initial deflection of two half-waves mode.
Case (3): Initial deflection of combined one half-wave and two half-waves modes.

The calculated results for three cases are plotted in Fig. 4.48.

The ultimate strength increases with the increase of the aspect ratio from 0.7 to 1.3. However, the ultimate strength drops when the aspect ratio exceeds 1.3. To consider the reason of this strength dropping, a flat plate, that is a plate with no initial deflection is considered.

The buckling mode changes from one half-wave mode to two half-waves mode at the aspect ratio of $\sqrt{2}$. When a flat plate of which aspect ratio is equal to $\sqrt{2}$ buckles, there exist two possibilities, which are into the one half-wave mode and the two half-waves mode, respectively. The buckling strength in these two case are the same, but the ultimate strength is different each other. It is not possible to perform elastoplastic large deflection analysis on a flat plate simulating bifurcation phenomenon at buckling. However, if such an analysis is possible to perform, the ultimate strength may jump from the broken line to the chain line as indicated by a fine vertical dashed line in Fig. 4.48 at $a/b = \sqrt{2}$.

Here, if it is assumed that the plate with initial deflection of a buckling mode collapses with the same half-waves number as the buckling mode, and that the ultimate strength for one half-wave mode collapse can be extrapolated as that for multi half-waves mode, Fig. 4.49 is obtained. This figure indicates that the ultimate strength abruptly varies at the aspect ratio at which the buckling mode changes from m half-waves mode to $m + 1$ half-waves mode.

On the other hand, for Case (3) with initial deflection of $A_{01}/t = A_{02}/t = 0.01$, the ultimate strength is plotted by solid circle marks in Fig. 4.48. In this case, the change in collapse mode from one half-wave

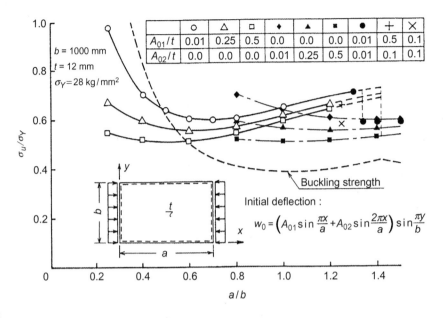

FIG. 4.48

Ultimate compressive strength of rectangular plate with simple initial deflection [13].

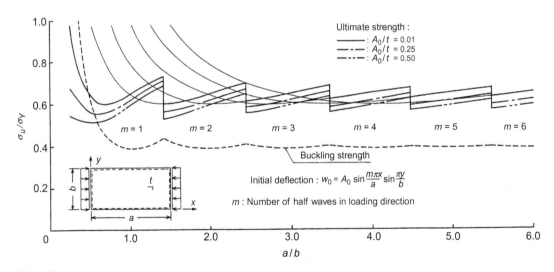

FIG. 4.49

Ultimate compressive strength of rectangular plate with initial deflection of buckling mode [13].

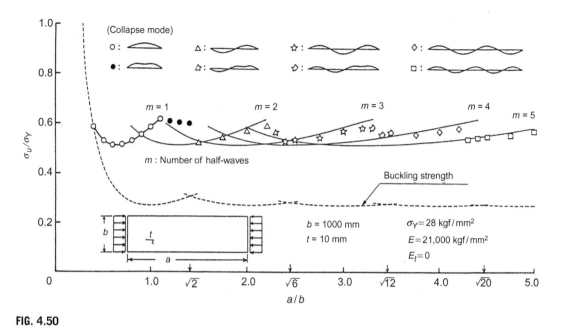

FIG. 4.50

Ultimate compressive strength of rectangular plate with initial deflection of thin-horse mode [13].

mode to two half-waves mode takes place at the aspect ratio of 1.34, which is smaller than $\sqrt{2}$ at which the buckling mode changes. The solid and the chain lines in Fig. 4.48 indicate the ultimate strength for Case (1) and Case (2), respectively. Then, there exists two ultimate strength within a certain range of the aspect ratio. It depends on the combination of the components of initial deflection expressed by Eq. (4.44) which ultimate strength is attained.

Here, the plate of $a/b = 1.25$ is considered as Case (3) giving the initial deflection as follows:

Case (a): $A_{01}/t = 0.5$, $A_{02}/t = 0.1$
Case (b): $A_{01}/t = A_{02}/t = 0.1$

The ultimate strength for Case (a) is plotted in Fig. 4.48 by + and that for Case (b) by ×, respectively. It is known that Case (a) with larger magnitude of initial deflection shows higher ultimate strength compared to Case (b) with smaller magnitude of initial deflection. It should be noticed that not only the magnitude of initial deflection but also its shape dominate the compressive ultimate strength.

In general, initial deflection in plating in actual structure is of a very complex mode and it is not clear which component dominates the buckling/yielding collapse behavior. Here, to examine this problem, an alternative series of nonlinear FEM analysis is performed on a rectangular plate varying the aspect ratio, a/b, between 0.4 and 5.0. The breadth and the thickness are set as $b = 1000$ mm and $t = 10$ mm. The plate is assumed to be accompanied by idealized initial deflection shown in Fig. 2.12 and Table 2.3 taking its maximum magnitude as 1% of the plate thickness.

The calculated ultimate compressive strength is plotted in Fig. 4.50 together with collapse modes. It is seen that the ultimate strength abruptly changes at the aspect ratios a little smaller than those at which the buckling mode changes. It is also seen that around these aspect ratios, deflection mode is not a clear m half-waves mode but a mixture of two components of the primary and the secondary buckling modes. On the other hand, the solid lines with $m = 2$ to $m = 5$ are illustrated on the basis of the solid line with $m = 1$ expanding its horizontal coordinate 2, 3, 4, and 5 times that for $m = 1$.

It is seen that the ultimate strength represented by marks is a little lower than the solid lines but is higher than the minimum value of the solid lines. Therefore, it may be possible to estimate the ultimate compressive strength of a rectangular plate with this minimum value. Varying the slenderness ratio of the plate, the minimum value of the ultimate compressive strength is calculated for the cases with and without welding residual stresses, and simple formulas are proposed in Ueda and Yao [10,12] to evaluate the ultimate strength in terms of slenderness ratio and the magnitude of initial deflection.

4.5.4 BUCKLING/PLASTIC COLLAPSE BEHAVIOR: WIDE PLATES

4.5.4.1 Wide rectangular plates in ship structures

In ship structures, there exist wide plates subjected to uni-axial or bi-axial thrust. For example, a wood-chip carrier has a longitudinal stiffening system at its deck and bottom/inner bottom plating whereas a transverse stiffening system at its side shell plating. The side shell plating is partitioned by two or three side stringers and plenty of transverse frames. The partitioned local panel of side shell plating is an ultra-wide rectangular plate, and is subjected to combined thrust and in-plane bending loads acting on its longer sides when a hull girder is subjected to longitudinal bending.

Another example is local panels in bottom and inner-bottom plating. They are subjected to bi-axial thrust produced by longitudinal bending moment and thrust due to pressure loads on side shell plating. Buckling/plastic collapse behavior of a wide plate is a little different from that of a short or a long rectangular plate. In this section, buckling/plastic collapse behavior of a wide plate is explained on the basis of the results of nonlinear FEM analyses.

A series of elastoplastic large deflection analysis is performed by the FEM on ultra-wide rectangular plate subjected to in-plane compression on its longer side. The length of the plate is fixed as $a = 800$ mm, and the breadth, b, is changed between 1600 and 6400 mm. The thickness, t, is varied as 10, 12, 16, 20, and 24 mm. The yield stress of the material is assumed to be 313.6 MPa.

4.5.4.2 Fundamental behavior during buckling/plastic collapse

Fig. 4.51A and B shows the average stress-central deflection and the average stress-average strain relationships, respectively, for an ultra-wide plate of which breadth is $b = 4800$ mm. In all cases of the slenderness ratios, the plate undergoes elastic buckling of a sinusoidal mode with one half-wave, but collapses in a cylindrical mode at the middle part of its width.

Fig. 4.52A shows how the deflection mode changes from a sinusoidal mode to a cylindrical mode. According to the change in deflection mode in the postbuckling and postultimate strength range, the stress distribution changes as indicated in Fig. 4.52B, where σ_x represents the membrane stress in the loading direction. It is seen that stress in a central part is uniform and becomes almost zero as deflection increases. Fig. 4.52A shows also the spread of yielded region. It can be seen how the yielding spreads as the deflection increases. It is known that a cylindrical mode has been changed to a so-called roof mode. This corresponds to the fact that a plastic hinge line is formed along the centerline parallel to a longer side, and the in-plane stress in the loading direction becomes almost zero at the central part of

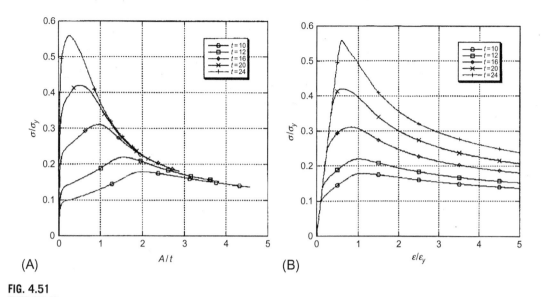

FIG. 4.51

Collapse behavior of ultra-wide rectangular plate subjected to uni-axial thrust (a = 800 mm; b = 4800 mm; σ_Y = 313.6 MPa). (A) Average stress-central deflection relationship. (B) Average stress-average strain relationship.

the plate. At the same time, in-plane stress in the direction of a shorter side is also zero in the central cylindrical part of the plate.

Fig. 4.53 shows the deflection components developed in the plate shown in Fig. 4.52. The solid lines are obtained by elastoplastic large deflection analysis, and the dashed line without considering yielding. Yielding starts at open circle marks on the solid lines. In this case, elastic and elastoplastic deflection components do not differ so much, and so their deflection modes. This implies that deflection of a cylindrical mode is obtained also by elastic large deflection analysis. Although it is not shown in Fig. 4.53, deflection components of three half-waves mode in the loading direction increase with the increase in the average strain, which results in the formation of a plastic hinge line along the centerline parallel to the longer side.

The aspect ratio of the plate considered in Figs. 4.51–4.53 is $a/b = 1/8$, and this plate may be called the ultra-wide plate. Here, an ordinary wide plate of which aspect ratio, a/b, is 1/2 is considered. The average stress-central deflection and the average stress-average strain relationship for this plate are shown in Fig. 4.54A and B, respectively. Thickness is changed as t = 10, 12, 16, 20, and 24 mm. For the plate of t = 12 mm, developments of deflection components are shown in Fig. 4.55 and changes in deflection mode, yielded region, and stress distribution are illustrated in Fig. 4.56.

It is seen that yielding starts at the four corners of the plate as in the case of a short thin plate. At this loading step, high in-plane normal stress is produced along two sides parallel to the shorter sides (see Fig. 4.56) and high bending shear stress in the corner regions (see Fig. 4.28).

For this aspect ratio, developments of elastic and elastoplastic deflection components are different. This feature is different from that observed in the case of an ultra-wide plate shown in Fig. 4.52. However, the resulting elastic and elastoplastic deflection modes are both of a cylindrical mode. The difference in the deflection mode is the shape of deflection in the loading direction as indicated in Figs. 4.36 and 4.37 of a short plate.

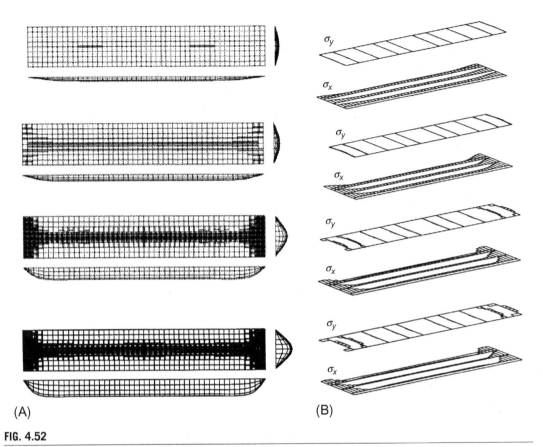

FIG. 4.52

Change in deflection mode, yielded region, and stress distribution during buckling/plastic collapse
($a \times b \times t = 800 \times 4800 \times 16$ mm). (A) Deflection mode and yielding. (B) Membrane stress distribution.

Varying the aspect ratio of thin and thick plates, average stress-central deflection and average stress-average strain relationships are plotted in Fig. 4.57A and B, respectively. Thickness of the thin plate is taken as 12 mm and that of the thick plate as 24 mm. In all cases, the plate undergoes elastic buckling and the buckling strength saturates to a certain value as the aspect ratio of the plate becomes low. Here, the buckling stress of a rectangular plate subjected to uni-axial thrust in the direction of shorter sides is expressed as:

$$\sigma_{cr} = \frac{\pi^2 E}{12(1-\nu^2)} \left(\frac{b}{a} + \frac{a}{b}\right)^2 \left(\frac{t}{b}\right)^2 = \frac{\pi^2 E t^2}{12(1-\nu^2)a^2} \left(1 + \frac{a^2}{b^2}\right)^2 \tag{4.46}$$

where a, b, and t are length, breadth, and thickness of the plate.

For a specified a and t, the buckling strength by Eq. (4.46) reduces to:

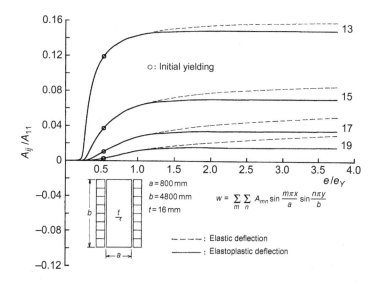

FIG. 4.53

Change in deflection components during buckling/plastic collapse ($a \times b \times t = 800 \times 6400 \times 16$ mm).

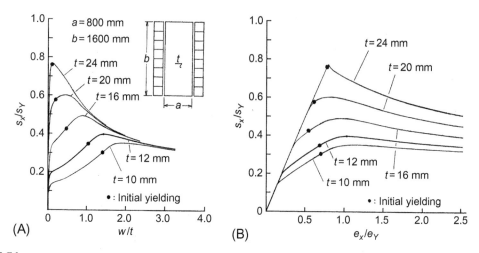

FIG. 4.54

Collapse behavior of ultra-wide rectangular plate subjected to uni-axial thrust ($a = 800$ mm; $b = 1600$ mm; $\sigma_Y = 313.6$ MPa). (A) Average stress-central deflection relationship. (B) Average stress-average strain relationship.

$$\sigma_{cr} = \frac{\pi^2 E t^2}{12(1-\nu^2)a^2} \tag{4.47}$$

when the aspect ratio, a/b, approaches zero. This is the reason why the buckling strength approaches to a certain value as the breadth increases in Fig. 4.57.

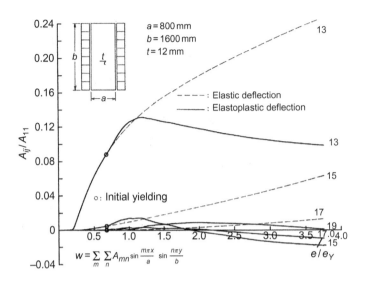

FIG. 4.55

Change in deflection components during buckling/plastic collapse.

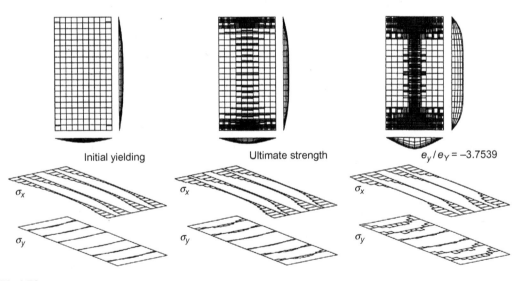

FIG. 4.56

Change in deflection mode, yielded region, and stress distribution during buckling/plastic collapse ($a \times b \times t = 800 \times 1600 \times 12$ mm).

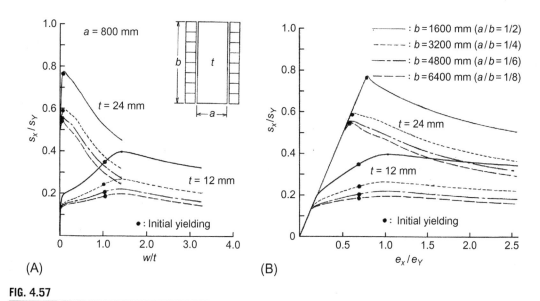

FIG. 4.57

Collapse behavior of ultra-wide rectangular plate subjected to uni-axial thrust (a = 800 mm; b = 1600 mm; σ_Y = 313.6 MPa). (A) Average stress-central deflection relationship. (B) Average stress-average strain relationship.

Another feature observed in Figs. 4.51, 4.54, and 4.57 is that the ultimate strength is obtained soon after yielding starts in the case of thick plates. This is because high bending strain is produced when deflection starts to develop owing to thicker thickness in the wide range of cylindrically curved central part of the plate. The high bending strain causes yielding in the compression side of bending as well as unloading in the tension side of bending. In the compression side of bending, stress does not increase because of the yielding. At the same time, in the tension side of bending stress decreases due to unloading. This is considered to be the reason of abrupt drop of capacity soon after the start of yielding.

4.6 POSTULTIMATE STRENGTH BEHAVIOR OF RECTANGULAR PLATE UNDER UNI-AXIAL THRUST

4.6.1 PLATES FOR ANALYSIS

In Section 4.5, buckling/plastic collapse behavior of a rectangular plate under uni-axial thrust was explained up to the ultimate strength. However, when a collapse behavior of a plated structure is considered as a whole, the postultimate strength behavior of an individual structural component has to be considered since it affects very much on the buckling/plastic collapse behavior of the whole structures. From this point of view, postultimate strength behavior of rectangular plates subjected to uni-axial thrust is examined on the basis of the nonlinear FEM analyses.

A series of elastoplastic ultra-large deflection analysis is performed [9,11]. Eighteen plates are chosen for analysis from the panels of deck plating on which mode and magnitude of initial deflection were measured [14,15]. The measured initial deflection is approximated by Eq. (4.48), that is

$$w_0 = \sum_{i=1}^{m} A_{0i} \sin \frac{i\pi x}{a} \sin \frac{\pi y}{b} \qquad (4.48)$$

For the present analysis, m is taken as 11. The coefficients of deflection components for the analyzed plates are given in Table 4.4 together with their sizes. Substituting the coefficients in Table 4.4 into Eq. (4.48), initial deflection of a thin-horse (or hungry-horse) mode is calculated. Typical modes of initial deflection are shown in Fig. 4.58. The plate is assumed to be simply supported along four sides and forced displacements are given on the shorter sides.

At the beginning, a series of analysis is performed on 18 panels giving the initial deflection of a thin-horse mode produced using the coefficients of deflection components in Table 4.4. Another series of analysis is then performed giving the initial deflection of a buckling mode of which magnitude is indicated in Table 4.4. The initial deflection for both cases can be compared in Fig. 4.58.

When a thick plate undergoes plastic buckling, the aspect ratio of one-half wave of the buckling deflection giving the lowest buckling strain is 0.7 [2]. Thick plates which exactly satisfy this condition are plates No. 10 and 11. For other thick plates, the buckling mode of which aspect ratio of the one half-wave is nearest to 0.7 is assumed.

Including the case when elastic buckling takes place, assumed buckling modes are as follows:

Plates No. 01–04: four half-waves
Plates No. 05–07: seven half-waves
Plates No. 08–09 and 12–14: three half-waves
Plates No. 10–11 and 15–18: five half-waves

The plate is divided into 10 elements in its breadth direction and into $(m + 1) \times 10$ elements in the loading direction, where m is the number of half-waves of the buckling mode. Material of all the plates is the high tensile strength steel of which yielding stress, σ_Y, is 313.6 MPa. The Young's modulus, E, is taken as 205.8 GPa. The material is assumed to be elastic-perfectly plastic.

4.6.2 COLLAPSE BEHAVIOR BEYOND ULTIMATE STRENGTH

Firstly, collapse behavior of thin plate is considered. Deflection mode and yielded region of plate No. 18 with initial deflection of a buckling mode and of a thin-horse mode are shown in Fig. 4.59A and B, respectively. Deflection is shown by multiplying 6. Regarding the yielded region, horizontal lines represent the yielding from the face and vertical lines that from the back toward the thickness in the element, and their numbers represent the numbers of yielded layers. In the case of elements with full of lines, all the layers are yielded. The average stress-average strain and average stress-deflection relationships for these two cases are compared in Fig. 4.60A and B, respectively.

Fig. 4.59A indicates that the plate with initial deflection of a buckling mode deflects in a sinusoidal mode above the buckling strength, but the deflection mode gradually changes to a periodical roof mode. In this case, deflection mode is periodical and the strength reduction beyond the ultimate strength is gradual.

Table 4.4 Coefficients of Measured Initial Deflection (in mm)

No.	$w_{0\,max}$	A_{01}	A_{02}	A_{03}	A_{04}	A_{05}	A_{06}	A_{07}	A_{08}	A_{09}	$A_{0\,10}$	$A_{0\,11}$
01	0.7859	0.7463	0.0257	0.1240	−0.0008	0.0656	0.0229	−0.1117	−0.0159	−0.0162	−0.0647	0.0077
02	1.2970	1.3460	0.0902	0.2236	0.0363	0.0868	0.0816	−0.1024	0.0361	0.0138	−0.0256	0.0276
03	1.5262	1.2350	0.3393	0.3435	−0.1293	0.0609	0.0148	0.0310	0.0310	0.0333	−0.0747	0.0185
04	1.6301	1.2392	−0.4825	0.2680	−0.0495	−0.0286	0.0286	0.0459	−0.1104	−0.0562	−0.0043	0.0077
05	−0.4163	−0.3543	−0.0972	0.0051	0.0433	0.0316	0.0169	0.0938	0.0183	0.0485	0.0324	0.0232
06	0.4624	0.3520	−0.1582	0.0950	0.0065	0.0973	−0.0135	0.0713	−0.0122	0.0555	0.0081	0.0281
07	0.6548	0.4787	−0.0400	0.1955	0.0585	0.1073	−0.0214	0.1294	0.0260	0.0985	0.0285	0.0897
08	3.0083	3.0924	−0.3364	0.3619	−0.0562	−0.0166	−0.0762	0.0099	−0.0584	−0.0598	−0.0300	0.0132
09	3.3261	3.6868	0.0492	0.7494	0.0545	0.3818	0.0155	0.1419	−0.0485	−0.0413	0.0134	0.0916
10	3.1418	3.3394	0.1875	0.2967	−0.0595	0.1944	−0.1278	0.1433	−0.0299	0.0699	−0.0135	0.0421
11	3.8194	4.1613	−0.4340	0.9214	0.0133	0.5593	0.0151	0.2425	−0.0042	0.0833	0.0216	0.0935
12	2.4127	2.6681	−0.2861	0.6864	−0.0967	0.5339	0.0076	0.2338	−0.0584	0.1050	−0.0444	0.0730
13	3.7661	3.9553	0.2182	1.2040	0.0949	0.3592	0.0248	−0.0228	0.0803	−0.1374	−0.0163	−0.0153
14	4.1092	4.5350	0.1825	1.4192	−0.3868	0.6556	−0.1380	0.3410	−0.04289	0.1547	−0.0919	0.1354
15	3.0660	1.8620	0.3439	1.0984	0.3603	0.5491	0.2338	0.1545	0.0026	0.1640	−0.0081	−0.0597
16	3.3584	2.7932	−0.7274	0.9647	0.1594	0.7269	−0.1866	0.1623	0.3599	0.2548	0.1927	0.0336
17	3.8388	3.5429	0.1604	1.5268	0.0882	0.7295	0.1600	0.3718	0.1658	0.0716	0.0383	0.0391
18	3.8267	4.5272	−0.0463	1.1956	0.0386	0.6717	0.0967	0.2407	0.0345	0.0554	−0.0024	−0.0092

Notes: No. 01–04: $a \times b \times t = 2400 \times 800 \times 34.5$ mm; No. 05–07: $a \times b \times t = 4200 \times 835 \times 22$ mm; No. 08–09: $a \times b \times t = 2100 \times 800 \times 19$ mm; No. 10–11: $a \times b \times t = 2800 \times 800 \times 19$ mm; No. 12–14: $a \times b \times t = 2400 \times 835 \times 15$ mm; No. 15–18: $a \times b \times t = 4200 \times 835 \times 13$ mm.

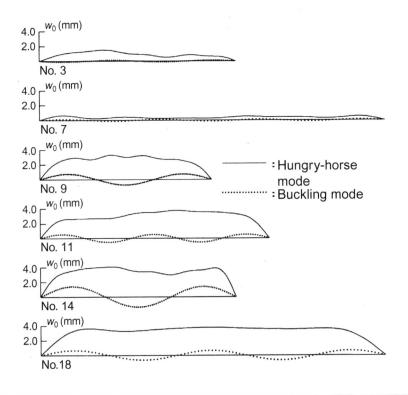

FIG. 4.58

Measured initial deflection and buckling component.

Also in the case of a plate with initial deflection of a thin-horse mode shown in Fig. 4.59B, deflection of a buckling mode becomes dominant above the buckling load ($\sigma_{cr}/\sigma_Y = 0.5751$) as can be seen in Fig. 4.59B. The deflection component of the buckling mode grows until the load reaches to its maximum value, the ultimate strength. In this case, however, the deflection mode is not perfectly periodical because of the existence of deflection components other than the buckling mode although they start to decrease above the buckling load [10]. Beyond the ultimate strength, deflection is localized at the left end of the plate with the concentration of yielding, and the load carrying capacity rapidly decreases. It can be observed in Fig. 4.59B that deflection of a buckling mode disappears and elastic unloading has taken place in the remaining part of the plate.

Fig. 4.61 shows the average stress-average strain relationships in individual parts of the plate divided into five. In the case of a plate with initial deflection of a buckling mode, ultimate strength is attained simultaneously at all the five half-waves parts of the deflection, and similar behavior is observed in these five half-waves parts in the postultimate strength range. Contrary to this, in the case of a plate with initial deflection of a thin-horse mode, only parts A and B attain their ultimate strength, but other parts undergo elastic unloading before they reach their own ultimate strength as indicated in Fig. 4.61. Also part B undergoes unloading soon after the ultimate strength has been attained. With the increase of the local deflection and the yielding in part A, the yielding spreads into the left-hand side of part B, and the average strain in part B again starts to increase.

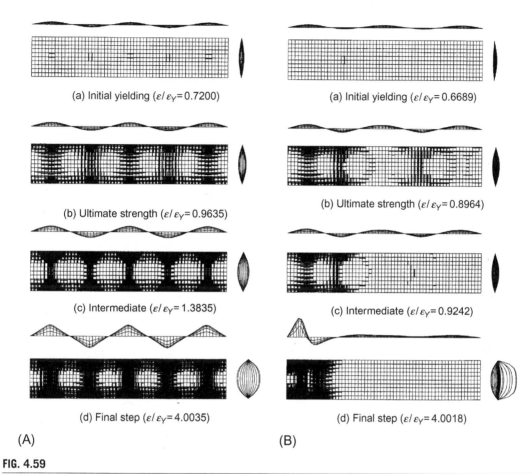

(a) Initial yielding ($\varepsilon/\varepsilon_Y = 0.7200$)

(a) Initial yielding ($\varepsilon/\varepsilon_Y = 0.6689$)

(b) Ultimate strength ($\varepsilon/\varepsilon_Y = 0.9635$)

(b) Ultimate strength ($\varepsilon/\varepsilon_Y = 0.8964$)

(c) Intermediate ($\varepsilon/\varepsilon_Y = 1.3835$)

(c) Intermediate ($\varepsilon/\varepsilon_Y = 0.9242$)

(d) Final step ($\varepsilon/\varepsilon_Y = 4.0035$)

(d) Final step ($\varepsilon/\varepsilon_Y = 4.0018$)

(A)

(B)

FIG. 4.59

Change in deflection mode and yielding in thin plate (No. 18). (A) Initial deflection of a buckling mode. (B) Initial deflection of a thin-horse mode.

The average strain of the whole plate is the average of those in five parts. Consequently, the average strain of the plate with initial deflection of a thin-horse mode is smaller compared to that of the plate with initial deflection of a buckling mode because of the negative strain produced by elastic unloading in parts B, C, D, and E. This is the reason why the capacity rapidly decreases beyond the ultimate strength in the case of a plate with initial deflection of a thin-horse mode.

Deflection mode and spread of yielded region of No. 03 panel with initial deflection of a buckling mode are shown in Fig. 4.62A, and those with initial deflection of a thin-horse mode in Fig. 4.69B. This plate undergoes plastic buckling at the critical strain of $\varepsilon_{cr}/\varepsilon_Y = 4.0241$ with four half-waves according to the Plastic Deformation Theory [7], where ε_Y is yield strain.

Similar behavior is observed as in the case of thin plate, although concentration of plastic deformation takes place also in the case with initial deflection of a buckling mode. Some thick plates show such behavior.

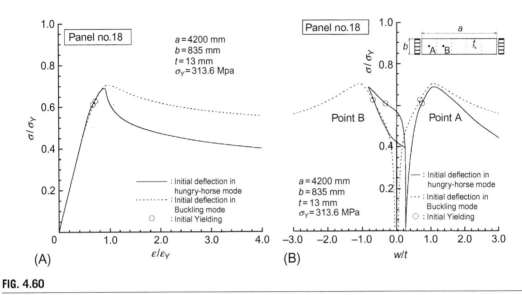

FIG. 4.60

Collapse behavior of thin plate with different initial deflection (No. 18). (A) Average stress-average strain relationships. (B) Average stress-deflection relationships.

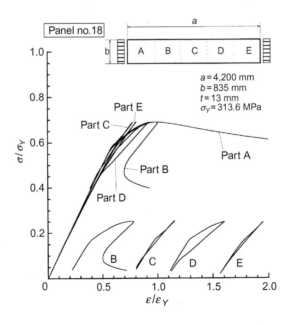

FIG. 4.61

Average stress-average strain relationships in individual buckling half-wave parts (No. 18; with initial deflection of thin-horse mode).

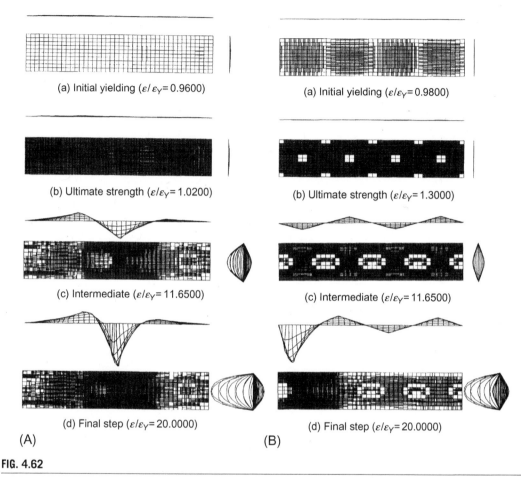

(a) Initial yielding ($\varepsilon/\varepsilon_Y = 0.9600$)

(a) Initial yielding ($\varepsilon/\varepsilon_Y = 0.9800$)

(b) Ultimate strength ($\varepsilon/\varepsilon_Y = 1.0200$)

(b) Ultimate strength ($\varepsilon/\varepsilon_Y = 1.3000$)

(c) Intermediate ($\varepsilon/\varepsilon_Y = 11.6500$)

(c) Intermediate ($\varepsilon/\varepsilon_Y = 11.6500$)

(d) Final step ($\varepsilon/\varepsilon_Y = 20.0000$)

(d) Final step ($\varepsilon/\varepsilon_Y = 20.0000$)

(A)   (B)

FIG. 4.62

Change in deflection mode and yielding in thin plate (No. 03). (A) Initial deflection of a buckling mode.
(B) Initial deflection of a thin-horse mode.

4.6.3 DEFLECTION MODES BEYOND ULTIMATE STRENGTH

Representative deflection modes of the compressed plates at the last step are summarized in Fig. 4.63. The left column shows the plates with initial deflection of a buckling mode, and the right column those with initial deflection of a hungry-horse mode. In all cases, the plates are at the last incremental step far beyond the ultimate strength as indicated in Fig. 4.60A.

In the case of plates with initial deflection of a buckling mode, collapse mode is in a periodical roof mode except Plates No. 01–04 with very large thickness. In the case of plate No. 06, unloading is going to take place at the third and the sixth half-wave from the left. Such unloading is not detected in other cases. On the other hand, when the plates are accompanied by initial deflection of a thin-horse mode,

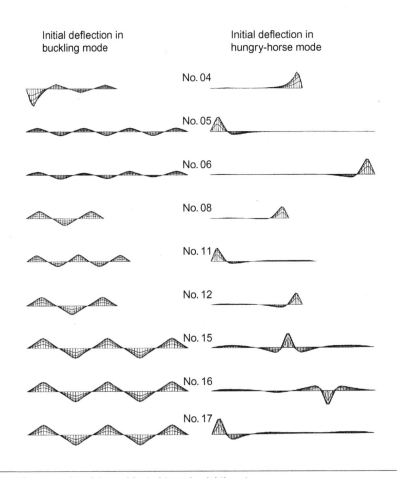

FIG. 4.63

Collapse modes of rectangular plates subjected to uni-axial thrust.

deflection of a buckling mode develops above the buckling load in the case of thin plates. Beyond the ultimate strength, however, deflection and yielding are localized at a certain one half-wave of the deflection mode, and the remaining part undergoes elastic unloading. The location where plastic deformation is localized is in many cases the end of the plate, but in some cases other part of the plate.

The collapse modes of plates which are not shown in Fig. 4.63 are as follows:

Plate No. 01: same as Plate No. 04
Plate No. 02: same as Plate No. 03
Plate No. 07: same as Plate No. 05
Plate No. 09: same as Plate No. 08
Plate No. 10: same as Plate No. 11
Plate No. 13 and 14: same as Plate No. 12

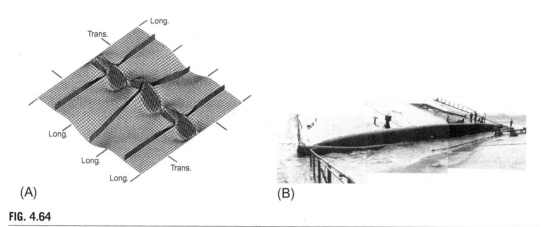

FIG. 4.64

Concentration of plastic deformation beyond the ultimate strength. (A) Stiffened plate. (B) Hull girder.

4.6.4 CONCENTRATION OF PLASTIC DEFORMATION BEYOND ULTIMATE STRENGTH

Plate in real structure is not accompanied by initial deflection of a buckling mode but of a thin-horse mode. Therefore, concentration of plastic deformation takes place beyond the ultimate strength as indicated in Fig. 4.63. Similar collapse behavior can be observed in the case of stiffened plate as illustrated in Fig. 4.64A. Even when a hull girder collapses under longitudinal bending, concentration of plastic deformation takes place in one cross-section as indicated in Fig. 4.64B, and the hull girder collapses in a jack-knife mode. Concentration of plastic deformation is accompanied by rapid reduction in capacity beyond the ultimate strength. This is very important when the safety of structures is considered.

4.7 BUCKLING/PLASTIC COLLAPSE BEHAVIOR OF RECTANGULAR PLATES UNDER UNI-AXIAL CYCLIC LOADING

4.7.1 BUCKLING/PLASTIC COLLAPSE BEHAVIOR OF SHORT PLATES UNDER SINGLE CYCLIC LOADING

The average stress-average strain and average stress-central deflection relationships for the plate in single cyclic loading are indicated in Fig. 4.65A and B, respectively. The size of the plate is $a \times b \times t = 800 \times 1000 \times 10$ mm. Bi-linear stress-strain relationship is assumed with:

Yield stress σ_Y: 30 kgf/mm^2
Young's modulus E: 21,000 kgf/mm^2
Strain hardening rate H': 210 kgf/mm^2

The loading starts from compression, and the unloading starts at $\varepsilon/\varepsilon_Y = -0.6, -1.0, -1.5, -2.0, -2.5,$ and -3.0 in Cases A, B, C, D, E, and F. Initial deflection of

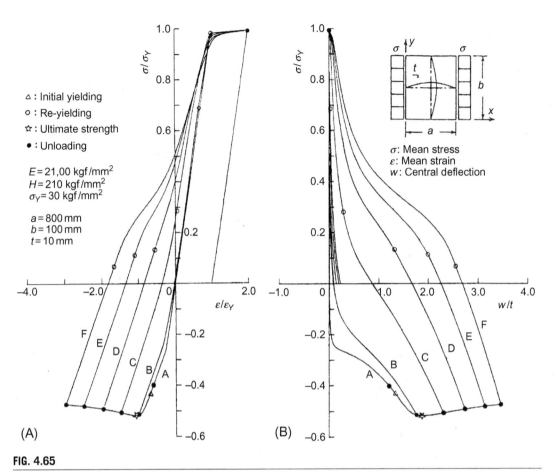

FIG. 4.65

Behavior of short thin plate subjected to tension after compressive collapse. (A) Average stress-average strain relationship. (B) Average stress-central deflection relationship.

$$w_0 = A_0 \sin \frac{\pi x}{a} \sin \frac{\pi y}{b} \tag{4.49}$$

is given.

With the increase in the applied strain, initial yielding occurs at the four corners of the plate at the point of open triangular mark on the average stress-average strain and average stress-central deflection curves. Then, with further increase in the applied strain, the ultimate strength is attained at the point of open star mark. At this stage, yielded region spreads mainly along the sides parallel to the load as indicated in Fig. 4.32A. At solid circle marks, unloading starts by reversing the sign of the forced displacement.

In Case A, the unloading starts in the elastic range before yielding occurs. In this case, behavior is completely reversible. In Case B, unloading starts just before the ultimate strength is attained. In this case, some strain and deformation remain when the average stress has returned to zero although they are not so large. After unloading, no residual stress exists in the plate in Case A, whereas small residual stress is observed in Case B.

In Cases C through F, the unloading starts after the ultimate strength has been attained. In these cases, at the unloading points, the yielded zone spreads also in the central part of the plate due to high bending stress. When the average stress returned to zero, considerable residual stress is observed in the plate according to the magnitude of residual deflection. This may be attributed to the nonuniform distribution of remaining plastic strains over the plate and toward the thickness direction.

Here, if the compressive load is again applied after the unloading, the behavior is elastic and reversible up to the starting point of unloading indicated by solid circular mark. In this case, the points of solid circular marks represent the re-yielding points, and the following behavior under increasing compressive strain is the same as that without unloading.

If the tensile load is applied after unloading in the compression range, elastic behavior continues until re-yielding takes place at the points of open circle marks in all cases. As the tensile load increases after re-loading, the in-plane stiffness, which is the slope of the average stress-average strain curve, gradually decreases with the spread of yielded region due to reversed bending, but it recovers with the decrease in lateral deflection. Then, general yielding occurs in the plate, and the stiffness suddenly decreases.

The forced displacements are again reversed in the tension range at solid circle mark point on the average stress-average strain and average stress-central deflection curves. The behavior beyond this point is elastic until the average tensile stress returns to zero. At this stage, the residual stress are very small in all cases. This may be because the plastic strains in the plate have become almost uniform due to tensile loading after general yielding.

Fig. 4.66A and B shows the average stress-average strain and average stress-central deflection relationships when loading starts in tension, respectively. G, H, and I represent the cases that the unloading starts at $\varepsilon/\varepsilon_Y = 0.5$, 2.0, and 3.0. In these cases, the plate tends to become flat with the increase in the applied tensile load.

In Case G, the unloading starts in the elastic range, and the behavior during unloading is completely reversible. In Cases H and I, the yielding takes place all over the plate at once at the open triangular mark point on the average stress-average strain and average stress-central deflection curves. The forced displacements are then reversed at the points of solid circle marks in the tension range. At these points, almost uniform plastic strains are produced in the plate, and the residual stress in the plate is infinitesimally small when the tensile load is removed. The residual deflection is smaller than the initial deflection in all cases. That is, the smaller residual deflection is obtained with higher applied tensile strain before unloading.

As the compressive load increases after unloading in the tension range, the deflection gradually increases until the load reaches around the buckling load. Above the buckling load, deflection rapidly increases, and re-yielding takes place at the points of open circle marks. Then with further increase in the applied compressive strain, the ultimate strength is attained at the points of open star marks on the average stress-average strain and average stress-central deflection curves. At the points of solid circle marks, the forced displacements are again reversed, and the plate behaves elastically until the compressive stress becomes zero. At this stage, considerable residual stress and deflection again remain in the plate. This may be attributed to the nonuniform distribution of plastic strains remaining in the plate.

Thick plates also show the similar collapse behavior although yielding takes place before buckling occurs under compression.

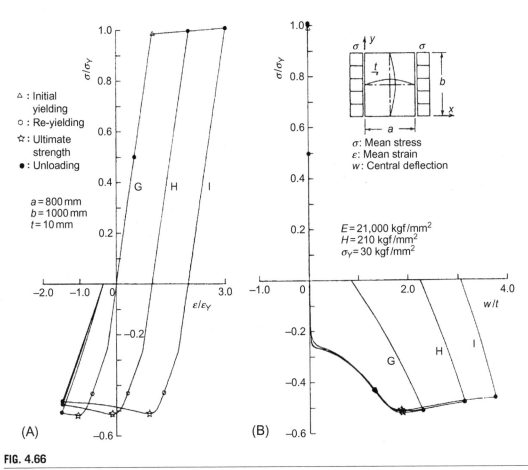

FIG. 4.66

Behavior of short thin plate subjected to compression after tensile yielding. (A) Average stress-average strain relationship. (B) Average stress-central deflection relationship.

4.7.1.1 Residual deflection after compressive collapse

The deflection at the start and the end of unloading in the compression range are plotted in Fig. 4.67 against the applied compressive strain increment. It is observed that the residual deflection increases with the increase in the applied strain increment. However, the increasing rate of residual deflection gradually decreases, and the recovered deflection during unloading becomes nearly constant.

4.7.2 BUCKLING/PLASTIC COLLAPSE BEHAVIOR OF SHORT PLATES UNDER MULTICYCLIC LOADING

At the beginning, results of analysis on the same plate indicated in Fig. 4.65 are shown. In this case, the cyclic loading starts from compression with the constant strain range of $-2.0 \leq \varepsilon/\varepsilon_Y \leq 2.0$ as indicated in Fig. 4.68A and B. As the loading starts from zero strain point, the first cycle is different

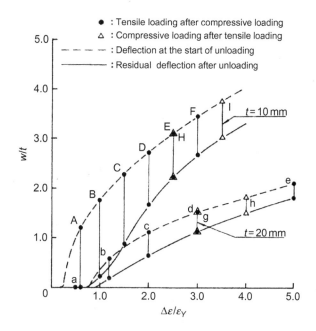

FIG. 4.67

Residual deflection after unloading in compression range.

from the following cycles. These loops seem to converge to the respective certain loops after several cycles under the constant strain amplitude. However, within the analyzed number of cycles, complete convergence of the hysteresis loop has not been obtained.

As in the case of single cyclic loading, similar behavior is observed in the case of thick plate although yielding starts before buckling.

Another example is the case of $a \times b \times t = 2750 \times 1000 \times 10$ mm. The material is the same with that for the plate in Fig. 4.65, and initial deflection of thin-horse mode is assumed as:

$$w_0 = \sum_{i=1}^{m} A_{0i} \sin \frac{i \pi x}{a} \sin \frac{\pi y}{b} \qquad (4.50)$$

where

$A_{01}/t = 0.037213$	$A_{02}/t = -0.002$	$A_{03}/t = 0.01$	$A_{04}/t = 0.000744$
$A_{05}/t = 0.003721$	$A_{06}/t = -0.000213$	$A_{07}/t = 0.001063$	$A_{08}/t = 0.00000002$
$A_{09}/t = -0.00000008$	$A_{0\,10}/t = -0.000048$	$A_{0\,11}/t = -0.000242$	

The results of analysis are given in Fig. 4.69A and B, which show the average stress-average strain and the change in deflection mode with the increase in cyclic numbers. The aspect ratio, a/b, of this plate is 2.75 and the plate may buckle in three half-waves mode in the loading direction.

In the first cycle, the deflection is of an almost regular three half-waves mode when the compressive ultimate strength is attained. Beyond the ultimate strength, however, it changes to an irregular three

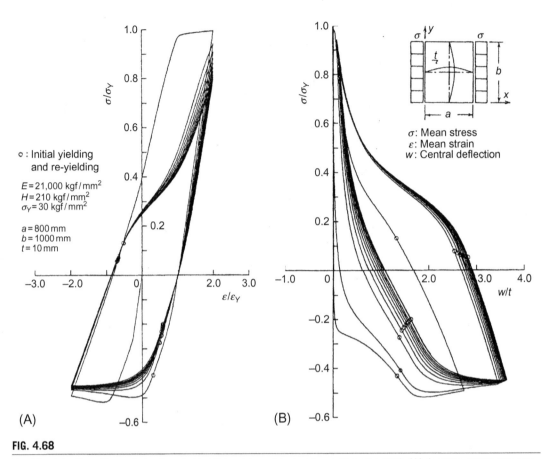

FIG. 4.68

Buckling/plastic collapse behavior of short thin plate subjected to multi cyclic load. (A) Average stress-average strain relationship. (B) Average stress-central deflection relationship.

half-waves mode with the increase in the applied compressive strain. During this process, yielding and deflection concentrate in the half-wave at the left end of the plate. Thus, deflection mode changes from three half-waves mode to two half-waves mode during unloading in the first cycle. This is the reason why in-plane stiffness changes at around σ/σ_Y is about -0.03 during unloading. When unloading has started, the behavior is at the beginning elastic, but soon re-yielding starts.

The behavior during the following tensile loading is almost the same as that of short plate in Fig. 4.68. After unloading in the tension range, residual deflection is concentrated near the left end of the plate.

It is observed that capacity both in tension and compression ranges decrease with the increase in the cyclic numbers within the analyzed cycles. Consequently, average stress-average strain relationship seems to converge to a certain hysteresis loop, although complete convergence is not attained within the analyzed cycles.

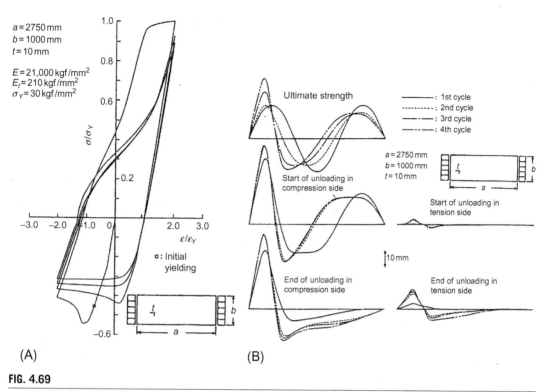

FIG. 4.69

Buckling/plastic collapse behavior of long plate under multi cyclic loading (Case *N*). (A) Average stress-average strain relationships. (B) Change in deflection mode.

EXERCISES

4.1 A simply supported rectangular plate ($a \times b \times t$) is subjected to longitudinal thrust; see Fig. 4.1. In this case, buckling stress is expressed as:

$$\sigma_{cr} = \frac{\pi^2 kE}{12(1 - v^2)} \left(\frac{t}{b}\right)^2 \tag{E4.1}$$

where

$$k = \left(\frac{a}{bm} + \frac{bm}{a}\right)^2 \tag{E4.2}$$

m is a number of half-waves of the buckling mode. Derive Eqs. (E4.1), (E4.2), and show how buckling coefficient, *k*, varies with respect to aspect ratio of the plate.

4.2 In the above case, buckling mode changes from m half-waves to $m + 1$ half-waves at the aspect ratio of $\sqrt{m(m + 1)}$. Prove this.

4.3 Derive Eqs. (4.5), (4.6) applying the *Principle of Virtual Work* or *Principle of Minimum Potential Energy*.

4.4 Explain what deflection components are produced in the postultimate strength range of a short plate.

4.5 Explain how deflection components varies as the thrust load increases in the case of a long plate.

4.6 Fig. 4.8 indicated the buckling coefficients for rectangular plates under uni-axial thrust with different out-of-plane boundary conditions. Explain why the buckling coefficient for Case B converges to that of Case A.

4.7 Explain how the welding residual stress in the loading direction affects the buckling strength of rectangular plate subjected to uniaxial thrust.

4.8 In the case of thin plate, welded plate sometimes buckles with no external compressive loads. Explain why this happens.

4.9 Following the derivation in Fujikubo and Yao [6], derive Eq. (4.26), which expresses the bi-axial buckling strength considering the influence of plate-stiffener interactions.

4.10 In the case of a simply supported rectangular plate subjected to bi-axial thrust, the buckling strength interaction relationship is expressed as follows:

$$\sigma_{xcr} + \left(\frac{a}{mb}\right)^2 \sigma_{ycr} = \frac{\pi^2 E}{12(1 - v^2)} \left(\frac{t}{b}\right)^2 \left(\frac{mb}{a} + \frac{a}{mb}\right)^2 \tag{E4.3}$$

where a, b, and t are length, breadth, and thickness of the rectangular plate; see Fig. 4.70. m is the number of half-waves of the buckling mode in the longitudinal direction. Derive Eq. (E4.3).

4.11 Consider a continuous stiffened plate indicated in Fig. 4.12. $a \times b \times tt = 2550 \times 850 \times 15$ mm and the stiffener is an angle-bar of $h \times b_f \times t_w/t_f = 250 \times 90 \times 12/17$ mm. For the local panel

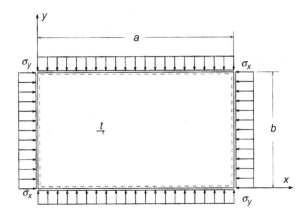

FIG. 4.70

Rectangular plate subjected to bi-axial thrust.

of this stiffened plate, calculate the buckling stress and draw buckling strength interaction curves for the following three cases:

(a) case with no interaction between plate and stiffeners

(b) case with interaction between plate and stiffeners

(c) case with interaction between plate and stiffeners as well as welding residual stress

4.12 Explain what is the secondary buckling. What is the physical meaning of the secondary buckling?

4.13 Explain about snap-through as a dynamic phenomenon which takes place during loading and unloading accompanied by secondary buckling.

4.14 Assuming initial and total deflections by Eqs. (4.36), (4.37), prove that bending stress components are expressed by Eq. (4.39).

4.15 Assuming initial and total deflections by Eqs. (4.38), (4.39), prove that Airy's stress function is expressed as:

$$F = \frac{E}{32}(A^2 - A_0^2)\left(\frac{a^2}{b^2}\cos\frac{2\pi x}{a} + \frac{b^2}{a^2}\cos\frac{2\pi y}{b}\right) - \frac{1}{2}\sigma y^2 + g(y) \qquad \text{(E4.4)}$$

Then, show that in-plane stress components are expressed by Eq. (4.38).

4.16 Draw the distributions of in-plane and bending stress components expressed by Eqs. (4.38), (4.39).

4.17 Explain the buckling/plastic collapse behavior of thin plate subjected to uni-axial thrust.

4.18 Explain the buckling/plastic collapse behavior of thick plate subjected to uni-axial thrust.

4.19 Explain the differences in developments of elastic and elastoplastic deflection components in the postbuckling range

4.20 Consider a simply supported thin plate subjected to uni-axial thrust. Explain why deflection mode changes from sinusoidal mode to roof mode beyond the ultimate strength.

4.21 Explain how the in-plane boundary condition affect the buckling/plastic collapse behavior of a simply supported thin/thick plate subjected to uni-axial thrust.

4.22 Explain how the welding residual stress affect the elastoplastic behavior of plate subjected to uni-axial tensile/compressive load when buckling does not occur.

4.23 Explain how the welding residual stress affect the elastoplastic behavior of plate subjected to uni-axial thrust load when buckling occurs.

4.24 Explain how the ultimate strength varies with respect to the aspect ratio of a simply supported rectangular plate under uni-axial thrust.

4.25 Explain how deflection mode changes in case of a simply supported wide plate subjected to uni-axial thrust.

4.26 Explain the influence of mode of initial deflection on the postultimate strength behavior of a simply supported rectangular plate subjected to uni-axial thrust.

4.27 Why rapid reduction in the capacity takes place beyond the ultimate strength when a simply supported rectangular plate with initial deflection of a hungry thin-horse is subjected to uni-axial thrust.

4.28 Explain about the residual deflection when a plate is subjected to cyclic loading.

4.8 APPENDIX: APPLICATION OF METHOD OF LEAST SQUARES TO DERIVE DEFLECTION COMPONENTS FROM FEM RESULTS

Deflection field in a simply supported plate can be expressed with $n \times m$ deflection components as:

$$w = \sum_{i=1}^{m} \sum_{j=1}^{n} A_{ij} \sin \frac{i\pi x}{a} \sin \frac{j\pi y}{b} \tag{4.51}$$

On the other hand, by the FEM analysis, deflections at nodal points are calculated. At the nodal point k locate at (x_k, y_k), deflection in z-direction, w_k, is obtained.

Here, to specify the deflection components, A_{ij}, in Eq. (4.51) for the calculated deflection by FEM, the *Method of Least Squares* can be applied.

According to the *Method of Least Squares*, the quantity representing the magnitude of error is defined as follows:

$$R = \sum_{k=1}^{kn} \left(w_k - \sum_{i=1}^{m} \sum_{j=1}^{n} A_{ij} \sin \frac{i\pi x_k}{a} \sin \frac{j\pi y_k}{b} \right)^2 \tag{4.52}$$

where kn is a number of nodal points.

The coefficients of deflection components, A_{ij}, are determined from the condition that real A_{ij}s minimize the error expressed by Eq. (4.52). This condition is expressed as:

$$\frac{\partial R}{\partial A_{ij}} = 0 \tag{4.53}$$

Substitution of Eq. (4.52) into Eq. (4.53) reduces to:

$$\frac{\partial R}{\partial A_{ij}} = \frac{\partial}{\partial A_{ij}} \sum_{k=1}^{kn} \left(w_i - \sum_{i=1}^{m} \sum_{j=1}^{n} A_{ij} \sin \frac{i\pi x_i}{a} \sin \frac{j\pi y_i}{b} \right)^2$$

$$= \sum_{k=1}^{kn} 2 \sin \frac{i\pi x_k}{a} \sin \frac{j\pi y_k}{b} \left(w_k - \sum_{i=1}^{m} \sum_{j=1}^{n} A_{ij} \sin \frac{i\pi x}{a} \sin \frac{j\pi y}{b} \right)^2 = 0 \tag{4.54}$$

Eq. (4.52) becomes a linear simultaneous equation with respect to A_{ij} as follows:

$$[H]\{A_{ij}\} = \{Q\} \tag{4.55}$$

where

$$[H] = \begin{bmatrix} h_{11} & h_{12} & a_{13} & EEEEEEEEE & h_{1k} \\ h_{21} & h_{22} & a_{23} & EEEEEEEEE & h_{2k} \\ h_{31} & h_{32} & a_{33} & EEEEEEEEE & h_{3k} \\ E & E & E & EEEEEEEEE & E \\ E & E & E & EEEEEEEEE & E \\ E & E & E & EEEEEEEEE & E \\ h_{k1} & h_{k2} & a_{k3} & EEEEEEEEE & h_{kk} \end{bmatrix}$$

$$\{Q\} = \begin{Bmatrix} q_1 \\ q_2 \\ q_3 \\ E \\ E \\ E \\ q_k \end{Bmatrix}$$

$$\Downarrow$$

$$[H] = \begin{bmatrix} h_{11} & h_{12} & a_{13} & \cdots & h_{1k} \\ h_{21} & h_{22} & a_{23} & \cdots & h_{2k} \\ h_{31} & h_{32} & a_{33} & \cdots & h_{3k} \\ \vdots & \vdots & \vdots & \cdots & \vdots \\ h_{k1} & h_{k2} & a_{k3} & \cdots & h_{kk} \end{bmatrix}$$

$$\{Q\} = \begin{Bmatrix} q_1 \\ q_2 \\ q_3 \\ \vdots \\ q_k \end{Bmatrix}$$

and

$$k_{MN} = \sum_{k=1}^{kn} \sin\frac{p\pi x_k}{a} \sin\frac{q\pi x_k}{a} \sin\frac{r\pi y_k}{b} \sin\frac{s\pi y_k}{b}$$

$$q_M = \sum_{k=1}^{kn} w_k \sin\frac{q\pi x_k}{a} \sin\frac{r\pi y_k}{b}$$

$$M = (p-1)n + q, \quad N = (r-1)n + s$$

4.9 APPENDIX: APPLICABILITY OF FEM CODE TO BUCKLING/PLASTIC COLLAPSE ANALYSIS OF PLATES SUBJECTED TO CYCLIC LOADING

The applicability of the FEM code, "ULSAS," to buckling/plastic collapse behavior of plates under cyclic loading is demonstrated here. Analysis is performed on a column specimen tested by Fujita et al. [16]. The size of the specimen is $L \times B \times H = 1000 \times 45.80 \times 36.95$ mm. The specimen was tested under a simply supported condition.

In the experiment, the compressive load is first applied. After the compressive ultimate strength has been attained, the axial compressive load is removed. Then, compressive load is again applied until the end shortening reaches around 1.1 times the height of the cross-section. Then, the axial compressive load is again removed, and the tensile axial load is continuously applied until the end shortening reaches about a quarter of the height of the cross-section. And the, unloading in the tension side and the re-loading in the compression side is performed.

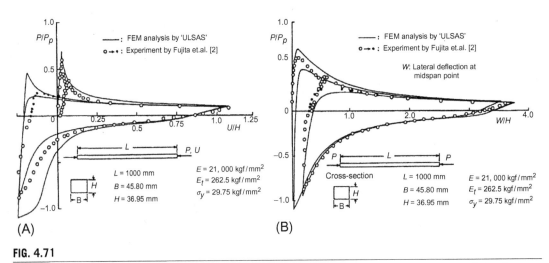

FIG. 4.71

Comparison of measured and calculated buckling/plastic collapse behavior of column under cyclic loading (R36C specimen). (A) Load-end shortening relationship. (B) Load-deflection relationship.

A quarter of the specimen is analyzed with 10 elements. The material is assumed to follow the kinematic hardening law with bilinear stress-strain relationship.

Fig. 4.71 shows the comparison of measured and calculated results for R36C specimen, in which (A) represents load-end shortening relationship and (B) load-lateral deflection relationships. The open circles are the measured results and the solid lines calculated results applying ULSAS. The analysis is performed up the 2.5 cycles, although the experimental results are shown only up to the beginning of the second cycle for clarity.

In the analysis, relatively sharp peak load appears at the beginning of the second cycle, although this is not observed in the measured results. The residual deflection at the end of the first cycle is greater than the calculated results. However, it may be said that the calculated results show good correlations with the measured results except near the peak loads at the beginning of the first and the second cycle in compression. Similar correlations are observed also for other specimens in Yao et al. [17]. It is concluded that ULSAS is applicable to the analysis on cyclic buckling/plastic collapse behavior of plates.

REFERENCES

[1] Timoshenko S, Gere J. Theory of elastic stability. McGraw-Hill Kogakusha, Ltd; 1961.
[2] Ueda Y. Elastic, elastic-plastic and plastic buckling of plates with residual stresses. PhD Dissertation. Lehigh University; 1962.
[3] Masubuchi K. Analysis of welded structures. Pergamon Press; 1980, 496 pp.
[4] Ueda Y, Yamakawa T. Inelastic buckling strength of plates with residual stress under bending. Trans Kansai Soc Naval Arch 1967;125:29–33 [in Japanese].

[5] Terazawa K, Ueda Y. Buckling strength of plates with residual stress under shear. Trans Kansai Soc Naval Arch 1965;117:33–37 [in Japanese].

[6] Fujikubo M, Yao T. Elastic local buckling strength of stiffened plate considering plate/stiffener interaction and welding residual stress. Mar Struct 1999;12:543–564.

[7] Yao T, Fujikubo M, Ko JY. Large deflection behaviour of rectangular plates under inplane compression. Trans West Jpn Soc Naval Arch 1995;89:179–190 [in Japanese].

[8] Murakami C, Yao T. On the snap-through phenomenon at the secondary buckling of rectangular plate under thrust. In: Proc the 14th int offshore and polar engineering conf, Toulon, France; 2004. p. 445–452.

[9] Yao T, Nikolov P. Buckling/plastic collapse of plates under cyclic loading. J Soc Naval Arch Jpn 1990;168:451–464.

[10] Ueda Y, Yao T, Nakacho K, Tanaka Y, Handa K. Compressive ultimate strength of rectangular plates with initial imperfections due to welding (3rd report): prediction method of compressive ultimate strength. J Soc Naval Arch Jpn 1973;154:361–370 [in Japanese].

[11] Yao T, Nikolov P. Numerical experiment on buckling/plastic collapse behaviour PF plates under cyclic loading. In: Fukumoto Y, Lee G, editors. Stability and ductility of steel structures under cyclic loading. Boca Raton, Ann Arbor, London: CRC Press; 1992. p. 203–214.

[12] Ueda Y, Yao T, Nakamura K. Compressive ultimate strength of rectangular plates with initial imperfections due to welding (1st report): effects of the shape and magnitude of initial deflection. J Soc Naval Arch Jpn 1970;148:235–244 [in Japanese].

[13] Ueda Y, Yao T. The influence of complex initial deflection modes on the behaviour and ultimate strength of rectangular plate in compression. J Constr Steel Res 1985;5:265–302.

[14] Yao Y, Fujikubo M, Yanagihara D, Zha Y, Murase T. Post-ultimate strength behaviour of rectangular panel under thrust. J Soc Naval Arch Jpn 1998;183:351–359.

[15] Yao T, Fujikubo M, Yanagihara D, Murase T. Post-ultimate strength behaviour of long rectangular plate subjected to uni-axial thrust. In: Proc 11th ISOPE conf, Stavanger, Norway, vol. IV; 2001. p. 390–397.

[16] Fujita Y, Nomoto T, Yuge K. Behaviour of deformation of structural members under compressive and tensile loads (1st report): on the buckling of a column subjected to repeated loading. J Soc Naval Arch Jpn 1984;156:346–354 [in Japanese].

[17] Yao T, Fujikubo M, Ko JY. Secondary buckling of thin plates with initial deflection. J Soc Naval Arch Jpn 1994;176:309–318 [in Japanese].

BUCKLING/PLASTIC COLLAPSE BEHAVIOR AND STRENGTH OF RECTANGULAR PLATES SUBJECTED TO COMBINED LOADS

5.1 COLLAPSE BEHAVIOR AND STRENGTH OF CONTINUOUS PLATES UNDER COMBINED LONGITUDINAL/TRANSVERSE THRUST AND LATERAL PRESSURE LOADS

5.1.1 MODEL FOR ANALYSIS

In the case of ship and ship-like floating structures, the main load acting on deck plating is an uni-axial thrust. On the other hand, bottom plating is subjected to combined transverse thrust and lateral pressure as well as longitudinal thrust, and side shell plating to combined shear, bending, and partly lateral pressure.

When a ship bottom plating is subjected to combined transverse thrust and lateral pressure, the buckling deflection as well as deflection due to lateral pressure can be assumed to be symmetrical with respect to two centerlines in each span and bay. At the same time, lateral deflection in each local panel is in the same direction under the action of lateral pressure load. Such a buckling/plastic collapse behavior can be analyzed fundamentally using a double-span double-bay model as indicated in Fig. 5.1.

Along the four sides of the model, symmetry condition is imposed. In the present analysis, longitudinal stiffeners and transverse frames are not modeled, but the plate is assumed to be simply supported along the lines where these members are attached. Considering the continuity of plating, the in-plane displacements of the sides in their perpendicular directions are assumed to be uniform. The breadth of the plate, b, is fixed as 800 mm, while the aspect ratio, a/b, and the thickness, t, are varied systematically.

The same model can be used for continuous plate subjected to combined longitudinal or biaxial thrust and lateral pressure loads when the buckling mode in the longitudinal direction is of an odd half-waves mode. If the buckling mode is of an even half-waves mode in the loading direction, triple $(1/2 + 1 + 1/2)$ span model can be used imposing continuous condition along the two centerlines in the transverse direction.

The initial deflection of a thin-horse mode is assumed, which is expressed as

$$w_0 = \alpha \left| \sum_m A_{0m} \sin \frac{m\pi x}{a} \sin \frac{\pi y}{b} \right| \tag{5.1}$$

Buckling and Ultimate Strength of Ship and Ship-like Floating Structures. http://dx.doi.org/10.1016/B978-0-12-803849-9.00005-8

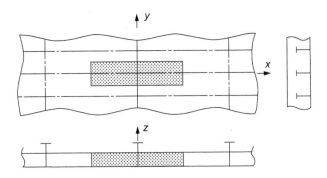

FIG. 5.1

Double span/double bay model [1].

	Table 5.1 Coefficients of Components in Initial Deflection of Idealized Thin-Horse Mode					
a/b	A_{01}/t	A_{03}/t	A_{05}/t	A_{07}/t	A_{09}/t	$A_{0\,11}/t$
1.0	1.1158	0.1377	−0.0123	−0.0043	0.0039	−0.0011
2.0	1.1421	0.2287	0.0326	−0.0109	−0.0049	0.0027
3.0	1.1458	0.3079	0.1146	0.0327	0.0000	−0.0074
4.0	1.1439	0.3385	0.1579	0.0743	0.0293	0.0062
5.0	1.1271	0.3483	0.1787	0.0995	0.0537	0.0256

1.00	1.02
1.01	1.03

Type (a)

1.00	0.90
0.90	1.00

Type (b)

1.00	0.80
0.80	1.00

Type (c)

1.00	0.60
0.60	1.00

Type (d)

FIG. 5.2

Assumed distribution of initial deflection in adjacent panels [1].

where m is taken as 11. Considering the symmetry condition of deflection mode with respect to the center line of each local panel, odd terms are set as zero. The magnitudes of deflection coefficients which result in $w_{0\,max}/t = 1.0$ are shown in Table 5.1. According to the measured results, α is different panel by panel. In the present analysis, on the basis of the observed fact, four typical patterns are assumed as indicated in Fig. 5.2.

The maximum magnitude of initial deflection, $w_{0\,max}$, is taken as 0.01 t for the elastic large deflection analysis to investigate into the elastic buckling behavior. On the other hand, for elastoplastic large deflection analysis to investigate into the collapse behavior, the maximum magnitude of initial deflection is taken as

$$w_{0\,max} = \eta \beta^2 t \qquad (5.2)$$

where β is the slenderness ratio of the plate which is defined as

$$\beta = \frac{b}{t}\sqrt{\frac{\sigma_Y}{E}} \qquad (5.3)$$

σ_Y and E are the yield stress and Young's modulus of the material, respectively. The coefficient, η, in Eq. (5.2) is taken as 0.025, 0.05, and 0.1; see Fig. 2.14 in Chapter 2.

The material properties assumed in the analysis are as follows:

Yield stress: $\sigma_Y = 313.6$ MPa
Young's modulus: $E = 205.8$ GPa
Poisson's ratio: $\nu = 0.3$
Strain hardening rate: $H' = E/65$

Both elastic and elastoplastic large deflection analyses are performed. The applied lateral pressure ranges from 0 to 60 m water-head. This range corresponds to a possible range when the sum of hydrostatic pressure and hydrodynamic pressure is considered.

5.1.2 INFLUENCE OF LOADING SEQUENCE ON BUCKLING/PLASTIC COLLAPSE BEHAVIOR

When collapse analysis is performed on rectangular plate subjected to combined loads, the loading sequence may affect the calculated results of buckling/plastic collapse behavior and the ultimate strength. To examine the influence of loading sequence on the calculated results, seven loading sequences are considered as indicated in Table 5.2 for combined longitudinal thrust and lateral pressure loads.

Firstly, elastic large deflection analysis is performed on continuous plating of a "bulk carrier" following the four loading sequences, Cases 1 through 4, in Table 5.2. For this analysis, initial deflection

Table 5.2 Assumed Loading Sequences [1]

Case	First Stage	Second Stage	Analysis
1	WP up to 30 m-wh	Thrust	Elastic
2	Thrust with $\Delta u = 0.025$ mm + WP up to 30 m-wh	Thrust	Elastic
3	Thrust with $\Delta u = 0.0125$ mm + WP up to 30 m-wh	Thrust	Elastic
4	Thrust with $\Delta u = 0.00625$ mm + WP up to 30 m-wh	Thrust	Elastic
5	WP up to 30 m-wh	Thrust	Elastoplastic
6	Thrust with $\Delta u = 0.0125$ mm + WP up to 30 m-wh	Thrust	Elastoplastic
7	Thrust with $\Delta u = 0.00625$ mm + WP up to 30 m-wh	Thrust	Elastoplastic

Notes: δu, applied in-plane displacement increment for thrust loading; WP, applied water pressure; m-wh, meters water-head.

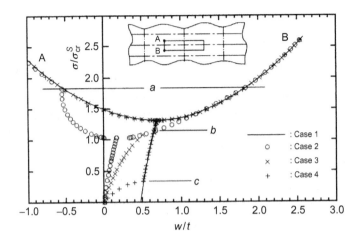

FIG. 5.3

Influence of loading sequence on elastic large deflection behavior [2].

of Type (a) shown in Fig. 5.2 is assumed. In Case 1, lateral pressure is firstly applied up to 30 m water-head, and then the thrust load. On the other hand, in Cases 2, 3, and 4, lateral pressure and thrust loads are applied simultaneously changing the ratio of thrust load to pressure load until lateral pressure reaches 30 m water-head, and then only thrust load is continuously applied.

The average stress-deflection relationships for these four cases are shown in Fig. 5.3. The average stress is nondimensionalized by elastic buckling strength, σ_{cr}^s, under simply supported condition, and deflection by panel thickness. The deflection is measured at the centers, A and B, of the adjacent panels. In Case 1, central deflection nearly equal to the half plate thickness is produced in each panel by lateral pressure of 30 m water-head. Then, deflection increases in the adjacent panels with the same magnitude, but bifurcation takes place when the average stress reaches around 1.36 times the elastic buckling stress, σ_{cr}^s.

In Case 2, bifurcation stress is nearly equal to σ_{cr}^s, and the pressure loading ends at the stress level of a in Fig. 5.3. Above this stress level, average stress-deflection relationship coincides with that of Case 1. On the other hand, in Cases 3 and 4, pressure loading ends at the stress levels of b and c in Fig. 5.3, which are below the bifurcation stress. After this, the average stress-deflection curve coincides with that of Case 1. It should be noticed that the sequence of loads does not affect the elastic large deflection behavior after the pressure load has reached the specified level.

Fig. 5.4 shows the results of elastoplastic large deflection analysis on the same plate in Cases 5, 6, and 7 in Table 5.2. Also for this analysis, initial deflection of Type (a) is imposed. In Case 5, yielding starts during the pressure loading before thrust is applied. In Case 6, the pressure loading ends at the stress level a after the maximum capacity has been attained. After reaching the level a, the behavior is fundamentally the same as that of Case 5, and the second load peak is attained. Similar behavior is observed also in Case 7, although the pressure loading ends before the maximum capacity is attained.

In all cases, behavior after the pressure loading has ended is almost the same, but some differences are observed among the three. This may be because of the different plastic deformations accumulated through different strain histories. Similar calculations are performed on other cases, and larger

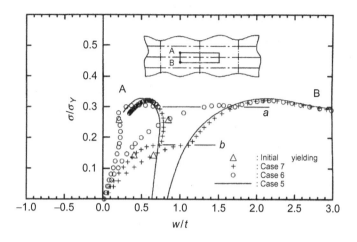

FIG. 5.4

Influence of loading sequence on elastoplastic large deflection behavior [1].

differences are observed in some cases in the collapse behavior depending on different yielding states. However, the ultimate strength does not differ so much in all cases as indicated in Fig. 5.4.

From the results indicated above, it can be concluded that the calculated ultimate strength is not so much affected by the loading sequence when a continuous plating is subjected to combined lateral pressure and thrust.

Hereafter, lateral pressure is firstly applied to the specified level, and then thrust load is applied in the case of combined thrust and lateral pressure loads.

5.1.3 INFLUENCE OF LATERAL PRESSURE ON ELASTIC BUCKLING STRENGTH OF CONTINUOUS PLATE UNDER COMBINED LONGITUDINAL/TRANSVERSE THRUST AND LATERAL PRESSURE

Ship bottom plating in hogging condition is subjected to combined longitudinal/transverse thrust and lateral pressure. Under the action of longitudinal/transverse thrust, local panel buckles fundamentally in a simply supported mode. On the other hand, under lateral pressure, deflection is of a clamped mode is produced because of the continuity of plating. This implies that lateral pressure may increase the local panel buckling strength since deflection of a clamped mode, which is different from the buckling mode, resists against the occurrence of the buckling deflection of a simply supported mode.

Here, a series of finite element method (FEM) elastic large deflection analysis is performed on ship bottom plating, which is modeled as a continuous plating; see Fig. 5.1 [1]. The shaded part is analyzed introducing symmetry condition along the four sides of the analyzed part. The longitudinal and the transverse stiffeners are not modeled, but the plate is assumed to be simply supported along the stiffener lines. The size of the local panel is $a \times b \times t = 2400 \times 800 \times 13.5$ mm. Two cases are considered as initial deflection. One is of a buckling mode expressed as

$$w_0 = A_{0m} \sin \frac{m \pi x}{a} \sin \frac{\pi y}{b} \tag{5.4}$$

where A_{0m}/t is taken as 0.01, and the other is of an idealized initial deflection of a thin-horse mode expressed as follows:

$$w_0 = \left| \sum_{i=1}^{11} A_{0i} \sin \frac{i\pi x}{a} \sin \frac{\pi y}{b} \right| \tag{5.5}$$

where A_{0i} is given as follows:

$$A_{01}/t = 0.11438 \longrightarrow A_{01}/t = 0.011438$$
$$A_{03}/t = 0.03079 \longrightarrow A_{03}/t = 0.003079$$
$$A_{05}/t = 0.01146 \longrightarrow A_{05}/t = 0.001146$$
$$A_{09}/t = 0.00327 \longrightarrow A_{07}/t = 0.000327$$
$$A_{0;11}/t = 0.00073 \longrightarrow A_{0;11}/t = 0.000073$$

The maximum magnitude of the latter initial deflection is also 0.01 times the plate thickness. In this case, the maximum magnitude of initial deflection in the adjacent four panels are changed one by one by 1% to make the occurrence of buckling numerically possible for the continuous plating (Type (a) in Fig. 5.2).

The relationship between average stress and two central deflection at adjacent panels are plotted in Fig. 5.5A and B for longitudinal thrust and transverse thrust, respectively [2]. The dotted lines shows the relationships when initial deflection is of a buckling mode, and the solid lines that when initial

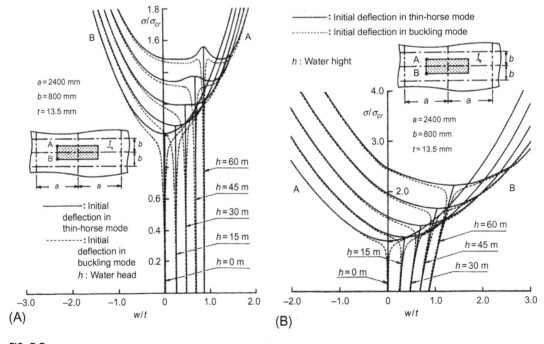

FIG. 5.5

Influence of lateral pressure on compressive buckling behavior [2]. (A) Longitudinal thrust. (B) Transverse thrust.

deflection is of a thin-horse mode. Buckling behavior similar to the bifurcation phenomenon is clearly observed in the latter case. This buckling behavior is at the same time elastic large deflection behavior, which was explained in detail in Chapter 4. In the present chapter, only the increase in buckling strength due to lateral pressure is highlighted.

An alternative series of FEM analysis is also performed for a continuous plating without stiffener changing the breadth to thickness ratio, b/t_p [4]. The breadth of the panel is kept as $b = 800$ mm, and the aspect ratio, a/b, is taken as 3.0, 4.0, 5.0, and 6.0.

The continuous plate is assumed to be simply supported along the longitudinal stiffener lines and the transverse frame lines. Stiffeners and frames are not modeled and symmetry condition or periodically continuous condition is imposed along the four boundaries of the shaded part in Fig. 5.1.

On the basis of the calculated results, empirical formulas to evaluate the buckling strength under combined longitudinal/transverse thrust and lateral pressure is derived as follows [4].

For longitudinal thrust:

$$\sigma_{cr}^w = \left[1 + \frac{(qb^4/Et_p^4)^{1.6}}{576}\right]\sigma_{cr}^s \quad (a/b \geq 2.0) \tag{5.6}$$

For transverse thrust:

$$\sigma_{cr}^w = \left[1 + \frac{(qb^4/Et_p^4)^{1.75}}{160(a/b)^{0.95}}\right]\sigma_{cr}^s \quad (a/b \geq 2.0) \tag{5.7}$$

where σ_{cr}^s is the buckling stress when the plate is simply supported.

In Eq. (5.6), the increase of buckling strength due to lateral pressure is represented by a function of (qb^4/Et_p^4) only. Neither the aspect ratio, a/b, nor half-wave number, m, is included. This may be because the buckling strength under longitudinal thrust is almost the constant value regardless of the aspect ratio when it is larger than 2.0. On the other hand, Eq. (5.7) for transverse thrust includes the aspect ratio a/b.

The buckling strength calculated by Eqs. (5.6), (5.7) is compared with the results of FEM analysis in Fig. 5.6A and B for various combination of the aspect ratio, slenderness ratio, and pressure level. As a whole, very good correlations are observed between both results, which demonstrates the accuracy of Eqs. (5.6), (5.7).

Here, Figs. 5.7 and 5.8 indicate the deflection and in-plane stress in the transverse direction produced by lateral pressure, q. Results of several cases are shown with different thickness and aspect ratio. It is known that deflection mode is almost the same and flat at the middle part of the plate when the aspect ratio is 2.0 or more, and their magnitudes are almost the same. On the other hand, for the plate of aspect ratio being 1.0, the plate is not long enough to produce similar deflection described above. This difference in deflection mode produces different in-plane stress in the panel. This is the reason why application of Eqs. (5.6), (5.7) is restricted for the range of $a/b \geq 2.0$ [4].

As the lateral pressure is increased, deflection of a clamped mode is more magnified. This dominant deflection component as well as the action of lateral pressure delay the development of deflection of a simply supported buckling mode caused by transverse thrust. It is worth noting that the buckling strength is increased with the increase in the applied lateral pressure to a certain pressure level, but further increase of the lateral pressure has an opposite effect on the buckling strength in the case of a

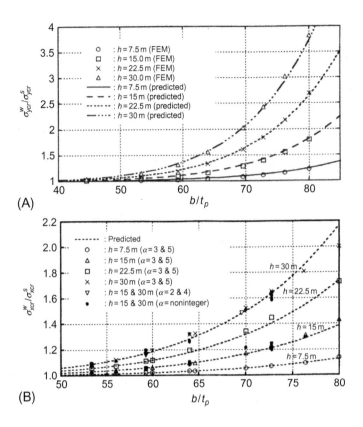

(A)

(B)

FIG. 5.6

Influence of lateral pressure on local buckling strength of continuous plating under thrust [4]. (A) Transverse thrust ($a/b = 3.0$). (B) Longitudinal thrust.

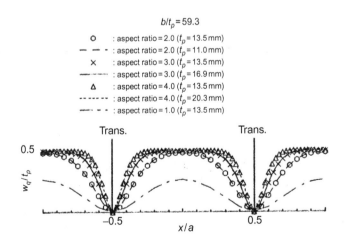

FIG. 5.7

Deflection shape in the loading direction produced by lateral pressure [4].

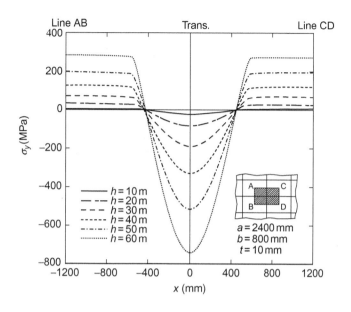

FIG. 5.8

Distribution of membrane stress produced by lateral pressure in transverse direction ($a/b = 3.0$) [5].

thin plating shown in Fig. 5.5B. This phenomenon cannot be explained by the above-mentioned feature alone. This is known from the distribution of membrane stress in the transverse direction plotted in Fig. 5.8.

Due to the continuity of the analyzed model at its spans or bays, the edges of the analyzed model remain straight. As a result, when large deflection is produced by lateral pressure, tensile membrane stress is produced in the central part of the panel whereas compressive stress is produced near the edges to satisfy self-balancing equilibrium condition of forces in the transverse direction. It is obvious that the tensile stress in the central part of the panel delays the occurrence of buckling. At the same time, compressive stress near the edges may accelerate the occurrence of local buckling near the edges. In fact, buckling deflection in the case of $h = 60$ m in Fig. 5.8B has been found to take place locally near the shorter edges of the panel. This is the reason why the buckling strength does not increase so much under a high lateral pressure especially in the case of a thin panel.

Another problem is the overestimation of the elastic buckling strength when the aspect ratio is below 2.0. To solve this problem, another series of elastic large deflection analysis is performed by FEM, and Eq. (5.7) is revised on the basis of the results of FEM analysis. That is [5]

$$\sigma_{cr}^q = R_q \cdot \sigma_{cr}^s \tag{5.8}$$

where

$$\sigma_{cr}^s = \frac{E\pi^2}{12(1 - \nu^2)} \left(1 + \frac{a^2}{b^2}\right)^2 \left(\frac{t}{a}\right)^2 \tag{5.9}$$

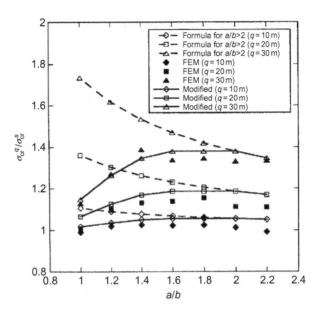

FIG. 5.9

Variation of elastic buckling strength with respect to aspect ratio for continuous plating subjected to combined transverse thrust and lateral pressure [5].

$$
R_q = \begin{cases}
R_{q2} + (R_{q2} - R_{q1})\left(\frac{2b}{a} - 2\right) & (1 \le a/b \le 1.5) \\
R_{q2} & (1.5 < a/b < 2.0) \\
1 + \dfrac{(qb^4/Rt^4)^{1.75}}{160(a/b)^{0.95}} & (2.0 \le a/b)
\end{cases}
\tag{5.10}
$$

$$
R_{q1} = 1 + \frac{(qb^4/Rt^4)^2}{1560}, \quad R_{q2} = 1 + \frac{(qb^4/Rt^4)^{1.75}}{160 \cdot 2^{0.95}}
\tag{5.11}
$$

The estimated buckling strength by Eqs. (5.8) through (5.11) is plotted by solid lines in Fig. 5.9. Relatively good correlations are observed between the FEM results and the new predictions.

5.1.4 BUCKLING/PLASTIC COLLAPSE BEHAVIOR UNDER COMBINED TRANSVERSE THRUST AND LATERAL PRESSURE

5.1.4.1 Buckling/plastic collapse behavior

Firstly, a series of elastoplastic large deflection analysis is performed on the bottom plating of VLCC ($a \times b \times t = 4200 \times 840 \times 19$ mm) to examine influences of the magnitude and the pattern of initial deflection on buckling/plastic collapse behavior. In Fig. 5.10A and B, average stress-central deflection relationships are plotted for three cases assuming the magnitude of initial deflection, η, in Eq. (5.2) as $0.025, 0.05$, and 0.1. Lateral pressure is not applied in this analysis. Fig. 5.10A and B shows the results of analysis when Types (a) and (d) initial deflection in Fig. 5.2 are assumed, respectively.

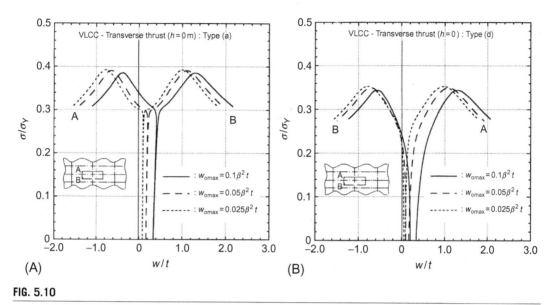

FIG. 5.10

Influence of magnitude of initial deflection on collapse behavior of continuous plating subjected to transverse thrust ($a/b = 5.0$; no lateral pressure) [1]. (A) Initial deflection of Type (a). (B) Initial deflection of Type (d).

In both cases, the difference in the ultimate strength is not so large within the assumed range of the magnitude of initial deflection. However, when initial deflection of Type (a) is prescribed, deflection of a buckling mode does not develop so much until the average stress reaches the buckling stress. This is because the magnitude of initial deflection is almost the same and in the same direction in the adjacent four panels, which constrain the growth of deflection of a buckling mode. Because of this, magnitude of deflection has hardly varied until the average stress reaches very near to the buckling stress.

On the contrary, when initial deflection of Type (d) is prescribed, the magnitude of initial deflection in the adjacent panels differ by 40%. Because of this, deflection of a simply supported buckling mode grows from the start of thrust loading; see Fig. 5.10B. Also in this case, similar collapse behavior is observed as in the case with Type (a) initial deflection.

Fig. 5.11A and B shows average stress-average strain relationships for the same bottom plating but subjected to combined transverse thrust and lateral pressure of 30 m water-head. The parameter, η, representing the magnitude of initial deflection is taken as 0.025 and 0.1 in Fig. 5.11A and B, respectively. Four types of initial deflection in Fig. 5.2 are assumed.

When initial deflection is small ($\eta = 0.025$), the ultimate strength except the case of Type (a) initial deflection is almost the same, while Type (a) initial deflection results in 8% higher ultimate strength. A similar tendency is observed also in the case of larger initial deflection of $\eta = 0.1$ although the difference in ultimate strength becomes larger among Types (b), (c), and (d); see Fig. 5.11B. Similar results are obtained for the cases with different pressure loads including zero pressure.

The average stress-central deflection relationships are illustrated in Fig. 5.12 for the continuous plating with local panel of $a \times b \times t = 2400 \times 800 \times 10$ mm subjected to combined transverse thrust

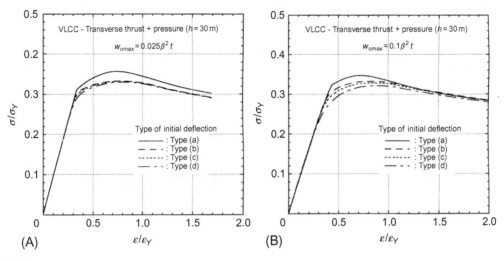

FIG. 5.11

Influence of pattern of initial deflection on collapse behavior of continuous plating subjected to combined transverse thrust and lateral pressure ($a/b = 5.0$; $h = 30$ m) [1]. (A) Maximum magnitude of 0.025 $\beta^2 t$. (B) Maximum magnitude of 0.1 $\beta^2 t$.

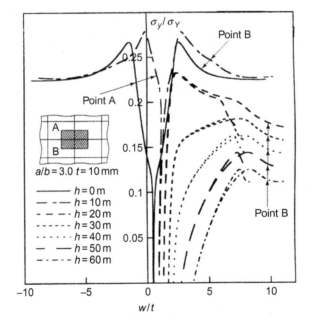

FIG. 5.12

Average stress-central deflection relationships for rectangular plating subjected to transverse thrust and lateral pressure ($a/b = 3.0$) [5].

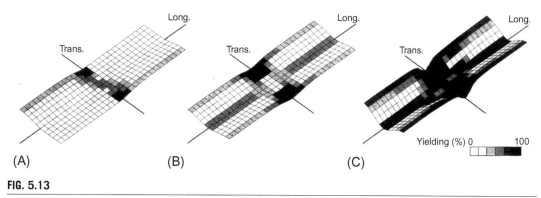

FIG. 5.13

Typical collapse mode and spread of yielding at ultimate strength ($a/b = 3.0$) [3]. (A) $h = 10$ m water-head. (B) $h = 20$ m water-head. (C) $h = 30$ m water-head.

and lateral pressure. Initial deflection of Type (b) is considered and the pressure is varied between 0 and 60 m water-head at every 10 m. Typical collapse modes for the cases in Fig. 5.12 are indicated in Fig. 5.13.

As shown in Fig. 5.13A, when the lateral pressure is relatively low ($h = 10$ m), the collapse mode is similar to that of an isolated plate which is simply supported along all sides. The spread of yielding is slightly different in the adjacent bays and spans. This is due to the different magnitude of initial deflection in adjacent panels as indicated in Fig. 5.2. Under the higher lateral pressure, the plate collapses as if it were clamped along all sides; see Fig. 5.13B and C.

5.1.4.2 Influence of lateral pressure on ultimate compressive strength

The ultimate compressive strength of continuous plating subjected to combined transverse thrust and lateral pressure is plotted against lateral pressure in Fig. 5.14. The bold solid lines represent the ultimate strength by the present analysis. On the other hand, the ultimate strength of an isolated rectangular plate of which four sides are clamped is also analyzed and the obtained ultimate strength is indicated by thin dashed lines in the same figure.

The solid marks at $h = 0$ m in Fig. 5.14 indicate the ultimate strength of continuous plating under pure transverse thrust. These strengths are generally higher than the ultimate strengths of simply supported isolated plates as will be seen later. This is because the rotation at the edges of each local rectangular plate at buckling is elastically constrained along all sides as a result of the continuity of plating and the presence of initial deflection of a thin-horse mode.

In the case of a thin plate of $t = 10$ mm, with the increase in lateral pressure, the ultimate strength of continuous plating increases due to the increase of elastic buckling strength, and the collapse mode gradually changes to a clamped mode from a simply supported mode. For the water-head of $h = 15–25$ m, the ultimate strength of a continuous plate almost coincides with that of an isolated plate clamped along all sides. For further increase in lateral pressure, the ultimate strength monotonously decreases. The ultimate strength of an isolated clamped plate shown by the dashed lines is more rapidly decreases than that of a continuous plate, with an increase in lateral pressure. This is because the internal forces and deformation produced by lateral pressure are closer to those of a clamped plate

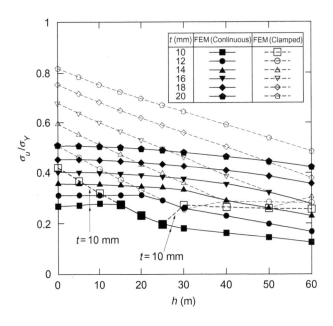

FIG. 5.14

Influence of lateral pressure on compressive ultimate of continuous plating subjected to combined transverse thrust and lateral pressure ($a \times b \times t/b = 2400 \times 800 \times 10.0$ mm) [5].

under transverse thrust than those of a continuous plate that behaves like a simply supported plate. As a result, beyond a certain level of lateral pressure, a clamped mode deflection is more enhanced and the collapse mode changes from a simply supported mode to a clamped mode.

Here, in Fig. 5.14, abrupt increase is observed in the ultimate compressive stress at the water-head of $h = 30$ m in the case of the isolated plate of $t = 10$ mm with clamped edges. Fig. 5.15 shows the related average stress-central deflection relationships. Under the water-head of $h = 20$ m and below, a clear peak of thrust load exists, and with further increase in the deflection, the postultimate strength capacity once decreases but again increases due to the development of membrane actions and strain hardening effect of $H' = E/65$. On the other hand, for the lateral pressure of $h = 30$ m and above, the peak stress at about $w/t = 2.0-2.5$ vanishes and the first peak stress is attained at about $w/t = 25$. The first peak of each average stress-central deflection curve in Fig. 5.15 is plotted as the ultimate strength in Fig. 5.14. The jump appeared in Fig. 5.14 between $h = 25$ m and $h = 30$ m is caused by the change in collapse behavior explained above.

It should also be noticed that the water-head of $h = 30$ m almost coincides with the plastic collapse pressure, $h_c = 29.3$ m, calculated by applying the rigid plastic mechanism analysis. For the plate of $t = 12$ mm, a similar jump is observed at $h = 40$ m, while plastic collapse pressure for this plate is $h_c = 42.2$ m. These results indicate that the plastic collapse pressure, h_c, gives the water-head at which bending stiffness has been almost lost and the membrane action becomes dominant. It should be noticed that the membrane strain at the ultimate strength under high pressure load, say at $w/t = 25$ m,

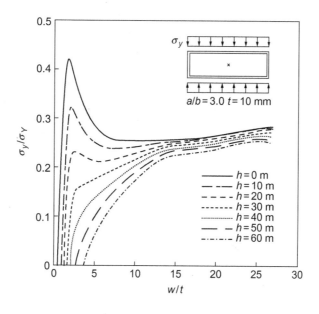

FIG. 5.15

Average stress-central deflection relationship of isolated clamped plate subjected to combined transverse thrust and lateral pressure ($a \times b \times t/b = 2400 \times 800 \times 10.0$ mm) [5].

is far beyond the yielding strain, that is, of the order of 100 times the yielding strain. The calculated behavior with such large strain neglecting fracture may not be realistic.

5.1.4.3 Estimation of ultimate strength of continuous plating subjected to combined transverse thrust and lateral pressure

Considering the buckling/plastic collapse behavior explained above and giving prior attention to the physical background of formulations, a method to estimate the ultimate compressive strength of continuous plating subjected to combined transverse thrust and lateral pressure was proposed [5]. For this loading condition, there exist two fundamental collapse modes.

One is similar to a simply supported mode shown in Fig. 5.13A but is subjected to the elastic constraint along all sides due to the continuity of plating and the presence of symmetric initial deflection of a thin-horse mode. The other is a clamped mode observed typically in thin plates under high lateral pressure; see Fig. 5.13B and C. The former mode is herein simply called a simply supported mode and the latter a clamped mode. The ultimate strength formulas for the respective collapse modes are derived first. Then, the ultimate strength of continuous plating is estimated by coupling these formulas.

It has been found that the lateral pressure has basically two opposite influences on the buckling/plastic collapse behavior and strength of continuous plating, that is the increased elastic buckling strength and the earlier yielding due to bending deformation both caused by lateral pressure.

Simply supported plate under transverse thrust

Fig. 5.16 shows the simply assumed stress distribution of a rectangular plate under transverse thrust when the ultimate strength is attained. Here, the stress, σ_{ue}^s, at the end parts is approximated by that of a square plate of which side length is b. On the other hand, the ultimate strength of the middle part, σ_{um}^s, is approximated by that of a both-ends simply supported column of length b with a unit width. The superscript s indicates a simply supported condition.

The results of FEM analysis on simply supported plates with various aspect ratios and slenderness ratios are shown by solid lines with solid marks in Fig. 5.17. On the basis of the calculated results on square plates ($a/b = 1.0$), the ultimate strength of the end parts can be expressed as

$$\frac{\sigma_{ue}^s}{\sigma_Y} = \frac{2.4}{\beta} - \frac{1.4}{\beta^2} \quad (\leq 1.0) \tag{5.12}$$

where β is given by Eq. (5.3). This formula gives a little higher ultimate strength compared to that proposed by Faulkner [6] with 2.0 and 1.0 instead of 2.4 and 1.4, respectively.

On the other hand, the ultimate strength of the middle part can be estimated by the stress at the intersection point of the elastic thrust-deflection curve and the rigid-plastic thrust-deflection curve as described in Section 5.4. It is given as

$$\frac{\sigma_{um}^s}{\sigma_Y} = \frac{0.06}{\beta} - \frac{0.6}{\beta^2} \quad (\leq 1.0) \tag{5.13}$$

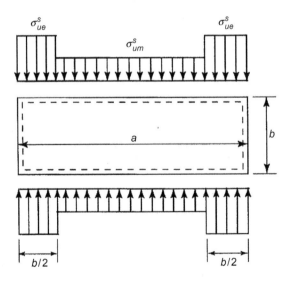

FIG. 5.16

Assumed stress distribution in rectangular plate under transverse thrust at ultimate strength [5].

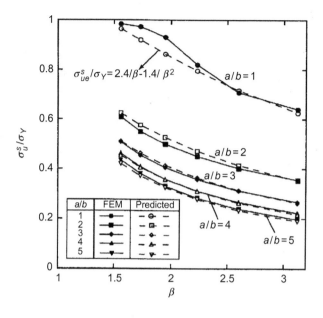

FIG. 5.17

Comparison of calculated and predicted ultimate strength of simply supported rectangular plate under transverse thrust [3].

The ultimate strength is represented as the weighted average of σ_{ue}^s and σ_{um}^s as

$$\frac{\sigma_u^s}{\sigma_Y} = \frac{b}{a}\frac{\sigma_{ue}^s}{\sigma_Y} + \left(1 - \frac{b}{a}\right)\frac{\sigma_{um}^s}{\sigma_Y} \tag{5.14}$$

The ultimate strength calculated by Eq. (5.14) is plotted by dashed lines with open marks in Fig. 5.17. Comparing the calculated results with those by the FEM, it is known that Eq. (5.14) gives good prediction of the ultimate strength of a simply supported rectangular plate subjected to transverse thrust.

Continuous plate which collapses in simply supported mode under combined transverse thrust and lateral pressure

It has been shown that lateral pressure increases the elastic buckling strength of continuous plating. According to Eq. (5.8), the increased buckling strength can be expressed as follows using the increasing rate, R_q.

$$\sigma_{cr} = \sigma_{cr}^s R_q \tag{5.15}$$

R_q is given by Eqs. (5.9)–(5.11). Here, the original definition of the slenderness ratio, β, of the plate is related to a square root of the ratio, σ_Y/σ_{cr}^s, and is expressed as

$$\beta = \frac{b}{t}\sqrt{\frac{\sigma_Y}{E}} = \pi\sqrt{\frac{k}{12(1-\nu^2)}} \times \sqrt{\frac{\sigma_Y}{\sigma_{cr}^s}} \qquad (5.16)$$

So, the buckling strength influenced by lateral pressure can be regarded as that free from lateral pressure but of the plate with an equivalent slenderness ratio. Considering the relationship of Eq. (5.15), this equivalent slenderness ratio, or reduced slenderness ratio can be expressed as

$$\beta_q = \frac{b}{t}\sqrt{\frac{\sigma_Y}{ER_q}} \qquad (5.17)$$

Eq. (5.17) implies that a continuous plate with the slenderness ratio, β, subjected to combined transverse thrust and lateral pressure can be regarded as a continuous plate with the reduced slenderness ratio, β_q, having the same elastic buckling strength under no lateral pressure. In other words, the original plate thickness, t, is replaced by the increased thickness, $\sqrt{R_q} \times t$, that gives the same elastic buckling strength.

Replacing β in Eqs. (5.12)–(5.14) by β_q, the ultimate strength of a continuous plate is calculated and is plotted in Fig. 5.18A by thin dashed and dotted lines. On the other hand, bold solid lines with solid marks in the same figure represent the results of the FEM analysis. It is seen that both results show good correlations only when lateral pressure is low. This is because the influence of lateral pressure to reduce the yielding strength is not considered in the above estimation, although the influence to increase the elastic buckling strength is included.

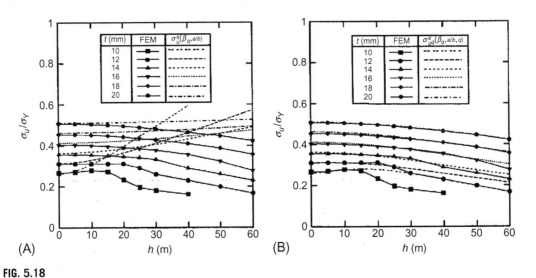

FIG. 5.18

Comparison of calculated and predicted ultimate strength of continuous plate under combined transverse thrust and lateral pressure [5]. (A) Prediction considering only increase of buckling strength. (B) Prediction considering both increase of buckling strength and decrease of yielding strength.

According to the theory of rigid-plastic mechanism analysis [7], the reduction of the ultimate strength due to lateral pressure can be expressed as a function of the following parameter:

$$\gamma_q = \frac{qb^2}{\sigma_Y t^2} = \frac{1}{\beta^2} \frac{qb^4}{Et^4} \tag{5.18}$$

γ_q corresponds to the ratio of the bending moment produced by lateral pressure to the fully plastic bending moment of the plate per unit width. Using γ_q, the ultimate strength of a continuous plate under lateral pressure is approximated as

$$\sigma_{uq}^s = \frac{\sigma_u^s}{1 + A\gamma_q^B} \tag{5.19}$$

By curve-fitting through the results of the FEM analysis, the parameters, A and B, are defined as

$$A = 0.067 - 0.0121\frac{a}{b} \quad (\geq 0), \qquad B = 0.154 + 0.577\frac{a}{b} \tag{5.20}$$

The ultimate strength predicted by Eq. (5.19) is compared with that by the FEM in Fig. 5.18B. The predicted strength is generally in good agreement with the FEM result. For very thin plate, however, Eq. (5.20) gives a little higher ultimate strength. This is because the plate collapses in a clamped mode. The ultimate strength formulas for such collapse mode are derived in the following.

Clamped plate under transverse thrust

Following the same approach with the simply supported plate under transverse thrust, the ultimate strength is approximated as

$$\frac{\sigma_u^c}{\sigma_Y} = \frac{b}{a}\frac{\sigma_{ue}^c}{\sigma_Y} + \left(1 - \frac{b}{a}\right)\frac{\sigma_{um}^c}{\sigma_Y} \tag{5.21}$$

where σ_{ue}^c and σ_{um}^c are the ultimate strength of the end parts and the central part of the clamped plate as indicated in Fig. 5.16. On the basis of the calculated results by the FEM on a square plate all sides clamped as well as analytical solutions for a both-ends clamped column given in Section 6.4, the following formulas are derived.

$$\frac{\sigma_{ue}^c}{\sigma_Y} = 1.27 - 0.18\beta \quad (\leq 1.0) \tag{5.22}$$

$$\frac{\sigma_{um}^c}{\sigma_Y} = \frac{3.48}{\beta^2} - \frac{0.262}{\beta} \quad (\leq 1.0) \tag{5.23}$$

Continuous plate which collapses in clamped mode under combined transverse thrust and lateral pressure

In the case of a clamped plate, the deflection mode produced by lateral deflection is similar to the buckling mode under transverse thrust. So, the lateral pressure has little influence on the increase of the elastic buckling strength. As shown in Fig. 5.14, the ultimate strength in a clamped mode monotonously decreases with the increase in the lateral pressure. This may indicate that the influence of lateral pressure on the ultimate strength of a clamped plate can be expressed in terms of only the parameter, γ_q.

Performing the curve fitting through the results of the FEM analysis, the ultimate strength formula of a continuous plate which collapses in a clamped mode is expressed as follows:

$$\sigma_{uq}^c = \sigma_u^c(1 - F\gamma_q + G\gamma_q^2) \tag{5.24}$$

where

$$F = 0.0821 + 0.019\frac{a}{b}, \quad G = 0.00495 + 0.000836\frac{a}{b} \tag{5.25}$$

Procedure and results of estimation

The ultimate strength of a continuous plate subjected to combined transverse thrust and lateral pressure is finally estimated by

$$\sigma_{uyq}(\beta_q, a/b, q) = \text{Min}[\sigma_{uq}^s, \sigma_{uq}^c] \tag{5.26}$$

Taking the lower strength between σ_{uq}^s and σ_{uq}^c, the change of the collapse mode is automatically considered.

The predicted ultimate strength is compared with the FEM results in Fig. 5.19A through D. The predicted ultimate strength is quite in good agreement with that obtained by the FEM analysis in the range of lateral pressure under consideration. A comparison of the ultimate strength between proposed formulas and the FEM analysis is summarized in Fig. 5.20. Good agreement is observed also in this figure.

5.1.5 BUCKLING/PLASTIC COLLAPSE BEHAVIOR UNDER COMBINED LONGITUDINAL THRUST AND LATERAL PRESSURE

5.1.5.1 Model of continuous plating for analysis

A model of continuous plating subjected to combined longitudinal thrust and lateral pressure is shown in Fig. 5.21. Under the longitudinal thrust, the number of half-waves of a buckling mode could be even or odd number depending on the aspect ratio of the local panel and the magnitude of lateral pressure. When the buckling half-wave number is even, double $(1/2 + 1/2)$ span model can be used imposing continuous condition along two boundaries in the transverse direction. On the other hand, when it is odd, triple $(1/2 + 1 + 1/2)$ span model has to be used imposing symmetry or continuous boundary condition along the two boundaries in the transverse direction. In the present analysis, triple span model is used with continuous boundary condition for all cases. On the other hand, symmetry condition is imposed along the two boundaries parallel to the x-axis. The continuity of plating is guaranteed by assuming that the in-plane displacements of the four sides of the model in their perpendicular in-plane directions are uniform.

Initial deflection of a thin-horse mode is assumed which is expressed by Eq. (5.1), that is

$$w_0 = \alpha \left| \sum_m A_{0m} \sin \frac{m\pi x}{a} \sin \frac{\pi y}{b} \right| \tag{5.1}$$

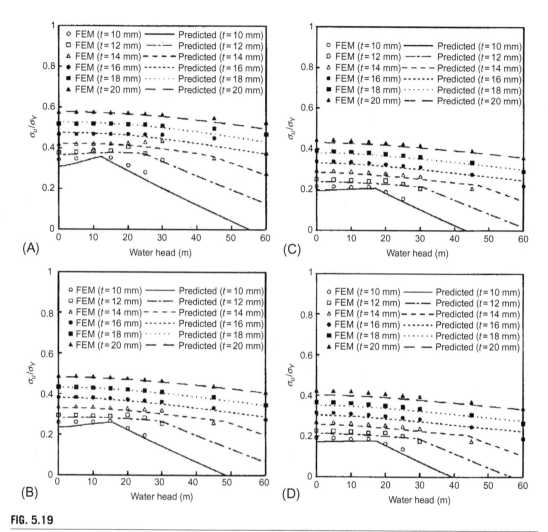

FIG. 5.19

Comparison of predicted and calculated ultimate strength of continuous plate under combined transverse thrust and lateral pressure [5].

The maximum magnitude of initial deflection is assumed as follows:

$$w_{0\,max} = 0.05\beta^2 t \tag{5.27}$$

where β is a slenderness ratio of the local panel and is defined as

$$\beta = \frac{b}{t}\sqrt{\frac{\sigma_Y}{E}} \tag{5.28}$$

σ_Y and E are the yield stress and Young's modulus of the material, respectively.

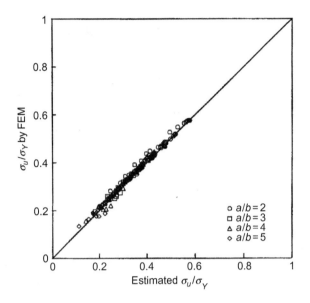

FIG. 5.20

Comparison of predicted and calculated ultimate strength of continuous plate under combined transverse thrust and lateral pressure [5].

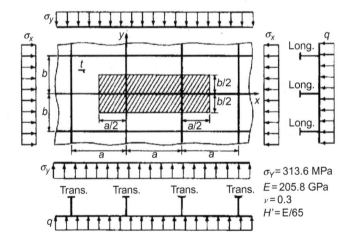

FIG. 5.21

Triple span double bay model of continuous plating subjected to combined longitudinal thrust and lateral pressure.

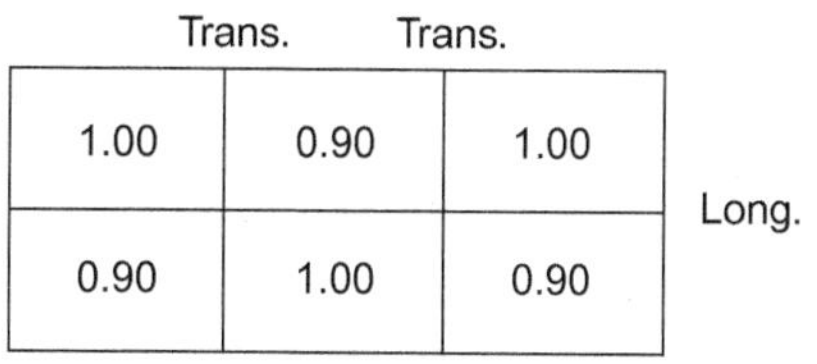

Trans.	Trans.		
1.00	0.90	1.00	Long.
0.90	1.00	0.90	

FIG. 5.22

Assumed distribution of the maximum magnitude of initial deflection in individual panels.

The aspect ratio of the local panel, a/b, is chosen as 3.0 and 5.0. The coefficients of initial deflection for these panels are given in Table 5.1. These coefficients give $w_{0\,max}/t = 1.0$. Considering the scatter of maximum magnitude of initial deflection in individual panels, the value of α in Eq. (5.1) is assumed as indicated in Fig. 5.22.

The assumed material properties are as follows:

Yield stress: $\sigma_Y = 313.6$ MPa
Young's modulus: $E = 205.8$ MPa
Poisson's ratio: $\nu = 0.3$
Strain hardening ratio: $H' = E/65$

A series of nonlinear FEM analyses is performed. The pressure load ranging 0 through 60 m water-head is firstly applied up to the individual specified value and then the longitudinal thrust is applied keeping the lateral pressure as its specified value. However, the upper limit of the water-head is set as 1.5 times the plastic collapse pressure of an all sides clamped rectangular plate under pure pressure load.

5.1.5.2 Collapse behavior and ultimate strength

The collapse mode and spread of yielding for the continuous plate of which aspect ratio and the thickness are $a/b = 3.0$ and $t = 12$ mm are shown in Fig. 5.23 for typical cases. When lateral pressure is relatively low ($h = 10$ m), the collapse mode is similar to that of an isolated local plate of which four sides are simply supported. In this case, the collapse mode is in three half-waves mode in the longitudinal direction although overall deflection due to lateral pressure is also superimposed as indicated in Fig. 5.23A.

When lateral pressure is intermediate ($h = 20$ m), collapse mode is of a superposition of an overall mode produced by lateral pressure and three half-waves buckling mode, although deflection of a buckling mode is relatively small. As a result, deflection is in the same direction in all panels. Yielding is limited near the end region of each panel. When lateral pressure is high, collapse is in an overall mode produced mainly by lateral pressure. In this case, deflection of a buckling mode is small and so-called roof mode is formed, which is a typical plastic collapse mode under lateral pressure. Yielding spreads in the end region and along the sides and the centerline in the longitudinal direction.

Varying the slenderness ratio of the local panel, a series of collapse analysis is performed, and the calculated ultimate compressive strength is plotted by solid lines with solid marks in Fig. 5.24. On the

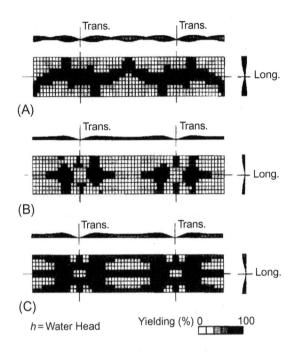

FIG. 5.23

Typical collapse modes of continuous plating subjected to combine longitudinal thrust and lateral pressure [8]. (A) $h = 10$ m (ultimate strength). (B) $h = 20$ m (ultimate strength). (C) $h = 30$ m (ultimate strength).

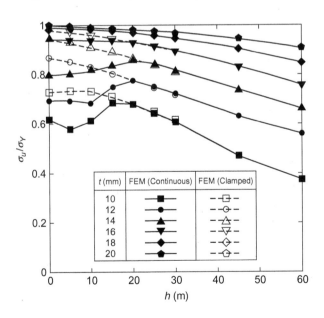

FIG. 5.24

Influence of lateral pressure on longitudinal ultimate strength of continuous plating subjected to combined longitudinal thrust and lateral pressure [8].

other hand, dashed lines with open marks in Fig. 5.24 represent the ultimate strength of an isolated local plate under all sides clamped condition.

As in the case of a plate subjected to combined transverse thrust and lateral pressure, the lateral pressure has two opposite influences on buckling/plastic collapse behavior of continuous plating. One is to increase the elastic buckling strength and the other is to reduce yielding strength. A little complicated ultimate strength curves in the lower pressure range is due to the opposite two factors, and the monotonously decreasing ultimate strength in the higher pressure range is due to earlier start of yielding by higher lateral pressure. It should be noticed that the ultimate strength of continuous plating agrees well with that of an isolated clamped plate in the higher pressure range.

5.1.5.3 *Estimation of ultimate compressive strength*
Ultimate strength under simply supported condition
A series of collapse analysis is performed by the FEM on continuous plating subjected to combined longitudinal thrust and lateral pressure. The aspect ratio, a/b, of the local panel is set as 3.0 and 5.0. The calculated ultimate strength is plotted against slenderness ratio of the local panel by solid marks in Fig. 5.25 when no lateral load is applied. On the other hand, the solid line in the same figure is the predicted ultimate strength by the following equations [8].

$$\frac{\sigma_u^s}{\sigma_Y} = \begin{cases} 1.0 & (\beta \leq 1.73) \\ 0.1 + \frac{1.571}{\beta} & (1.73 < \beta) \end{cases} \tag{5.29}$$

where β is the slenderness ratio of the local panel defined by Eq. (5.28). As is seen in Fig. 5.25, the predicted ultimate strength is in good agreement with the FEM results. Eq. (5.29) gives the compressive ultimate strength of a simply supported local panel subjected to longitudinal thrust.

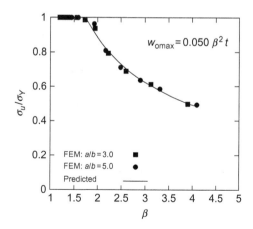

FIG. 5.25

Ultimate strength of continuous plating subjected to longitudinal thrust [8].

When continuous plating is subjected to combined longitudinal thrust and lateral pressure, slenderness ratio, β, in Eq. (5.29) could be replaced by reduced slenderness ratio, β_q, as in the case of continuous plating subjected to combined transverse thrust and lateral pressure explained in Section 5.1.4. Physical background of this method is that the slenderness ratio is equivalent to the ratio, $\sqrt{\sigma_Y/\sigma_{cr}^s}$, and σ_{cr}^s can be replaced by the buckling strength of continuous plating, σ_{cr}^q subjected to combined longitudinal thrust and lateral pressure, which is expressed as follows; see Eq. (5.6):

$$\sigma_{cr}^q = R_q \sigma_{cr}^s \tag{5.30}$$

where R_q is expressed from Eq. (5.6) as follows:

$$R_q = 1 + \frac{(qb^4/Et^4)^{1.6}}{576} \tag{5.31}$$

Thus, the reduced slenderness ratio can be expressed as follows:

$$\beta_q = \frac{b}{t}\sqrt{\frac{\sigma_Y}{R_q E}} = \frac{\beta}{\sqrt{R_q}} \tag{5.32}$$

If β_q by Eq. (5.32) is used instead of β in Eq. (5.29), it is considered that continuous plating with slenderness ratio of β subjected to lateral pressure, q, is equivalent to simply supported thicker plate of $t_q = t\sqrt{R_q}$ with slenderness ratio of β_q. Replacing β in Eq. (5.29) by β_q, the effect of increased elastic buckling strength on ultimate strength is considered.

Here, according to the theory of rigid-plastic large deflection analysis [7], the reduction in the ultimate strength due to lateral pressure may be represented by the following parameter:

$$\gamma_q = \frac{qb^2}{\sigma_Y t^2} = \frac{qb^4/Et^4}{\beta} \tag{5.33}$$

Physical meaning of γ_q is the ratio of bending moment produced by lateral pressure to the fully plastic bending moment of the plate per unit width. The reduction in the ultimate strength due to lateral pressure could be expressed in terms of γ_q with good accuracy. However, under the combined longitudinal thrust and lateral pressure, large scatter is observed in the predicted results. So, β_q is used instead of γ_q in order to get better agreement with the FEM results.

The main reason behind this is the fact that both buckling mode under transverse thrust and deflection mode produced by lateral pressure is of an overall mode under the combined transverse thrust and lateral pressure, whereas they are different each other under the combined longitudinal thrust and lateral pressure. On the other hand, when combined longitudinal thrust and lateral pressure act on continuous plating, the deflection produced by lateral pressure is of an overall mode, whereas that by buckling is of a several half-waves mode in the longitudinal direction. Such a difference in collapse mode cannot be accounted for by using only the parameter γ_q. This is the reason why reduction of the ultimate strength due to lateral pressure could be better expressed in terms of β_q, which depends on (qb^4/Et^4) as well as γ_q.

The ultimate compressive strength of continuous plating which collapses in a simply supported mode under combined longitudinal thrust and lateral pressure is finally expressed in the following form:

$$\sigma_{uq}^s = \sigma_u^s(1 + 0.0527\beta_q - 0.0223\beta_q^2) \tag{5.34}$$

The first term, σ_u^s, includes the effect of increased elastic buckling strength, and the second term that of reduced yielding strength. σ_u^s can be calculated by Eq. (5.29).

Ultimate strength under clamped condition

With an increase in the lateral pressure, collapse mode changes from a simply supported mode to a clamped mode as explained above. On the basis of the calculated results by the FEM analysis, the ultimate strength of a clamped rectangular plate subjected to longitudinal thrust can be expressed as follows:

$$\frac{\sigma_u^c}{\sigma_Y} = 0.8 + 0.277\beta - 0.096\beta^2 \quad (\leq 1.0) \tag{5.35}$$

On the other hand, when a clamped plate is subjected to combined longitudinal thrust and lateral pressure, elastic buckling strength is increased by lateral pressure since deflection modes by longitudinal thrust and lateral pressure loads are different each other. This results in the increase of the ultimate strength when the plate is thin as shown in Fig. 5.24 for the case of $t = 10$ mm in low lateral pressure range. This influence of lateral pressure on elastic buckling strength of a clamped plate can be estimated by replacing β in Eq. (5.29) by β_q in Eq. (5.32) but taking the following R_q instead of that by Eq. (5.31).

$$R_q = \begin{cases} 1.0 - 0.0041\alpha_q + 0.00227\alpha_q^2 & (\alpha_q \leq 10.0) \\ 0.727 + 0.0474\alpha_q - 0.0000845\alpha_q^2 & (10.0 < \alpha_q) \end{cases} \tag{5.36}$$

where

$$\alpha_q = \frac{qb^4}{Et^4} \tag{5.37}$$

R_q defined by Eq. (5.36) is an increasing ratio of elastic buckling strength of a clamped rectangular plate subjected to combined longitudinal thrust and lateral pressure, q.

Performing curve fitting through the results of FEM analysis, the ultimate compressive strength formula for continuous plating which collapses in a clamped mode can be represented as follows:

$$\sigma_{uq}^c = \sigma_u^c(1 - 0.014\alpha_q + 0.000156\alpha_q^2 - 0.0000007\alpha_q^3) \tag{5.38}$$

Similar to Eq. (5.34) for a simply supported plate, the first term of the above equation includes the effect of lateral pressure on elastic buckling strength of the plate, and the second term represents the reduction ratio of the ultimate strength due to earlier yielding caused by lateral pressure. The reduction of the ultimate strength due to lateral pressure in the case of a clamped plate is well represented by the parameter, α_q. It should be noted that these three parameters, α_q, β_q, and γ_q, are dependent each other.

Ultimate strength under combined longitudinal thrust and lateral pressure

The ultimate compressive strength of continuous plating subjected to combined longitudinal thrust and lateral pressure is finally estimated as

$$\sigma_{uxq} = \text{Min}[\sigma_{uq}^s, \sigma_{uq}^c] \tag{5.39}$$

By taking the lower value between σ_{uq}^s and σ_{uq}^c, the change of the collapse mode from a simply supported mode to a clamped mode can be automatically taken into account.

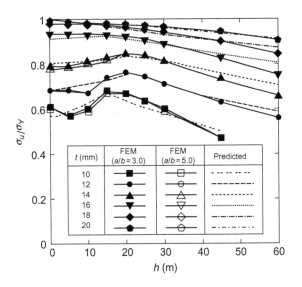

FIG. 5.26

Comparison of calculated and predicted ultimate strength of continuous plating subjected to combined longitudinal thrust and lateral pressure [8].

The predicted ultimate strength is plotted by thin lines with no mark in Fig. 5.26. Relatively good correlation is observed between calculated and predicted ultimate strength.

5.1.6 COLLAPSE BEHAVIOR UNDER COMBINED BI-AXIAL THRUST AND LATERAL PRESSURE

A series of elastoplastic large deflection analysis is performed for continuous plating subjected to combined bi-axial thrust and lateral pressure [8]. The pressure is firstly applied up to the specified level, and then forced displacements are exerted proportionally both in longitudinal and transverse directions. The ultimate strength interaction relationships are plotted in Fig. 5.27 for the plate of $a/b = 3.0$. The horizontal and the vertical axes represent σ_{ux}/σ_Y and σ_{uy}/σ_Y, respectively. Here, σ_{ux} and σ_{uy} are the average stresses in longitudinal and transverse direction, respectively, at the ultimate strength state, where the resultant stress, $\sqrt{\sigma_x^2 + \sigma_y^2}$, attains its maximum value.

Normalizing σ_{ux} and σ_{uy} by the ultimate strength, σ_{uxq} under combined longitudinal thrust and lateral pressure and σ_{uyq} under combined transverse thrust and lateral pressure, respectively, a new configuration of interaction relationships are obtained in terms of R_x and R_y as illustrated in Fig. 5.28, where

$$R_x = \frac{\sigma_{ux}}{\sigma_{uxq}}, \quad R_y = \frac{\sigma_{uy}}{\sigma_{uyq}} \tag{5.40}$$

Fig. 5.28 indicates that each interaction relationship consists of approximately two linear lines; a line inclined from horizontal along which the continuous plating behaves as if it were under combined

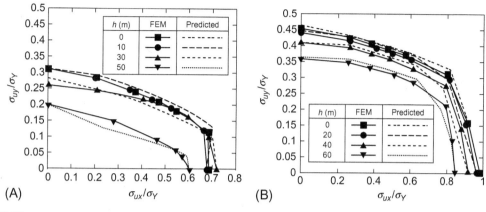

FIG. 5.27

Ultimate strength interaction relationships of continuous plating subjected to combined bi-axial thrust and lateral pressure ($a/b = 3.0$) [8]. (A) Thin panel ($t = 12$ mm). (B) Thick panel ($t = 18$ mm).

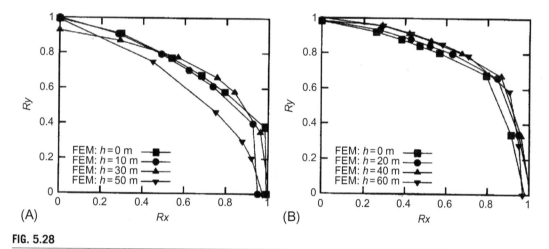

FIG. 5.28

Normalized ultimate strength interaction relationships of continuous plating subjected to combined bi-axial thrust and lateral pressure ($a/b = 3.0$) [8]. (A) Thin panel ($t = 12$ mm). (B) Thick panel ($t = 18$ mm).

transverse thrust and lateral pressure, and a line near to vertical along which it behaves as in the case under combined longitudinal thrust and lateral pressure.

On the basis of the calculated results by the FEM, the interaction relationship of the ultimate compressive strength in terms of R_x and R_y for continuous plating subjected to combined bi-axial thrust and lateral pressure can be expressed as follows:

$$R_y = f(R_x) + R_Q \tag{5.41}$$

where

$$f(R_x) = \begin{cases} \text{MIN}[1 - 0.199R_x - 0.407R_x^2, \ 26.405 - 26.405R_x] & 1.837 \le \beta \\ \text{MIN}[1 - 0.193R_x - 0.194R_x^2, \ 3.946 - 3.946R_x] & \beta < 1.837 \end{cases} \tag{5.42}$$

$$R_Q = \gamma_1 \gamma_q + \gamma_2 \gamma_q^2 \tag{5.43}$$

where

$$\gamma_1 = \begin{cases} -0.292 + 0.259\beta - 0.051\beta^2 & 1.837 \le \beta \\ 3.281 - 3.474\beta + 0.922\beta^2 & \beta < 1.837 \end{cases} \tag{5.44}$$

$$\gamma_2 = \begin{cases} 0.103 - 0.091\beta + 0.0183\beta^2 & 1.837 \le \beta \\ -1.453 + 1.630\beta - 0.458\beta^2 & \beta < 1.837 \end{cases} \tag{5.45}$$

R_Q is a parameter which reflects the influence of lateral pressure. As is known from Fig. 5.28, the interaction relationships for thin plates ($t = 12$ mm) are a little different from those for thick plates ($t = 18$ mm). This is the reason why two relationships are defined for $f(R_x)$ in Eq. (5.42) depending on the slenderness ratio, β. Comparison between the calculated and the predicted ultimate strength is performed in Fig. 5.29 as well as in Fig. 5.28. Good correlations are observed between the calculated and the predicted ultimate strength.

5.2 PLATES UNDER COMBINED UNI-AXIAL THRUST AND BENDING
5.2.1 LOADING CONDITIONS

When a ship's hull girder is subjected to longitudinal bending, continuous plating in side shell plating partitioned by longitudinal stiffeners is subjected to combined uni-axial thrust and bending. For hull

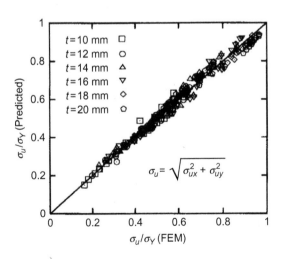

FIG. 5.29

Comparison of calculated and predicted ultimate strength of continuous plating subjected to combined bi-axial thrust and lateral pressure [8].

girders constructed in a longitudinal stiffening system, this combined load can be approximated by uni-axial thrust since bending stress is not dominant compared to compressive stress (average stress) except near the neutral axis of hull girder bending.

On the other hand, in the case of a middle height part of side shell plating of bulk carriers or wood-chip carriers with transversely stiffened side shell plating, the influence of bending cannot be ignored when the buckling strength is considered. Especially in the case of wood-chip carriers, their side shell plating is stiffened by vertical frames up to the upper deck. This side shell plating has two or three side stringers in its upper region, and is partitioned by these stringers and transverse frames. A partitioned local panel of the side shell plating is a wide rectangular plate, and is subjected to combined thrust and in-plane bending loads acting on its longer sides when a hull girder is subjected to sagging bending moment.

A wood-chip carriers has no top side tank but has double bottom and hopper side tanks. So, the neutral axis of the cross-section lies at a lower level of the cross-section. Because of this, under the sagging condition, high compressive stress is produced at the deck and the upper part of side shell plating. In this case, the buckling/plastic collapse behavior and strength of the local panel at the upper part of side shell plating have to be assessed under the action of the combined in-plane bending and thrust loads.

Series of elastic/elastoplastic large deflection FEM analyses are performed changing the aspect ratio of the panel and the ratio of thrust to in-plane bending loads. On the basis of the calculated results, characteristics of the collapse behavior and the ultimate strength as well as elastic buckling behavior are explained [9].

5.2.2 WIDE RECTANGULAR PLATES FOR ANALYSIS

The length of the plate is fixed as $a = 800$ mm, and the breadth, b, is varied as 1600, 2800, 4000, and 5200 mm. The thickness is varied as 12, 16, 20, and 24 mm. The material is assumed to be elastic-perfectly plastic with the yield stress of 313.6 MPa.

The plate is assumed to be simply supported along four sides. Initial deflection of one half-wave mode is given of which magnitude is 0.01 times the thickness. Influence of welding residual stress is not considered.

5.2.3 METHOD TO APPLY COMBINED THRUST AND BENDING LOADS

A series of elastic and elastoplastic ultra-large deflection analyses is performed by FEM. The combined loads are applied on its wider sides by forced displacement and rotation [10]. Considering the symmetry condition, only the right-hand side half of the rectangular plate is analyzed. On the loading side, fictitious loading point is provided at $y = y_0$ as illustrated in Fig. 5.30. Here, displacement in the horizontal direction and the rotation angle of this point are denoted as u_0 and θ_{z0}, respectively. Then, the displacements at nodal point, i, can be expressed through coordinate translation matrix as

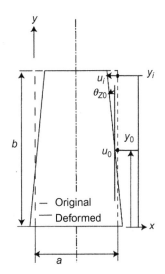

FIG. 5.30

Loading condition for combined thrust and bending.

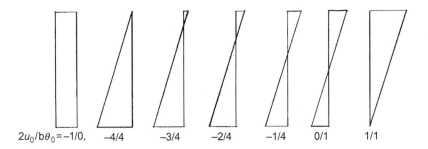

$$2u_0/b\theta_0 = -1/0, \quad -4/4 \quad -3/4 \quad -2/4 \quad -1/4 \quad 0/1 \quad 1/1$$

FIG. 5.31

Applied strain distribution for combined bending and thrust loading.

$$
\begin{Bmatrix} u_i \\ v_i \\ w_i \\ \theta_{xi} \\ \theta_{yi} \\ \theta_{zi} \end{Bmatrix}
=
\begin{bmatrix}
1 & 0 & 0 & 0 & 0 & y_0 - y_i \\
0 & 1 & 0 & 0 & 0 & 0 \\
0 & 0 & 1 & 0 & 0 & 0 \\
0 & 0 & 0 & 1 & 0 & 0 \\
0 & 0 & 0 & 0 & 1 & 0 \\
0 & 0 & 0 & 0 & 0 & 1
\end{bmatrix}
\begin{Bmatrix} u_0 \\ v_i \\ w_i \\ \theta_{xi} \\ \theta_{yi} \\ \theta_{z0} \end{Bmatrix}
\tag{5.46}
$$

Setting u_0 as free, pure bending load can be applied, whereas pure compression load can be applied by setting θ_{z0} as zero. For the combined loading, $2u_0/b\theta_{z0}$ is varied as $-1/0$, $-4/4$, $-3/4$, $-2/4$, $-1/4$, $0/1$, and $1/1$. The applied strain distributions corresponding to individual ratios of $2u_0/b\theta_{z0}$ are illustrated in Fig. 5.31.

5.2.4 **COLLAPSE BEHAVIOR UNDER PURE BENDING**

Fig. 5.32 shows the moment-curvature relationships of rectangular plates of which aspect ratio is 0.2 (a = 800 mm, b = 4000 mm). Two cases are shown of which thickness, t, is taken as 12 and 20 mm, respectively. The bending moment in Fig. 5.32 is divided by fully plastic bending moment and the curvature by that corresponding to initial yielding. The slope of the moment-curvature curve represents a flexural stiffness of the plate against in-plane bending. In the case of thin plate (t = 12 mm), the plate undergoes elastic buckling at the point where flexural rigidity rapidly changes. The thick plate also undergoes elastic buckling, but yielding by out-of-plane bending takes place soon after buckling.

In both cases, initial yielding takes place at points indicated by an open circle. The spreads of yielding and deflection modes at the ultimate strength and at the last incremental step are illustrated in Figs. 5.33 and 5.34 for cases of thin plate (t = 12 mm) and thick plate (t = 20 mm), respectively. It is seen that yielded region spreads only in the compression side of bending at the ultimate strength. Beyond the ultimate strength, yielded region spreads also in the tension side of bending.

In the case of a thick plate (t = 20 mm), the ultimate strength is attained soon after the initial yielding with small lateral deflection. Beyond the ultimate strength, capacity decreases with the increase in lateral deflection. Plastic hinge line can be seen along the centerline of the plate. However, plastic mechanism is not completely but formed starting from the compression side of bending in both cases.

Fig. 5.35 shows how the stress distribution changes during progressive collapse in the postultimate strength range. The top, the middle and the bottom are at the initial yielding, at the ultimate strength, and at the last incremental step. In the case of a thin plate (t = 12 mm) shown in Fig. 5.35A, relatively

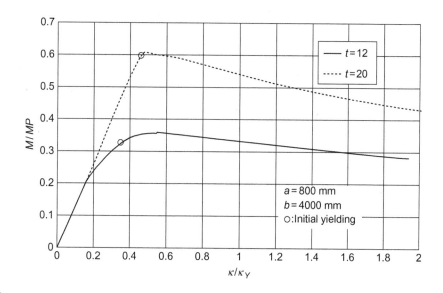

FIG. 5.32

Moment-curvature relationships for wide rectangular plates in pure bending [9].

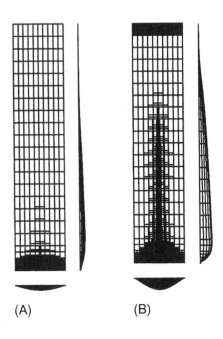

FIG. 5.33

Collapse mode of thin plate (t = 12 mm) [9]. (A) Ultimate strength. (B) Last step.

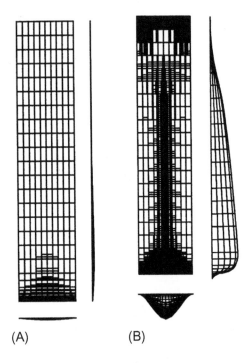

FIG. 5.34

Collapse mode of thick plate (t = 20 mm) [9]. (A) Ultimate strength. (B) Last step.

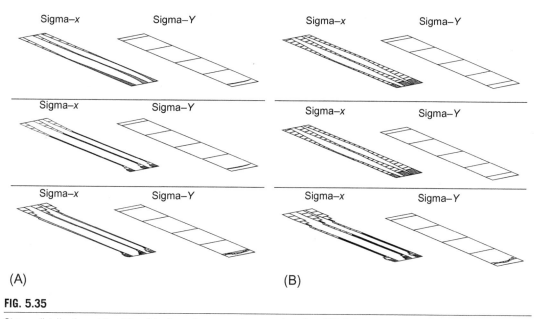

FIG. 5.35

Stress distribution at buckling/plastic collapse under pure bending [9]. (A) Thin plate (t = 12 mm). (B) Thick plate (t = 20 mm).

large deflection due to buckling is produced when initial yielding takes place. This is the reason why the distribution of σ_x is not linear and the compressive stress is very low in the wide region of the compression side of bending. At the same time, large deflection produces σ_y which is in self-equilibrium in a cross-section. At the last incremental step, distribution of σ_x near the compressed side is not uniform according to the magnitude and the sign of σ_y under the yielding condition:

$$\sigma_x^2 - \sigma_x\sigma_y + \sigma_y^2 + 3\tau_{xy}^2 - \sigma_Y^2 = 0 \tag{5.47}$$

In the case of a thick plate (t = 20 mm) shown in Fig. 5.35B, stress distribution is almost linear when yielding starts and at the ultimate strength. This is because buckling deflection has not yet developed at the ultimate strength as well as at the initial yielding. At the last incremental step, edge stress in a tension side is uniform and is equal to the yielding stress of the material. In the region adjacent to this tension side, some compressive stress is produced in y-direction as is shown in Fig. 5.35.

5.2.5 COLLAPSE BEHAVIOR UNDER PURE THRUST

Fig. 5.36 shows average stress-average strain relationships of the same plates in Section 5.2.4 under pure thrust. The buckling takes place in an elastic range in the case of a wide plate subjected to uni-axial thrust in the direction of the shorter side. As explained in Chapter 4, the buckling mode is of a half-wave sinusoidal mode, although deflection mode changes to a cylindrical mode with the increase in applied compressive load as indicated in Figs. 5.37 and 5.38.

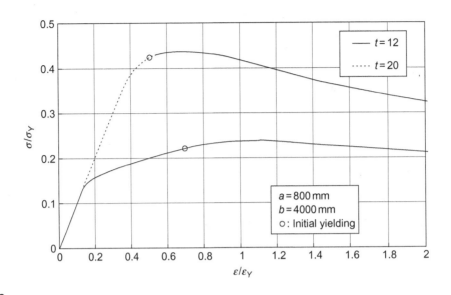

FIG. 5.36

Average stress-average strain relationships for wide rectangular plates in pure compression [9].

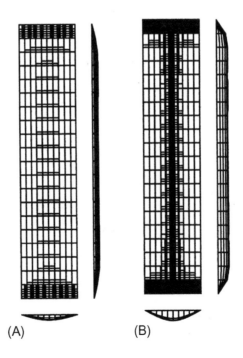

FIG. 5.37

Collapse mode of thin plate ($t = 10$ mm) [9]. (A) Ultimate strength. (B) Last step.

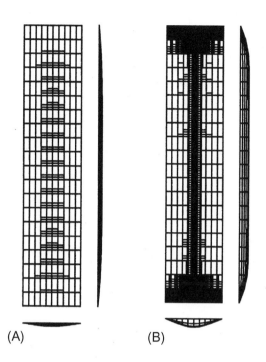

FIG. 5.38

Collapse mode of thick plate (t = 20 mm) [9]. (A) Ultimate strength. (B) Last step.

In the case of a thicker plate, yielding starts soon after buckling and the ultimate strength is attained. On the other hand, yielding takes place after large deflection of a cylindrical mode has been developed in the case of a thinner plate.

The spreads of yielding at the ultimate strength and at the last incremental step are also indicated in Figs. 5.37 and 5.38, for thin plate (t = 12 mm) and thick plate (t = 20 mm), respectively, together with deflection modes. In the case of a thin plate, ultimate strength is attained by yielding at shorter sides due to large membrane stress whereas it is attained by yielding along the central line parallel with the longer sides due to bending in the case of thick plate. Compared to the case of pure bending, plastic mechanism is clearly formed at the last incremental step regardless of the plate thickness.

5.2.6 COLLAPSE BEHAVIOR UNDER COMBINED THRUST AND BENDING LOADS

Average stress-average strain relationships for wide rectangular plates subjected to combined uni-axial thrust and bending loads are plotted in Fig. 5.39A and B, which are for thin plate (t = 12 mm) and thick plate (t = 20 mm), respectively. On the other hand, bending moment-curvature relationships for the same cases are plotted in Fig. 5.40A and B for thin and thick plates.

The relationships between bending moment and average stress, that is loading paths under forced displacement and rotation angle, for the same cases are also illustrated in Fig. 5.41A and B.

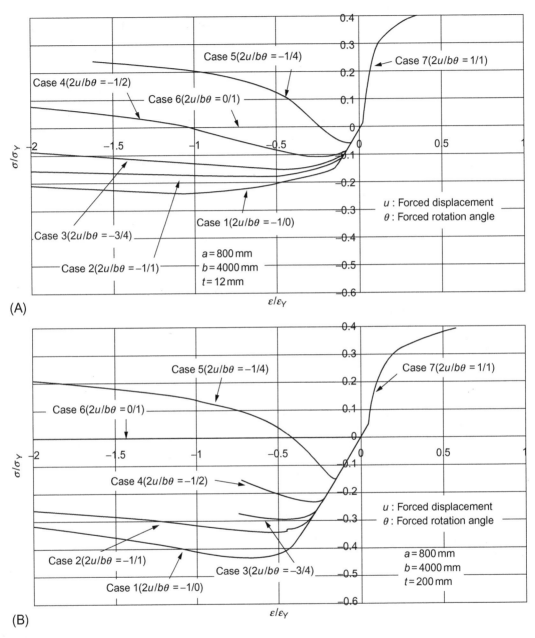

FIG. 5.39

Average stress-average strain relationships for wide rectangular plates under combined thrust and bending [9]. (A) Thin plate (t = 12 mm). (B) Thick plate (t = 20 mm).

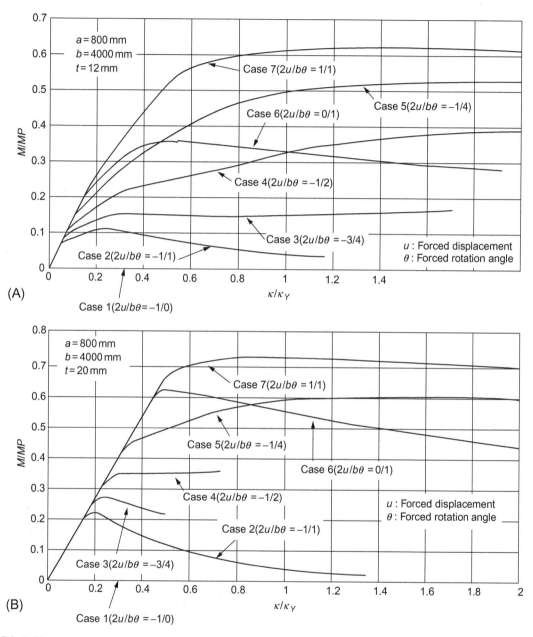

FIG. 5.40

Bending moment-curvature relationships for wide rectangular plates under combined thrust and bending [9]. (A) Thin plate ($t = 12$ mm). (B) Thick plate ($t = 20$ mm).

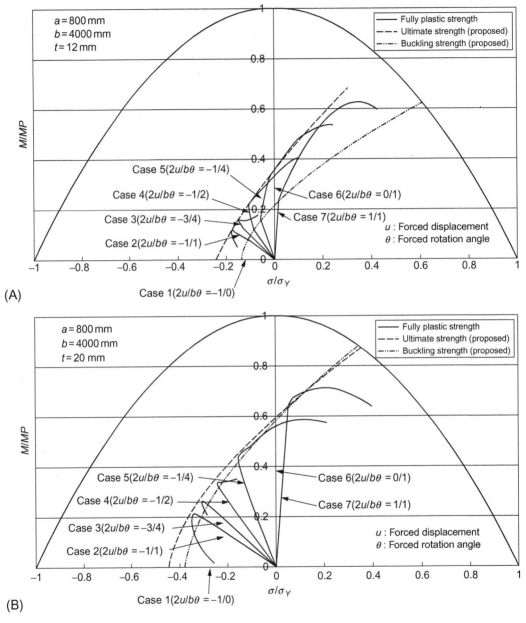

FIG. 5.41

Loading paths under combined thrust and bending [9]. (A) Thin plate ($t = 12$ mm). (B) Thick plate ($t = 20$ mm).

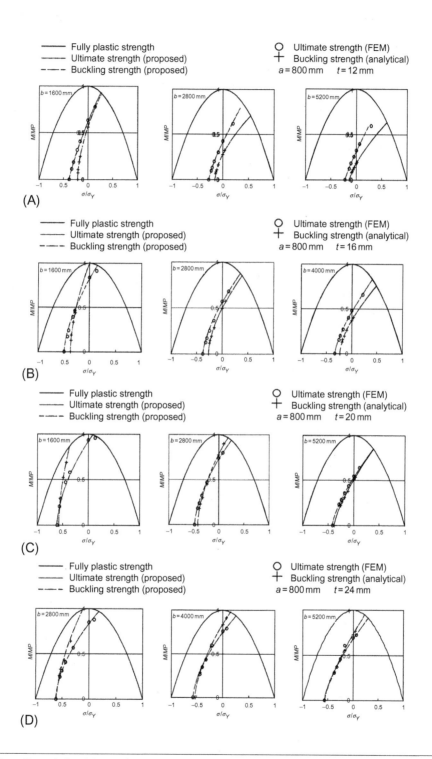

FIG. 5.42

Strength interaction relationships under combined thrust and bending [9]. (A) $t = 12$ mm. (B) $t = 16$ mm. (C) $t = 20$ mm. (D) $t = 24$ mm.

In Figs. 5.40 and 5.41, the kinked points on the straight line from the coordinate origin correspond to the occurrence of buckling. It is seen that the compressive buckling and ultimate strength decreases with the increase in bending moment and vice versa except the case of pure bending. Regarding the capacity against thrust, it decreases once its maximum value is attained as indicated in Figs. 5.40 and 5.41. On the other hand, bending capacity decreases with the decrease in compression capacity beyond the ultimate strength only when thrust load is dominant under combined bending and thrust, say Case 2 ($2u_0/b\theta_{z0} = -1/1$) for thin plate and Cases 2 ($2u_0/b\theta_{z0} = -1/1$) and 3 ($2u_0/b\theta_{z0} = -3/4$) for thick plate. This may be because large deflection is produced over the breadth direction under thrust dominant loading whereas one side becomes yielded in tension and deflection in the tension side of bending is reduced with increasing forced rotation under in-plane bending dominant loading. In the case of ordinary wood-chip carrier, the ratio, $2u/b\theta_{z0}$, is roughly between -3 and -4 at the upper region of the side shell plating. This implies that the loading path in this case locates near the horizontal axis in the loading path diagram shown in Fig. 5.42. It is considered that both capacities against thrust and bending loads decrease beyond the ultimate strength.

The broken and chain lines in Fig. 5.42 are the interaction curves for elastic buckling strength and ultimate strength, respectively, which are calculated by the formula explained in the following section.

5.2.7 APPROXIMATE FORMULAS TO EVALUATE BUCKLING/ULTIMATE STRENGTH OF RECTANGULAR PLATES SUBJECTED TO COMBINED THRUST AND BENDING LOADS

On the basis of the calculated results, strength interaction relationships in terms of thrust (average stress) and bending moment can be formulated as follows:

$$(a_1 - x)^{a_2} + y^{a_3} = a_4 \tag{5.48}$$

For the buckling strength interaction relationships, parameters in Eq. (5.48) are given as follows:

$$\begin{cases} x = \sigma/\sigma_{cr} \\ y = M/M_{cr} \end{cases} \tag{5.49}$$

$$\begin{cases} a_1 = 0.76 \\ a_2 = 1.0 \\ a_3 = 0.08 \times \log(\alpha - 0.15) + 1.8 \\ a_4 = 1.77 \end{cases} \tag{5.50}$$

where $\alpha = a/b$ is the aspect ratio of the plate. σ_{cr} and M_{cr} are buckling strength under pure thrust load and bending load, respectively. σ is positive when compressive load acts and is negative under tensile load.

On the other hand, when ultimate strength interaction relationships are considered, the parameters in Eq. (5.48) are as follows:

$$\begin{cases} x = \sigma/\sigma_Y \\ y = M/M_Y \end{cases} \tag{5.51}$$

$$\begin{cases} a_1 = 1.36 \\ a_2 = 0.98 \\ a_3 = a_5 \times \beta/\alpha + a_6 \\ a_4 = 2.32 \end{cases} \tag{5.52}$$

where

$$\beta = \frac{b}{t} \cdot \sqrt{\frac{\sigma_Y}{E}} \tag{5.53}$$

$$\begin{cases} a_5 = -0.125 \times \log \frac{\alpha - 0.1}{1000} - 1.44 \\ a_6 = 0.35 \times \log \frac{\alpha - 0.1}{1000} + 5.2 \end{cases} \tag{5.54}$$

The interaction curves are calculated applying Eqs. (5.48) through (5.54) and are illustrated in Fig. 5.42 together with the FEM results. The solid lines represent the fully plastic strength interaction relationship. On the other hand, the chain lines and dashed lines are the interaction relationships predicted by the proposed formulas. The predicted interaction relationships show good correlations with the FEM results.

5.3 PLATES UNDER COMBINED UNI-AXIAL THRUST AND SHEAR LOADS

5.3.1 MODEL FOR ANALYSIS

To examine the ultimate strength of a rectangular plate subjected to combined thrust and shear loads, the plate shown in Fig. 5.43 is analyzed [11]. The four sides are assumed to be kept straight under combine loading. The breadth of the plate, b, is fixed as 1000 mm and the thickness, t, and the length, a, are varied. Elastic-perfectly plastic material is assumed with yield stress, σ_Y, of 235.2 MPa.

Initial deflection of a thin-horse mode is assumed in the following form:

$$w_0 = \sum_m A_{0m} \sin \frac{m\pi x}{a} \sin \frac{\pi y}{b} \tag{5.55}$$

The maximum magnitude of initial deflection is set as

$$w_{0\,max} = 0.1 \left(\frac{b}{t} \sqrt{\frac{\sigma_Y}{E}} \right)^2 t \tag{5.56}$$

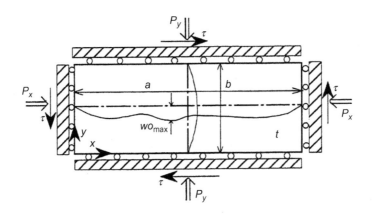

FIG. 5.43

Model for collapse analysis under combined thrust and shear loads [11].

5.3.2 ULTIMATE STRENGTH UNDER PURE SHEAR

In Fig. 5.44, ultimate shear strength obtained by the FEM analysis is plotted by open marks against $\sqrt{\tau_Y/\tau_{cr}^e}$ ratio of the plate, where τ_{cr}^e is elastic shear buckling stress. Results for three cases are plotted, which are a/b being 1.0, 3.0, and 5.0. The solid line is the buckling strength with Johnson's plasticity correction.

The ultimate strength under pure shear load is relatively high when it is compared to the compressive ultimate strength. This is because of the formation of tension field after buckling under the condition that all sides are kept straight. It is seen in Fig. 5.44 that the plate with higher aspect ratio shows higher ultimate strength. This is partly because initial deflection of a thin-horse mode of which maximum magnitude is 10% of the plate thickness resists against development of the deflection of a shear buckling mode. Another cause is that plate with higher aspect ratio shows lower elastic buckling strength which results in higher $\sqrt{\tau_Y/\tau_{cr}^e}$ ratio.

In the case of a square plate ($a/b = 1.0$), the formation of deflection of a buckling mode is clearly observed above the shear buckling strength, whereas almost no deflection of a buckling mode is observed when the aspect ratio, a/b is 3.0 and 5.0.

5.3.3 ULTIMATE STRENGTH UNDER COMBINED THRUST AND SHEAR

The ultimate strength of a rectangular plate of which aspect ratio, a/b, is 3.0 is calculated under combined transverse thrust and shear loads as well as combined longitudinal thrust and shear loads. For this analysis, three cases are considered, which are

with no shear load

with shear load of $\tau = 0.25\,\tau_Y$

with shear load of $\tau = 0.5\,\tau_Y$

Calculated ultimate strength is plotted by open marks in Figs. 5.45 and 5.46 for the cases of combined longitudinal thrust and shear loads and combined transverse thrust and shear loads,

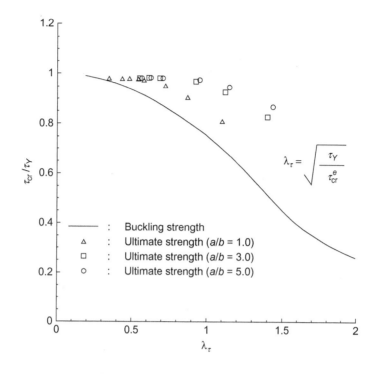

FIG. 5.44

Ultimate strength under pure sher load [11].

respectively. The solid, dashed, and dotted lines in these figures are the buckling strength calculated by the buckling interaction equation given in the Class NK rule [12] as follows:

$$\frac{\sigma_x}{\sigma_{xcr}^e} + \frac{\sigma_y}{\sigma_{ycr}^e} + \left(\frac{\tau}{\tau_{cr}^e}\right)^2 = 1 \tag{5.57}$$

where σ_{xcr}^e, σ_{ycr}^e, and τ_{cr}^e are elastic buckling strength of the plate subjected to σ_x alone, σ_y alone, and τ alone, respectively.

In Fig. 5.45, buckling strength and ultimate strength are plotted against the slenderness ratio, λ_e, for different intensity of applied shear stress. For the buckling strength, Johnson's plastic correction is performed in the following form:

$$\sigma_{cr} = \sigma_Y \left(1 - \frac{\sigma_Y}{4\sigma_{cr}^e}\right) \tag{5.58}$$

where σ_{cr}^e and σ_Y are elastic buckling stress and yield stress, respectively. It is seen that buckling strength curve intersects the horizontal axis. In the range of slenderness ratio above the intersecting point, the plate buckles by the specified shear stress alone.

It is seen in Fig. 5.46 that the ultimate strength is below the buckling strength in the range of slenderness ratio, λ_e, less than $\sqrt{2}$ when applied shear stress, τ, is 0.0 and 0.25 times τ_Y. On the other hand, the ultimate strength is almost the same with buckling strength when τ is 0.5 times τ_Y, although

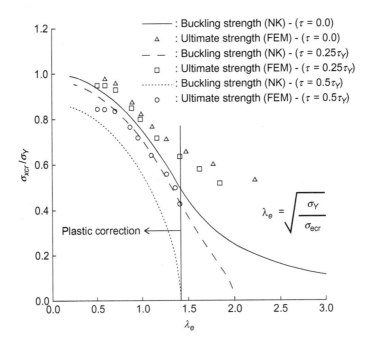

FIG. 5.45

Influence of shear stress on compressive ultimate strength of rectangular plate subjected to combined longitudinal thrust and shear loads ($a/b = 3.0$) [11].

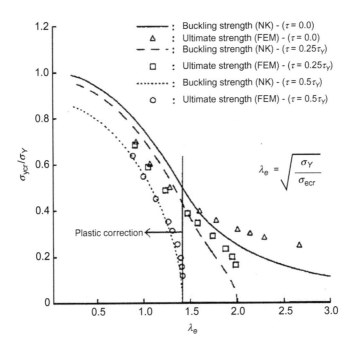

FIG. 5.46

Influence of shear stress on compressive ultimate strength of rectangular plate subjected to combined transverse thrust and shear loads ($a/b = 3.0$) [11].

this has no physical meaning. However, it can be said that yielding starts as the average transverse stress approaches to the buckling strength in the wide range along the centerline in the longitudinal direction, and the ultimate strength is soon attained; see Fig. 6.45. With the increase in the applied shear stress, further reduction is observed in the yielding strength and so the ultimate strength.

EXERCISES

5.1 Explain why lateral pressure increase the buckling strength of a rectangular plate subjected to longitudinal/transverse thrust.

5.2 Explain the influence of lateral pressure on the ultimate strength of a rectangular plate subjected to combined longitudinal/transverse thrust.

5.3 Explain the fundamental idea of how to estimate the ultimate strength of a rectangular plate subjected to combined transverse thrust.

5.4 Explain how the influence of lateral pressure on elastic buckling strength of a rectangular plate can be introduced when a simplified method is developed to estimate the ultimate strength of a rectangular plate subjected to combined longitudinal/transverse thrust.

5.5 Explain the difference in buckling behavior of a rectangular plate subjected to longitudinal/transverse thrust when the plate is accompanied with initial deflection of a buckling mode and of a thin-horse mode, together with the reason.

5.6 Explain how the loading sequence affects the buckling collapse behavior when combined loads act.

5.7 Explain how the loading sequence affects the buckling/plastic collapse behavior when combined loads act.

5.4 APPENDIX: ULTIMATE STRENGTH OF A STRIP SUBJECTED TO AXIAL THRUST

The ultimate strength, σ_{um}^s, of the middle part of a simply supported rectangular plate under transverse thrust can be evaluated modeling this part as a both-ends simply supported strip of length b with a unit width under thrust.

The elastic thrust-deflection relationship of a strip having the maximum initial deflection of w_0 is given by [13]

$$w = \frac{w_0}{1 - P/P_{cr}} \tag{5.59}$$

where w is the maximum deflection at the mid-span, and $P_{cr} = P_{cr}^s(= \pi^2 E t^3/12b^2)$ is the flexural buckling strength of a both-ends simply supported strip with a unit width. t is the plate thickness and E is Young's modulus of the material.

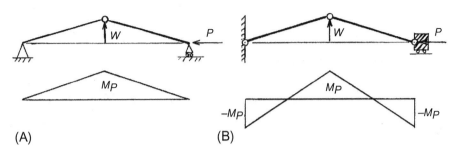

FIG. 5.47

Plastic mechanism of a strip under axial compression. (A) Simply supported strip. (B) Clamped strip.

On the other hand, the fully plastic interaction relationship of the cross-section of a strip can be expressed as follows:

$$\frac{M}{M_P} + \left(\frac{P}{P_P}\right)^2 = 1 \tag{5.60}$$

where $M_P = \sigma_Y t^2/4$ and $P_P = \sigma_Y t$ are the fully plastic bending moment and the fully plastic axial force, respectively.

Here, bending moment at mid-span point of a both-ends simply supported strip (see Fig. 5.47A) can be expressed as

$$M = wP \tag{5.61}$$

The relationship between thrust, P, and deflection, w, of a strip having a plastic hinge at mid-span point can be derived by substituting Eq. (5.61) into Eq. (5.60) as follows:

$$\frac{wP}{M_P} + \left(\frac{P}{P_P}\right)^2 = 1 \tag{5.62}$$

The ultimate strength is determined as the axial stress at the intersection of the elastic thrust-deflection curve, Eq. (5.59), and the plastic thrust-deflection curve, Eq. (5.62). Assuming the maximum initial deflection of $w_0 = \eta\beta^2 t$ (see Eq. 5.3), the ultimate strength of the both-ends simply supported strip, σ_{um}^s, is determined by solving the following equation:

$$\beta^2 = \frac{1 - \bar{\sigma}^2}{4\eta\bar{\sigma} + (12/\pi^2)\bar{\sigma}(1 - \bar{\sigma}^2)} \tag{5.63}$$

where $\bar{\sigma} = \sigma_{um}^s/\sigma_Y$. Performing the curve fitting through the calculated results of $\bar{\sigma}$-w relationship, Eq. (5.13) can be derived.

On the other hand, the ultimate strength, σ_{um}^c, of the middle part of an all-sides clamped rectangular plate under transverse thrust is evaluated considering a both-ends clamped strip of length b and a unit width. The elastic thrust-deflection relationship is represented by Eq. (5.59) but using the flexural buckling strength, $P_{cr}^c = \pi^2 E t^3/3b^2$ of the both-ends clamped strip instead of P_{cr}^s for P_{cr}.

When both ends are clamped (see Fig. 5.47B), bending moment at mid-span point of the strip can be expressed as

$$M = \frac{1}{2}wP \tag{5.64}$$

The relationship between thrust, P, and deflection, w, of a strip having a plastic hinge at mid-span point and both ends can be derived by substituting Eq. (5.64) into Eq. (5.60) as follows:

$$\frac{wP}{2M_P} + \left(\frac{P}{P_P}\right)^2 = 1 \tag{5.65}$$

The ultimate strength is determined as the axial stress at the intersection of the elastic thrust-deflection curve, Eq. (5.59), and the plastic thrust-deflection curve, Eq. (5.65). Assuming the maximum initial deflection of $w_0 = \eta\beta^2 t$ (see Eq. 5.3), the ultimate strength of the both-ends clamped strip, σ_{um}^c, is determined by solving the following equation:

$$\beta^2 = \frac{1 - \overline{\sigma}^2}{2\eta\overline{\sigma} + (3/\pi^2)\overline{\sigma}(1 - \overline{\sigma}^2)} \tag{5.66}$$

where $\overline{\sigma} = \sigma_{um}^c/\sigma_Y$. Performing the curve fitting through the calculated results of $\overline{\sigma}$-w relationship, Eq. (5.23) can be derived.

REFERENCES

[1] Yao T, Fujikubo M, Varghese B, Yamamura K, Niho O. Buckling/plastic collapse strength of wide rectangular plate under combined pressure and thrust. J Soc Naval Arch Jpn 1997;182:561–570.
[2] Fujikubo M, Yao T, Khedmati M, Harada M, Yanagihara D. Estimation of ultimate strength of continuous stiffened panel under combined transverse thrust and lateral pressure part 1: continuous plate. Mar Struct 2005;18:383–410.
[3] Fujikubo M, Yao T, Khedmati M. Estimation of ultimate strength of ship bottom plating under combined transverse thrust and lateral pressure. J Soc Naval Arch Jpn 1999;186:621–630.
[4] Fujikubo M, Yao T, Varghese B, Zha Y, Yamamura K. Elastic local buckling strength of stiffened plates considering plate/stiffener interaction and lateral pressure. In: Proc. 8th international offshore and polar engineering conference, Montreal, Canada; 1998. p. 292–299.
[5] Yao T, Niho O, Fujikubo M, Vargese B, Mixutani K. Buckling/plastic collapse strength of ship bottom plating. J Soc Naval Arch Jpn 1997;181:309–321 [in Japanese].
[6] Faulkner D. A review of effective plating for use in the analysis of stiffened plating in bending and compression. J Ship Res 1975;19:1–17.
[7] Okada H, Oshima K, Fukumoto Y. Compressive strength of long rectangular plates under hydrostatic pressure. J Soc Naval Arch Jpn 1979;146:270–280 [in Japanese].
[8] Khedmati M. Ultimate strength of ship structural members and systems considering local pressure effects [Doctoral thesis]. Hiroshima: Hiroshima University; 2000.
[9] Miyachi S, Iijima K, Yao T. Collapse behaviour and strength of wide rectangular plate under combined bending and thrust loads. In: Proc. 20th technical exchange and advisory meeting on marine structures, Seoul, Korea; 2006. p. 319–326.

[10] Fujikubo M, Pei Z. Progressive collapse analysis of ship's hull girder in longitudinal bending using idealized structural unit method. J Jpn Soc Naval Arch Ocean Eng 2005;1:187–196 [in Japanese].

[11] Fujikubo M, Yao T, Varghese B. Buckling and ultimate strength of plates subjected to combined loads. In: Proc. 7th int. offshore and polar eng. conf., Honolulu, USA; 1997. p. 380–387.

[12] Class N. Rules and guidance for the survey and construction of steel ships; 1993.

[13] Timoshenko S, Gere J. Theory of elastic stability. McGraw-Hill Kogakusha, Ltd; 1961.

BUCKLING/PLASTIC COLLAPSE BEHAVIOR AND STRENGTH OF STIFFENED PLATES

6.1 BUCKLING COLLAPSE BEHAVIOR AND STRENGTH OF STIFFENED PLATES

Ship structure is composed of thin plates on which a number of stiffeners are provided to increase their strength and stiffness. It is well known that the buckling strength of a stiffened plate increases with the increase in the flexural stiffness of the stiffener, but reaches its maximum limiting value when the flexural stiffness of the stiffener exceeds a certain value. The ratio of the flexural stiffness of a stiffener to that of a plate at this state is defined as the minimum stiffness ratio, $\gamma_{\min}^B$ [1,2].

As the load increases, a stiffened plate undergoes overall buckling when the stiffness ratio, γ, is lower than $\gamma_{\min}^B$, whereas it undergoes local plate buckling when γ is higher than $\gamma_{\min}^B$. This is shown in Fig. 6.1, which is the case of the simplest stiffened plate with one stiffener.

The length, the breadth, and the thickness of the plate are 1400, 600, and 4 mm, respectively. The angle-bar stiffener is provided along the centerline of the plate as indicated in Fig. 6.1 of which flange width is 30 mm, and its thickness of the web and the flange is 4 mm. A series of eigenvalue analysis and experiments were carried out varying the height of the stiffener [3]. The eigenvalue analysis was performed applying the finite strip method (FSM).

When the stiffener height is low, buckling of an overall mode (Mode 1 in Fig. 6.1) takes place, and the buckling strength increases with the increase in the stiffener height.

Above a certain size of the stiffener, which gives minimum stiffness ratio, $\gamma_{\min}^B$, the buckling mode changes to a local mode (Mode 2 in Fig. 6.1). As is known from Fig. 6.1 in this case, web of the stiffener deflects due to the bending moment from the plate along the plate-stiffener intersection line. Against the rotation of the plate along intersection line, resistance decreases as the stiffener becomes higher. Because of lower resistance from the stiffener to the plate, local buckling strength gradually deceases with the increase in the stiffener height.

With further increase in the stiffener height, buckling mode changes from a local buckling of the plate to a lateral buckling of the stiffener, which is shown as Mode 3 in Fig. 6.1. The buckling strength in Mode 3 decreases with the increase in the stiffener height.

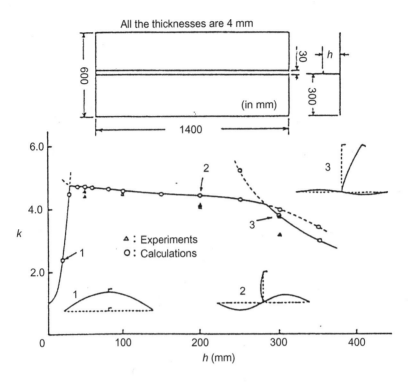

FIG. 6.1

Buckling strength of stiffened plate with a one-sided stiffener.

In all ranges of the stiffener height, the calculated buckling strength shows good agreement with measured buckling strength.

6.2 BUCKLING/PLASTIC COLLAPSE BEHAVIOR AND STRENGTH OF CONTINUOUS STIFFENED PLATES
6.2.1 MODELING OF CONTINUOUS STIFFENED PLATE FOR FEM ANALYSIS

Deck plating and bottom/inner bottom plating as stiffened plates are the main structural members in ship's hull girder under longitudinal bending. Because of this, many research works have been performed up to now both theoretically and experimentally on buckling/plastic collapse of stiffened plates subjected to longitudinal thrust. Bottom/inner bottom plating are subjected also to transverse thrust and lateral pressure loads in general.

Here, longitudinal stiffeners of the same size are attached to the deck plating with equal distances in general. The bottom and the inner bottom plating also have the same structure. Such stiffened plate is

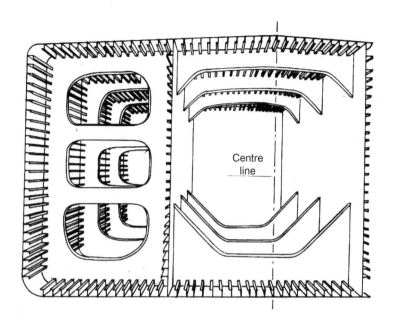

FIG. 6.2

Structure of single hull tanker.

hereafter called continuous stiffened plate. Fig. 6.2 indicates an example of continuous stiffened plating composing the structure of a single hull tanker.

As an example, a stiffened plate shown in Fig. 6.3 is considered [4]. This is a case that longitudinal girders and transverse frames are provided on plating. Between adjacent longitudinal girders, two longitudinal stiffeners are attached. When buckling/plastic collapse analysis is performed on such continuous stiffened plate, it is not necessary to analyze the whole plating, but it is enough if a part of the plating is analyzed on the basis of the geometrical considerations. For the stiffened plating in Fig. 6.3, extent of modeling could be as follows.

6.2.1.1 Modeling extent in longitudinal direction

The deflection of stiffened plate after buckling has taken place is in general represented as the sum of local deflection and overall deflection. If it is assumed that the plate is simply supported along the intersection lines of plate with longitudinal girders and transverse frames, local and overall deflection are represented by the dotted and solid lines in Fig. 6.3.

Regarding the extent of modeling in the longitudinal direction, two modelings can be considered depending on the local buckling mode of the plate [4]. If the number of half-waves of the local plate buckling mode in longitudinal direction is an odd number, the extent of modeling can be taken as *adg-beh* or *beh-cfi* imposing symmetry conditions on each side of the model, that is

$$\left. \begin{array}{l} u\text{: uniform} \\ \theta_y = \theta_z = 0 \end{array} \right\} \tag{6.1}$$

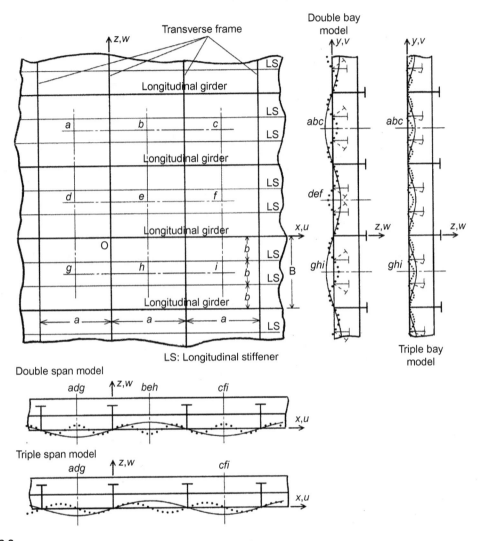

FIG. 6.3

Extent of modeling for continuous stiffened plate [4].

On the other hand, when the number of half-waves in the longitudinal direction is an even number, extent of the model has to be *adg-cfi*. In this case, periodically continuous condition has to be imposed on *adg* and *cfi*, which is expressed as follows:

$$\left.\begin{array}{l} u: \text{ uniform along } adg \text{ and } cfi \\ v_{adg} = v_{cfi}, \quad w_{adg} = w_{cfi} \\ \theta_{xadg} = \theta_{xcfi}, \quad \theta_{yadg} = \theta_{ycfi}, \quad \theta_{zadg} = \theta_{zcfi} \end{array}\right\}$$
(6.2)

The second and the third conditions are imposed at the two corresponding points with same y-coordinate along *adg* and *cfi*.

The model of which longitudinal extent is *adg-beh* or *beh-cfi* is hereafter called as double span model, and that with *adg-cfi* as triple span model.

6.2.1.2 Modeling extent in transverse direction

The extent of modeling in the transverse direction depends on configuration of the cross-section of stiffeners as well as loading conditions. When the stiffener has a symmetric cross-section such as a flat-bar or a tee-bar, the extent of modeling can be *abc-def* or *def-ghi* in Fig. 6.10 owing to the symmetrical deflection in plating. In this case, symmetry condition is imposed on *abc*, *def* or *ghi*, that is

$$
\left.
\begin{array}{ll}
v: & \text{uniform} \\
\theta_x = \theta_z = 0 &
\end{array}
\right\}
\tag{6.3}
$$

This boundary condition can be applied for stiffened plating subjected to longitudinal thrust, transverse thrust, lateral pressure, and their combinations.

On the other hand, when an angle-bar stiffener is attached, cross-section of the stiffener is not symmetrical. Even in this case, boundary condition expressed by Eq. (6.1) can be applied if the stiffened plate is subjected to only thrust load and no yielding occurs. When lateral pressure loads act, the force corresponding to lateral pressure passes through the shear center of the stiffener's cross-section, which is the intersecting point of mid-thickness lines of the web and the flange; see Fig. 6.4 [5]. In this case, if the stiffener is free, only translation takes place. However, the stiffener is attached to the plate and intersection line of the plate and the stiffener cannot move sideways. Consequently, the stiffener rotates as indicated in Fig. 6.4.

So, when lateral pressure acts on stiffened plate, the extent of modeling should be *abc-ghi*. The boundary condition for this case should be periodically continuous condition, which can be represented as follows:

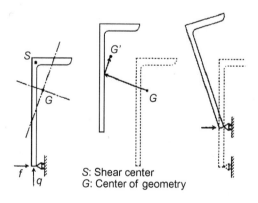

S: Shear center
G: Center of geometry

FIG. 6.4

Displacement and rotation of angle-bar stiffener in stiffened plating [5].

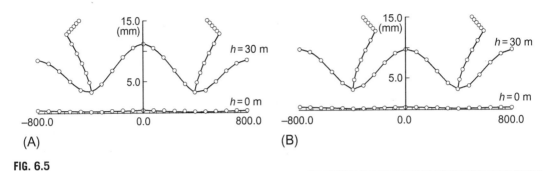

FIG. 6.5

Comparison of deflection mode between double and triple-bay models for continuous stiffened plate subjected to lateral pressure [5]. (A) Double-bay model. (B) Triple-bay model.

$$
\left.
\begin{array}{l}
v \text{ uniform along } abc \text{ and } ghi \\
u_{abc} = u_{ghi}, \qquad w_{abc} = w_{ghi} \\
\theta_{yabc} = \theta_{yghi}, \qquad \theta_{xabc} = \theta_{xghi}, \qquad \theta_{zabc} = \theta_{zghi}
\end{array}
\right\}
\tag{6.4}
$$

With this boundary condition, cross-section deforms as indicated in Fig. 6.12B. If symmetry condition is imposed instead of periodically continuous condition, the structure and its deformation shall be as indicated in Fig. 6.5A. Modeling with extent of *abc-def* or *def-ghi* is denoted as a double-bay model and that with *abc-ghi* as a triple-bay model, respectively. Triple span model gives more rational deflection mode.

When number of stiffeners becomes very large, buckling/plastic collapse behavior of this stiffened plate can be approximated by that of a longitudinal stiffener with attached plating. Also in this case, it is not necessary to perform collapse analysis on a whole stiffened plate. The modeling for this case is shown in Fig. 6.6, where longitudinal girder is not considered. It is enough if the analysis is performed on double span or triple span and double bay or triple bay model as shown in the figure. In-plane displacement perpendicular to the boundary is assumed to be uniform considering that the model is a part of continuous plating.

6.2.2 COLLAPSE BEHAVIOR OF STIFFENED PLATES UNDER LONGITUDINAL THRUST

6.2.2.1 Models for FEM collapse analysis

As fundamental models, stiffened panels from bottom plating of a bulk carrier and deck plating of a tanker (VLCC) are selected [4]. As indicated in Fig. 6.3, the spacing between adjacent transverse frames is denoted as a and that between adjacent longitudinal girders B. The spacing between adjacent longitudinal stiffeners is denoted as b.

For bulk carrier model, aspect ratio of the local plate is taken as $a/b = 3.0$, and that for VLCC model as $a/b = 5.0$. A local plate ($a \times b$) is defined as a part of stiffened plate partitioned by longitudinal girders/stiffeners and transverse frames. Six thicknesses are selected for each aspect ratio of the local plate, among which the thicknesses of existing ships are included. The assumed dimensions of the local plates are given in Table 6.1.

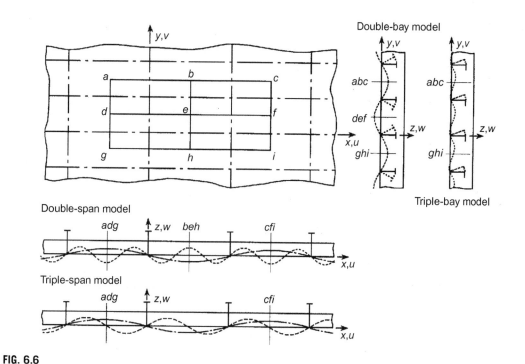

FIG. 6.6

Extent of modeling for continuous stiffened plate with plenty of stiffeners [4].

Table 6.1 Thickness of Local Plates (in mm)							
Ship Type	**$a \times b$**	**B10**	**B15**	**B20**	**B25**	**B30**	**B35**
Bulk carrier	2550×850	33	22	16	13	11	9.5
VLCC	4750×950	37	25	18.5	15.0	12.5	11.0
Note: *For example, B15 implies $\beta = b/t \cdot \sqrt{E/\sigma_Y} = 1.5$.*							

As for stiffeners, flat-bar, angle-bar, and tee-bar stiffeners are selected. Dimensions of the stiffeners are summarized in Table 6.2, and their shapes are shown in the illustration attached to Table 6.2 together with definitions of height, breadth, and thicknesses. Four sizes of stiffeners are considered for each type, which are denoted as S1, S2, S3, and S4.

The calculation models are denoted as "*pqSrBsNt*," where

- "*p*" stands for type of stiffener
    $p = F$: flat-bar; $p = A$: angle-bar; $p = T$: tee-bar
- "*q*" stands for aspect ratio of local plate
    $q = 3$: $a/b = 3.0$ $q = 5$: $a/b = 5.0$

Table 6.2 Types and Sizes of Longitudinal Stiffeners (in mm)

	Size 1	Size 2	Size 3	Size 4
Flat-bar	150×17	250×25	350×35	500×35
Angle-bar	$150 \times 90 \times 9/12$	$250 \times 90 \times 10/15$	$400 \times 100 \times 12/17$	$600 \times 150 \times 15/20$
Tee-bar	$138 \times 9 + 90 \times 12$	$235 \times 10 + 90 \times 15$	$383 \times 12 + 100 \times 17$	$580 \times 15 + 150 \times 20$

Flat-bar $\quad h \times t$
Angle-bar $\quad h \times b_f \times t_w / t_f$
Tee-bar $\quad h \times t_w + b_f \times t_f$

Flat-bar — Angle-bar — Tee-bar

- "r" stands for size of stiffener

  $r = 1$: S1-size; $\quad r = 2$: S2-size; $\quad r = 3$: S3-size; $\quad r = 4$: S4-size
- "s" stands for slenderness ratio of local plate

  $s = 10$: $a/b = 1.0$; $\quad s = 15$: $a/b = 1.5$; $\quad s = 20$: $a/b = 2.0$;

  $s = 25$: $a/b = 2.5$; $\quad s = 30$: $a/b = 3.0$; $\quad s = 35$: $a/b = 3.5$
- "n" stands for number of longitudinal stiffeners

  $t = n$: number of stiffener is n; $\quad t = $ inf: infinite number of stiffeners

For example, "A5S2B30N4" stands for stiffened plate with four angle-bar stiffeners of S2 size of which local plate has an aspect ratio of 5.0 and slenderness ratio of 3.0. Keeping the size ($a \times b$) of the local plate unchanged, the number of stiffeners is varied as 1, 2, 4, 8, and infinity.

It should be noticed that the distance between adjacent stiffeners are the same. Therefore, the aspect ratio of the whole stiffened plate varies as follows:

$$\text{BC series: } a/B = 1.50(N = 1); 1.00(N = 2); 0.60(N = 4); 0.33(N = 8)$$

$$\text{VLCC series: } a/B = 2.50(N = 1); 1.67(N = 2); 1.00(N = 4); 0.56(N = 8)$$

In summary, as for local panels, two aspect ratios and six slenderness ratios are considered, and as for stiffeners, three types, four sizes, and five numbers. Thus, number of calculated models is altogether $720 (= 2 \times 6 \times 3 \times 4 \times 5)$ [5].

6.2.2.2 Modeling for FEM collapse analysis

In the present analysis, triple span/triple bay model is used for stiffened panels with two, four, and eight stiffeners. In this case, modeling extent in the longitudinal direction is $1/2 + 1 + 1/2$ spans, and is $1/2 + 1 + 1/2$ bays in the transverse direction.

When the number of stiffeners is infinite, spacing between adjacent longitudinal stiffeners is considered as one bay, and longitudinal girder is not considered. On the other hand, when number of stiffener is one, the triple span/double full bay model is used.

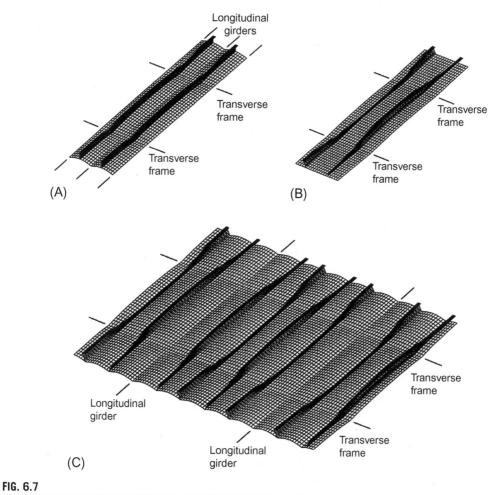

FIG. 6.7

Typical FEM models of stiffened panels [4]. (A) $N = 1$. (B) $N = $ infinite. (C) $N = 4$.

Fig. 6.7A–C shows stiffened plate models with one stiffener, with infinite stiffeners and four stiffeners, respectively. Models are accompanied by initial deflection, which shall be explained later in Section 6.2.2.4.

Longitudinal girders and transverse frames are not modeled, but the plate is assumed to be simply supported along the lines of their attachment. Stiffener's web is also assumed to be simply supported along the intersection lines with web of transverse frames. A multiple point constraint condition is imposed for displacements along the boundary cross-sections where periodically continuous boundary conditions are imposed.

The material is assumed to be elastic/perfectly plastic. Material properties assumed in the analyses are as follows:

Yield stress: $\sigma_Y = 313.6$ MPa
Young's modulus: $E = 205.8$ GPa
Poisson's ratio: $\nu = 0.3$
Strain hardening rate: $H' = 0$

6.2.2.3 Nonlinear FEM analysis

A series of elastoplastic large deflection analyses is performed using the nonlinear FEM code, MSC.Marc [6]. Bilinear shell element (Element No. 75) is used. This element is a four-noded thick-shell element with three translations and three rotations per node as degrees of freedom. Shear deformation in the thickness direction is considered in this element. Eleven integration points are provided toward thickness direction. Meshing of the local plate is the same regardless of the aspect ratio, and one half-wave region of the elastic buckling mode is represented by 10 × 10 elements; see Figs. 6.7 and 6.8. Stiffener web and flange are divided into six elements toward its depth and breadth directions, respectively. This division is the same regardless of the stiffener size.

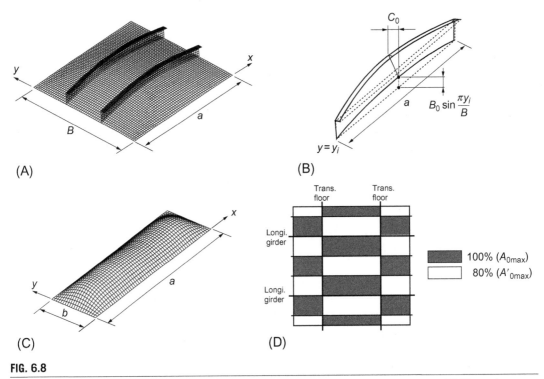

(A)

(B)

(C)

(D)

FIG. 6.8

Components of assumed initial deflection in stiffened plate [4]. (A) Overall buckling mode. (B) Flexural buckling/tripping mode. (C) Thin-horse mode. (D) Distribution of magnitude.

6.2.2.4 Initial deflection in FEM model

The initial deflection consists of three components, which are (i) overall buckling mode in stiffened plate; (ii) overall buckling and tripping modes in stiffeners; and (iii) thin-horse mode in local plates, which are indicated in Fig. 6.8A-C, respectively.

The measured results [7] show that initial deflection of an overall buckling mode is in an opposite direction at 7 boundaries among 20 measured boundaries of adjacent spans as described in Chapter 2. As for initial deflection of stiffeners in a horizontal direction (tripping mode), they are in an opposite direction at 4 spans among 20. However, the direction of initial deflection in the adjacent spans is set as opposite. This is because the buckling/ultimate strength is lower when initial deflections in the adjacent spans are in an opposite direction compared to the case of the same direction. Thus, initial deflection of an overall buckling mode in stiffened plate is expressed as

$$w_{0ov} = B_0 \sin \frac{\pi x}{a} \sin \frac{\pi y}{B} \tag{6.5}$$

The origin of the coordinate system is taken at an intersection point of a longitudinal girder and a transverse frame; see point O in Fig. 6.3.

Then, initial deflection of the ith longitudinal stiffener located at $y = y_i$ is expressed as follows:

$$w_{0si} = B_0 \sin \frac{\pi x}{a} \sin \frac{\pi y_i}{B} \tag{6.6}$$

Initial deflection of a tripping mode is expressed as

$$v_{0si} = C_0 \frac{z}{h_s} \sin \frac{\pi x}{a} \tag{6.7}$$

where $h_s = t_p/2 + h$ and z is measured from the mid-thickness plane of the plate.

On the other hand, initial deflection of a thin-horse mode in local plate is expressed as follows:

$$w_{0thin} = \left| A_{0\,max} \sum_{m=1}^{11} A_{0m} \sin \frac{\pi m x}{a} \sin \frac{\pi y}{B} \right| \tag{6.8}$$

Consequently, initial deflection in plate is expressed as the sum of deflections of an overall mode and of a thin-horse mode as follows:

$$w_{0p} = w_{0ov} + w_{0thin} = B_0 \sin \frac{\pi x}{a} \sin \frac{\pi y}{B}$$
$$+ \left| A_{0\,max} \sum_{m=1}^{11} A_{0m} \sin \frac{\pi m x}{a} \sin \frac{\pi y}{B} \right| \tag{6.9}$$

Measured results indicate that B_0 lies in the range of $-0.0007a$ and $0.0006a$ and C_0 in the range of $-0.00125a$ and $0.00135a$, respectively [7]. In the present analyses, B_0 and C_0 are assumed as follows:

$$B_0 = C_0 = 0.001 \times a \tag{6.10}$$

where a is indicated in Fig. 6.3. As for the second term in Eq. (6.9), the initial deflection of a thin-horse mode in local plate, deflection components given in Table 6.3 are used.

Table 6.3 Coefficients Making Initial Deflection of Thin-Horse Mode [7]

	A_{01}	A_{02}	A_{03}	A_{04}	A_{05}	A_{06}
$a/b = 3.0$	1.1458	−0.0616	0.3079	0.0229	0.1146	−0.0065
$a/b = 5.0$	1.1271	−0.0697	0.3483	0.0375	0.1787	−0.0199
	A_{07}	A_{08}	A_{09}	A_{10}	A_{11}	
$a/b = 3.0$	0.0327	0.0000	0.00000	−0.0015	−0.0074	
$a/b = 5.0$	0.0995	0.0107	0.0537	−0.0051	0.0256	

Coefficients of initial deflection $A_{0m}(m = 1, \ldots, 11)$ in Table 6.3 makes initial deflection of a thin-horse mode of which maximum magnitude is 1.0. Then, maximum magnitude of initial deflection of a thin-horse mode, $A_{0\,max}$, is assumed to be:

$$A_{0\,max} = \text{Smaller}[0.1\beta^2 t_p,\ 6\ \text{mm}] \tag{6.11}$$

where $0.1\beta^2 t_p$ is the magnitude of average initial deflection proposed by Smith et al. [8] based on the measured results, and 6 mm is the maximum allowable magnitude specified in Japan Shipbuilding Quality Standards [9]. In addition to this, 20% difference is given in the magnitude of initial deflection of a thin-horse mode in the adjacent local panels alternatively as indicated in Fig. 6.8D. That is, the maximum magnitude of initial deflection of nonshaded panels in Fig. 6.8D is set as

$$A'_{0\,max} = 0.8 \times A_{0\,max} \tag{6.12}$$

This is to give irregularity in the initial deflection and to obtain numerically stable solution.

When number of stiffeners is infinite, Eqs. (6.5), (6.6) reduce to:

$$w_{0s} = B_0 \sin\frac{\pi x}{a} \tag{6.13}$$

and so the first term of Eq. (6.9).

Fig. 6.9 shows initial deflection of a typical stiffened plate, model T3S4B35N2, defined as above. The initial deflection is magnified by 50 in this figure. Influence of welding residual stress is not considered in the present analysis.

6.2.2.5 Ultimate strength and collapse behavior of stiffened plates under longitudinal thrust [5]

The ultimate strength of a total of 720 cases are summarized in Tables 9.1–9.3 together with calculated results applying different methods in Section 9.5 in Chapter 9. For stiffened plate with tee-bar stiffeners, the ultimate strength obtained by nonlinear FEM analyses is plotted with various marks in Figs. 6.10 and 6.11. In each figure, the ultimate strength is plotted against slenderness ratio of the local plate.

Comparing the marks in figures, it is known that the ultimate strength is affected by the number of stiffeners when stiffener size is small and the slenderness ratio of the local plate is low. In the case of aspect ratio being $a/b = 3.0$, variation is observed in the ultimate strength depending on the number of stiffeners only when the stiffeners of size S1 are provided. On the other hand, when aspect ratio of

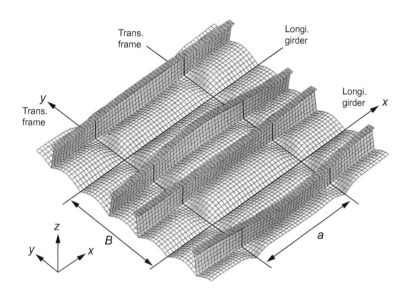

FIG. 6.9

Initial deflection given on T3S4B35 [4].

the local plate is 5.0, variation in the ultimate strength is observed also when stiffeners of size S2 are provided. These are the cases that overall buckling dominates the collapse behavior.

Here, two typical cases are considered that the ultimate strength varies and does not vary depending on the number of stiffeners. As the former example, T3S1B10 series (stiffened plate with tee-bar stiffeners of size S1 are attached; slenderness ratio and aspect ratio of the local plate are $\beta = 1.0$ and $a/b = 3.0$, respectively) is selected, and as the latter example, F3S2B25 series (stiffened plate with flat-bar stiffeners of size S2 are attached; slenderness ratio and aspect ratio of the local plate are $\beta = 2.5$ and $a/b = 3.0$, respectively). Fig. 6.12A and B shows respective average stress-average strain relationships [4]. On the other hand, collapse modes for these two cases are shown in Figs. 6.13 and 6.14, respectively, for the models with 1, 8, and infinite number of stiffeners [4].

Fig. 6.13 indicates that collapse mode of stiffened plate, T3S1B10 series, is in an overall buckling mode regardless of the number of stiffeners. Overall buckling strength mainly depends on the aspect ratio of the whole stiffened plate if plate thickness as well as stiffener size are the same.

In the present model, local buckling strength is the same regardless of the number of stiffeners since the aspect ratios of all local panels are the same. Contrary to this, overall buckling strength varies with the number of stiffeners because aspect ratio of the whole plate varies as the number of stiffeners increases.

In all cases indicated in Fig. 6.13A-C, stiffened panels collapse by the occurrence of overall buckling. In the postultimate strength range after overall buckling has taken place, stiffeners undergoes tripping at the mid-span point of the stiffener which locates in the compression side of overall bending.

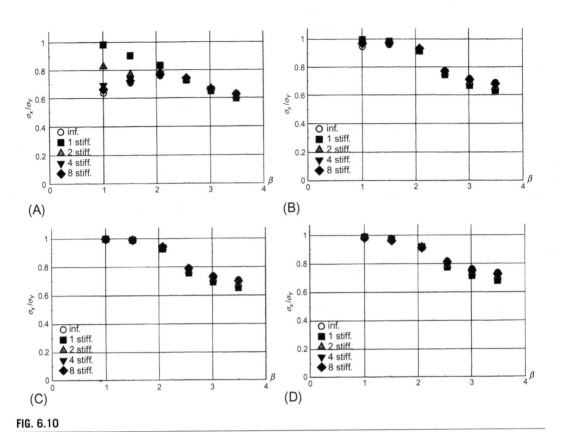

FIG. 6.10

Ultimate compressive strength of stiffened panels ($a/b = 3.0$; tee-bar stiffeners). (A) T3S1-series.
(B) T3S2-series. (C) T3S3-series. (D) T3S4-series.

On the other hand, Fig. 6.14A-C indicates that, plate collapses by local buckling and plastic deformation is concentrated at the mid-span region of local plates in the case of F3S2B25 series model. The collapse is dominated by local collapse of plate, although overall bending deformation takes place in stiffeners beyond the ultimate strength. Such collapse is denoted as plate induced (PI) failure, while the collapse of T3S1B10 series models is denoted as stiffener induced (SI) failure [7]. When PI failure takes place, the ultimate strength is almost the same regardless of the number of stiffeners.

In the actual ship structure, size of stiffeners are determined from the condition that overall buckling of a stiffened plate does not occur before local plate buckling takes place. In addition to this, thickness of deck and bottom plating is relatively thick.

Hereafter, on the basis of the above observation, continuous stiffened plate with many (more than two) stiffeners is considered, which is subjected to combined uni-/bi-axial thrust and lateral pressure. The model for FEM analyses is that shown in Fig. 6.6. Models combining double span/triple span and double bay/triple bay are used depending on the stiffener type and aspect ratio of local plate as well as load combinations.

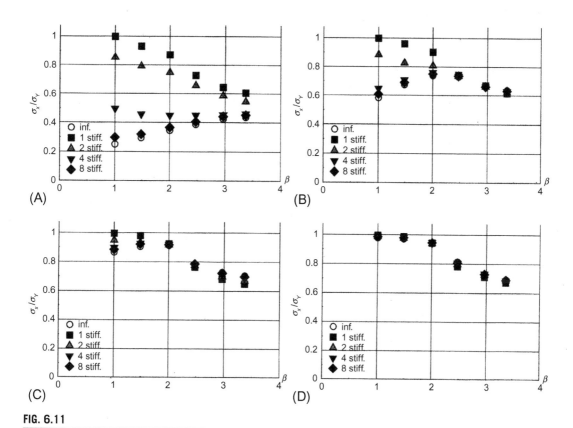

FIG. 6.11

Ultimate compressive strength of stiffened panels ($a/b = 5.0$; tee-bar stiffeners). (A) T5S1-series.
(B) T5S2-series. (C) T5S3-series. (D) T5S4-series.

6.2.3 COLLAPSE BEHAVIOR AND STRENGTH OF CONTINUOUS STIFFENED PLATES UNDER COMBINED LONGITUDINAL THRUST AND LATERAL PRESSURE LOADS

6.2.3.1 Modeling of continuous stiffened plates

In Section 6.2.2, collapse behavior of stiffened panels subjected to longitudinal thrust is explained. In ship and ship-like floating structures, however, stiffened panels in bottom or side shell plating are subjected to lateral pressure in addition to longitudinal thrust which is produced by longitudinal hull girder bending.

Here, continuous stiffened plate is considered subjected to combined longitudinal thrust and lateral pressure loads [8,9]. In the nonlinear FEM analyses performed here, a triple span/double bay model is used for stiffened panels with flat-bar and tee-bar stiffeners, and a triple span/triple bay model is used for those with angle-bar stiffeners; see Fig. 6.6. Along the boundary, periodically continuous condition is imposed. Owing to this modeling, any local buckling mode can be simulated. To keep the plate continuity condition along boundaries, in-plane displacement of the sides in their perpendicular

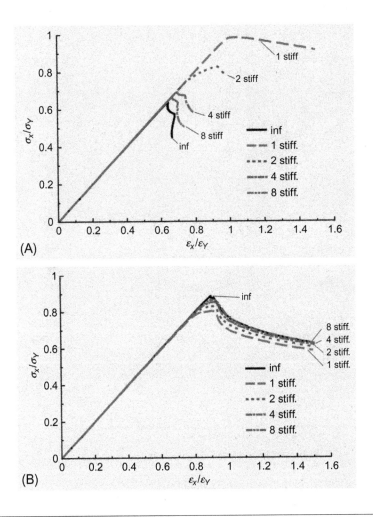

FIG. 6.12

Average stress-average strain relationships of stiffened panels with flat-bar stiffeners ($a/b = 3.0$) [4]. (A) Stiffeners of Size 2 and local plate with slenderness ratio of $b/t \cdot \sqrt{\sigma_Y/E} = 3.0$ (T3S1B10). (B) Stiffeners of Size 1 and local plate with slenderness ratio of $b/t \cdot \sqrt{\sigma_Y/E} = 1.0$ (F3S2B25).

direction is assumed to be uniform. In both models, the plate is assumed to be simply supported along the intersection lines of transverse members and plate, and the transverse members are not modeled.

Dimensions of the local plate are as follows:

Length: $a = 2400, 4000$ mm
Breadth: $b = 800$ mm
Thickness: $t_p = 13, 15, 20$ mm

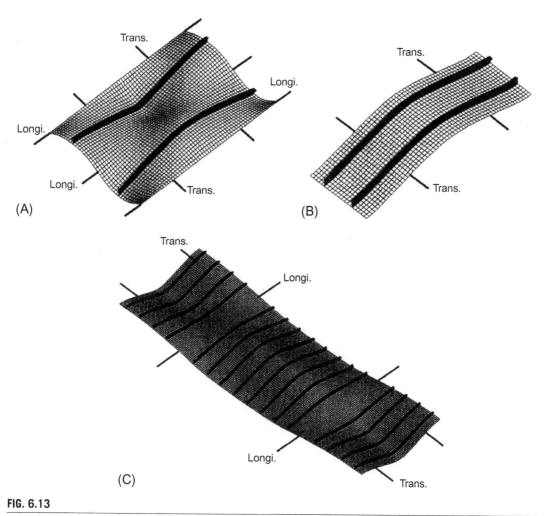

FIG. 6.13

Typical overall collapse modes of stiffened plate: T3S1B10 [4]. (A) One stiffener (two stiffened plates).
(B) Infinite number of stiffeners. (C) Eight stiffeners.

The material properties are as follows:

Yield stress: $\sigma_Y = 313.6$ MPa
Young's modulus: $E = 205.8$ GPa
Poisson's ratio: $v = 0.3$
Strain hardening rate: $H' = E/65$

The geometry of the stiffener cross-sections is indicated in Table 6.4. In each group, angle- and tee-bar stiffeners have the same moment of inertia of the cross-section.

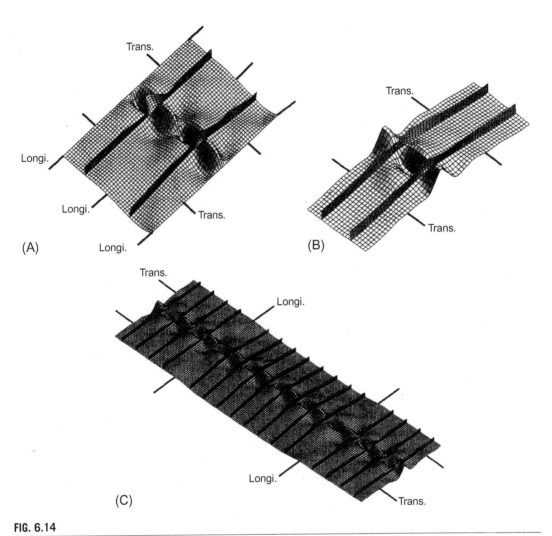

FIG. 6.14

Typical local collapse modes of stiffened plate: F3S2B25 [4]. (A) One stiffener. (B) Infinite number of stiffeners. (C) Eight stiffeners.

As for initial deflection, the sum of overall and local buckling modes is assumed as follows:

$$w_0 = w_{0s} + w_{0p} = A_0 \sin \frac{\pi x}{a} + B_0 \sin \frac{m\pi x}{a} \sin \frac{\pi y}{b} \tag{6.14}$$

At the same time, initial distortion is given to the stiffener, which is expressed as

$$v_0 = C_0 \frac{z}{h} \sin \frac{\pi x}{a} \tag{6.15}$$

Table 6.4 Geometry of Cross-Sections of Stiffeners (in mm)

Type	Model	Shape	h	t_w	b_f	t_f	h/t_w	I_s
1	F1	Flat	150	17	–	–	8.82	478.13
1	T1	Tee	150	9	90	12	16.67	478.13
1	A1	Angle	150	9	90	12	16.67	478.13
2	F2	Flat	250	19			13.16	2470.0
2	F2	Tee	250	10	90	15	25.00	2420.0
2	F2	Angle	250	10	90	15	25.00	2420.0
3	F2	Flat	350	35			10.00	12,500.0
3	F2	Tee	400	12	100	17	33.33	10,500.0
3	F2	Angle	400	12	100	17	33.33	10,500.0

Notes: h, height of stiffener; t_w, thickness of stiffener web; b_f, breadth of stiffener flange; t_f, thickness of stiffener flange; I_s, second moment of inertia of stiffener; $\sigma_Y = 313.6$ MPa, yield stress of stiffener.

The magnitude of initial deflection is assumed as follows:

$$A_0/a = 0.001, \quad B_0/t_p = 0.01, \quad C_0/a = 0.001 \tag{6.16}$$

m is a number of local buckling half-waves in the longitudinal direction and t_p is the plate thickness. a and b are the length and the breadth of local plate as indicated in Fig. 6.26.

Pressure loads are applied from a plate side and are varied as 0, 10, 20, 30, 45, and 60 m water-head. As was explained in Section 5.1, loading sequence affects very little on the ultimate strength under combined thrust and lateral pressure loads. In the present analyses, pressure load is firstly applied to a specified level, and then thrust load is applied.

6.2.3.2 Buckling/plastic collapse behavior under combined longitudinal thrust and lateral pressure

Collapse behavior of stiffened plates with flat-bar, angle-bar, and tee-bar stiffeners are similar although there exist some differences in the ultimate strength. Here, only the results of analysis on stiffened plates with tee-bar stiffeners are explained here.

Fig. 6.15 shows average stress-average strain relationships for Model F2 series with plate thicknesses of 13, 15, and 20 mm. For the cases with plate thickness of 13, 15, and 20 mm, deflection modes and spreads of yielding are illustrated in Figs. 6.16–6.18. In each figure, (A), (B), and (C) are the cases with 0, 20, and 45 m water-head.

When the plate is thin and no pressure load acts, local plate undergoes buckling of a three half-waves mode; see Fig. 6.16A. Soon after deflection of a buckling mode begins to develop, yielding starts at the top of the mid-span point of the stiffener which locates in the compression side of overall bending. The ultimate strength is attained by the occurrence of overall flexural buckling. Beyond the ultimate strength, plastic deformation concentrates near the mid-span region of the center span as indicated in Fig. 6.16A. In the remaining part, elastic unloading takes place during the concentration process of plastic deformation at the mid-span region. Similar collapse behavior is observed when the plate thickness is 15 mm; see Fig. 6.17A.

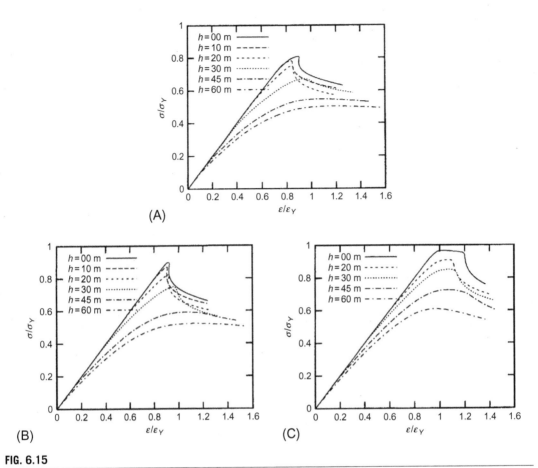

FIG. 6.15

Average stress-average strain relationships of continuous stiffened plating under combined longitudinal thrust and pressure loads (Model T2). (A) $t = 13$ mm. (B) $t = 15$ mm. (C) $t = 20$ mm.

When the plate thickness is 20 mm, general yielding takes place before local plate buckling occurs as indicated in the middle figure in Fig. 6.18A. Collapse mode beyond the ultimate strength is almost the same with that of thin plate. However, yielding in the postultimate strength range takes place at the compression side of overall bending, that is at compressed stiffener in both ends of the model and at compressed plate at the mid-span of the central span; see the lowest figure in Fig. 6.18A. It should be noticed that the top of the stiffener located in the tension side is not yielded in the middle span.

In Fig. 6.15C, plateau is observed in the average stress-average strain curve when $t = 20$ mm under no lateral pressure. This indicates that the peak load is obtained due to general yielding, and plastic overall buckling has taken place at the point where capacity starts to rapidly decrease. In cases of thinner stiffened panels, the ultimate strength is attained by the occurrence of elastic overall buckling.

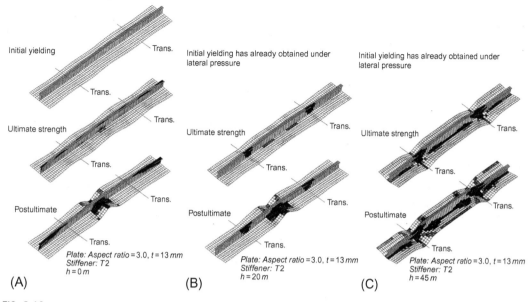

FIG. 6.16

Collapse modes and spreads of yielding under combined longitudinal thrust and pressure loads (Model T2; $t = 13$ mm). (A) $h = 0$ m. (B) $h = 20$ m. (C) $h = 45$ m.

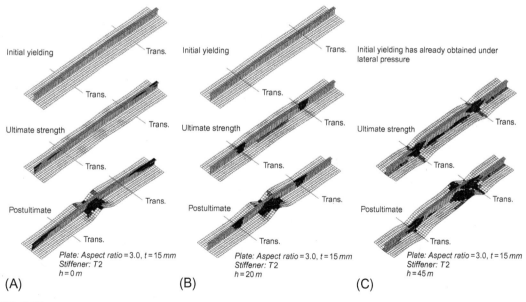

FIG. 6.17

Collapse modes and spreads of yielding under combined longitudinal thrust and pressure loads (Model T2; $t = 15$ mm). (A) $h = 0$ m. (B) $h = 20$ m. (C) $h = 45$ m.

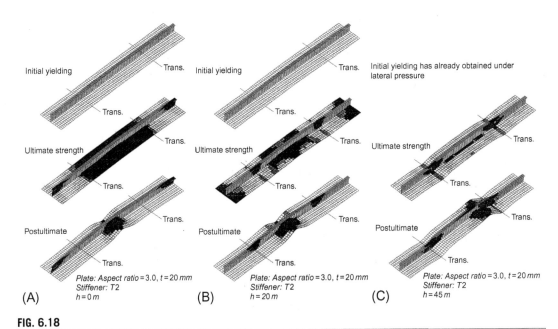

FIG. 6.18

Collapse modes and spreads of yielding under combined longitudinal thrust and pressure loads (Model T2; $t = 20$ mm). (A) $h = 0$ m. (B) $h = 20$ m. (C) $h = 45$ m.

When lateral pressure acts together with longitudinal thrust, collapse behavior is similar to the case of no lateral pressure if the water-head is below 10 m. However, with the increase of the applied lateral pressure, overall deflection mode changes from flexural buckling mode (simply supported mode) to clamped mode as can be seen in Figs. 6.16–6.18.

With the development of clamped mode deflection, yielding starts at the stiffener top at the span ends. This is because high compressive stress is produced at this location due to the clamped mode deflection of the stiffener.

Collapse mode in the postultimate strength range is almost the same regardless of the plate thickness when lateral pressure is around 20 m water-head. In this range of water-head, plate deflects to the opposite direction on both sides of the stiffener at the middle of the model. This indicates that deflection in panels is dominated by the occurrence of buckling. When pressure becomes higher than 30 m water-head, on the other hand, deflection of the plate is to the same direction on both sides of the stiffener. This indicates that deflection in panels is dominated by high lateral pressure.

When pressure load is 10 and 20 m water-head, plastic hinge (or concentrated plastic deformation) is located at the mid-span region of the center span. On the other hand, it is located near the span end region of the center span.

The initial yielding strength and the ultimate strength are plotted in Fig. 6.19 against water head for each plate thickness. Above a certain water head, yielding starts during pressure loading, and this is the reason why initial yielding strength curves intersect the horizontal axis.

In the case of stiffened plates with angle-bar stiffeners, the collapse behavior is almost the same with that of stiffened plates with tee-bar stiffeners. The flat-bar stiffeners have also similar effects on collapse behavior of stiffened plates although the in-plane stiffness beyond buckling under the action of

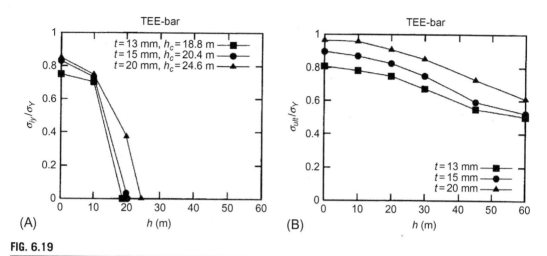

FIG. 6.19

Influence of lateral pressure load on compressive strength of stiffened plates (Model T2). (A) Initial yielding strength. (B) Ultimate strength.

lateral pressure is a little lower than stiffened plates with tee-bar and angle-bar/tee-bar stiffeners. This is because horizontal displacement at the top of stiffeners by local plate buckling is not constrained in the case of a flat-bar stiffener, whereas it is constrained by flange at the top of the web in the case of an angle-bar stiffener.

6.2.3.3 Influence of lateral pressure on collapse behavior and ultimate strength

On continuous stiffened panels in ship bottom plating, distributed pressure load acts from the plate side. On the other hand, pressure loads from the stiffener side may also act on continuous stiffened plate from liquid cargo or as inertia forces. When pressure load is high, hinge-induced (HI) collapse may take place in addition to stiffener induced (SI) failure and plate induced (PI) failure explained in Section 6.2.2.5. SI, PI, and HI failure modes are schematically illustrated in Fig. 6.20. In Fig. 6.20C, pressure load is acting from a stiffener side, but can be set also from a plate side.

Here, a series of nonlinear FEM analyses is performed varying the uniformly distributed pressure load between −40 and +40 m water heads. For the local plate analyzed here, initial deflection of the following form is assumed in the local plate:

$$w_{0p} = \sum_{m=3}^{7} B_{0m} \sin \frac{m\pi x}{a} \sin \frac{\pi y}{b} \tag{6.17}$$

where magnitude of each deflection component is set as 0.005 times the plate thickness. Initial deflection of a flexural buckling mode and that of a sideways mode are also assumed as follows:

$$w_{0s} = A_0 \sin \frac{\pi x}{a} \tag{6.18}$$

$$\phi_0 = C_0 \sin \frac{\pi x}{a} \tag{6.19}$$

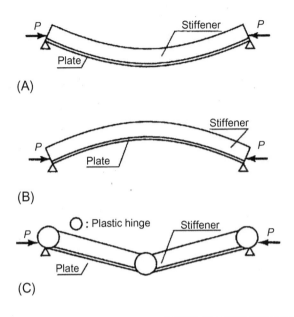

FIG. 6.20

Typical failure mode of continuous stiffened plate subjected to combined longitudinal thrust and lateral pressure [9]. (A) Stiffener-induced (SI) failure. (B) Plate-induced (PI) failure. (C) Hinge-induced (HI) failure.

The magnitudes of the two initial deflections are set as $A_0/a = C_0/a = 0.001$. Welding residual stress is not considered in this analysis.

Here, for the case of $a/b = 3.0$ and $t = 25$ mm with tee-bar stiffeners of type 1, the ultimate compressive strength is plotted by solid circles against lateral pressure in Fig. 6.21. The lines are the estimations obtained by the method explained later in Section 6.3.3.1. For the cases when lateral pressure is $q = -33, -15, -5, 5, 15$, and 32 m water-heads, deformation and spread of yielding are illustrated in Fig. 6.22A through F. Thickness of the shade represents the extent of yielded layer toward the thickness direction. Except the case in Fig. 6.22E, yielded region in the plate is restricted only to the region near the transverse members if any.

Fig. 6.22C and D indicates that buckling of a flexural mode has occurred when lateral pressure load is low. The direction of flexural deflection coincides with that of the initial deflection. In these cases, ultimate strength is attained when yielding spreads in the mid-span region of the stiffener which comes in the compression side of overall bending. A typical SI failure is observed.

Lateral pressure from the plate side produces tensile stress in the stiffener in the mid-span region, whereas compressive stress is produced in the same region by lateral pressure from the stiffener side. Because of this, lateral pressure from the plate side delays the compressive yielding in the mid-span region and increases the compressive ultimate strength as the pressure increases to a certain level as indicated in Fig. 6.20 (SI failure). On the other hand, lateral load from the stiffener side accelerates the compressive yielding in the mid-span region and this reduces the compressive ultimate strength as the pressure load increases; see Fig. 6.21 (SI failure).

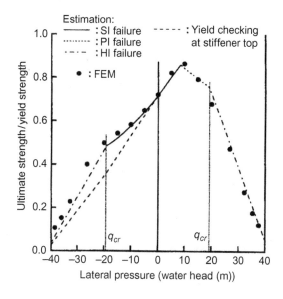

FIG. 6.21

Influence of lateral pressure on ultimate compressive strength of continuous stiffened plate subjected to combined longitudinal thrust and lateral pressure [9].

When pressure from the plate side becomes high, PI failure takes place as shown in Fig. 6.22E. It is seen that plate in the compression side of overall bending has collapsed in the mid-span region of a center span. With further increase in the lateral pressure load from the plate side, plastic hinges are formed at mid-span point and both ends of the central span as indicated in Fig. 6.22F. This is a typical HI failure.

In the HI and PI failures, compressive ultimate strength monotonously decreases as the pressure load increases as shown in Fig. 6.21.

When lateral pressure load acts from the stiffener side, tensile stress is produced in the plate by lateral pressure, and PI failure does not occur. In this case, failure mode changes from SI mode to HI mode as the pressure increases as indicated in Fig. 6.21. Also in this case, compressive ultimate strength monotonously decreases with the increase of lateral pressure.

Here, two cases are considered when PI failure occurs ($q = 15$ m) and when SI failure occurs ($q = -15$ m). When lateral pressure is from the plate side ($q = 15$ m), plate is yielded at the span ends; see Fig. 6.22E. On the other hand, no yielding is observed in the plate when lateral pressure is from the stiffener side ($q = -15$ m); see Fig. 6.22B.

Fig. 6.23 shows the bending moment distributions in the longitudinal direction for two cases of $q = \pm 15$ m before longitudinal thrust is applied and at the ultimate strength. When the pressure load acts from the plate side ($q = 15$ m), the bending moments at the supporting points (supporting points) decreases as the longitudinal thrust increases, and approaches to zero, that is a both-ends simply supported condition. On the other hand, when pressure load acts from the plate side ($q = -15$ m), the

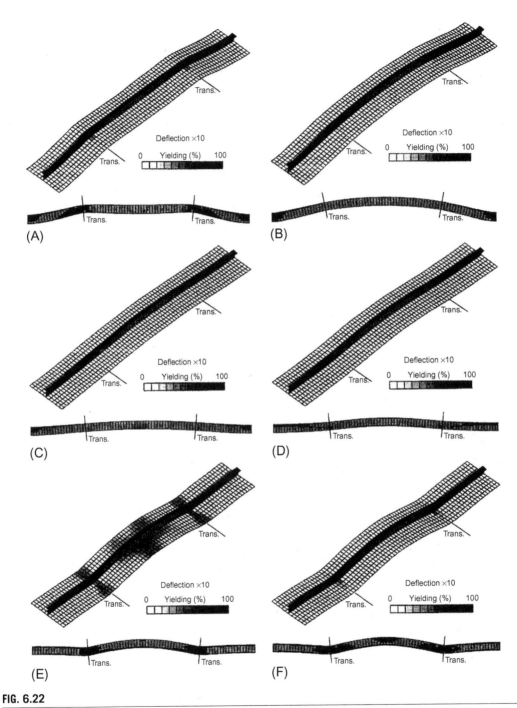

FIG. 6.22

Collapse mode and spread of yielding at ultimate strength of continuous stiffened plate subjected to combined longitudinal thrust and lateral pressure (type 1; tee-bar; $a \times b \times t_p = 2400 \times 800 \times 25$ mm; continued) [9]. (A) $q = -33$ m. (B) $q = -15$ m. (C) $q = -5$ m. (D) $q = 5$ m. (E) $q = 15$ m. (F) $q = 32$ m.

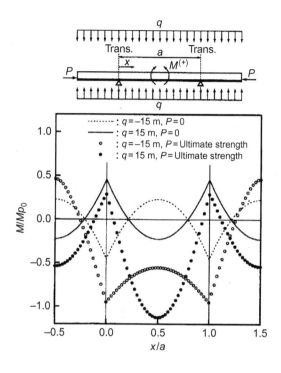

FIG. 6.23

Bending moment diagram in longitudinal direction (type 1; tee-bar; $a \times b \times t_p = 2400 \times 800 \times 25$ mm) [9].

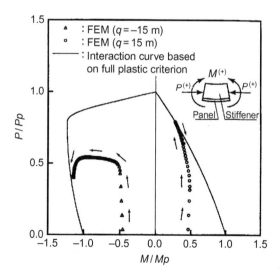

FIG. 6.24

Interaction relationships between axial force and bending moment (type 1; tee-bar; $a \times b \times t_p = 2400 \times 800 \times 25$ mm) [9].

bending moment at the supporting points increases to a fully plastic bending moment, and HI failure mode is formed; see Fig. 6.23.

Fig. 6.24 shows fully plastic interaction relationship between axial force and bending moment at the cross-section of continuous stiffened plate with tee-bar stiffeners of type 1 and $a \times b \times t_p$ = 2400 × 800 × 25 mm together with loading paths at supporting points for $q = \pm 15$ m. When the pressure is from the plate side ($q = 15$ m), the bending moment attains its maximum value when the axial force is about a half of the fully plastic axial force. After this, bending moment gradually decreases with the increase in the axial force, and the maximum axial force (the compressive ultimate strength) is attained when the plate in the central mid-span region becomes fully plastic.

On the other hand, when lateral pressure acts from the stiffener side ($q = -15$ m), the bending moment at the supporting points rapidly increases with a slight increase in the axial force after the yielding spreads near mid-span points in the left and right spans. This is because deflection of an overall flexural buckling mode rapidly develops as the axial force approaches the buckling strength. Then, the ultimate compressive strength has been attained, and the loading path asymptotically approaches to the fully plastic interaction curve.

It can be concluded that bending moment distribution is of a both ends clamped mode at the ultimate strength when the pressure load acts from the stiffener side, whereas that is of a both ends simply supported mode when the pressure load acts from the plate side. It can be also said that the ultimate strength is attained with small yielding when lateral pressure from the stiffener side is relatively low, for example, $q = -5$ m, while it is attained with large yielding when lateral pressure load from the stiffener side is relatively high. This is because developed clamped mode deflection resists against the development of simply supported buckling mode deflection.

6.2.4 COLLAPSE BEHAVIOR AND STRENGTH OF CONTINUOUS STIFFENED PLATES UNDER TRANSVERSE THRUST

6.2.4.1 Model for analysis

Three types of stiffeners are considered, which are a flat-bar stiffener, an angle-bar stiffener, and a tee-bar stiffener. For the stiffened plates with flat-bar and tee-bar stiffeners, a triple span/double bay model is used for the buckling/plastic collapse analysis of continuous stiffened plate, that is, part "*abhg*" or "*bcih*" in Fig. 6.6. In this model, symmetry condition is imposed along the longitudinal boundaries, *ag/bh* or *bh/ci*, whereas periodically continuous condition along transverse lines, *ac* and *gi*. In the case of stiffened panels with angle-bar stiffeners, triple span/triple bay model, "*acgi*" in Fig. 6.6, is used imposing periodically continuous condition along all sides of the model.

Considering the continuity condition of plating, the in-plane displacements of the model along its sides in their perpendicular direction are assumed to be uniform. The transverse frames are not modeled, but the plate is assumed to be simply supported along the lines of transverse frames.

For comparison purpose, continuous plating without stiffeners is also analyzed. For this case, the plate is assumed to be simply supported also along the plate/longitudinal stiffener web intersection lines.

Dimensions of the local plate are taken as follows:

Length: $a = 2400, 4000$ mm
Breadth: $b = 800$ mm
Thickness: $t = 10, 13, 15, 20,$ and 25 mm

Cross-sectional geometries of analyzed stiffeners are indicated in Table 6.4 in Section 6.2.3.1. In each group denoted as 1, 2, and 3, the stiffeners have the same moment of inertia. Hence, a flat-bar stiffener has a thicker web than tee-bar and angle-bar stiffeners.

As for initial deflection, three types of initial deflection are superposed, which are

(1) Initial deflection of thin-horse mode in local plates.

$$w_{0p} = \alpha_0 \left| \sum_{i=1}^{n} A_{0i} \sin \frac{i\pi x}{a} \sin \frac{\pi y}{b} \right| \tag{6.20}$$

(2) Initial deflection of overall flexural buckling mode:

$$w_{0s} = B_{0s} \sin \frac{\pi x}{a} \tag{6.21}$$

(3) Sideways initial deflection of stiffeners due to angular rotation about plate-stiffener web intersection line as defined in Section 6.2.3.3:

$$\phi_0 = C_0 \frac{z}{h_w + t_p/2} \sin \frac{\pi x}{a} \tag{6.22}$$

The magnitude of deflection components in Table 2.3 in Chapter 2 are used as A_{0i} in Eq. (6.7), for the initial deflection of a thin-horse mode. The maximum magnitude of this initial deflection is taken as

$$w_{0p \, max} = 0.05\beta^2 t \tag{6.23}$$

where

$$\beta = \frac{b}{t} \sqrt{\frac{\sigma_Y}{E}} \tag{6.24}$$

and σ_Y and E are the yield stress and Young's modulus, respectively. Eq. (6.23) represents the average level of initial deflection proposed by Smith [10]; see Fig. 2.14 (Chapter 2). α_0 in Eq. (6.20) is a parameter to account for variation of the magnitude of initial deflection bay by bay and span by span. In the analysis here, the difference of 10% is given in adjacent panels.

As for initial deflection of overall flexural buckling mode and sideways mode, it is assumed that $B_{0s} = C_0 = 0.001a$. Mechanical properties of the material are set as follows:

Yield stress: $\sigma_Y = 313.6$ MPa
Poisson's ratio: $\nu = 0.3$
Young's modulus: $E = 205.8$ GPa
Strain hardening rate: $H' = E/65$

6.2.4.2 Buckling/plastic collapse behavior under transverse thrust

For typical continuous stiffened panels, average stress-average strain relationships and average stress-deflection relationships are summarized in Fig. 6.25. Local plate is $a \times b = 2400 \times 800$ mm, and the stiffener is a tee-bar of $h \times t_w + b_f \times t_f = 250 \times 15 + 90 \times 16$ mm. In Fig. 6.25, results of two cases, $t = 10$ mm and $t = 20$ mm, are plotted, each with and without stiffeners.

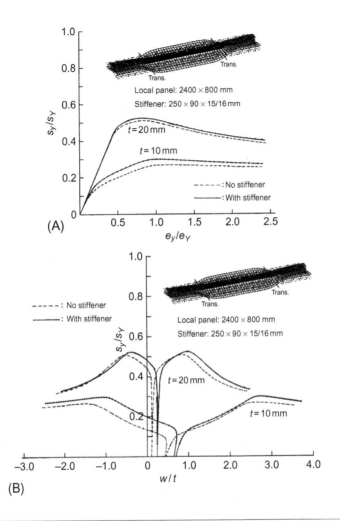

FIG. 6.25

Progressive collapse behavior of continuous plating and stiffened plating under transverse thrust (local plate: 2400 × 800 mm; stiffener: 250 × 10 + 100× 15 mm). (A) Average stress-average strain relationships. (B) Average stress-deflection relationships.

The maximum initial deflection in panels calculated by Eq. (6.23) is 4.88 and 2.44 mm for $t = 10$ mm and $t = 20$ mm, respectively, and that in stiffener is $a/1000 = 2.4$ mm. These initial deflections are in the same direction, and resist against development of deflection of a simply supported buckling mode. Comparing the cases with and without stiffeners, the former has larger initial deflection since initial deflection of an overall buckling mode is superposed to the plate initial deflection. This is the reason why buckling/ultimate strength of stiffened plate is higher than that of plate alone since plate in stiffened plate has larger initial deflection different from the buckling mode. However, the buckling/plastic collapse behavior is fundamentally the same regardless of the stiffener.

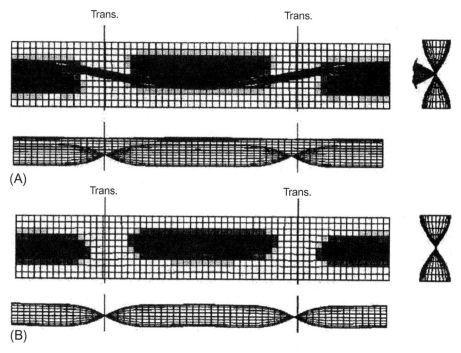

FIG. 6.26

Collapse modes and equivalent stress distributions of continuous plating and continuous stiffened plating under transverse thrust (local plate: 2400 × 800 × 10 mm; stiffener: 250 × 10 + 100 × 15 mm) [10]. (A) Continuous stiffened plate. (B) Continuous plate.

The local plate undergoes buckling with a sinusoidal mode of one half-wave. With further increase in the compressive load in the transverse direction, deflection mode changes to a cylindrical mode with a long flat part in the middle of the plate.

Fig. 6.26A and B shows deflection mode and stress distribution at the ultimate strength for stiffened plate and plate alone, respectively, when the plate thickness is $t = 10$ mm. As for stress, von Mises equivalent stress is shown. Thicker region is of a higher stress level, and the thinner region of a lower stress level. As mentioned above, deflection modes and stress distributions for two cases with and without stiffeners are quite similar. This indicates that the ultimate strength estimation method explained in Section 5.1.4.3 for continuous plating subjected to transverse thrust could be applied to the ultimate strength estimation of continuous stiffened plate subjected to transverse thrust considering the influence of stiffeners, provided that longitudinal stiffeners and transverse frames are strong enough so that plate locally collapses.

The ultimate strength of continuous stiffened plate under transverse thrust is plotted against slenderness ratio of the local plate in Fig. 6.27A and B together with that of continuous plate. In both figures, stiffener is of a tee-bar type, but (A) is of type 2 and (B) of type 3. The circles are the results of FEM analysis and the lines are the estimations by simple formulas explained later in Section 6.3.4.

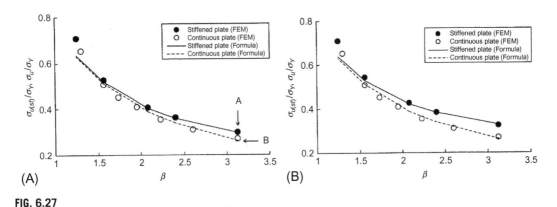

FIG. 6.27

Ultimate strength of continuous plating and continuous stiffened plating under transverse thrust
(local plate: 2400 × 800 mm) [11]. (A) Tee-bar stiffener of type 2. (B) Tee-bar stiffener of type 3.

The continuous stiffened plate shows higher ultimate strength than continuous plating. This is due to the torsional resistance of stiffener against the development of plate deflection in addition to the larger initial deflection explained before.

The increase in the ultimate strength becomes larger as the thickness of plate becomes thinner and the stiffener size larger. The points A and B in Fig. 6.27A indicate the ultimate strength of continuous stiffened plate and continuous plate of which collapse behaviors are shown in Figs. 6.25 and 6.26.

Fig. 6.28 summarized the ultimate strength of continuous stiffened panels under transverse thrust. In each figure denoted as (A), (B), and (C), ultimate strength is compared between stiffened panels with stiffeners of different types but the same flexural rigidities. An angle-bar stiffener has higher flexural/torsional stiffness than a tee-bar stiffener. This is the reason why stiffened plate with angle-bar stiffeners show higher ultimate strength than that with tee-bar stiffeners, although the difference is small. On the other hand, a flat-bar stiffener makes the ultimate strength the highest. This is because the thickness of flat-bar stiffener is the largest to have the same flexural stiffness under the condition that the web height is the same with angle-bar and tee-bar stiffeners.

The height of the flat-bar stiffener of type 3 (350 mm) is lower than that of angle-bar or tee-bar stiffener (400 mm), but the thickness is 35 mm, which is extremely thick. Because of this, ultimate strength of stiffened plate with type 3 flat-bar stiffener is very high.

6.2.5 BUCKLING/PLASTIC COLLAPSE BEHAVIOR UNDER COMBINED TRANSVERSE THRUST AND LATERAL PRESSURE

The same model is used as that in Section 6.2.4. In the analysis, pressure load is firstly applied to a specified pressure level and then transverse thrust load is applied. Typical collapse modes are illustrated in Fig. 6.29A and B.

When lateral pressure is relatively low, the collapse occurs in a simply supported mode as in the case of stiffened plates subjected pure transverse thrust; see Fig. 6.29A. As the lateral pressure increases, collapse mode changes from a simply supported mode to a clamped mode as indicated in Fig. 6.29B.

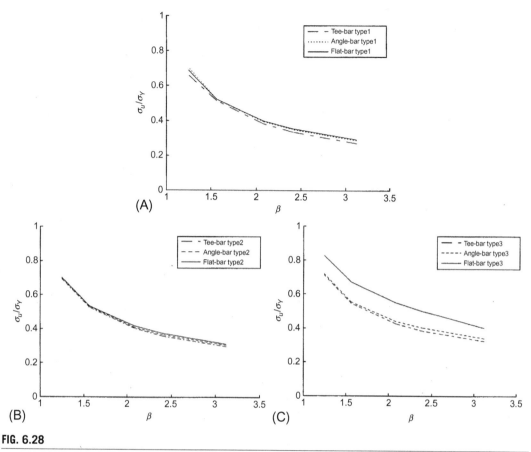

FIG. 6.28

Ultimate strength of continuous plating and continuous stiffened plating under transverse thrust (local plate: 2400 × 800 mm) [11]. (A) Type 1 stiffener. (B) Type 2 stiffener. (C) Type 3 stiffener.

The ultimate strength of stiffened plate with tee-bar stiffeners subjected to transverse thrust is plotted against applied pressure in Fig. 6.30A and B for thin plate ($t = 10$ mm) and thick plate ($t = 20$ mm), respectively. The ultimate strength of continuous plate without stiffeners is also plotted in the figure by dashed lines. The ultimate strengths for stiffened plates in Fig. 6.31A and B are indicated by A and B in Fig. 6.30A, respectively.

For a thin plate ($t = 10$ mm) without stiffener indicated by dashed line in Fig. 6.30A, the ultimate strength increases with the increase in the applied pressure up to about 15 m water head. This is because of the increase in the elastic buckling strength owing to lateral pressure as explained in Section 6.2.3.

For further increase in the applied lateral pressure, collapse mode changes from a simply supported mode to a clamped mode as indicated in Fig. 6.29B, and the ultimate strength starts to decrease with the increase in the applied lateral pressure. It should be noticed in Fig. 6.30A that the difference in ultimate strength between stiffened and unstiffened plates is larger under relatively small lateral

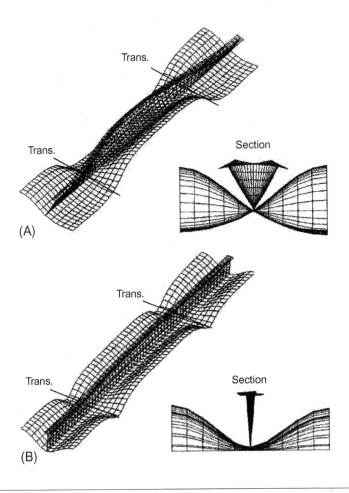

FIG. 6.29

Typical collapse modes of continuous stiffened plate subjected to combined transverse thrust and lateral pressure (local plate: 2400 × 800 × 10 mm; stiffener: 400 × 12 + 100 × 17 mm) [11]. (A) Water head of 5 m. (B) Water head of 15 m.

pressure where the local plates collapse in a simply supported mode. This is because the torsional deformation of the stiffener and its enhancing effect on the elastic buckling strength of local plates are dominant when deflection of a simply supported mode develops than the case when deflection of a clamped mode develops.

In the case of a thicker plate ($t = 20$ mm) in Fig. 6.30B, the increase in the elastic buckling strength due to lateral pressure is small, and the ultimate compressive strength monotonously decreases with the increase in the applied lateral pressure. Thicker continuous plate indicated by dashed line in Fig. 6.30B shows the same collapse behavior with continuous stiffened plates.

Here, q_{pl} in Fig. 6.30A indicates the lateral pressure corresponding the plastic collapse strength under pure lateral pressure obtained by the rigid plastic mechanism analysis on an all sides clamped plate. q_{pl} is given as follows [12].

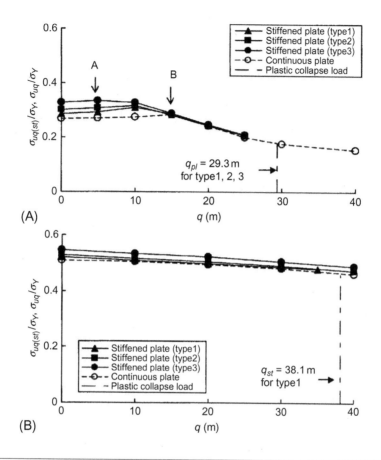

FIG. 6.30

Ultimate strength of stiffened plate with tee-bar stiffener subjected to combined transverse thrust and lateral pressure (local plate: 2400 × 800 mm; tee-bar stiffener) [11]. (A) Thin plate (t = 10 mm). (B) Thick plate (t = 20 mm).

$$q_{pl} = \frac{48\alpha^4}{3\alpha^2 + 2(1 - \sqrt{1 + 3\alpha^2})} \frac{M_P^{PL}}{a^2} \tag{6.25}$$

where $\alpha = a/b$ is an aspect ratio of the local plate and M_P^{PL} is the fully plastic bending moment of the plate per unit width, which is expressed as $t_p^2 \sigma_Y / 4$. On the other hand, q_{st} in Fig. 6.30B is the lateral pressure corresponding to the plastic mechanism of a both ends clamped beam with three hinges, which is expressed as

$$q_{st} = \frac{16}{L} M_P^{ST} \tag{6.26}$$

where M_P^{ST} is the fully plastic bending moment of the stiffener cross-section with attached plating of a full width.

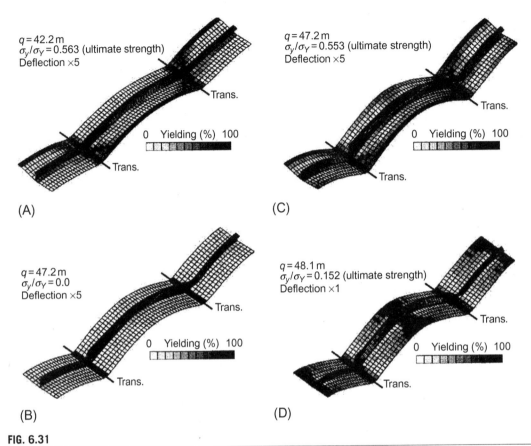

FIG. 6.31

Collapse modes and spreads of yielding of continuous stiffened plate subjected to combined transverse thrust and lateral pressure (local plate: 2400 × 800 × 20 mm; tee-bar stiffener: 1590 × 9 + 90 × 12 mm) [11]. (A) h = 42.2 m (ultimate strength). (B) h = 42.2 m (zero thrust). (C) h = 47.2 m (ultimate strength). (D) h = 48.1 m (ultimate strength).

It is known from Fig. 6.30 that the ultimate strength under transverse thrust of a stiffened plate is equivalent or higher than that of continuous plate as long as the applied lateral pressure is below the plastic collapse pressure, q_{pl} or q_{st}.

As shown by the dashed lines in Fig. 6.30, the continuous plate with rigid supports (no lateral deflection along stiffener lines) can sustain a significant magnitude of thrust load even when the lateral pressure exceeds q_{pl} or q_{st}. This is mainly due to a membrane action in the plate and a material strain-hardening effect. For the case of stiffened plate, however, when lateral pressure is increased up to the value close to whichever the smaller of q_{pl} or q_{st}, the thrust load cannot be increased any more or even when it is possible, the behavior under the action of thrust load becomes unstable.

Fig. 6.31 shows some typical collapse mode of thicker plate ($t = 20$ mm) with tee-bar stiffeners of type 1 subjected to lateral pressure higher than q_{st} of 38.1 m water head. There exists thrust capacity of $\sigma_y/\sigma_Y = 0.563$ even for the lateral pressure of $h = 42.2$; see Fig. 6.31A. With further increase in the lateral pressure, however, tripping deformation as well as plastic-hinge rotation takes place at the both end supports as shown in Fig. 6.31B and C. In Fig. 6.31B, only lateral pressure is applied whereas in Fig. 6.31C, transverse thrust load is applied in addition to lateral pressure load. $\sigma_y/\sigma_Y = 0.553$ is the transverse thrust capacity under the lateral pressure of $h = 42.2$ m. When lateral pressure of $h = 48.1$ m is applied, the three-point hinge mechanism has been completely developed as indicated in Fig. 6.31D, and the capacity for transverse thrust is reduced down to $\sigma_y/\sigma_Y = 0.152$.

The above results indicate that the stiffened panels may have a certain capacity even when the applied lateral pressure exceeds plastic mechanism collapse strength, q_{pl} or q_{st}. However, q_{pl} or q_{st} should be regarded as the maximum pressure capacities from the practical design viewpoint.

6.2.6 COLLAPSE BEHAVIOR OF CONTINUOUS STIFFENED PLATES UNDER BI-AXIAL THRUST

A series of nonlinear FEM analyses is performed on continuous stiffened panels explained in Section 6.2.3.1 applying bi-axial thrusts [13]. Results of analyses on three typical cases are shown in Fig. 6.32A–C together with the results of analyses on continuous plates without stiffeners. Deflection modes are also shown for some cases.

Fig. 6.32A shows the case with thin local plate. In this case, the ultimate strength of continuous stiffened plate is higher than that of continuous plate for all bi-axial stress ratios. This is partly because the elastic buckling strength of local plate is increased due to constraint against plate deflection by the stiffener. In addition to this, when longitudinal thrust is dominant, stiffener carries further thrust load after the plate has been locally collapsed. For this case, collapse modes for $\sigma_y/\sigma_x = 1/2$ and $\sigma_y/\sigma_x = 0$ are indicated. In the former case, transverse thrust is the dominant load, while longitudinal thrust is dominant in the latter case.

In the case shown in Fig. 6.32B, ultimate strength of continuous stiffened plate is a little higher than that of continuous plate when transverse thrust is dominant. This is because local plate buckling has occurred as indicated in the figure, and the stiffener increased the local buckling strength. On the other hand, in the range where longitudinal thrust is dominant, ultimate strength of continuous stiffened plate is lower than that of continuous plate. This is because overall flexural buckling has taken place in the case of continuous stiffened plate before the plate collapses locally as indicated in the figure.

Fig. 6.32C shows the case when slenderness ratio as a beam-column is low. In this case, local plate buckling takes place in the range of the stress ratio, σ_x/σ_y, being lower than 1.0. In this range, continuous stiffened plate and continuous plate show almost the same ultimate strength. When σ_x/σ_y ratio becomes higher than 1.0, overall flexural buckling takes place before local plate buckling occurs. Because of this, the ultimate strength of continuous stiffened plate becomes very low compared to that of continuous plate.

It can be concluded that there exist two collapse modes in continuous stiffened plates subjected to bi-axial thrust, which are local plate collapse between stiffeners and overall flexural collapse as a beam-column. The boundary of the two collapse modes is indicated in Fig. 6.32 by dotted lines.

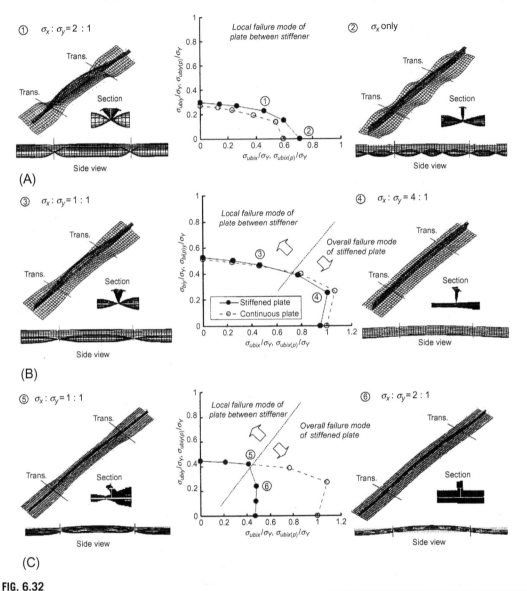

FIG. 6.32

Ultimate strength interaction relationships of continuous stiffened panels and continuous plates subjected to bi-axial thrust [13]. (A) Type 2 tee-bar stiffener; $a/b = 3.0$; $t_p = 10$ mm. (B) Type 2 tee-bar stiffener; $a/b = 3.0$; $t_p = 20$ mm. (C) Type 1 tee-bar stiffener; $a/b = 5.0$; $t_p = 20$ mm.

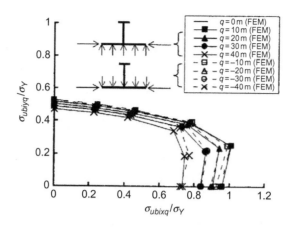

FIG. 6.33

Ultimate strength interaction relationships of continuous stiffened panels subjected to combined bi-axial thrust and lateral pressure [13].

6.2.7 COLLAPSE BEHAVIOR OF CONTINUOUS STIFFENED PLATES UNDER COMBINED BI-AXIAL THRUST AND LATERAL PRESSURE LOADS

A series of nonlinear FEM analyses is performed on continuous stiffened panels explained in Section 6.2.3.1 applying bi-axial thrusts after applying lateral pressure load to a specified level [13]. Results of analyses on continuous stiffened plate with $a/b = 3.0$, $t = 10$ mm and type 2 tee-bar stiffener are plotted in Fig. 6.33.

It is known from the figure that the ultimate strength interaction curve shrinks toward the origin of the coordinate system as the pressure increases. The reduction of the ultimate strength due to lateral pressure is more significant when longitudinal thrust is dominant. This is because of the earlier occurrence of yielding in the stiffener caused by large flexural deflection of the stiffener. As far as in the analyzed range, direction of the pressure load does not affect the compressive ultimate strength so much whether it is from the plate side or from the stiffener side.

6.3 SIMPLIFIED METHOD TO EVALUATE COMPRESSIVE ULTIMATE STRENGTH OF CONTINUOUS STIFFENED PLATES SUBJECTED TO COMBINED BI-AXIAL THRUST AND LATERAL PRESSURE

6.3.1 MODELING FOR ULTIMATE STRENGTH EVALUATION

Regarding simple formulas to evaluate the ultimate strength of stiffened plates subjected to thrust, Faulkner proposed to use Eulerian buckling strength of stiffener with attached plating applying Johnson-Ostenfeld plasticity correction as the compressive ultimate strength of stiffened plate [14]. For the breadth of attached plate, he proposed a simple expression considering the influence of welding residual stress. In Faulkner's method, stiffener is modeled by a both-ends simply supported column,

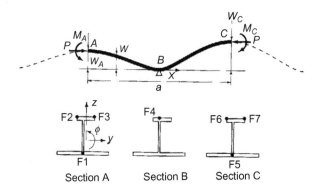

FIG. 6.34

Double-span model and possible points of initial yielding in cross-section [15].

and the interaction between adjacent panels is not considered. The influences of initial deflection and welding residual stress are not explicitly considered but implicitly included.

On the other hand, Carlsen proposed simple formulas to evaluate the ultimate compressive strength of stiffened plate on the basis of Perry-Robertson formula [7]. In his method, two types of collapse modes are considered for continuous stiffened plate, which are SI (stiffener induced) failure and PI (plate induced) failure. The SI failure occurs when the stiffener is in the compression side of the overall bending and the stiffener collapses. On the other hand, PI failure occurs when the plate is in the compression side of the overall bending and the plate collapses. In the PI failure, shift of neutral axis of the cross-section after local plate has buckled is taken into account. PI and SI failure modes are schematically illustrated in Fig. 6.20A and B appeared in Section 6.2.3.3.

In the following, double-span model is explained to evaluate the ultimate strength of continuous stiffened plate subjected to combined bi-axial thrust and lateral pressure. In the model, interaction between adjacent spans and that between stiffener web and local plate are considered. Fig. 6.34 indicates the proposed model and the possible initial yielding points in the cross-section of typical stiffeners. In the proposed method, the ultimate strength is determined from the initial yielding condition.

The continuous stiffened plate shown in Fig. 6.6 is considered, which is subjected to combined longitudinal thrust, transverse thrust, and lateral pressure as well as each component alone. As for pressure load, uniformly distributed pressure load is considered which acts from the plate side (q) and the stiffener side ($-q$), respectively.

6.3.2 CONTINUOUS STIFFENED PLATES SUBJECTED TO LONGITUDINAL THRUST

6.3.2.1 Modeling and assumptions

In the case of ship and ship-like floating structure, the collapse behavior of continuous stiffened plate subjected to longitudinal thrust can be summarized as follows [15]:

(1) In the postbuckling range of the local plate, overall deflection of a flexural buckling mode of the stiffener starts to increase rapidly after the yielding starts in the stiffener, and then the ultimate strength is attained.

(2) The interaction between adjacent spans cannot be ignored. For example, if initial deflection in the stiffener is in the same direction in the adjacent spans, the ultimate strength increases. In some cases, yielding is concentrated near the end of the span, that is cross-section B in Fig. 6.34, and not in the mid-span as Perry-Robertson formula.

(3) When flat-bar stiffener of a high depth/thickness ratio is provided, interactive buckling between plate and stiffener may occur and the ultimate strength is attained. In this case, torsional buckling of a stiffener may occur.

To simulate the above-mentioned collapse behavior, a double-span beam-column model indicated in Fig. 6.34 is considered. In the transverse direction, a double-bay model is considered. The stiffener is modeled as a beam-column with flexural/torsional stiffness and the attached plate of the effective width, b_e. The influence of interaction between plate and stiffener web on local panel buckling is taken into account and the spring of k_ϕ is put along the stiffener web/plate intersection line as the resistance by plate against rotation of the stiffener. Symmetry condition is imposed at both ends (mid-span points) of the model.

As for the initial deflection, flexural and sideways modes are assumed in the stiffener and for plate initial deflection of a panel buckling mode. Flat-bar, angle-bar, and tee-bar stiffeners are considered of which cross-sections are indicated in Fig. 6.35.

Two coordinate systems are considered. The x, y, z-system is for total cross-section and the origin of the coordinate system is placed at the center of geometry of the whole cross-section with attached plating, while x', y', z'-system is for only stiffener, and the coordinate origin is located at the center of geometry of the stiffener cross-section.

Regarding the welding residual stress, rectangular distributions shown in Fig. 6.36 are assumed. According to Eqs. (2.2), (2.3) in Chapter 2, parameters in Fig. 5.36 are expressed as

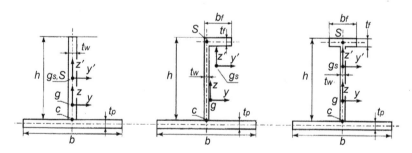

g : Centroid of total effective cross-section
gs : Centroid of stiffener cross-section
S : Shear center of stiffener
c : Plate-stiffener intersection point

FIG. 6.35

Cross-section of stiffeners [15].

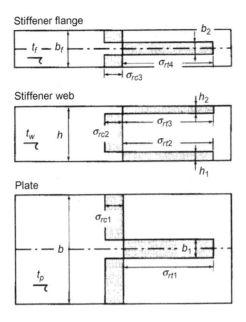

FIG. 6.36

Assumed welding residual stress in cross-section of stiffeners [16].

$$
\begin{cases}
b_1 = t_w + 0.52\Delta Q_{max}(2t_p + r_w) \\
b_2 = t_w + 0.52\Delta Q_{max}(2t_f + r_w) \\
h_1 = (t_w/t_p) \times 0.26\Delta Q_{max}(2t_p + r_w) \\
h_2 = (t_w/t_f) \times 0.26\Delta Q_{max}(2t_f + r_w)
\end{cases}
\tag{6.27}
$$

where ΔQ_{max} is given by Eqs. (2.4), (2.5) in Chapter 2. In cases of flat-bar, angle-bar, and rolled tee-bar stiffeners, $h_2 = b_2 = 0$, whereas in the case of fabricated tee-bar stiffener, h_2 and b_2 are calculated by Eq. (6.27). Tensile residual stress is set equal to the yield stress of the material. Compressive residual stresses, σ_{rc1}, σ_{rc2}, and σ_{rc3}, are assumed to be the same in magnitude, and the equilibrium condition of forces is considered all over the cross-section to derive compressive residual stress; see Eq. (2.8) in Chapter 2.

6.3.2.2 Fundamental equations as beam-column model

For the derivation of fundamental equations, the following points are assumed [15].

(1) Axial force-deflection and axial force-distortion relationships are derived considering that the deflection and the distortion are independent.

(2) According to the results of nonlinear FEM analyses, deflection of the stiffener is relatively small when the ultimate strength is attained. On this basis, effective width is considered to be uniform toward the longitudinal direction, and the location of neutral axis to be the same.

When symmetry condition is imposed on both ends, A and C, shear force at these cross-sections is zero. Consequently, vertical reaction force at support, B, is also zero. Denoting the initial deflection and the total deflection as w_{0s} and w, respectively, the equilibrium condition of bending moment of the beam-column can then be expressed as follows:

$$EI_z \frac{d^2(w - w_{0s})}{dx^2} = -(Pw - Pw_A + M_A) \tag{6.28}$$

where

E: Young's modulus
I_y: second moment of inertia of the cross-section for vertical bending
P: axial force
w_A: deflection at cross-section A
M_A: bending moment at cross-section A

As for initial deflection, the following expression is used allowing the different magnitudes and directions in the adjacent spans.

$$w_{0s} = \begin{cases} -W_{0s1} \sin \frac{\pi x}{a} & x \leq 0 \\ -W_{0s2} \sin \frac{\pi x}{a} & x \geq 0 \end{cases} \tag{6.29}$$

When $W_{0s1} = -W_{0s2}$, initial deflection is of an Eulerian buckling mode of a both-ends simply supported column.

Applying Eq. (6.28) in the regions, AB and BC, and considering the continuity condition of deflection and slope at point B as well as symmetry conditions at mid-span points, A and C, relationship between axial force and bending moment is derived. Then, the bending moments at cross-sections, A, B, and C, are obtained in the following forms:

$$\begin{cases} M_A = \dfrac{P}{1 - P/P_{cr}} \left(W_{0s1} - \dfrac{W_{0s1} + W_{0s2}}{\pi} \dfrac{\alpha}{\sin \alpha} \right) \\[3mm] M_B = -\dfrac{P}{1 - P/P_{cr}} \dfrac{W_{0s1} + W_{0s2}}{\pi} \dfrac{\alpha}{\tan \alpha} \\[3mm] M_C = \dfrac{P}{1 - P/P_{cr}} \left(W_{0s2} - \dfrac{W_{0s1} + W_{0s2}}{\pi} \dfrac{\alpha}{\sin \alpha} \right) \end{cases} \tag{6.30}$$

where $\alpha = (\pi/2) \cdot \sqrt{P/P_{cr}}$ and P_{cr} is the elastic flexural or flexural/torsional buckling strength of a both-ends simply supported column, which is given as P_{crb} for flat-bar and tee-bar stiffeners and as P_{crbt} for angle-bar stiffener; see Section 6.4.

To represent sideways deflection of the stiffener top or distortion, torsional angle is defined. As for initial torsion, the following expression is used:

$$\phi = \phi_0 \sin \frac{\pi x}{a} \tag{6.31}$$

Then, torsional angle is expressed in terms of axial force as follows:

$$\phi = \frac{\phi_0}{1 - P/P_t} \sin \frac{\pi x}{a} \tag{6.32}$$

where P_t is elastic torsional or flexural/torsional buckling strength, which is given by P_{crt} for flat-bar and tee-bar stiffeners and by P_{crbt} for angle-bar stiffener; see Section 6.4.

6.3.2.3 Effective width of local plate and effective thickness of stiffener after buckling

After the plate has undergone local buckling, stress in the plate becomes not uniform as illustrated in Fig. 6.37A. As local buckling deflection develops in the plate, deflection is produced also in the stiffener web in the case of a flat-bar stiffener. In this case, the sideways deflection in the stiffener web is the same magnitude as that in the plate. This also cause non-uniform stress distribution in the stiffener web as shown in Fig. 6.37B. As deflection develops, the in-plane stiffness of the plate decreases as indicated in Fig. 6.38.

Details of the derivation of effective width and effective thickness are introduced in Tanaka et al. [4].

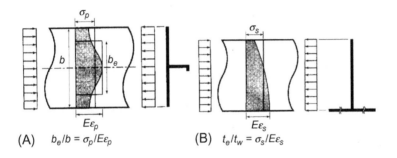

FIG. 6.37

Stress distribution in plating beyond local buckling [15]. (A) Local plate. (B) Flat-bar stiffener.

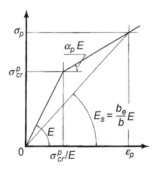

FIG. 6.38

Change in in-plane stiffness of plate after buckling [15].

6.3.2.4 Condition to determine ultimate strength

The ultimate strength is defined as the axial force, P, when the yielding condition expressed by the following equation has been satisfied at any of the seven points indicated in Fig. 6.34 [15,16].

$$\Gamma = \sigma_Y - (\sigma_P + \sigma_B + \sigma_{BT} + \sigma_{WRS}) = 0 \tag{6.33}$$

where

$$
\begin{cases}
\sigma_P = \dfrac{P}{A}: & \text{axial stress} \\[2mm]
\sigma_B = -\dfrac{M_y z}{I_y}: & \text{bending stress} \\[2mm]
\sigma_{BT} = Ey'(z_c' - z_s')\left(\dfrac{d^2\phi}{dx^2} - \dfrac{d^2\phi_0}{dx^2}\right): & \text{warping stress} \\[2mm]
\sigma_{WRS}: & \text{welding residual stress}
\end{cases}
\tag{6.34}
$$

The fourth term in the right-hand side represents warping stress, which can be ignored when flat-bar stiffeners are provided. In Eq. (6.34), M_y, which is either of M_A, M_B, or M_C in Eq. (6.30), includes the axial force, P. Therefore, Eq. (6.33) is a nonlinear equation with respect to P, and cannot be solved as a closed form solution. In the actual calculation, increasing P step by step, initial yielding condition, Eq. (6.33), is checked at seven points shown in Fig. 6.34 in the three cross-sections of a stiffener.

6.3.2.5 Validation of the proposed evaluation method

A series of nonlinear FEM analysis is performed on continuous stiffened panels shown in Table 6.5. Three types of stiffeners are considered, which are flat-bar, angle-bar, and tee-bar stiffeners, and three sizes in each type. As for local plate, two sizes are considered, which are $a \times b = 2400 \times 800$ mm and $a \times b = 4000 \times 800$ mm. plate thickness, t, is varied as 10, 13, 15, 20, and 25 mm. Yield stress of the material is 313.6 MPa for both plate and stiffeners.

Initial deflection of a buckling mode is given to the plate of which maximum magnitude is set as 0.01 times the plate thickness, t. As for initial deflection in a stiffener, flexural buckling mode of 0.001 times a is given and the sideways deflection of the same magnitude at the stiffener top. Welding residual stress is calculated applying Eq. (6.27) together with Eqs. (2.4), (2.5) in Chapter 2. The calculated compressive residual stress is $\sigma_{cri}/\sigma_Y = 0.07 - 0.18$ for stiffened panels with flat-bar stiffeners, where i is 1 and/or 2; see Fig. 6.36. For stiffened panels with angle-bar and rolled tee-bar stiffeners, $\sigma_{cri}/\sigma_Y = 0.05 - 0.11$, and those with fabricated tee-bar stiffeners, $\sigma_{cri}/\sigma_Y = 0.09 - 0.24$. These values are quite possible in actual ship and ship-like floating structures [18].

Table 6.5 Section Properties of Stiffeners for FEM Analyses (in mm) [15]			
	Type 1	**Type 2**	**Type 3**
Flat-bar	150×17	250×19	350×35
Angle-bar	$150 \times 90 \times 9/12$	$250 \times 90 \times 10/15$	$400 \times 100 \times 12/17$
Tee-bar	$150 \times 9 + 90 \times 12$	$250 \times 10 + 90 \times 15$	$400 \times 12 + 100 \times 17$
Notes: *flat-bar, $h \times t_w$; angle-bar, $h \times b_f \times t_w/t_f$; tee-bar, $h \times t_w + b_f \times t_f$.*			

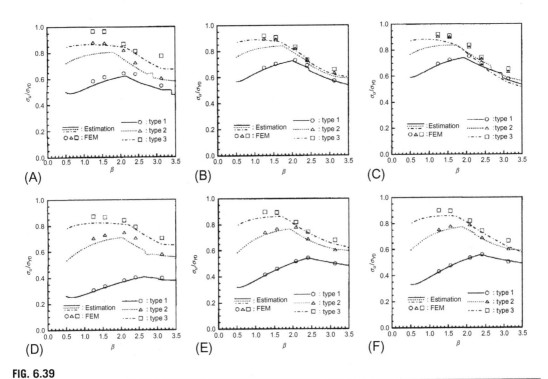

FIG. 6.39

Comparison of calculated and predicted ultimate strength of continuous stiffened plates subjected to longitudinal thrust [15]. (A) Flat-bar stiffener ($a/b = 3.0$). (B) Angle-bar stiffener ($a/b = 3.0$). (C) Tee-bar stiffener ($a/b = 3.0$). (D) Flat-bar stiffener ($a/b = 5.0$). (E) Angle-bar stiffener ($a/b = 5.0$). (F) Tee-bar stiffener ($a/b = 5.0$).

For all the combinations of local plating and stiffeners, estimated ultimate compressive strength is compared with the FEM results in Fig. 6.39. For the tee-bar, welding residual stress of only the rolled one is considered. It is observed that the proposed method gives good estimation of the ultimate strength of continuous stiffened panels subjected to longitudinal thrust. It is also known that the better estimation is obtained for the stiffened panels of which local plate has aspect ratio of $a/b = 5.0$ than for those with $a/b = 3.0$. This is because assumed collapse mode as a beam-column more likely takes place when aspect ratio, that is, the length of the beam-column, is longer.

6.3.3 CONTINUOUS STIFFENED PLATE SUBJECTED TO COMBINED LONGITUDINAL THRUST AND LATERAL PRESSURE

6.3.3.1 *Method to evaluate compressive ultimate strength*

Initial yielding condition at points, F1–F7, is considered to determine the ultimate compressive strength. The initial yielding condition in this case is expressed as [16]

$$\Gamma = \sigma_Y - (\sigma_P + \sigma_B + \sigma_{BT} + \sigma_{WRS} + \sigma_Q) = 0 \tag{6.35}$$

The above equation is fundamentally the same as Eq. (6.33), but σ_Q, which represents bending stress by lateral pressure load, is newly added.

When SI failure takes place, σ_Q is given as a bending stress at the mid-span point of a both ends clamped beam subjected to combined axial thrust and lateral pressure load as follows:

$$\sigma_Q = -\frac{qba^2}{24} \cdot \frac{6(\alpha \operatorname{cosec} \alpha - 1)}{\alpha^2} \cdot \frac{z_t}{I_y} \tag{6.36}$$

where

$$\alpha = \frac{\pi}{2}\sqrt{\frac{P}{P_{cr}}} \tag{6.37}$$

and P_{cr} is elastic flexural buckling strength under both ends simply supported condition. I_y is second moment of inertia of the full cross-section, and z_t represents z coordinate at points F2 and F3 in Fig. 6.34 measured from the neutral axis of the cross-section.

On the other hand, when PI failure occurs, σ_Q is given as a bending stress at the mid-span point of a both ends simply supported beam subjected to combined axial thrust and lateral pressure load as follows:

$$\sigma_Q = -\frac{qba^2}{8} \cdot \frac{2(\alpha \sec \alpha - 1)}{\alpha^2} \cdot \frac{z_p}{I_y} \tag{6.38}$$

where z_p represents z coordinate at point F5 in Fig. 6.34 measured from the center of geometry of the cross-section.

As the lateral pressure load increases without axial force, a plastic hinge is firstly formed at a mid-span point when lateral pressure load reaches to:

$$q_{cr} = \frac{12I_y}{a^2 b z_t}\sigma_Y \tag{6.39}$$

Then, with further increase in the lateral pressure load, plastic hinges are formed also at both clamped ends, and a plastic mechanism is formed when lateral pressure reaches to:

$$q_{max} = C_H \frac{16}{a^2 b} M_P \tag{6.40}$$

When HI failure takes place under the action of combined longitudinal thrust and lateral pressure, the compressive ultimate strength is approximated by linear interpolation equation expressed as follows:

$$\sigma_u = -\frac{\sigma_u^{cr}}{q_{max} - q_{cr}}(q - q_{cr}) + \sigma_u^{cr} \quad (q_{cr} \geq q \geq q_{max}) \tag{6.41}$$

The theoretical value of C_H in Eq. (6.40) is 1.0 when a stiffener with attached plating undergoes pure flexural buckling. However, to account for the influence of local distortion of the stiffener or local plate buckling, C_H is set as 0.95.

Here, in some cases when SI failure takes place, the ultimate strength is attained after the yielding spreads toward the height direction of the stiffener web. To take this into account, the coordinate, z_t, where initial yielding condition is examined, is shifted toward the plate. That is, for negative value of lateral pressure, q, z_t is changed to z_t' which is expressed as

$$z_t' = z_t - C_S \frac{q}{q_{cr}}(z_t - z_s) \quad (z_t' \geq z_s)$$ (6.42)

z_s is a distance between the center of geometry and shear center of the full cross-section. For angle-bar and tee-bar stiffeners, C_S is taken as 1.0, whereas for flat-bar stiffener, C_S is taken as 2/3.

6.3.3.2 Validation of proposed evaluation method

The estimated ultimate strength is plotted by solid, dashed, and chain lines for SI, PI, and HI failures, respectively, in Fig. 6.21 in Section 6.2.3.3. The predicted ultimate strength is in good agreement with the FEM results represented by solid circles. The dashed line is for SI failure using z_t instead of modified z_t' defined by Eq. (6.42). This gives a little lower estimate of the compressive ultimate strength, which indicates that the use of z_t results in conservative ultimate strength.

For the continuous stiffened panels with flat-bar and tee-bar stiffeners and the aspect ratio of the local plate being $a/b = 3.0$, the predicted ultimate strengths are compared with the FEM results in Fig. 6.40A through F. In general, it can be said that better estimation is obtained in the case when pressure load acts from the plate sides. This is because assumed condition to determine the compressive ultimate strength better corresponds to the actual collapse behavior.

More concretely, spread of yielded region toward the height direction of the stiffener web is not so much when lateral pressure is from the plate side, while yielding spreads toward the height direction under the action of lateral pressure from the stiffener side when ultimate strength is attained.

6.3.4 CONTINUOUS STIFFENED PLATES SUBJECTED TO TRANSVERSE THRUST [10]

According to the results of FEM analyses, the buckling/plastic collapse behavior of continuous stiffened plates subjected to transverse thrust is basically the same with that of continuous plating under transverse thrust as described in Section 6.2.4. The only difference is the presence of the influence of stiffener to increase buckling strength of the local plate in the case of stiffened plate.

The local buckling strength of continuous stiffened plate under transverse thrust can be derived from Eq. (4.26) in Chapter 4 setting σ_x as zero. The resulting buckling strength under transverse thrust can be expressed as

$$\sigma_{ycr}^{ST} = k_y \sigma_{ycr}$$ (6.43)

where

$$k_y = \frac{\kappa_5 - \sqrt{\kappa_5^2 - 4\kappa_3\kappa_6}}{2\kappa_3}$$ (6.44)

$$\sigma_{ycr} = \frac{E\pi^2}{12(1-\nu^2)} \left(\frac{t}{a}\right)^2 \left(\frac{a}{b} + \frac{b}{a}\right)^2$$ (6.45)

κ_3, κ_5, and κ_6 are given in Ref. [4.6] in Chapter 4. As for k_y, simpler formula is proposed by Class NK [17], which is expressed as follows:

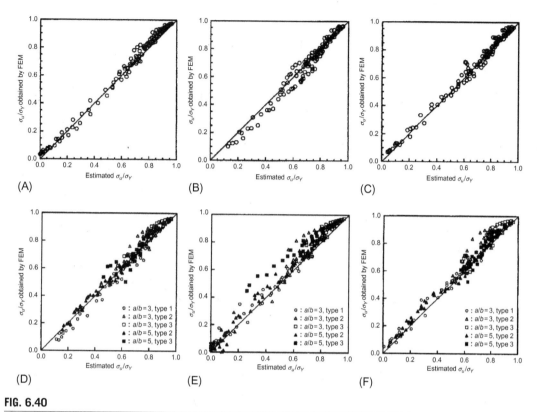

FIG. 6.40

Comparison of calculate and predicted ultimate strength of continuous stiffened plates subjected to combined longitudinal thrust and lateral pressure [16]. (A) Flat-bar ($+q$). (B) Angle-bar ($+q$). (C) Tee-bar ($+q$). (D) Flat-bar ($\pm q$). (E) Angle-bar ($\pm q$). (F) Tee-bar ($\pm q$).

$$k_y = c \left(\frac{t_w}{t}\right)^3 + 1 \tag{6.46}$$

where

for flat-bar: $c = 0.12 - 0.02(a/b)$
for angle-bar: $c = 0.98 - 0.14(a/b)$
for tee-bar: $c = \begin{cases} 0.6 & (2 \le a/b \le 3) \\ 1 - 0.133(a/b) & (3 \le a/b \le 5) \end{cases}$

When t_w/t is greater than 1.0, it is set that $t_w/t = 1.0$.

k_y calculated by Eq. (6.44) is compared with that calculated by Eq. (6.46) in Fig. 6.41. It is seen that prediction by Eq. (6.46) is quite accurate, and Eq. (6.46) can be used instead of Eq. (6.44).

It was shown in Section 5.1.4.3 in Chapter 5 that the ultimate strength formula for continuous plates under transverse thrust is expressed as a function of the slenderness ratio of the local plate expressed by:

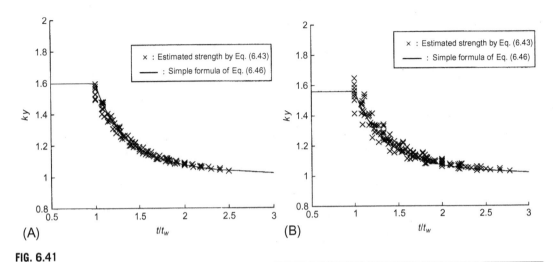

FIG. 6.41

Comparison of exact and approximate buckling coefficients for transverse thrust [11]. (A) Tee-bar stiffener. (B) Angle-bar stiffener.

$$\beta = \frac{b}{t}\sqrt{\frac{\sigma_Y}{E}} \tag{6.47}$$

The above expression comes from $\sqrt{\sigma_Y/\sigma_{cr}}$. From this viewpoint, a new slenderness parameter can be determined as follows:

$$\beta^{ST} = \frac{b}{t}\sqrt{\frac{\sigma_Y}{Ek_y}} \tag{6.48}$$

Then, it can be considered that this continuous stiffened plate is regarded equivalent with a thicker continuous plate with the thickness of $t\sqrt{k_y}$.

On the basis of the equivalence from the buckling viewpoint, the ultimate strength of continuous stiffened plate subjected to transverse thrust can be expressed by the same equation as Eq. (5.14) in Chapter 5, that is

$$\frac{\sigma_{us}^{ST}}{\sigma_Y} = \frac{b}{a}\frac{\sigma_{ue}^{s}}{\sigma_Y} + \left(1 - \frac{b}{a}\right)\frac{\sigma_{um}s}{\sigma_Y} \tag{6.49}$$

where

$$\frac{\sigma_{ue}^{s}}{\sigma_Y} = \frac{2.4}{\beta^{ST}} - \frac{1.4}{\beta^{ST\,2}} \leq 1.0 \tag{6.50}$$

$$\frac{\sigma_{um}^{s}}{\sigma_Y} = \frac{0.06}{\beta^{ST}} + \frac{0.6}{\beta^{ST\,2}} \leq 1.0 \tag{6.51}$$

Eqs. (6.50), (6.51) are derived from Eqs. (5.12), (5.13) in Chapter 5 by replacing β by β^{ST}.

The predicted ultimate strengths of continuous stiffened plates as well as continuous plates are compared with the results of FEM analysis in Fig. 6.27 in Section 6.2.4.2. Very good correlations are observed between predicted and calculated ultimate strengths for continuous stiffened panels as well as continuous plates. It should be noticed that the ultimate strength is underestimated especially when the local plate is thin if the influence of the stiffener on buckling strength is not considered.

6.3.5 CONTINUOUS STIFFENED PLATES SUBJECTED TO COMBINED TRANSVERSE THRUST AND LATERAL PRESSURE

It was pointed out in Chapter 5 that lateral pressure has two opposite effects on the buckling/plastic collapse behavior of continuous plate, which are to increase the elastic buckling strength and to reduce initial yielding strength. The effect of increasing the elastic buckling strength on the ultimate strength can be taken into account by the same approach described in Section 6.3.4. That is, the plate slenderness ratio is modified to that for the equivalent thicker plate having the same elastic buckling strength.

For the continuous stiffened plate subjected to combined transverse thrust and lateral pressure load, the coupled effects of stiffener and lateral pressure are considered by introducing the following new slenderness parameter on the basis of Eqs. (5.15), (6.43), that is

$$\beta_q^{ST} = \frac{b}{t}\sqrt{\frac{\sigma_Y}{Ek_yR_q}} \tag{6.52}$$

R_q represents the increasing ratio of the elastic buckling strength of continuous plating owing to lateral pressure and was given in Section 5.1.3 as follows:

$$R_q = \begin{cases} R_{q2} + (R_{q2} - R_{q1})\left(\dfrac{2b}{a} - 2\right) & (1 \leq a/b << 1.5) \\[2mm] R_{q2} & (1.5 < a/b < 2.0) \\[2mm] 1 + \dfrac{(qb^4/Rt^4)^{1.75}}{160(a/b)^{0.95}} & (2.0 \leq a/b) \end{cases} \tag{5.10}$$

$$R_{q1} = 1 + \frac{(qb^4/Rt^4)^2}{1560}, \quad R_{q2} = 1 + \frac{(qb^4/Rt^4)^{1.75}}{160 \cdot 2^{0.95}} \tag{5.11}$$

In addition to these, the plastic collapse load for the continuous stiffened plate due to pure lateral pressure, q_{pl} and q_{st}, as described in Section 6.2.5, are expressed as follows:

$$q_{pl} = \frac{48\alpha^4}{3\alpha^2 + 2(1 - \sqrt{1 + 3\alpha^2})} \frac{M_P^{PL}}{a^2} \tag{6.25}$$

$$q_{st} = \frac{16}{L}M_P^{ST} \tag{6.26}$$

where $\alpha = a/b$, and M_P^{PL} and M_P^{ST} are explained in Section 6.3.5.

It is considered that when the lateral pressure load, q, is higher than q_{pl} or q_{st}, then the compressive ultimate strength is zero, that is $\sigma_{uq}^{ST} = 0$.

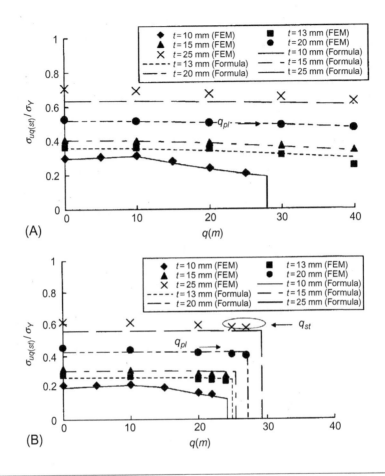

FIG. 6.42

Comparison of calculated and predicted compressive ultimate strength (stiffened plate with type 2 tee-bar stiffeners) [11]. (A) $ab = 3.0$. (B) $a/b = 5.0$.

The predicted compressive ultimate strengths are compared with results of FEM analyses in Fig. 6.42A and B for the cases of continuous stiffened panels with type 2 tee-bar stiffeners. Good agreements are observed between the predicted and calculated ultimate strengths.

At the end, a limitation in the application of the ultimate strength formulas explained above is examined since they are based on the results of FEM analyses for the limited range of geometrical parameters of stiffened panels. According to a series of comparison between predicted and FEM strengths, applicable ranges of the dimensions are as follows:

Plate slenderness ratio: $1.2 \leq \beta(= (b/t)\sqrt{\sigma_Y/E}) \leq 3.5$
Stiffener column slenderness ratio: $0.2 \leq \gamma(= \sqrt{P_Y/P_{cr}}) \leq 1.2$
Plate aspect ratio: $1.0 \leq a/b \leq 5.0$
Lateral pressure: $-40 \, \text{m} \leq q \leq 40 \, \text{m}$ water head

6.3.6 CONTINUOUS STIFFENED PLATES SUBJECTED TO BI-AXIAL THRUST [14]

Firstly, the following parameters are defined as

$$R_x = \frac{\sigma_{ubix}}{\sigma_{ux}}, \quad R_y = \frac{\sigma_{ubiy}}{\sigma_{uy}} \tag{6.52}$$

where σ_{ux} and σ_{uy} are the ultimate strength when only longitudinal thrust acts and transverse thrust acts, respectively. Figs. 6.43 and 6.44 show the interaction relationships of the ultimate strength in terms of R_x and R_y for continuous stiffened panels with type 2 tee-bar stiffeners subjected to bi-axial thrust. The dashed line is a circle and the chain line represents the von Mises' yield surface. Lines of $R_x = 1.0$ and $R_y = 1.0$ are also drawn by chain lines.

The ultimate strength of continuous stiffened panels with type 2 tee-bar stiffeners is calculated varying the slenderness ratio as a beam column but fixing the plate thickness as $t = 10$ mm and $t = 20$ mm. The calculated results are plotted in Figs. 6.43 and 6.44, respectively.

The collapse mode can be grouped into two, which are overall collapse by flexural buckling of stiffeners and local plate collapse. The former collapse occurs when longitudinal thrust is dominant, and the latter when transverse thrust is dominant. The same conclusion was obtained from Fig. 6.32 in Section 6.2.6.

On the basis of the above observations, it can be concluded that the interaction relationship asymptotically approaches to a circle with the increase in plate slenderness ratio and with decrease in column slenderness ratio. On the other hand, it asymptotically approaches to $R_x = 1.0$ and $R_y = 1.0$ when the plate slenderness ratio decreases and the column slenderness ratio increases.

It is considered that the interaction curve approaches to the fully plastic strength interaction curve. However, it does not completely reach the fully plastic interaction curve as indicated in Figs. 6.43

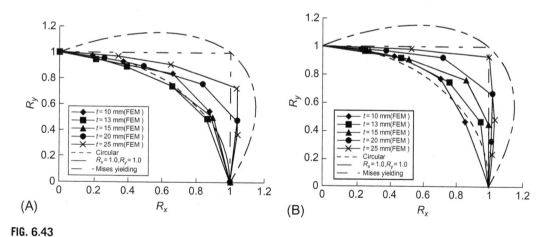

(A)

(B)

FIG. 6.43

Ultimate strength interaction relationships under bi-axial thrust (continuous stiffened panels with type 2 tee-bar stiffeners; $t = 10$ mm) [13]. (A) $a/b = 3.0$. (B) $a/b = 5.0$.

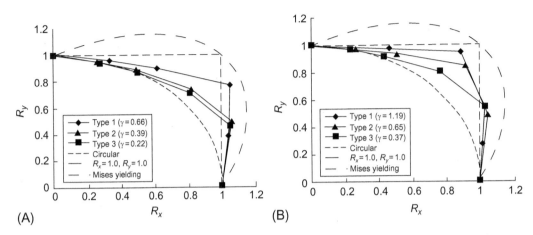

FIG. 6.44

Ultimate strength interaction relationships under bi-axial thrust (continuous stiffened panels with type 2 tee-bar stiffeners; $t = 20$ mm) [13]. (A) $a/b = 3.0$. (B) $a/b = 5.0$.

and 6.44. This may be partly because of the initial deflection, and partly because of the occurrence of the buckling. This is the reason why $R_x = 1.0$ and $R_y = 1.0$ are introduced.

Accordingly, the following interaction relationship are proposed [18].

(a) When $\beta \geq 1.0$:

$$R_x^\delta + c_1 R_y^\delta - c_2 R_y = 1.0 \tag{6.54}$$

where $\beta = (b/t) \cdot \sqrt{\sigma_Y/E}$, $R_x = \sigma_{ubix}/\sigma_{ux}$, $R_y = \sigma_{ubiy}/\sigma_{uy}$ and

$$\begin{cases} \delta = \frac{\gamma}{\beta-1}\left(\frac{a}{b} - 1\right) & \leq 2.0 \\ c_1 = 3.8 - 1.4\beta & \leq 1.0 \\ c_2 = 2.8 - 1.4\beta & \leq 0.0 \end{cases} \tag{6.55}$$

(b) When $\beta \leq 1.0$:

$$R_x = 1.0 \quad \text{or} \quad R_y = 1.0 \tag{6.56}$$

The interaction relationship under bi-axial thrust can be schematically illustrated as in Fig. 6.45.

6.3.7 CONTINUOUS STIFFENED PLATES SUBJECTED TO COMBINED BI-AXIAL THRUST AND LATERAL PRESSURE

Here, parameters similar to R_x and R_y defined in Section 6.3.6 are newly defined as

$$R_{xq} = \frac{\sigma_{ubixq}}{\sigma_{uxq}}, \quad R_{yq} = \frac{\sigma_{ubiyq}}{\sigma_{uyq}} \tag{6.57}$$

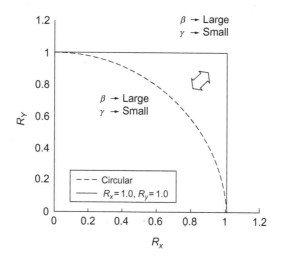

FIG. 6.45

Ultimate strength interaction relationships under bi-axial thrust [13].

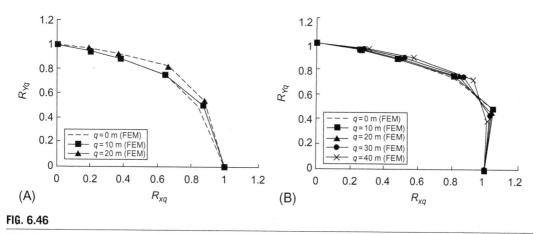

FIG. 6.46

Ultimate strength interaction relationships under combined bi-axial thrust and lateral pressure (continuous stiffened panels with type 2 tee-bar stiffeners) [13]. (A) $a/b = 3.0$. (B) $a/b = 5.0$.

where σ_{uxq} is the ultimate compressive strength of continuous stiffened plate subjected to combined longitudinal thrust and lateral pressure, while σ_{uyq} is that of continuous stiffened plate subjected to combined transverse thrust and lateral pressure. The interaction curve may be between two simple curves indicated in Fig. 6.45.

For continuous stiffened plates with type 2 tee-bar stiffeners subjected to combined bi-axial thrust and lateral pressure, calculated results are plotted in terms of R_{xq} and R_{yq} in Fig. 6.46A and B.

Fig. 6.46 indicates that the interaction relationships represented by R_{xq} and R_{yq} are almost the same regardless of the magnitude of lateral pressure. Consequently, the interaction for continuous stiffened plate subjected to bi-axial thrust can be used as the interaction for stiffened plate subjected to combined bi-axial thrust and lateral pressure by replacing R_x and R_y by R_{xq} and R_{yq}.

Rewriting Eqs. (6.56)–(6.61), the ultimate strength interaction relationship for continuous stiffened panels subjected to combined bi-axial thrust and lateral pressure is given as follows:

(a) When $\beta \geq 1.0$:

$$R_{xq}^\delta + c_1 R_{yq}^\delta - c_2 R_{yq} = 1.0 \tag{6.58}$$

where $\beta = (b/t_p) \cdot \sqrt{\sigma_Y/E}$, $R_{xq} = \sigma_{ubixq}/\sigma_{uxq}$, $R_{yq} = \sigma_{ubiyq}/\sigma_{uyq}$, and

$$\begin{cases} \delta = \frac{\gamma}{\beta-1}\left(\frac{a}{b} - 1\right) & \leq 2.0 \\ c_1 = 3.8 - 1.4\beta & \leq 1.0 \\ c_2 = 2.8 - 1.4\beta & \leq 0.0 \end{cases} \tag{6.59}$$

(b) When $\beta \leq 1.0$:

$$R_{xq} = 1.0 \quad \text{or} \quad R_{yq} = 1.0 \tag{6.60}$$

Applying the above-mentioned method, ultimate strength interaction relationships are derived and compared with the FEM results in Fig. 6.47. The same comparison is made by comparing calculated and predicted stress in the form of resultant stress:

$$\frac{\sigma_u}{\sigma_Y} = \sqrt{\left(\frac{\sigma_{ubixq}}{\sigma_Y}\right)^2 + \left(\frac{\sigma_{ubiyq}}{\sigma_Y}\right)^2} \tag{6.61}$$

The results are plotted in Fig. 6.48A–C for continuous stiffened panels with flat-bar, angle-bar, and tee-bar stiffeners, respectively. Relatively good correlations are observed between predicted and calculated results in Figs. 6.47 and 6.48.

6.3.8 CLOSED FORM FORMULAS TO EVALUATE ULTIMATE STRENGTH OF STIFFENED PLATES SUBJECTED TO COMBINED IN-PLANE LOADS AND LATERAL PRESSURE

In the method explained above, iterative calculation is required to obtain initial yielding strength, Eq. (6.35). On the basis of this method, Harada [18] and Harada et al. [19,20] derived closed form formulas to evaluate the ultimate strength of stiffened panels subjected to combined in-plane loads and lateral pressure introducing some assumptions as follows:

(1) As for the effective width, constant value is used which is the effective width when average strain reaches the yield strain.
(2) As for warping stress, constant value is used which is the warping stress when axial force is 40% of the general yielding axial force.
(3) Buckling strength of a closed form is used which includes influence of plate/stiffener web interaction.

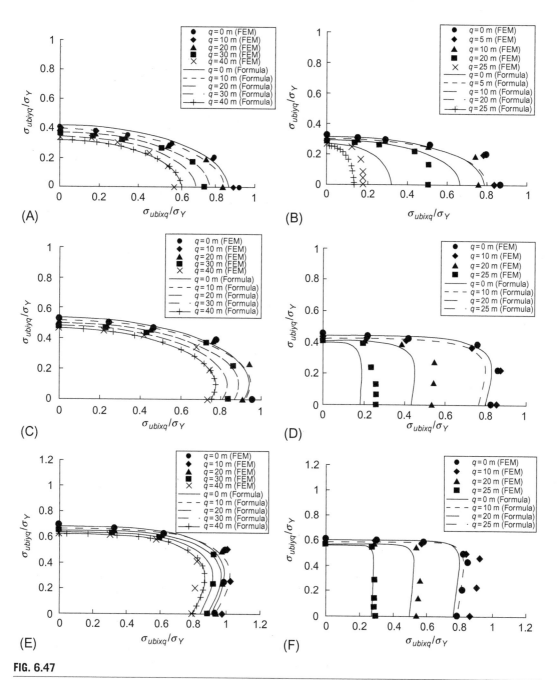

FIG. 6.47

Comparison of calculated and predicted ultimate strength interaction relationships under combined bi-axial thrust and lateral pressure [13]. (A) $a/b = 3.0$, $t_p = 15$ mm. (B) $a/b = 5.0$, $t_p = 15$ mm. (C) $a/b = 3.0$, $t_p = 20$ mm. (D) $a/b = 5.0$, $t_p = 20$ mm. (E) $a/b = 3.0$, $t_p = 25$ mm. (F) $a/b = 5.0$, $t_p = 25$ mm.

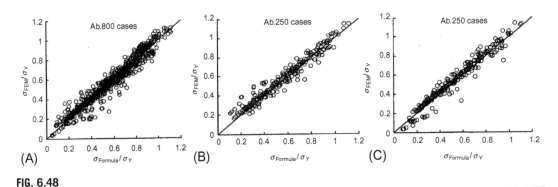

FIG. 6.48

Comparison of calculated and predicted ultimate strength of continuous stiffened plates under combined bi-axial thrust and lateral pressure [13]. (A) Tee-bar stiffeners. (B) Angle-bar stiffeners. (C) Flat-bar stiffeners.

(4) Point of stress check is moved toward the depth direction in stiffener web according to the slenderness ratio, $\gamma = \sqrt{A\sigma_Y/P_{crb}}$, of the stiffener with attached plating. F1, F3, F4, F6, and F7 are at the flange surface when slenderness ratio is less than 0.5. It is changed to the center of geometry of the stiffener cross-section without attached plating when slenderness ratio is greater than 0.55. Between 0.5 and 0.55, it is linearly interpolated with respect to slenderness ratio.

(5) Effective thickness of a flat-bar stiffener is derived from a simple condition that angle between deflected stiffener web and plate is unchanged from a right angle.

Thus, the ultimate strength formula is given as

$$\Gamma = c_1 R_y^\delta - c_2 R_y - c_3 R_x R_y + R_x^\delta - 1 = 0 \tag{6.62}$$

where

$$R_x = \frac{\sigma_{ux}*}{c_z \alpha_x \sigma_Y}, \quad R_y = \frac{\sigma_{uy}*}{c_z \alpha_y \sigma_Y} \tag{6.63}$$

$\sigma_{ux}*$ and $\sigma_{uy}*$ are the average stresses in x- and y-directions at the ultimate strength state under combined bi-axial in-plane loads and lateral pressure. Here, setting the stress ratio of applied bi-axial in-plane loads, η, as

$$\eta = \frac{\sigma_y}{\sigma_x} = \frac{\sigma_{uy}*}{\sigma_{ux}*} \tag{6.64}$$

Eq. (6.62) reduces to:

$$(c_1 \eta^\delta) R_{ux}^\delta - c_3 \eta R_{ux}^2 - c_2 \eta R_{ux} - 1 = 0 \tag{6.65}$$

The ultimate strength, R_{ux}, in a closed form is obtained from Eq. (6.65).

c_z in Eq. (6.58) represents the influence of shear load, and is given as follows:

$$c_z = \sqrt{1 - \left(\frac{\tau}{\tau_Y}\right)}, \quad \tau_Y = \frac{1}{\sqrt{3}}\sigma_Y \tag{6.66}$$

Other variables necessary for evaluation of the ultimate strength by Eq. (6.65) is given in Section 6.5.

EXERCISES

6.1 Explain three fundamental buckling modes of stiffened plate subjected to longitudinal thrust.

6.2 What is γ^B_{min}?

6.3 Explain the similarity and difference between collapse modes of stiffened plate and continuous plate both subjected to transverse thrust.

6.4 To increase the buckling strength of a square plate with four stiffeners, which is the best way among the four methods indicated in Fig. 6.49? It is assumed that stiffeners are strong enough and no overall buckling takes place.

6.5 What are PI (plate induced) failure, SI (stiffener induced) failure, and HI (plastic hinge induced) failure?

6.6 Explain about modeling extent when nonlinear FEM analysis is performed on stiffened plate.

6.7 Explain the fundamental idea of a simple method to evaluate the ultimate strength of stiffened plate subjected to longitudinal thrust.

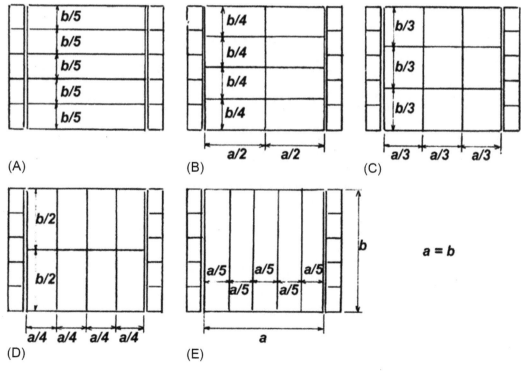

FIG. 6.49

Possible stiffening methods.

6.4 APPENDIX: BUCKLING STRENGTH OF COLUMN WITH ATTACHED PLATING UNDER AXIAL COMPRESSION

In the calculation of buckling strength of a column model in axial compression, plate is assumed to contribute only to the flexural stiffness of the column cross-section.

In the case of a flat-bar or tee-bar stiffener with symmetry cross-section, flexural buckling strength and torsional buckling strength are expressed as follows:

$$P_{crb} = \alpha_1 \tag{6.67}$$

$$P_{crt} = \alpha_4/\xi\alpha_3 \tag{6.68}$$

On the other hand, interactive flexural/torsional buckling takes place and the buckling strength is expressed as

$$P_{crbt} = \left(\beta_2^2 - \sqrt{\beta_2^2 - 4\beta_1\beta_2}\right) \Big/ 2\beta_1 \tag{6.69}$$

where

$$\begin{cases}
\alpha_1 = \pi^2 E I_y/a^2 \\
\alpha_2 = \pi^2 E I'_{yz}(z'_s - z'_c)^2/a^2 \\
\alpha_3 = I'_0/A_s - z'^2_s + z'^2_c \\
\alpha_4 = \pi^2 E I'_z(z'_s - z'_c)^2/a^2 + C' + a^2 k_\phi/\pi^2 \\
\beta_1 = \xi(\xi y'^2_s - \alpha_3) \\
\beta_2 = \xi(\alpha_1\alpha_3 - 2y'_s\alpha_2) + \alpha_4 \\
\beta_3 = \alpha_2^2 - \alpha_1\alpha_4
\end{cases} \tag{6.70}$$

and

$$\begin{cases}
\xi = A_s/A_e, \quad I_y = \int z^2 dA_e, \quad I'_y = \int z'^2 dA_s, \\
I'_z = \int y'^2 dA_s, \quad I'_{yz} = \int y'z' dA_s, \\
I'_0 = I'_y + I'_z + (y'^2_s + z'^2_s)A_s, \quad k_\phi = \frac{Et_p^3}{3b}
\end{cases} \tag{6.71}$$

In the above expression, A_s and A_e are cross-sectional area of the stiffener and the stiffener with attached plate with effective width, respectively. C' is Saint-Venant's torsional stiffness of the stiffener cross-section.

6.5 APPENDIX: PARAMETERS IN CLOSED FORM FORMULAS TO EVALUATE ULTIMATE STRENGTH OF STIFFENED PLATE SUBJECTED TO COMBINED BI-AXIAL IN-PLANE LOADS AND LATERAL PRESSURE

6.5.1 CONDITION TO EVALUATE THE ULTIMATE STRENGTH

Harada [18] and Harada et al. [19,20] consider four loading conditions combining compression/tension in x- and y- directions. For each case, variables in Eq. (6.62):

$$\Gamma = c_1 R_y^\delta - c_2 R_y - c_3 R_x R_y + R_x^\delta - 1 = 0 \tag{6.62}$$

are defined as follows.

6.5.1.1 Bi-axial compression

$$\delta = \frac{\gamma}{\beta - 1}\left(\frac{a}{b} - 1\right) \geq 2.0 \tag{6.72}$$

$$\begin{cases} c_1 = 3.8 - 1.4\beta \leq 1.0 \\ c_2 = 2.8 - 1.4\beta \leq 0.0 \\ c_3 = 0.0 \end{cases} \tag{6.73}$$

$$\begin{cases} \alpha_x = \sigma_{uxq}/\sigma_Y \\ \alpha_y = \sigma_{uyq}/\sigma_Y \end{cases} \tag{6.74}$$

Here, σ_{uxq} and σ_{uyq} are the ultimate strength only when load in x-direction and that in y-direction are applied, respectively. β and γ are slenderness ratios of local plate and stiffener with attached plating, which are defined as

$$\beta = \frac{b}{t}\sqrt{\frac{\sigma_Y}{E}}, \quad \gamma = \sqrt{\frac{A\sigma_Y}{P_{cr}}} \tag{6.75}$$

As for yield stress, σ_Y is defined as the area mean of yield stresses, σ_{YP} for plating and σ_{YS} for stiffeners as follows:

$$\sigma_Y = \frac{1}{A}(A_p\sigma_{YP} + A_s\sigma_{YS}) \tag{6.76}$$

A_s and A_p are cross-sectional areas of a stiffener and the attached plating, respectively, and A is the total area of the stiffener with attached plating as

$$A = A_p + A_s \tag{6.77}$$
$$A_p = bt \tag{6.78}$$

$$A_s = \begin{cases} ht_w & \text{(flat-bar)} \\ (h - t_f)t_w + b_ft_f & \text{(angle-bar)} \\ ht_w + b_ft_f & \text{(tee-bar)} \end{cases} \tag{6.79}$$

When local plate buckling has taken place, effective width, b_e, is used instead of full width, b. In the case of flat-bar stiffener, the effective thickness of the web, t_e, is also used instead of the full thickness, t_w. Then, A_p and A_s in Eqs. (6.76)–(6.78) have to be replaced by the effective area, A_e, defined by Eq. (6.101)–(6.103) in Section 6.5.5.

6.5.1.2 Combined compression in x-direction and tension in y-direction

$$\delta = 2.0 \tag{6.80}$$

$$\begin{cases} c_1 = 1.0 \\ c_2 = 0.0 \\ c_3 = 1.0 \end{cases} \tag{6.81}$$

$$\begin{cases} \alpha_x = \begin{cases} 1.0 & \text{when} \quad \beta \geq 2.0 \\ \sigma_{uxq}/\sigma_Y & \text{when} \quad \beta \leq 2.0 \end{cases} \quad \alpha_y = 1.0 \tag{6.82}$$

$\sigma_{uxq}*$ has to be equal to or less than σ_{uxq}.

6.5.1.3 Combined tension in x-direction and compression in y-direction

$$\delta = 2.0 \tag{6.83}$$

$$\begin{cases} c_1 = 1.0 \\ c_2 = 0.0 \\ c_3 = 1.0 \end{cases} \tag{6.84}$$

$$\begin{cases} \alpha_x = 1.0 \\ \alpha_y = 1.0 \end{cases} \tag{6.85}$$

$\sigma_{uyq}*$ has to be equal to or less than σ_{uyq}.

6.5.1.4 Bi-axial tension

$$\delta = 2.0 \tag{6.86}$$

$$\begin{cases} c_1 = 1.0 \\ c_2 = 0.0 \\ c_3 = 1.0 \end{cases} \tag{6.87}$$

$$\begin{cases} \alpha_x = 1.0 \\ \alpha_y = 1.0 \end{cases} \tag{6.88}$$

6.5.2 BUCKLING STRENGTH OF LOCAL PLATE

As shown in Fig. 4.3 in Chapter 4, the buckling strength of a local plate looks as if it decreases owing to initial deflection. Elastoplastic buckling strength of local plate is assumed to be given by

$$\frac{\sigma_{cr}^p}{\sigma_Y} = \frac{1}{2} \left\{ \frac{\sigma_{ecr}^p}{\sigma_Y} + 1 - \sqrt{\left(\frac{\sigma_{ecr}^p}{\sigma_Y} - 1 \right)^2 + \Delta} \right\} \tag{6.89}$$

where

$$\sigma_{ecr}^p = \frac{k_x \pi^2 E}{3(1 - v^2)} \left(\frac{t}{b} \right)^2 \tag{6.90}$$

$$k_x = c_x \left(\frac{t_w}{t} \right)^3 + 1 \tag{6.91}$$

$$c_x = \begin{cases} 0.07 & \text{for angle-bar and tee-bar} \\ 0.03 & \text{for tee-bar} \end{cases} \tag{6.92}$$

6.5.3 EFFECTIVE WIDTH OF THE LOCAL PLATE BEYOND THE OCCURRENCE OF LOCAL BUCKLING

$$\frac{b_e}{b} = (1 - \alpha_p)\frac{\sigma_{xcr}}{\sigma_{YP}} + \alpha_p \tag{6.93}$$

where

$$\alpha_p = \frac{1 + \{a/(m_1 b)\}^4}{3 + \{a/(m_1 b)\}^4} \tag{6.94}$$

6.5.4 EFFECTIVE THICKNESS OF FLAT-BAR STIFFENER BEYOND THE OCCURRENCE OF LOCAL PLATE BUCKLING

The effective thickness of a flat-bar stiffener web is given as

$$\frac{t_e}{t_w} = \frac{2\pi^2}{3}\left(\frac{h}{b}\right)^2\left(1 - \frac{b_e}{b}\right) \tag{6.95}$$

6.5.5 VARIABLES NECESSARY TO EVALUATE WARPING STRESS

$$\alpha = 1 - \frac{\sigma_w}{\sigma_Y} \tag{6.96}$$

where

$$\sigma_w = \begin{cases} \frac{pi^2 EW_{s2}b_f}{2ah_w}(0.5t_f - h_w) \times \left(1 - \frac{0.4\sigma_Y A}{P_{crt}}\right) & \text{(tee-bar, angle-bar)} \\ 0 & \text{(flat-bar)} \end{cases} \tag{6.97}$$

6.5.6 σ_{uxq}

The ultimate strength, σ_{uxq}, under combined longitudinal thrust and lateral pressure is calculated as follows.

6.5.6.1 When thrust load is dominant ($0 \leq q \leq q_{cr}$ or $-q_{cr} \leq q \leq 0$)

$$\sigma_{uxq} = \text{Minimum}[\sigma_{PI}, \quad \sigma_{SI}] \tag{6.98}$$

where

$$\sigma_{PI} = \frac{A_e}{A}\left\langle P_{cr}\left\{\frac{1}{A_e} + \frac{z_p}{I_e}\left(W_{s1} + \frac{5a^4 b|q|}{384EI_e}\right)\right\} + \alpha\sigma_Y - \frac{qa^2 b}{8}\frac{z_p}{I_e}\right.$$

$$-\sqrt{\left[P_{cr}\left\{\frac{1}{A_e}+\frac{z_p}{I_e}\left(W_{s1}+\frac{5a^4b|q|}{384EI_e}\right)\right\}+\alpha\sigma_Y-\frac{qa^2b}{8}\frac{z_p}{I_e}\right]^2}$$
$$-4\left(\alpha\sigma_Y-\frac{qa^2b}{8}\frac{z_p}{I_e}\right)\frac{P_{cr}}{A_e}\Bigg\rangle \tag{6.99}$$

$$\sigma_{SI}=\frac{A_e}{A}\Bigg\langle P_{cr}\left\{\frac{1}{A_e}+\frac{z_s}{I_e}\left(W_{s1}+\frac{a^4b|q|}{384EI_e}\right)\right\}+\alpha\sigma_Y-\frac{qa^2b}{24}\frac{z_s}{I_e}$$
$$-\sqrt{\left[P_{cr}\left\{\frac{1}{A_e}+\frac{z_s}{I_e}\left(W_{s1}+\frac{a^4b|q|}{384EI_e}\right)\right\}+\alpha\sigma_Y-\frac{qa^2b}{24}\frac{z_s}{I_e}\right]^2}$$
$$-4\left(\alpha\sigma_Y-\frac{qa^2b}{24}\frac{z_s}{I_e}\right)\frac{P_{cr}}{A_e}\Bigg\rangle \tag{6.100}$$

where A_e is the effective cross-sectional area introducing effective width and effective thickness beyond the local plate buckling, and is given as follows:

$$A_e=\begin{cases}A_{pe}+A_s & \text{(angle-bar, tee-bar)}\\ A_{pe}+A_{se} & \text{(tee-bar)}\end{cases} \tag{6.101}$$

$$A_{pe}=b_e t \tag{6.102}$$

$$A_{se}=h_w t_e \tag{6.103}$$

6.5.6.2 When pressure load is dominant ($q \geq q_{cr}$ or $-q_{cr} \geq q$)

$$\sigma_{uxq}=\sigma_{HI} \tag{6.104}$$

$$\sigma_{HI}=-\frac{\sigma_{HI}*}{q_{st}-q_{cr}}(|q|-q_{cr})+\sigma_{HI}* \tag{6.105}$$

$$\sigma_{HI}*=\text{Minimum}\ [\sigma_{PI}|_{q=q_{cr}},\quad \sigma_{SI}|_{q=q_{cr}}] \tag{6.106}$$

$$q_{st}=\frac{16}{a^2b}Z_p\sigma_Y,\quad q_{cr}=\frac{12}{a^2b}\frac{I}{s_{s0}}\sigma_Y \tag{6.107}$$

6.5.7 σ_{uyq}

The ultimate stress, σ_{uyq}, under combined transverse thrust and lateral pressure is given as

$$\sigma_{uyq}=\text{Minimum}\ [\sigma_{uyqq(ST)},\ \sigma_{uyq(C)}]\leq \sigma_{uy(ST)} \tag{6.108}$$

$\sigma_{uyqq(ST)}$ represents the ultimate strength of continuous stiffened plate, and is given as follows:

$$\sigma_{uyqq(ST)} = \frac{\sigma_{uyq(ST)}}{1 + A\gamma_q^B} \tag{6.109}$$

where

$$\sigma_{uyq(ST)} = \sigma_{YP}\left\{\left(\frac{2.4}{\beta_{sq}} - \frac{1.4}{\beta_{sq}^2}\right)\frac{b}{a} + \left(\frac{0.06}{\beta_{sq}} - \frac{0.6}{\beta_{sq}^2}\right)\left(1 - \frac{b}{a}\right)\right\} \tag{6.110}$$

$$\sigma_{uy(ST)} = \sigma_{YP}\left\{\left(\frac{2.4}{\beta_s} - \frac{1.4}{\beta_s^2}\right)\frac{b}{a} + \left(\frac{0.06}{\beta_s} - \frac{0.6}{\beta_s^2}\right)\left(1 - \frac{b}{a}\right)\right\} \tag{6.111}$$

and

$$\beta_{sq} = \frac{b}{t}\sqrt{\frac{\sigma_{YP}}{E\kappa_Y R_q}}, \quad \beta_s = \frac{b}{t}\sqrt{\frac{\sigma_{YP}}{E\kappa_Y}} \tag{6.112}$$

$$R_q = 1 + \frac{\{|q|E(b/t)^4\}^{1.75}}{160(a/b)^{0.95}}, \quad \kappa_Y = c_Y\left(\frac{t_w}{t}\right)^3 + 1 \tag{6.113}$$

$$c_Y = \begin{cases} 0.98 - 0.14(a/b) & \text{(angle-bar)} \\ 0.6 & \text{(tee-bar} \quad 2 \le a/b \le 3) \\ 1.0 - 0.133(a/b) & \text{(tee-bar} \quad 3 \le a/b \le 5) \\ 0.12 - 0.22(a/b) & \text{(flat-bar)} \end{cases} \tag{6.114}$$

Furthermore,

$$\gamma_q = \frac{|q|b^2}{\sigma_{YP}t^2} \tag{6.115}$$

$$\begin{cases} A = 0.067 - 0.0121(a/b) \le 0.0 \\ B = 0.154 + 0.577(a/b) \le 0.0 \end{cases} \tag{6.116}$$

On the other hand, $\sigma_{uyq(C)}$ in Eq. (6.108) represents the ultimate strength of a local plate which is clamped along all sides.

$$\sigma_{uyq(C)} = \begin{cases} \sigma_{uy(C)}(1 - F\gamma_q + G\gamma_q^2) & \text{(when } q < q_{pl} \text{ or } q < q_{st}) \\ 0 & \text{(when } q_{pl} \ge q \text{ or } q \ge q_{st}) \end{cases} \tag{6.117}$$

where

$$\sigma_{uy(C)} = \sigma_{YP}\left\{(1.27 - 0.18\beta)\frac{b}{a} + \left(\frac{3.48}{\beta^2} - \frac{0.262}{\beta}\right)\left(1 - \frac{b}{a}\right)\right\} \tag{6.118}$$

$$\beta = \frac{b}{t}\sqrt{\frac{\sigma_{YP}}{E}} \tag{6.119}$$

$$\begin{cases} F = 0.0821 + 0.0191(a/b) \\ G = 0.00495 + 0.000836(a/b) \end{cases} \tag{6.120}$$

q_{pl} and q_{st} in Eq. (6.118) represent fully plastic collapse strength of local plate and a stiffener with attached plating, respectively, which are given as

$$q_{pl} = \frac{12\sigma_{YP}t^2}{a^2} \frac{(a/b)^4}{3(a/b)^2 - 2(\sqrt{1 + 3(a/b)^2} - 1)} \tag{6.121}$$

$$q_{st} = \frac{16}{a^2b}Z_p\sigma_Y \tag{6.122}$$

REFERENCES

[1] Ueda Y, Yao T. Ultimate strength of compressed stiffened panels and minimum stiffness ratio of their stiffeners. Eng Struct 1983;5:97–107.

[2] Ueda Y, Yao T, Katayama M, Nakamine M. Minimum stiffness ratio of a stiffener against ultimate strength of a plate (2nd report). J Soc Nacal Arch Jpn 1978;143:308–315 [in Japanese].

[3] Yoshida K. Buckling analysis of plate structures by the strip elements. J Soc Naval Arch Jpn 1971;130:161–171 [in Japanese].

[4] Tanaka S, Yanagihara D, Yasuoka A, Harada M, Okazawa S, Fujikubo M, et al. Evaluation of ultimate strength of stiffened panels under longitudinal thrust. Mar Struct 2014, 36:21–50.

[5] Yao T, Fujikubo M, Yanagihara D. Consideration on FEM modelling for buckling/plastic collapse analysis of stiffened panels. Trans West Jpn Soc Naval Arch 1998;95:121–128 [in Japanese].

[6] MSC.Marc 2005r3. User's guide.

[7] Carlsen C. A parametric study of collapse of stiffened panels in compression. Struct Eng 1980;58B(2):33–40.

[8] Khedmati M. Ultimate strength of ship structural members and systems considering local pressure effects. [Doctoral thesis]. Hiroshima: Hiroshima University; 2000.

[9] Yanagihara D, Fujikubo M, Harada M. Estimation of ultimate strength of continuous stiffened plate under combined thrust and lateral pressure. J Soc Naval Arch Jpn 2003;194:161–170 [in Japanese].

[10] Smith C, Davidson P, Chapman J, Dowling P. Strength and stiffness of ship plating under in-plane compression and tension. Trans RINA 1987;130:277–296.

[11] Fujikubo M, Harada M, Yao T, Khedmati M, Yanagihara D. Estimation of ultimate strength of continuous stiffened plate under combined transverse thrust and lateral pressure, part 2: continuous stiffened plate. Mar Struct 2005;18:411–427.

[12] Kusuda T. Plastic analysis of plate structures subjected to transverse load. J Soc Naval Arch Jpn 1960;108:195–202 [in Japanese].

[13] Harada M, Fujikubo M, Yanagihara D. Estimation of ultimate strength of continuous stiffened plate under combined biaxial thrust and lateral pressure. J Soc Naval Arch Jpn 2004;196:189–198 [in Japanese].

[14] Faulkner D, Adamchak J, Snyder J, Vetter M. Synthesis of welded grillages to withstand compression and normal loads. Comput Struct 1973;3:221–246.

[15] Fujikubo M, Yanagihara D, Yao T. Estimation of ultimate strength of continuous stiffened panels under thrust. J Soc Naval Arch Jpn 1999;185:203–212 [in Japanese].

[16] Fujikubo M, Yanagihara D, Yao T. Estimation of ultimate strength of continuous stiffened panels under thrust (2nd report). J Soc Naval Arch Jpn 1999;186:631–638 [in Japanese].

[17] Class NK. Guidelines for tanker structures, guidelines for direct strength analysis; 2001.

[18] Harada M. Study on practical method to evaluate ultimate strength of stiffened panels. Doctoral thesis. Hiroshima University; 2004.

[19] Harada M, Fujikubo M, Yanagihara D. Development of a set of simple formulae for estimation of ultimate strength of a continuous stiffened plate under combined loads. J Jpn Soc Naval Arch Ocean Eng 2005;2:387–395 [in Japanese].

[20] Harada M, Fujikubo M, Yanagihara D. Development of a set of closed-form formulae for estimation of ultimate strength of a continuous stiffened plate under combined in-plane loads and lateral pressure. ClassNK technical bulletin 2007;25:11–21.

BUCKLING/PLASTIC COLLAPSE BEHAVIOR AND STRENGTH OF PLATE GIRDERS SUBJECTED TO COMBINED BENDING AND SHEAR LOADS

7.1 RESEARCH ON BUCKLING OF PLATE GIRDERS IN SHIP AND SHIP-LIKE FLOATING STRUCTURES

Plate girders consisting of web and flange plates have been main longitudinal strength members in single-hull ship structures. They are also main transverse strength members. The plate girders are usually subjected to combined shear and bending/compression loads. Basler performed a series of experiments to investigate into the collapse behavior of plate girders subjected to shear and bending loads and derived design formulas to evaluate ultimate strength [1–4]. Akita and Fujii [5,6] and Fujii [7] revised Basler's formulas, taking into account of the collapse of flange due to lateral load produced by the action of tension field in the web panel after shear buckling has occurred.

These formulas are, however, for plate girders with free flanges of a finite width such as deck or bottom girders or transverse rings in single hull tankers. In 1992, MARPOL convention [8] came into effect and all tankers had become to have double hull structures. Because of this, flange of bottom girders has become a part of continuous plating such as inner bottom plating or longitudinal bulkhead. The isolated free flange and continuous plating may show different buckling behavior as a flange of the girder during buckling/plastic collapse.

Regarding the collapse behavior and strength of plate girders of which flanges is a part of the continuous plating, Olaru et al. [9] and Olaru [10] performed a series of nonlinear finite element method (FEM) analysis to clarify its buckling/plastic collapse behavior and the ultimate strength. The influences of stiffeners and perforation in the web panel are also examined.

7.2 BUCKLING/PLASTIC COLLAPSE BEHAVIOR AND STRENGTH OF UNSTIFFENED PLATE GIRDERS

7.2.1 BASLER'S FINDINGS

7.2.1.1 *Plate girders subjected to pure bending*

In the early times, it was known that column members cannot carry further loads above their buckling strength. On the other hand, it was also known that plate members can carry further loads above the buckling strength when elastic buckling takes place. This was the reason why somewhat smaller safety factor was used for web buckling than that for column buckling, although the exact value of safety factor was not known. Because of this, a series of research works had started by Basler [2] at Lehigh University to clarify the load carrying capacity of plate girders beyond the web buckling.

Basler conducted a series of experiments on 27 plate girder specimens to clarify the collapse behavior of plate girders subjected to shear, bending, and their combinations. He explained the buckling collapse behavior of the web of a plate girder subjected to pure bending showing the measured results in Fig. 7.1 [1].

Fig. 7.1 indicates that

(1) buckling deflection in the web develops only in the compression side of bending;

(2) bending moment is not sustained at the web where deflection develops; and

(3) consequently, compressive stress does not increase in the region where large deflection is produced after web buckling.

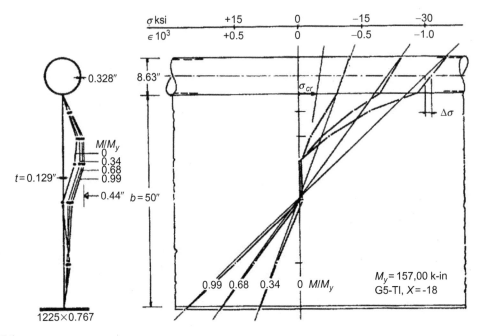

FIG. 7.1

Measured deflection and stress distribution [1].

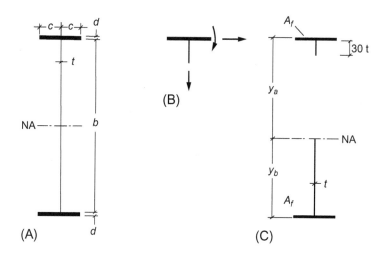

FIG. 7.2

Possible buckling mode of compression flange.

Test results indicated that the web elastic buckling does not directly lead to the failure of the plate girder in bending. Basler recognized the importance of the compression flange, and concluded that buckling of the compression flange could be a trigger to the overall collapse of the plate girder [1].

For the plate girder of which cross-section is shown in Fig. 7.2A, Basler considered three buckling modes as flange buckling, which are flexural buckling (with vertical deflection toward the web), lateral buckling (with horizontal deflection), and torsional buckling; see Fig. 7.2B. Regarding the flexural buckling, Basler discussed the limit of the slenderness ratio, b/t_w, where b and t_w are the depth and the thickness of web plating. As for the lateral buckling and torsional buckling, he discussed the minimum width to thickness ratio of the flange to exclude torsional buckling. Fig. 7.2C shows the effective cross-section of a buckled plate girder in bending.

On the basis of the observed results, he derived design formulas to evaluate the ultimate strength of a plate girder subjected to pure bending, which shall be explained later.

7.2.1.2 Plate girders subjected to shear load

It had been known that the formation of the tension field in web plating largely affects the ultimate strength of plate girders subjected to shear loads; see Fig. 7.3. From this point of view, Basler considered that the stiffener spacing: a, the girder depth: b, the web thickness: t, as well as the yield stress: σ_Y, and the Young's modulus: E of the material are the influential parameters which may affect the ultimate shear strength. He proposed the formula

$$V_u = V_P f(a,\ b,\ t,\ \sigma_Y,\ E) \tag{7.1}$$

where V_P is the fully plastic shear force, which is defined as

$$V_P = \frac{1}{\sqrt{3}} \sigma_Y bt \tag{7.2}$$

The tensile force produced by tension field in the web cannot be sustained by a flange which has relatively low flexural stiffness against bending in the plane of the web. Contrary to this, the vertical

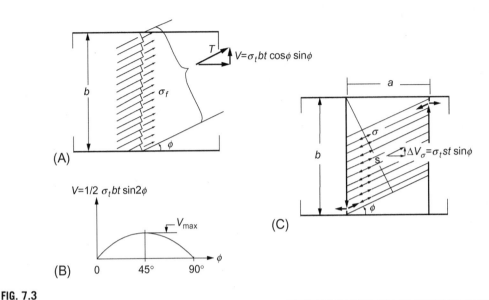

FIG. 7.3

Action of tension field in web of plate girder [3].

stiffener can carry the force from the tension field; see Fig. 7.3C. Consequently, "only a part of the web contains a pronounced tension field which gives rise to shear force" beyond the web buckling as Basler mentioned. On the basis of this concept, Basler developed the formulas to evaluate the ultimate shear strength of plate girders. The detail of his formulation is introduced later together with Fujii's formulas.

7.2.1.3 Plate girders subjected to combined bending and shear loads

For the collapse of plate girders subjected to combined bending and shear loads, Basler considered the ultimate strength interaction relationships. In deriving the ultimate strength interaction relationship, Basler assumed that the bending moment does to affect the interaction relationship until it reaches to a certain value, M_f, which is the bending moment produced by upper and lower flanges alone when they have yielded. Another measure is the fully plastic bending moment, M_P, which is attained under pure bending. Fig. 7.4 schematically shows the interaction relationship, where M_Y indicates the initial yielding strength of the plate girder cross-section in bending. The point Q_1 corresponds when the bending moment is equal to M_f, and the point Q_2 to M_P.

Fig. 7.4 shows the strength interaction relationships together with assumed stress distributions when a plate girder collapses under combined shear and bending loads. The interaction relationship is expressed as

$$\left(\frac{V}{V_u}\right)^n + \left(\frac{M - M_f}{M_P - M_f}\right)^m = 1 \qquad (7.3)$$

In Fig. 7.4, (3a) stands for the case of $m = 2$ and $n = 1$, and (3b) is the case of $m = n = 2$, respectively.

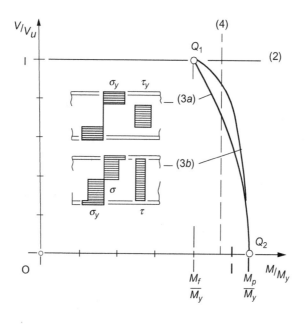

FIG. 7.4

Ultimate strength interaction relationship [4].

Basler further modified this interaction relationship from the viewpoint of tension field. He cut off the curve by the requirement [2]:

$$\frac{M}{M_u} \leq 1.0 \tag{7.4}$$

The test results by Basler are plotted on the interaction relationships in Fig. 7.4 with the modification by Eq. (7.4) [4]. It is seen that pretty good correlations are observed between predicted and measured ultimate strength. However, Fujii [7] pointed out that

(1) The contribution of a flange which resists against the action of tension field in a web is not considered.
(2) The direction of the tension field in a web is not properly estimated.
(3) The influence of shear force produced in the vertical stiffener is ignored when the equilibrium condition of forces is considered in the web panel.

These problems were solved by Fujii [5–7] as shall be explained in Section 7.2.2.

7.2.2 FUJII'S FORMULATIONS TO EVALUATE ULTIMATE STRENGTH OF PLATE GIRDERS

Assuming that the plate girder is so designed as not to undergo overall lateral buckling or local flange buckling and vertical stiffeners are strong enough, Fujii derived formulas to evaluate the ultimate strength of plate girders subjected to combined bending and shear loading [5–7].

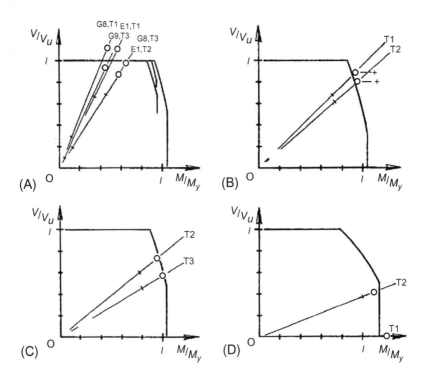

FIG. 7.5

Comparison of predicted and measured ultimate strength of plate girders subjected to combined bending and shear loads [4].

7.2.2.1 *Plate girders subjected to shear load*

Stress components

When the web of the plate girder is in pure shear stress state, the stress components in η-ξ coordinate system shown in Fig. 7.6 can be expressed as follows:

$$\left.\begin{aligned} \sigma_\xi &= \tau \sin 2\phi \\ \sigma_\eta &= -\tau \sin 2\phi \\ \tau_{\xi\eta} &= \tau \cos 2\phi \end{aligned}\right\} \tag{7.5}$$

Denoting the shear stress, τ, at local web buckling as τ_{cr}, the above stress components become as

$$\left.\begin{aligned} \sigma_\xi &= \tau_{cr} \sin 2\phi \\ \sigma_\eta &= -\tau_{cr} \sin 2\phi \\ \tau_{\xi\eta} &= \tau_{cr} \cos 2\phi \end{aligned}\right\} \tag{7.6}$$

The shear buckling stress, τ_{cr}, can be calculated as

$$\tau_{cr} = \frac{\pi^2 k_s E}{12(1 - v^2)} \left(\frac{t_w}{h}\right)^2 \tag{7.7}$$

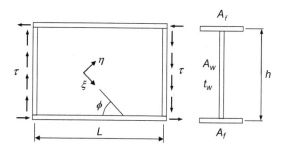

FIG. 7.6

Notation of plate girder and definition of coordinate system.

where

$$
k_s = \begin{cases} 8.98 + \frac{5.61}{\lambda^2} - \frac{1.99}{\lambda^3} & 1.0 \le \lambda \\ 8.39\lambda - 3.44 + \frac{2.31}{\lambda} + \frac{5.34}{\lambda^2} & \lambda < 1.0 \end{cases}
\tag{7.8}
$$

$$\lambda = \ell/h \quad \text{(aspect ratio of web panel)}$$

When τ_{cr} is greater than $0.5 \times \tau_Y$, the following τ_{cr}^p is used instead of τ_{cr}.

$$
\tau_{cr}^p = \tau_Y \left(1 - \frac{\tau_Y}{4\tau_{cr}}\right)
\tag{7.9}
$$

Equilibrium conditions

Until the web buckling takes place in shear, normal stresses in two perpendicular directions are the same in magnitude but opposite in sign as is known from Eq. (7.5), and increase with the increase in applied shear stress, τ. Beyond the web buckling in shear, a tension field is formed as indicated in Fig. 7.3, and the tensile stress, σ_1, in the direction of buckling wave can further increases; see Fig. 7.7. On the other hand, compressive stress, σ_2, which is in the perpendicular direction to the buckling wave, is considered not to increase as the buckling deflection develops. That is

$$
\sigma_2 = -\tau_{cr} \cos 2\phi
\tag{7.10}
$$

Here, the normal and shear stresses along the boundary of the web panel are considered; see Fig. 7.7. Considering the equilibrium of forces along the upper edge of the web panel, normal and shear stresses are derived as follows:

$$
\sigma_v = \sigma_1 \sin^2 \alpha + \sigma_2 \cos^2 \alpha
\tag{7.11}
$$
$$
\tau = (\sigma_1 - \sigma_2) \sin \alpha \cos \alpha
\tag{7.12}
$$

In the same manner, considering the equilibrium condition of forces along the vertical edge of the web panel, normal and shear stresses along this side are derived as follows:

$$
\sigma_w = \sigma_1 \cos^2 \alpha + \sigma_2 \sin^2 \alpha
\tag{7.13}
$$

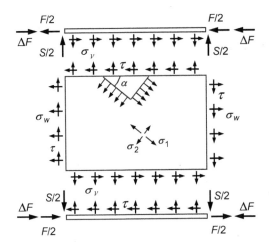

FIG. 7.7

Stresses produced in and along the boundaries of web panel.

$$\tau = (\sigma_1 - \sigma_2)\sin\alpha\cos\alpha \tag{7.14}$$

Assuming that the stresses are uniform along the side, the applied shear force can be expressed as

$$V = ht_w\tau = A_w(\sigma_1 - \sigma_2)\sin\alpha\cos\alpha \tag{7.15}$$

where $A_w = ht_w$ is the sectional area of web panel at intersection with a vertical stiffener.

Ultimate shear strength
Case with strong flange. When the flange has high flexural stiffness against pull-in action toward the web, all over the web panel may yields with uniform stress. Applying the Tresca's yield condition, the fully yielded condition can be expressed as

$$|\sigma_1 - \sigma_2| = \sigma_{wY} \tag{7.16}$$

where σ_{wY} is the yield stress of the web panel. Substituting Eq. (7.10) into Eq. (7.16), the following relationship is obtained:

$$\sigma_1 = \sigma_{wY} - \tau_{cr}\sin 2\alpha \tag{7.17}$$

Substituting Eq. (7.10), (7.17) into the first equation in Eq. (7.11), the following relationship is derived:

$$\frac{\sigma_v}{\sigma_{wY}} = \sin^2\alpha - \frac{v_{cr}}{2}\sin 2\alpha \tag{7.18}$$

where

$$v_{cr} = \frac{\tau_{cr}}{\tau_{wY}}$$

Considering the trigonometric relationships:

$$2 \sin^2 \alpha = 1 - \cos 2\alpha$$
$$\cos^2 2\alpha = 1 - \sin^2 2\alpha$$

the following expression is obtained from Eq. (7.15):

$$\sin 2\alpha = \frac{\left(1 - 2\frac{\sigma_v}{\sigma_{wY}}\right) v_{cr} + \sqrt{1 + v_{cr}^2 - \left(1 - 2\frac{\sigma_v}{\sigma_{wY}}\right)^2}}{1 + v_{cr}^2} \tag{7.19}$$

Substituting Eqs. (7.16), (7.19) into Eq. (7.15), the following relationship is derived:

$$v = \frac{V}{V_P} = \frac{\left(1 - 2\frac{\sigma_v}{\sigma_{wY}}\right) v_{cr} + \sqrt{1 + v_{cr}^2 - \left(1 - 2\frac{\sigma_v}{\sigma_{wY}}\right)^2}}{1 + v_{cr}^2} \tag{7.20}$$

where $V_P = bt\tau_{wY}$ is the fully plastic shear force. In the above expression, v_{cr} is constant when the size of the plate girder is specified, but σ_v/σ_{wY} varies with the increase in the applied shear load. Therefore, the maximum value of v is attained when

$$\frac{\partial v}{\partial \left(\frac{\sigma_v}{\sigma_{wY}}\right)} = 0 \tag{7.21}$$

Eq. (7.21) is satisfied when

$$\frac{\sigma_v}{\sigma_{wY}} = \frac{1 - v_{cr}}{2} \tag{7.22}$$

Substituting Eq. (7.22) into Eq. (7.20), the ultimate shear strength is obtained as follows:

$$v_u = \frac{V_u}{V_P} = 1.0 \tag{7.23}$$

At the same time, the angle of tension field, α, in this case is obtained from Eq. (7.19) as

$$\alpha = 45° \tag{7.24}$$

This is the case when all over the web plate is yielded with uniform stress.

Case with weak flange: wide spacing between vertical stiffeners. On the other hand, when the flange is relatively weak, and stiffener spacing is wide, the flange collapses before the web fully yields forming a plastic mechanism as a both ends clamped beam under the action of σ_v, as indicated in Fig. 7.8. Performing the *Rigid Plastic Mechanism Analysis*, σ_v at the plastic collapse of the flange is derived as

$$\sigma_{v\,max} = \frac{4A_f t_f \sigma_{fY}}{\ell^2 t_w} \tag{7.25}$$

where $A_f = b_f t_f$ and σ_{fY} are the cross-sectional area and the yield stress of the flange, respectively. Dividing Eq. (7.25) by σ_{wY}, new parameter is introduced, which is

$$\frac{\sigma_{v\,max}}{\sigma_{wY}} = 4\frac{ht_f}{\ell^2}\frac{A_f \sigma_{fY}}{A_u \sigma_{wY}} \equiv \frac{\varepsilon}{2} \tag{7.26}$$

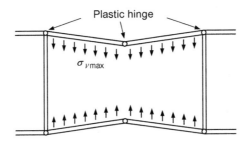

FIG. 7.8

Formation of plastic mechanism in flange.

Replacing $\sigma_v/\sigma wY$ in Eqs. (7.19), (7.20) with $\sigma_{v\,max}/\sigma wY$ defined by Eq. (7.26), the ultimate shear strength and the inclination angle of tension field when the flange collapses are evaluated as follows:

$$v_u = \frac{(1-\varepsilon)v_{cr} + \sqrt{1 + v_{cr}^2 - (1-\varepsilon)^2}}{1 + v_{cr}^2} \tag{7.27}$$

$$\alpha = \tan^{-1} \frac{v_{cr} + \sqrt{1 + v_{cr}^2 - (1-\varepsilon)^2}}{2 - \varepsilon} \tag{7.28}$$

Case with weak flange: narrow spacing between vertical stiffeners. When the spacing between vertical stiffener is narrow and flange is not so stiff, the breadth of tension field is limited to a narrow band as illustrated in Fig. 7.9. In this case, yielding condition is satisfied within the partial tension field whereas the flange collapses forming a plastic mechanism.

As illustrated in Fig. 7.9, the normal stress, σ_2, is the same all over the web panel and is represented by Eq. (7.10). On the other hand, σ_1' in the partial tension field is different from σ_1 in the triangular region beside the partial tension field.

Here, Fig. 7.9 indicates that the applied shear force can be expressed using τ and τ' as follows:

$$V = t_w(h - \ell \tan\alpha) \cdot \tau' + t_w \ell \tan\alpha \cdot \tau$$
$$= A_w\{(1 - \lambda \tan\alpha) \cdot \tau' + \lambda A_w \tan\alpha \cdot \tau\} \tag{7.29}$$

Substituting Eq. (7.12) into Eq. (7.29), it reduces to

$$V = A_w\{(1 - \lambda \tan\alpha)(\sigma_1' - \sigma_2) + \lambda A_w \tan\alpha(\sigma_1 - \sigma_2)\} \sin\alpha \cos\alpha \tag{7.30}$$

In the partial tension field, σ_1' is derived from the yielding condition; see Eq. (7.16), as follows:

$$\sigma_1' = \sigma_{wY} - \tau_{cr} \sin 2\alpha \tag{7.31}$$

On the other hand, σ_1 when the flange collapses forming a plastic mechanism is obtained from Eqs. (7.10), (7.11), and (7.26) as

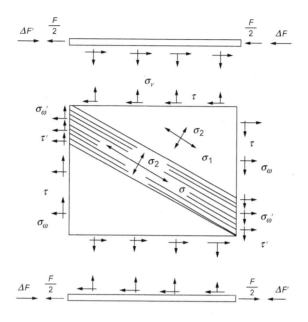

FIG. 7.9

Partial tension field in narrow vertical stiffener spacing.

$$\sigma_1 = \frac{\frac{\varepsilon}{2} + \frac{v_{cr}}{2}\sin 2\alpha \cos^2 \alpha}{\sin^2 \alpha}\sigma_{wY} \tag{7.32}$$

Substituting Eqs. (7.10), (7.31), and (7.32) into Eq. (7.30), the applied shear force is obtained in a nondimensionalized form as

$$v = \varepsilon\lambda + \lambda v_{cr}\sin 2\alpha + (1 - \lambda\tan\alpha)\sin 2\alpha \tag{7.33}$$

v in Eq. (7.33) takes its maximum value when $\partial V / \partial \alpha = 0$, which results in

$$\alpha = \tan^{-1}\frac{1 + \lambda v_{cr}}{\lambda} \tag{7.34}$$

Substituting Eq. (7.34) into Eq. (7.33), the ultimate shear strength in this case is obtained as follows:

$$v_u = \sqrt{\lambda^2 + (1 + \lambda v_{cr})^2} - (1 - \varepsilon)\lambda \tag{7.35}$$

This ultimate strength is under the condition:

$$\sigma_1 \leq \sigma_1' \tag{7.36}$$

This condition is rewritten using Eqs. (7.31), (7.32), and (7.34) as follows:

$$\varepsilon \leq 1 - \frac{\lambda + v_{cr}(1 + \lambda v_{cr})}{\sqrt{\lambda^2 + (1 + \lambda v_{cr})^2}} \tag{7.37}$$

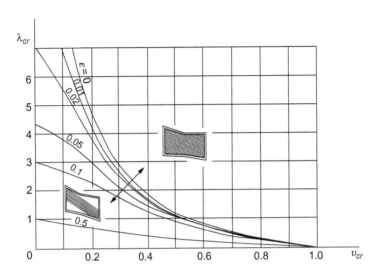

FIG. 7.10

Relationship between critical values and collapse mode [7].

or

$$\lambda \leq \frac{1}{\sqrt{\varepsilon + v_{cr}^2}} \frac{\sqrt{1 - \varepsilon} - v_{cr}}{1 + v_{cr}^2} \equiv \lambda_{cr} \tag{7.38}$$

The relationship between λ_{cr} and v_{cr} is plotted in Fig. 7.10 varying ε.

Summary of formulas for ultimate shear strength. The ultimate shear strength of plate girders is summarized using previous equations as follows:

1. $(1 - v_{cr}) < \varepsilon$

$$v_u = 1.0 \tag{7.23}$$

$$\alpha = 45° \tag{7.24}$$

2. $1 - \frac{\lambda + v_{cr}(1 + \lambda v_{cr})}{\sqrt{\lambda^2 + (1 + \lambda v_{cr})^2}} < \varepsilon \leq (1 - v_{cr})$

$$v_u = \frac{(1 - \varepsilon)v_{cr} + \sqrt{1 + v_{cr}^2 - (1 - \varepsilon)^2}}{1 + v_{cr}^2} \tag{7.27}$$

$$\alpha = \tan^{-1} \frac{v_{cr} + \sqrt{1 + v_{cr}^2 - (1 - \varepsilon)^2}}{2 - \varepsilon} \tag{7.28}$$

3. $\varepsilon \leq 1 - \dfrac{\lambda + v_{cr}(1+\lambda v_{cr})}{\sqrt{\lambda^2 + (1+\lambda v_{cr})^2}}$

$$v_u = \sqrt{\lambda^2 + (1 + \lambda v_{cr})^2} - (1 - \varepsilon)\lambda \tag{7.35}$$

$$\alpha = \tan^{-1} \frac{1 + \lambda v_{cr}}{\lambda} \tag{7.34}$$

7.2.2.2 Plate girders subjected to pure bending

As indicated in Fig. 7.1, the normal stress in the web panel does not increase beyond the buckling at the location where deflection develops in the case of a plate girder subjected to bending moment. Akita and Fujii idealized such stress distribution as illustrated in Fig. 7.11 [6]. That is, the stress is linearly distributed over the web plate until web buckling occurs under pure bending load. Then, bending moment is sustained only at the compression flange and the web near the compression flange after buckling takes place in bending. Such behavior is similar to the postbuckling behavior of thin plate subjected to compression. The web part, ch in Fig. 7.11, where compressive stress increases even after the buckling is called the "effective width."

Bending moment beyond buckling is expressed as

$$M = M_{cr} + EI'(\phi - \phi_{cr}) \tag{7.39}$$

where

M: Applied bending moment
M_{cr}: Bending moment at buckling
ϕ: Curvature
ϕ_{cr}: $= M_{cr}/EI$ Curvature at buckling
EI: Flexural stiffness of cross-section before buckling
EI': Flexural stiffness of cross-section after buckling
A_f: Cross-sectional area of a flange
A_w: Cross-sectional area of a web

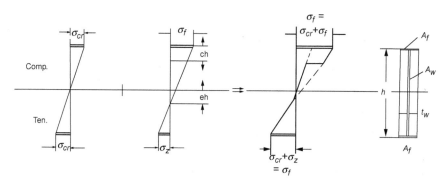

FIG. 7.11

Idealized stress distribution in web panel after buckling by bending [6].

and

$$I = \left(\frac{1}{2}A_f + \frac{1}{12}A_w\right)h^2 \tag{7.40}$$

$$I' = \left(\frac{1}{2} + 2e^2\right)h^2 A_f + \frac{1}{3}h^2 A_w \left[\frac{1}{4} + 3e^2 - \left(\frac{1}{2} + e - c\right)^3\right] \tag{7.41}$$

$$e = \left(\frac{1}{2} + c + 2\frac{A_f}{A_w}\right) - \sqrt{2\left(1 + 2\frac{A_f}{A_w}\right)\left(c + \frac{A_f}{A_w}\right)} \tag{7.42}$$

$$c = 1.21\frac{t_w}{h}\sqrt{E\sigma_Y} \tag{7.43}$$

Here, the stress in the compression flange can be expressed as

$$-\sigma_f = \frac{hM_{cr}}{2I} - \frac{h(1 + 2e)(M - M_{cr})}{2I'} \tag{7.44}$$

From the condition that $\sigma_f = \sigma_{fY}$, where (σ_{fY}) is the yield stress of the flange, the initial yielding moment after buckling is derived as follows:

$$M_1 = M_{cr} + (M_Y - M_{cr})\frac{I'}{I}\frac{1}{1 + 2e} \tag{7.45}$$

where $M_Y = 2I\sigma_{fY}/2I$ is the initial yielding strength when buckling does not take place. Akita and Fujii showed that the bending moment, M_1, is very near to the ultimate bending moment, and he proposed to consider M_1 as the ultimate bending moment when web panel undergoes buckling in bending [6].

He also proposed to consider the bending moment at which web buckling takes place after the compression flange has been yielded as the ultimate bending moment.

In summary, the ultimate bending moment can be expressed in a nondimensionalized form as follows:

1. $m_{cr} \leq 1.0$

$$m_u = m_{cr} + \beta(1 - m_{cr}) \tag{7.46}$$

2. $1 < m_{cr}$

$$m_u = m_{cr} \tag{7.47}$$

where

$$m_u = \frac{M_u}{M_Y}, \quad m_{cr} = \frac{M_{cr}}{M_Y}, \quad \beta = \frac{1}{1 + 2e}\frac{I'}{I} \tag{7.48}$$

and β represents the ratio of section modulus after buckling to that before buckling. β calculated by Eq. (7.48) varying the sectional area ratio, A_f/A_w, of flange to web is plotted in Fig. 7.12.

The elastic buckling strength ($m_{cr} \leq 1$) can be calculated by

$$m_{cr} = \frac{k\pi^2}{12(1 - v^2)}\left(\frac{t_w}{h}\right)^2\left(\frac{E}{\sigma_Y}\right) \tag{7.49}$$

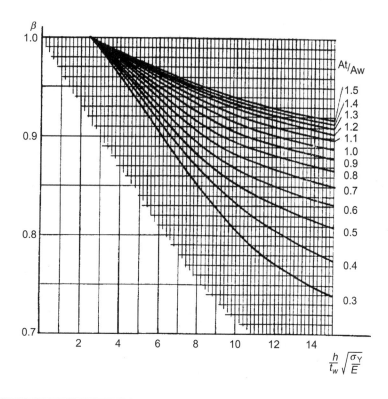

FIG. 7.12

Idealized stress distribution in web panel after buckling by bending [6].

where

$$k = 15.87 + \frac{1.87}{\lambda^2} + 8.6\lambda^2 \quad \lambda \leq \tfrac{2}{3}$$
$$k = 23.9 \quad \tfrac{2}{3} < \lambda$$

and $\lambda = \ell/h$ is the aspect ratio of the web panel.

For the elastoplastic buckling,

$$m_{cr} = f_s = \frac{M_Y}{M_P} \quad \text{when} \quad \frac{h}{t_w}\sqrt{\frac{\sigma_Y}{E}} \leq 2.42 \tag{7.50}$$

$$m_{cr} = \frac{\left(0.952\sqrt{k} - \frac{h}{t_w}\sqrt{\frac{\sigma_Y}{E}}\right)f_s + \left(\frac{h}{t_w}\sqrt{\frac{\sigma_Y}{E}} - 2.52\right)}{0.952\sqrt{k} - 2.42} \quad 2.42 < \frac{h}{t_w}\sqrt{\frac{\sigma_Y}{E}} \tag{7.51}$$

The buckling strength, m_{cr}, calculated by Eqs. (7.49), (7.50), and (7.51) is plotted in Fig. 7.13 together with the ultimate strength, m_y, calculated by Eqs. (7.46), (7.47).

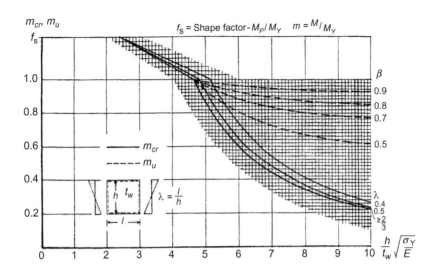

FIG. 7.13

Buckling strength and ultimate strength of plate girders under pure bending [6].

7.2.2.3 Plate girders subjected to combined bending and shear loads

Buckling strength interaction relationship

Akita and Fujii [6] assumed buckling interaction relationship of web panel under combined shear and bending as follows:

$$\left(\frac{M'_{cr}}{M_{cr}}\right)^2 + \left(\frac{V'_{cr}}{V_{cr}}\right)^2 = 1 \tag{7.52}$$

where M_{cr} and V_{cr} are buckling strength under pure bending and pure shear, respectively. Depending on the ratio of bending moment to shear force, bending moment, M'_{cr} and V'_{cr} at buckling are evaluated by Eq. (7.52). M'_{cr} and V'_{cr} are used as buckling strength instead of M_{cr} and V_{cr}, respectively, when the ultimate strength is calculated.

Fully plastic strength interaction relationship

When buckling does not occur in web and flange of a plate girder shown in Fig. 7.14, fully plastic bending moment and fully plastic shear force are expressed as

$$M_P = hA_f\sigma_{fY} + \frac{1}{4}hA_w\sigma_{wY} \tag{7.53}$$

$$V_P = A_w\tau_{wY} \tag{7.54}$$

The fully plastic stress distribution under combined bending and shear is shown in Fig. 7.14. Applying von Mises's yielding condition, the following equation is obtained in the web:

$$\sigma^2 + 3\tau^2 = \sigma_{wY}^2 \tag{7.55}$$

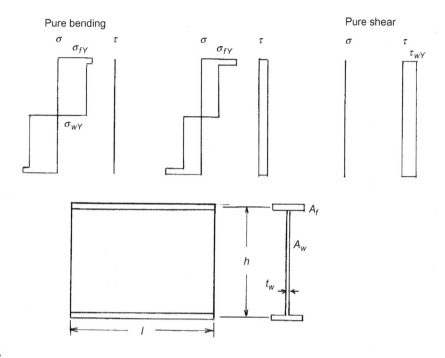

FIG. 7.14

Fully plastic stress distribution under combined bending and shear.

Here, the stresses, σ and τ, in Fig. 7.14 produce bending moment and shear force expressed as

$$M'_P = hA_f\sigma_{fY} + \frac{1}{4}hA_w\sigma \tag{7.56}$$

$$V'_P = A_w\tau \tag{7.57}$$

From Eq. (7.55), the shear yielding stress, τ_{wY}, of the web panel is given as

$$\tau_{wY} = \frac{1}{\sqrt{3}}\sigma_{wY} \tag{7.58}$$

With Eqs. (7.53), (7.54), and (7.56)–(7.58), the fully plastic strength interaction relationship is derived in the following form:

$$\{(\gamma + 1)m' - \gamma\}^2 + v'^2 = 1 \tag{7.59}$$

where

$$\gamma = 4\frac{A_f\sigma_{fY}}{A_w\sigma_{wY}} \tag{7.60}$$

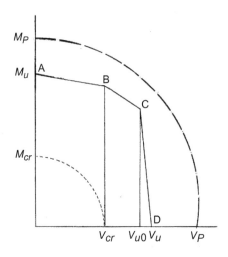

FIG. 7.15

Ultimate strength interaction relationships of plate girder under combined bending and shear [8].

and

$$m' = \frac{M'_P}{M_P}, \quad v' = \frac{V'_P}{V_P}$$

When buckling takes place, the ultimate strength interaction relationship is not simple as Eq. (7.52) or (7.59).

Ultimate strength interaction relationship

Fig. 7.15 schematically shows the strength interaction relationships of a plate girder subjected to combined bending and shear. The dashed line is the buckling strength interaction curve by Eq. (7.52), and the broken line the fully plastic strength interaction curve by Eq. (7.59). On the other hand, the solid line represents the ultimate strength interaction relationship considering buckling.

Point A represents the ultimate strength under pure bending. At this point, shear force is zero and bending moment is calculated by Eq. (7.45) or (7.46).

At point B, shear force is equal to the buckling strength, that is $v' = v'_{cr}$. On the other hand, Akita and Fujii defined the bending moment at point B as follows:

$$m' = \beta(1 - v'_{cr}) + m_{fP}v'_{cr} \tag{7.61}$$

where m_{fP} is the fully plastic bending moment sustained by flanges and is given as follows:

$$m_{fP} = \frac{1}{M_Y} h A_f \sigma_{fY} \tag{7.62}$$

The ultimate shear force at point C is evaluated by Eq. (7.27) or (7.35) setting $\varepsilon = 0$. For this calculation, lateral force acting on a flange is considered to be negligibly small. The shear force and the bending moment are given as [6]:

1. $1 - \dfrac{\lambda + v_{cr}(1+\lambda v_{cr})}{\sqrt{\lambda^2 + (1+\lambda v_{cr})^2}} < \varepsilon \le (1 - v_{cr})$

$$v_C = \frac{2v_{cr}}{1 + v_{cr}^2} \tag{7.63}$$

$$m_C = m_{fP} - 2m_{wP}\frac{1 - v_{cr}^2}{1 + v_{cr}^2} \tag{7.64}$$

2. $\varepsilon \le 1 - \dfrac{\lambda + v_{cr}(1+\lambda v_{cr})}{\sqrt{\lambda^2 + (1+\lambda v_{cr})^2}} < \varepsilon$

$$v_C = \sqrt{\lambda^2 + (1 + \lambda v_{cr})^2} - \lambda \tag{7.65}$$

$$m_C = m_{fP} - m_{wP}\left\{ 2\lambda v_{cr} + \frac{(1+\lambda^2)(v_C - v_{cr} - \lambda v_{cr}^2)}{\lambda + v_C} \right\} \tag{7.66}$$

where m_{wP} is the fully plastic bending moment sustained by web and is given as follows:

$$m_{wP} = \frac{1}{4M_Y} hA_w \sigma_{wY} \tag{7.67}$$

At point D, the bending moment is zero and the ultimate shear strength, v_u, is given by Eq. (7.23), (7.27), or (7.35), depending on the stiffness of flanges as well as stiffener spacing.

7.2.3 NUMERICAL EXPERIMENTS ON COLLAPSE BEHAVIOR OF PLATE GIRDERS

7.2.3.1 Test specimens for numerical simulation of collapse behavior

To confirm the validity of Fujii's formulas, Olaru et al. [9] performed a series of elastoplastic large deflection analyses on tested girders. Dimensions and material properties are summarized in Table 7.1.

As girder specimens collapsed in shear, G6-T1 in Busler's experiments [2] was selected. The FEM meshing for G6-T1 specimen is shown in Fig. 7.16. The web thickness excluding the central two

Table 7.1 Dimension and Yielding Stress of Analyzed Test Girders

Model	L	h_w	t_w	b_f^u	t_f^u	b_f^b	t_f^b	σ_Y^w	σ_Y^{fu}	σ_Y^{fb}
G6-T1	1905	1270	4.90	308	19.7	308	19.7	253	261	261
G2-T1	1905	1270	6.86	309	19.5	309	19.6	243	266	259
G1-T1	1905	1270	6.86	522	10.8	311	19.3	227	244	247
BS-3.2-0.0	500	1000	3.20	120	14.0	120	14.0	235	372	372
DB	2400	1990	11.0	800	18.0	800	15.0	314	314	314

Notes: L, web panel length (between vertical stiffeners) [mm]; h_w, t_w, height and thickness of web panel [mm]; b_f^u, t_f^u, width and thickness of upper flange [mm]; b_f^b, t_f^b, width and thickness of lower flange [mm]; σ_Y^w, yield stress of web [MPa]; σ_Y^{fu}, yield stress of upper flange [MPa]; σ_Y^{fb}, yield stress of lower flange [MPa].

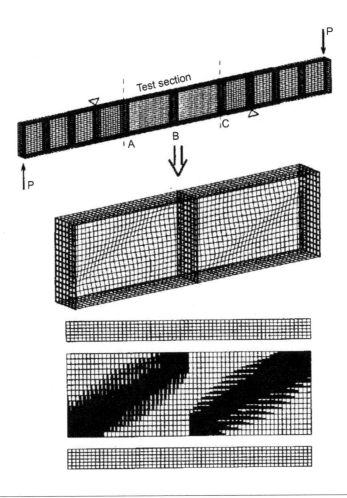

FIG. 7.16

Deformation and yielding at the ultimate strength (G6-T1 specimen) [9].

panels were increased so that shear collapse takes place in the central two panels. Basler applied two equal forces in opposite directions as indicated in Fig. 7.16. This loading condition produces bending moment. However, it was ignored considering that the produced bending moment is small. In the test, cover plates are provided at two supporting points.

As girder specimens collapsed in pure bending, G2-T1 and G1-T1 specimens were selected, of which FEM meshings are illustrated in Figs. 7.18 and 7.20, respectively. G1-T1 is a special specimen having wide and thin flange in the compression side of bending. Four points bending load was applied on these specimens.

As girder specimens subjected to combined bending and shear, BS-3.2-0.0 in Fig. 7.22 was selected which was tested in SR127 research project [11]. Three point bending load was applied on this test specimen.

The last model denoted as DB is a part of double bottom. Shear, bending, and combined shear and bending loads were applied on this specimen. The DB model is shown in Fig. 7.24. Width of the flange is taken equal to the spacing between longitudinal stiffeners. The edge line of the flange is actually the centerline of the panel between stiffeners. So, symmetry condition was imposed along the edges of upper and lower flanges. In addition to this, intersecting lines of the web and the flange are restrained against horizontal displacement.

All girder components including stiffeners and cover plates are modeled by four-noded isoparametric bi-linear degenerated shell elements with reduced integral.

Because of nonsymmetry as for loading condition and structural arrangements, full model was considered for Basler's specimens. On the other hand, for BS-3.2-0.0 specimen, only a half was analyzed introducing symmetry condition.

Measured yield stress of the material is used, which is indicated in Table 7.1. Since, with few exceptions, the material of the test specimens was mild steel, bi-linear stress-strain relationship was assumed in the analysis setting strain hardening rate as 0.14% of the Young's modulus. For SR-3.2-0.0 specimen, measured initial deflection was given as an initial condition.

7.2.3.2 Collapse behavior of G6-T1 specimen under shear

For G6-T1 specimen, the deflection and spread of yielding at the ultimate strength are illustrated in Fig. 7.16, and shear force-shear strain and shear force-deflection relationships are plotted in Fig. 7.17.

The web slenderness ratio, h/t_w, is 259, and the web panel undergoes elastic buckling at around $V/V_P = 0.2$. The buckling strength is well predicted by Fujii's formulas, Eqs. (7.7), (7.8). Beyond the buckling, in-plane stiffness, which is the slope of the shear force-shear strain curve, is reduced by about 25%. Yielding starts at an open circle on the curve.

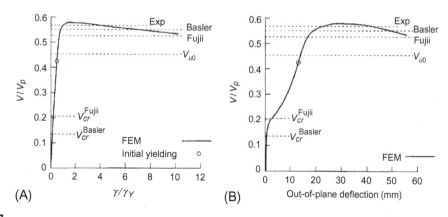

FIG. 7.17

Load-displacement relationship of G6-T1 during progressive collapse [9]. (A) Shear force-shear strain relationship. (B) Shear force-deflection relationship.

It is clearly seen in Fig. 7.16 that a partial tension field is formed in web panels, and the subsequent load above bucking load is carried by frame actions of the flanges and the stiffeners sustaining tension field. The ultimate shear strength is around 60% of the fully plastic shear strength, V_P.

At the ultimate strength state, tension field action causes a wide spread of yielding along the diagonal line. In this case, load carrying capacity does not decrease so much beyond the ultimate strength. Such postultimate strength plateau is often observed in the experiment [2].

Applying both methods by Basler and Fujii, the ultimate shear strength is evaluated and plotted together with the experimental result. As for Fujii's method, Eq. (7.35) is applicable for evaluation of shear ultimate strength of G6-T1 specimen. This is the case when a partial tension field is formed and a plastic mechanism is not formed in the flange.

7.2.3.3 Collapse behavior of G2-T1 and G1-T1 specimens under pure bending

In the case of a plate girder subjected to pure bending, the flange located in the compression side of bending plays an important role since the web panel in the compression side of bending does not carry compressive load except near the compression flange once the buckling has taken place in bending.

For G2-T1 specimen tested in pure bending, deformation and the spread of yielding are shown in Fig. 7.18, and the bending moment-rotation relationship in Fig. 7.19. Fig. 7.18 indicates that web panel buckled under bending and the compression flange with longer span in torsion. Sharp drop of the capacity beyond the ultimate strength observed in Fig. 7.19 is due to the occurrence of torsional buckling of the flange in the compression side of bending.

The ultimate strength evaluated by Basler's and Fujii's formulas is indicated together with the measured ultimate strength in Fig. 7.19. Relatively good agreement is observed between calculated and measured ultimate strength.

On the other hand, G1-T1 specimen has a wide and thin compression flange. This flange is similar to double bottom girder in proportion. For this specimen, the deformation and spread of yielding in the postultimate strength range are shown in Fig. 7.20, and the bending moment-rotation relationship in Fig. 7.21.

Very rapid drop of capacity is observed in Fig. 7.21 beyond the ultimate strength. In this case, the local torsional buckling of the compression flange becomes the primary cause of the girder failure. Large deformation and yielding are observed in the mid-span region of the longer compression flange.

7.2.3.4 Collapse behavior of BS-3.2-0.0 specimen under combined bending and shear

Three-point bending load was applied on BS-3.2-0.0 specimen tested in SR127 research panel [10]. Considering symmetry of the loading condition and the structure, a half is analyzed imposing a symmetry condition along the symmetry line. Deformation and spread of yielding in the postultimate strength range are illustrated in Fig. 7.22, and the load-relative shear displacement relationship in Fig. 7.23.

Fig. 7.23 indicates that the web buckles a slightly above analytical buckling strength, P_{cr}, and the in-plane stiffness is reduced by around 25%. The open circles in the figure are measured points during collapse test on BS-3.2-0.0. Very good correlation is observed between the measured and calculated results.

Above the load level indicated by P_u^{ex}, deflection rapidly increased in the experiment, and the maximum load was obtained at P_{max}^{ex}. FEM analysis shows that the capacity does not decrease after the ultimate strength has been attained.

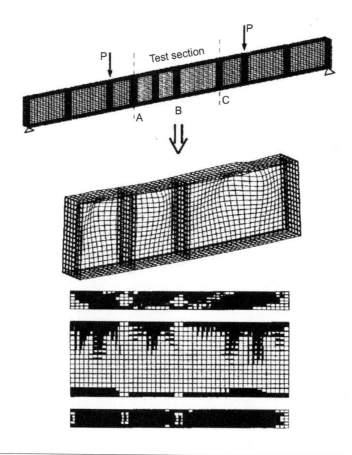

FIG. 7.18

Deformation and yielding in postultimate strength range (G2-T1 specimen) [9].

Such postultimate strength behavior is similar to that of G6-T1 indicating that flange buckling did not occur since bending moment was not so large. However, the presence of small bending moment reduce the ultimate shear strength, and the girder cannot attain its ultimate shear strength, V_u.

Fig. 7.24 shows the ultimate strength interaction relationships according to Fujii's formulas. On this interaction diagram, ultimate strength points obtained by experiment and FEM analysis are plotted together with that obtained by Basler's formulas. It can be said that Fujii's formulas give accurate estimation of the ultimate strength, whereas Basler's formulas overestimate the ultimate strength.

Fig. 7.24 indicates that the loading condition for BS-3.2-0.0 is that of a shear dominant case.

7.2.4 COLLAPSE BEHAVIOR OF PLATE GIRDERS IN DOUBLE BOTTOM STRUCTURE

7.2.4.1 Collapse under shear

As a typical girder in ship's double bottom structure, DM model in Table 7.1 was considered [9,11]. The boundary and loading conditions are shown in Fig. 7.25 and Table 7.2. Unlike the usual experimental

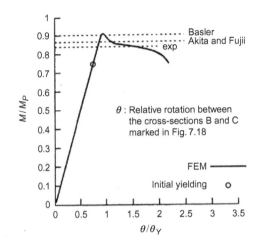

FIG. 7.19

Load-displacement relationship of G2-T1 during progressive collapse [9].

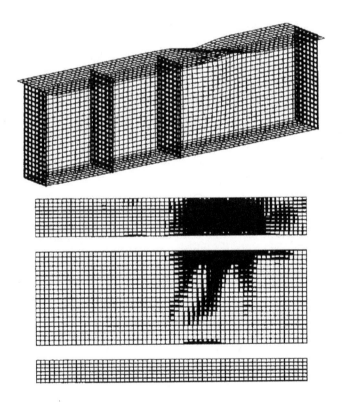

FIG. 7.20

Deformation and yielding in postultimate strength range (G1-T1 specimen) [9].

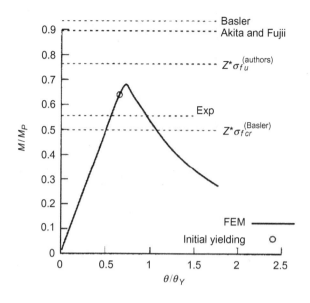

FIG. 7.21

Load-displacement relationship of G1-T1 during progressive collapse [9].

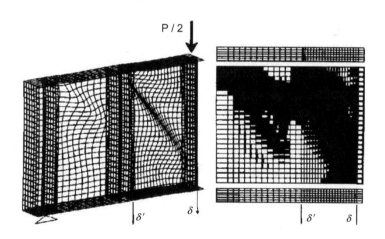

FIG. 7.22

Deformation and yielding in postultimate strength range (BS-3.2-0.0 specimen) [9].

set-ups for shear loading, the DM model is loaded in a clamped-ends beam condition as indicated in Fig. 7.25 to resemble pure shear condition making the bending moment at mid-span point as zero.

The boundary and the loading conditions are summarized in Table 7.2, where constrained degree of freedom is indicated by *. The applied forced displacement is denoted as δ, and Ci represents the

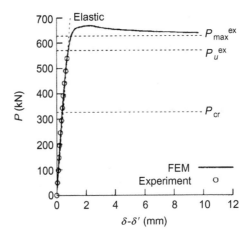

FIG. 7.23

Load-displacement relationship of BS-3.2-0.0 [9].

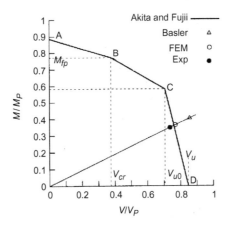

FIG. 7.24

Ultimate strength interaction relationship [9].

uniform displacement along the corresponding boundary line. Web panel is assumed to be simply supported along edges 1 and 4 and the constraint from the adjacent web panels is not considered. Lateral deflection in y-direction is constrained along lines 8 and 11 considering that the flange is a part of continuous plating, that is, bottom plating or inner bottom plating. Along edges 7, 9, 10, and 12, displacement in y-direction is kept uniform and symmetry condition was imposed considering the condition of continuous plating.

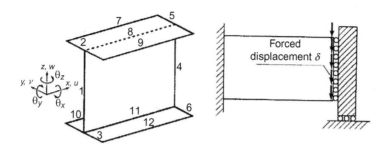

FIG. 7.25

Application of shear force on DB model [11].

Table 7.2 Boundary Conditions for FEM Analysis on DB Models [11]												
Edge	1	2	3	4	5	6	7	8	9	10	11	12
u	*	*	*	C1	C1	C1						
v	*			*			C2	*	C3	C4	*	C5
w	*	*	*	δ	δ	δ						
θ_x	*	*	*	*	*	*	*		*	*		*
θ_y	*			*	*							
θ_z		*	*		*	*	*	*	*	*	*	*

Deformation and spread of yielding in DB model in shear are shown in Fig. 7.26, and shear force-shear strain relationship in Fig. 7.27. The collapse behavior is almost the same as that of Basler's G6-G1 specimen in a sense that the ultimate strength is dependent mainly on the web behavior. However, in the DB model, flexural stiffness of the flange is low, which results in bending collapse of flange due to pull-in action of the tension field in the postultimate strength range. The collapse of the flange is considered as the cause of decrease in the capacity after the ultimate strength has been attained.

7.2.4.2 Collapse under pure bending

Under pure bending, the compression flange of the plate girder may buckle torsionally. On the other hand, the flange of the DB model in the compression side of bending is subjected to thrust and may locally buckle as a continuous panel under thrust. The difference in deformation of these two flanges are schematically shown in Fig. 7.28.

Two cases are analyzed here: the DB model and the same model but with free edge flange of the same size. Deformation and spread of yielding of former DB model in pure bending are shown in Fig. 7.29, and the moment-rotation angle relationships for two cases in Fig. 7.30.

The change in bending stress distribution is also shown in Fig. 7.29. It is known that compressive stress in the buckled part does not increase and the neutral axis moves toward the tension side of bending.

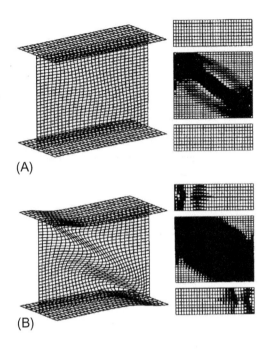

(A)

(B)

FIG. 7.26

Deformation and spread of yielding (DB model in shear) [9]. (A) Ultimate strength ($V/V_P = 0.68$).
(B) Postultimate strength range ($V/V_P = 0.55$).

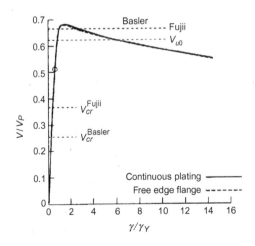

FIG. 7.27

Shear force-shear strain relationship (DB model in shear) [9].

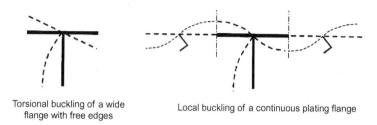

Torsional buckling of a wide
flange with free edges

Local buckling of a continuous plating flange

FIG. 7.28

Local buckling of wide flange [11].

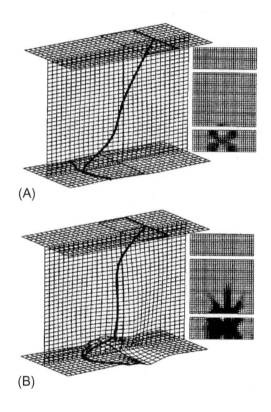

(A)

(B)

FIG. 7.29

Deformation and spread of yielding (DB model in bending) [9]. (A) Ultimate strength ($M/M_P = 0.67$).
(B) Postultimate strength range ($M/M_P = 0.51$).

Significant difference is observed in the bending moment-rotation angle relationship depending
on the condition of the compression flange. In the case of a continuous flange, local panel buckling
takes place and the compressive ultimate strength is attained in the flange. The continuous plating
condition evidently ensures the higher ultimate strength because the rotation of the compression flange

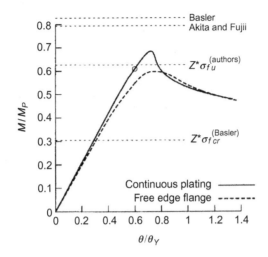

FIG. 7.30

Bending moment-rotation angle relationships (DB model in bending) [9].

is constrained owing to the continuity condition of the flange. On the other hand, decrease in the capacity beyond the ultimate strength is much rapid in the case of a continuous flange. This may be because of the concentration of buckling deformation at a certain cross-section of the compression flange and the resulting elastic unloading in the remaining part of the span. However, after a while, moment-rotation curves for the two cases become almost the same.

7.2.4.3 Collapse under combined shear and bending

Collapse behavior of the DB model was simulated changing the ratio of bending moment to shear force as follows:

$$m/v = 0.5: \text{shear dominant condition}$$
$$m/v = 1.0: \text{significant interaction}$$
$$m/v = 2.0: \text{bending dominant condition}$$

where $m = M/M_P$ and $v = V/V_P$ indicate the contributions of bending and shear, respectively, under the combined loading.

Collapse behavior in the case of $m/v = 0.5$ is almost the same as that of shear dominant case shown in Figs. 7.26 and 7.27. The presence of varying small bending moment only slightly affects the collapse behavior of compressed flange without producing major differences in collapse behavior of the girder.

Collapse behavior in the case of $m/v = 2.0$ is similar to that under pure bending shown in Figs. 7.29 and 7.30. Local buckling of the compression flange as continuous plating becomes the trigger to the overall collapse, although location where local buckling takes place changes to the cross-section where induced bending moment shows larger value.

For the case of $m/v = 1.0$, deformation and the spread of yielding are shown in Fig. 7.31, and the bending moment-rotation angle in Fig. 7.32. Two cases are shown in Fig. 7.32, which are DB model

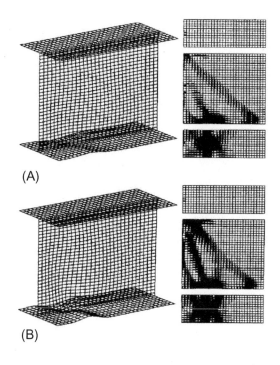

FIG. 7.31

Deformation and spread of yielding (DB model in combined bending and shear) [9]. (A) Ultimate strength ($M/M_P = 0.53$). (B) Postultimate strength range ($M/M_P = 0.42$).

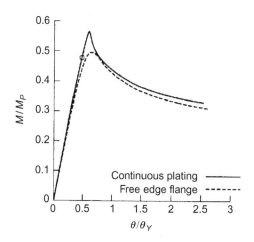

FIG. 7.32

Bending moment-rotation angle relationships (DB model in combined bending and shear) [9].

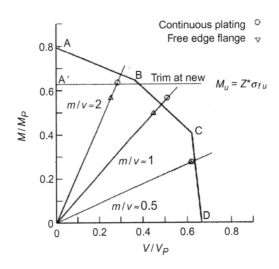

FIG. 7.33

Ultimate strength interaction relationship of DB model [11].

and the same model but with free edge flange of the same size. Similar tendency is observed between two bending moment-rotation angle curves as those under pure bending.

Fig. 7.31 indicates that yielding takes place in the web along the diagonal in the tension field. At the same time, yielding occurs in the web near the compression flange at the location where larger bending moment is produced.

Obtained results are plotted in the ultimate strength interaction diagram applying Fujii's formulas in Fig. 7.33. Good correlation is observed between ultimate strength by Fujii's formulas and those by FEM. It can be said that Fujii's formulas can be used for the estimation of the ultimate strength of girders in double bottom structure especially in shear.

Fujii recommended considering that the web panel is clamped along the stiffener lines when buckling strength is calculated. This suggestion seems to be valid. It is known that V_{u0} introduced by Fujii is very important to construct the ultimate strength interaction relationships.

7.3 BUCKLING/PLASTIC COLLAPSE BEHAVIOR AND STRENGTH OF STIFFENED GIRDERS IN SHEAR

Shear collapse of plate girders in double bottom structures is one of the dominant failure modes under lateral loads on bottom and/or inner bottom plating due to water pressure and/or cargo weight. The girder in double bottom structure is usually stiffened by stiffeners. In the case of longitudinal girders, horizontal stiffeners are provided, while on transverse girders (floors) vertical stiffeners. The web of these girders sometimes has perforations for piping and passing. In this section, influences of stiffeners and perforation on collapse behavior and ultimate strength of plate girders are explained on the basis of the FEM results.

7.3.1 FEM MODELS FOR ANALYSIS

The longitudinal and transverse girders taken from the double bottom structure of real ship structures are considered. The one-span girder between intersecting girders in the perpendicular direction is considered as a representative unit. Typical transverse and longitudinal girders are shown in Fig. 7.34. The numbers along the boundaries indicate the boundary conditions given in Table 7.1.

The longitudinal girders have two horizontal stiffeners, while the transverse girders have two vertical stiffeners. The stiffeners are of a flat bar-type and have snipped ends. For comparison purpose, longitudinal girder of the same size but without horizontal stiffeners is also analyzed. For the case of perforated web, an oval hole is provided at the center of the web as illustrated in Fig. 7.34.

Young's modulus, E, is taken as 206 GPa and the yielding stress, σ_Y, as 275 MPa. Bi-linear idealization of stress-strain relationships is assumed with strain-hardening rate, E_T, being 0.1% of the Young's modulus. Initial deflection of a buckling mode without stiffeners is prescribed on the web

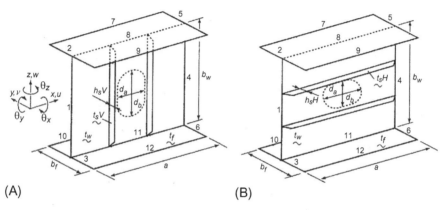

FIG. 7.34

Notation and lines of boundary condition for girder models [11]. (A) Transverse girders. (B) Longitudinal girders.

Model	a	b_w	t_w	b_f	t_f	h_{sV}	t_{sV}	h_{sH}	t_{sH}	d_a	d_b
Table 7.3 Dimension of FEM Models for Shear Loading [11]											
G11	2400	1600	11	800	20	–	–	–	–	–	–
V11	2400	1600	11	800	20	150	12	–	–	–	–
H11	2400	1600	11	800	20	–	–	150	12	–	–
PV11	3200	1780	11	800	18.5*, 16**	150	10	–	–	500	800
PH11	3200	1780	11	800	18.5*, 16**	–	–	180	10	800	500

Notes: a, length of web plate; b_w, breadth of web plate; t_w, thickness of web plate; b_f, breadth of flange; t_f, thickness of flange ("*," upper flange, "**," bottom flange); h_{sV}, height of vertical stiffener; t_{sV}, thickness of vertical stiffener; h_{sH}, height of horizontal stiffener; t_{sH}, thickness of horizontal stiffener; d_a, length of hole; d_b, breadth of hole.

panel taking its maximum magnitude as 10% of the web thickness. The welding residual stress is not considered. The dimensions of girders analyzed here are summarized in Table 7.3.

The girder models with web thickness of 11 mm shall hereafter denoted as follows:

G11: unstiffened web
V11: vertically stiffened web
H11: horizontally stiffened web
PV11: perforated and vertically stiffened web
PH11: perforated and horizontally stiffened web

Actual stiffener ends are snipped. Snipped ends are not realized in the FEM modeling. However, setting the rotation and the displacement free at the ends of stiffeners, almost the same condition is obtained as snipped ends.

7.3.2 GIRDERS WITH VERTICALLY STIFFENED WEB PANEL

Firstly, the G11 model is calculated to examine the fundamental collapse behavior of double bottom girders in shear free from influences of stiffeners and perforation. The average shear stress-shear strain relationship is plotted in Fig. 7.35 by a dotted line. Here, the average shear stress is calculated by dividing total shear force by the web cross-sectional area, and the average shear strain is calculated as relative displacement between web ends divided by total span length.

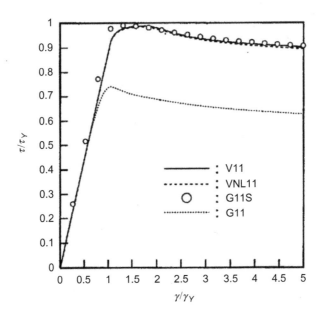

FIG. 7.35

Average shear stress-shear strain relationships of vertically stiffened girder models [11].

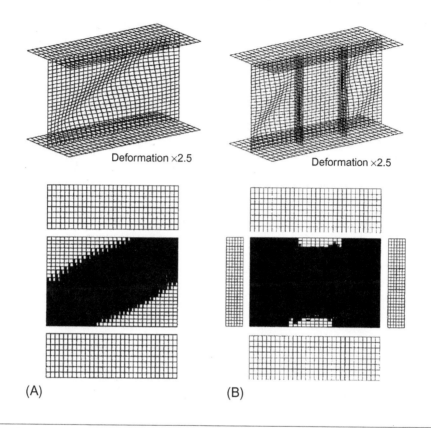

Deformation ×2.5   Deformation ×2.5

(A)   (B)

FIG. 7.36

Comparison of deformation and spread of yielding in post-ultimate strength range (G11 and V11) [11].
(A) G11 ($\gamma/\gamma_Y = 2.6$). (B) V11 ($\gamma/\gamma_Y = 5.2$).

For this slender web of G11, elastic shear buckling occurs at around $\tau/\tau_Y = 0.5$. With the increase in the lateral deflection, the tension field develops and the web can carry further load. Fig. 7.36A shows the spread of yielding in the postultimate strength range. Because of the tension field action, the reduction in load-carrying capacity beyond the ultimate strength is also small. Comparing to the other cases in Fig. 7.35, G11 shows the lowest load carrying capacity.

Average stress-average strain relationship of V11 is plotted by the solid line in Fig. 7.35. Its deformation and spread of yielding is illustrated in Fig. 36B. Unlike G11, three sub-panels divided by two equally spaced stiffeners are much shorter in length. Hence their elastic shear buckling strength is much higher than that of unstiffened web panel of G11 and exceeds shear yielding stress, τ_{wY}. Consequently, the ultimate strength of V11 is almost the same as shear yielding strength.

It is known from Fig. 7.36B that the collapse is more enhanced in the two end sub-panels rather than the central sub-panel. This is due to the additional bending moment action in the end parts. To know the influence of such nonuniform collapse behavior on the average stress-average strain relationships as a whole girder, and to examine the effectiveness of vertical stiffeners as a panel breaker, two additional calculations have been performed. One is VNL11, which is a virtual model derived from V11 by removing the vertical stiffeners but constraining lateral deflection of the web panel along the two stiffener lines. The other is G11S, which corresponds to one sub-panel of G11. The calculated average shear stress-shear strain relationships are plotted by dashed line and open circles in Fig. 7.35 for VNL11 and GS11, respectively.

It can be concluded that the considered vertical stiffener, which is typical in ship double bottom structure, has enough strength as a panel breaker even in the postultimate strength range. It is seen in Fig. 7.36B that the vertical stiffeners remain as almost straight despite their lower stiffness in comparison with web panel.

7.3.3 GIRDERS WITH HORIZONTALLY STIFFENED WEB PANEL

Deformation and spread of yielding of H11 model with h_{sH} at the ultimate strength and in subsequent two stages are illustrated in Fig. 7.37. At the ultimate strength shown in Fig. 7.37A, deflection in web panel is very small and so deflection of the stiffeners. All the web panel is yielded at this stage.

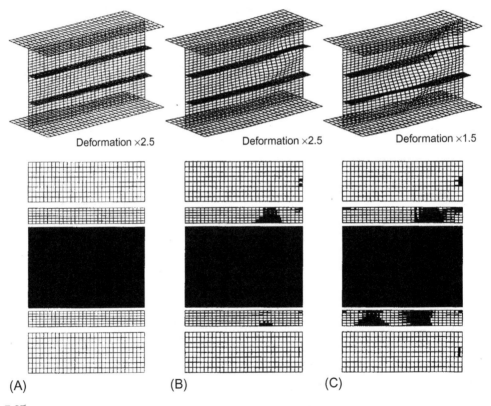

FIG. 7.37

Deformation and spread of yielding (H11 with $h_{sH} = 150$ mm) [11]. (A) $\gamma/\gamma_Y = 9.1$. (B) $\gamma/\gamma_Y = 10.3$. (C) $\gamma/\gamma_Y = 15.5$.

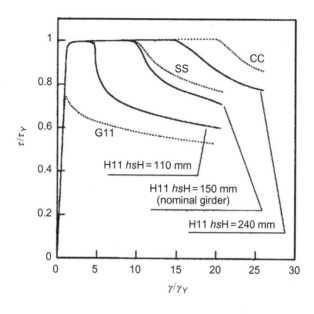

FIG. 7.38

Average shear stress-shear strain relationships (H11 models) [11].

At the strain level corresponding to Fig. 7.37B, plastic buckling has occurred in the web panel and lateral deflection has started to increase locally. Due to this deflection, also the stiffeners begin to deflect in a tripping mode, and yielding is observed at the deflected region. At strain level corresponding to Fig. 7.37C, deflection in web panel and stiffeners further increases. During the collapse process from that in Fig. 7.37B to that in Fig. 7.37C, deflection mode changes from local mode to global mode. Since the web panel is fully yielded, this mode change is quite rapid.

Average shear stress-shear strain relationships for H11 with h_{sH} = 110, 150, and 240 mm are plotted by the solid lines in Fig. 7.38. In the case of h_{sH} = 150 mm, capacity start to decrease at around γ/γ_Y = 10, which correspond to the start of plastic buckling at the web panel; see Fig. 7.37B. The capacity reduction in postultimate strength range is very rapid because of the failure of stiffeners. The solid curves show almost the same ultimate shear strength, which is equal to the fully plastic shear strength, regardless of the stiffener height. Wider plastic plateau is observed compared to the girders with vertical stiffeners. The width of plateau increases with the increase in the stiffener height.

To demonstrate the effectiveness of horizontal stiffeners, stress-strain relationship for G11 model is also plotted in Fig. 7.38. It is known that the ultimate shear strength which is about 75% of the fully plastic shear strength has increased up to the fully plastic shear strength owing to the two horizontal stiffeners.

The ductility of horizontally stiffened girders is measured by γ_{cr} defined as the shear strain at ultimate strength. To examine the effect of stiffener size on the magnitude of γ_{cr}, a series of FEM analyses has been performed varying the stiffener height, h_{sH}, and stiffener thickness, t_{sH}. Virtual models derived from H11 model by replacing two stiffeners with simply supported nodal lines (SS)

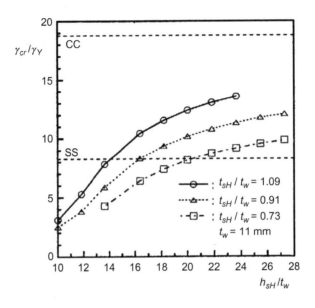

FIG. 7.39

Influence of stiffener size on critical shear strain, γ_{cr} (H11 models) [11].

and clamped nodal lines (CC) are also analyzed. In SS and CC models, in-plane displacement along the nodal lines is allowed, and only the out-of-plane deflection is constrained. The obtained average shear stress-shear strain relationships are plotted by dotted lines in Fig. 7.38. It is seen that CC model gives the upper limit of γ_{cr}.

The variations of critical shear strain, γ_{cr}, with respect to two parameters:

h_{sH}/t_w: ratio of stiffener height to web thickness, which is related to the ratio of flexural stiffness of stiffener to that of web panel

t_{sH}/t_w: ratio of stiffener thickness to web thickness, which is related to the ratio of torsional rigidity of stiffener to flexural stiffness of web panel

are shown in Fig. 7.39.

7.3.4 GIRDERS WITH VERTICALLY STIFFENED AND PERFORATED WEB PANEL

For PV11 model, deformation and spread of yielding are shown in Fig. 7.40 and average shear stress-shear strain relationship in Fig. 7.41. As seen in Fig. 7.40A, buckling does not occur until ultimate strength is attained, although yielding spreads around the hole in the middle sub-panel.

The ultimate shear strength is almost the same as fully yielded shear strength at the perforated cross-section in the middle. Beyond the ultimate strength buckling deformation develop only in the perforated middle sub-panel, which results in the reduction of capacity as the shear deformation increases.

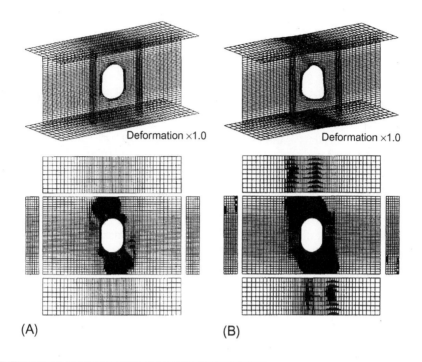

FIG. 7.40

Deformation and spread of yielding (PV11 model) [11]. (A) $\gamma/\gamma_Y = 1.2$ (ultimate strength). (B) $\gamma/\gamma_Y = 9.3$.

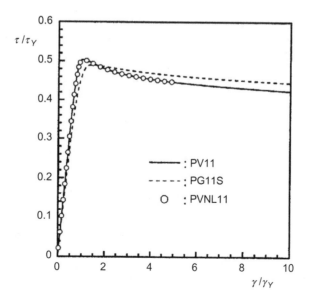

FIG. 7.41

Average shear stress-shear strain relationships (PV11 model) [11].

The dashed line in Fig. 7.41 is for PG11S, which is a girder model representing the isolated central sub-panel with perforation from PV11. Some differences are observed as the shear strain increases although the difference is not so significant. This indicates that the constraints from the sub-panels on both sides slightly increase the ultimate shear strength.

On the other hand, the open circles in Fig. 7.41 represent the result of collapse analysis on PVNL11 model, which is a virtual model obtained from PV11 model by removing stiffeners and constraining lateral deflection of the nodal lines to which stiffeners are attached with simply supported condition. The collapse behavior of PVNL11 is almost the same with that of PV11.

It can be concluded that the vertical stiffeners attached to the perforated web affect little on the ultimate shear strength and postultimate strength behavior of the girder.

7.3.5 GIRDERS WITH HORIZONTALLY STIFFENED AND PERFORATED WEB PANEL

Collapse behavior of PH11 is illustrated in Fig. 7.42, where deformation and spread of yielding are shown at the ultimate strength and in the postultimate strength state. It is seen that the yielding spreads at the vicinity of hole where stress concentrates. The ultimate strength is attained when perforated

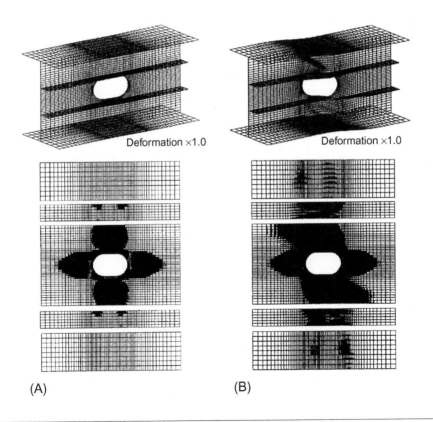

Deformation ×1.0   Deformation ×1.0

(A)   (B)

FIG. 7.42

Deformation and spread of yielding (PH11 model) [11]. (A) $\gamma/\gamma_Y = 2.5$ (Ultimate strength). (B) $\gamma/\gamma_Y = 9.4$.

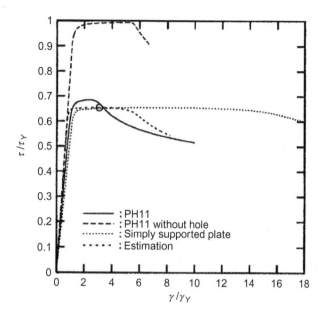

FIG. 7.43

Average shear stress-shear strain relationships (PH11 model) [11].

cross-section fully yields, and the ultimate shear strength is almost equal to the fully plastic shear strength of the perforated cross-section. After the shear strain further increases, plastic shear buckling takes place locally around the hole, and this immediately leads to the overall collapse including stiffener tripping as indicated in Fig. 7.42B.

Fig. 7.43 shows the average shear stress-shear strain relationships, and the results of collapse analysis on PH11 are plotted by the solid line. The rapid reduction of capacity around $\gamma/\gamma_Y = 3.0$ is because of the occurrence of local plastic buckling shown in Fig. 7.42B. The broken line in Fig. 7.43 is for H11 with no perforation. The ductility of PH11 is much lower than that of H11. This is because of the earlier occurrence of plastic local buckling near the hole in the case of PH11.

The dotted line in Fig. 7.43 shows the average shear stress-shear strain relationship for the sub-panel model, which is identical to the middle sub-span of PH11. The two sides of this model are assumed to be simply supported against out-of-plane deflection. The stiffness, that is, the slope of stress-strain curve, as well as the ultimate shear strength of this sub-panel model is a little different from those of a full-span model, PH11. This may be because constraints from the sub-panels on both sides of the perforated sub-panel in the case of PH11 are not considered in a single sub-panel model.

Using this curve for the collapsed part (central sub-panel) and the elastic stress-strain relationship during unloading for the remaining sub-panels on both sides, the average stress-strain relationship for the full span can be approximately estimated, which is plotted by short broken line in Fig. 7.43. This curve is denoted as "estimation." The ultimate shear strength of estimation is a little lower

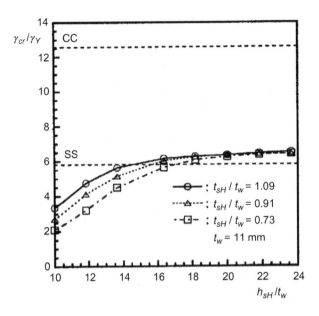

FIG. 7.44

Influence of stiffener size on critical shear strain γ/γ_Y (PH11 model) [11].

than PH11 model, but the capacity reduction beyond the ultimate strength can be simulated by the estimation.

The variation of critical shear strain, γ_{cr}, with respect to h_{sH}/t_w and t_{sH}/t_w is plotted in Fig. 7.44. Here, γ_{cr} is the shear strain of the middle sub-panel with perforation at the ultimate strength. All ductility curves saturate to the same level regardless of the height and the thickness of horizontal stiffeners. The saturated level is a little higher than that of SS model.

The influence of torsional stiffness of the stiffener on ductility is small compared to the case without perforation. This may be because stiffeners yield and their torsional stiffnesses decrease before the ultimate strength has been attained, which reduces the rotational constraint on the web panel.

EXERCISES

(1) How do the stress distribution in web panel of a plate girder vary after buckling has occurred under pure bending?

(2) What is the tension field in case of a plate girder subjected to pure shear load?

(3) How do the stress components change in the tension field in the web of a plate girder after buckling?

(4) How is the shear ultimate strength of a plate girder affected by its flanges?

(5) Derive fully plastic strength interaction relationship in terms of shear force and bending moment when a plate girder is subjected to combined shear and bending loads and buckling does not take place.

(6) What is the difference between single plate girder and bottom girder in ship and shiplike floating structures?

REFERENCES

[1] Basler K, Thürlimann B. Strength of plate girders in bending. Proc ASCE 1968;87(ST6):153–181.

[2] Basler K, Yen B, Mueller J, Thürlimann B. Web buckling tests on welded plate girders, report 251-11. Lehigh University; 1961. p. 7–48.

[3] Basler K. Strength of plate girders in shear. Proc ASCE 1961;87(ST7):151–180.

[4] Basler K. Strength of plate girders under combined shear and bending. Proc ASCE 1961;87(ST7):181–197.

[5] Akita Y, Fujii T. Minimum weight design of structures based on buckling strength and plastic collapse (1st report): plastic collapsing load of plate girders accompanied by web's shear buckling. J Naval Arch Jpn 1966;119:200–208 [in Japanese].

[6] Akita Y, Fujii T. Minimum weight design of structures based on buckling strength and plastic collapse (2nd report): strength of plate girders in postbuckling under combined bending and shear. J Naval Arch Jpn 1966;120:156–164 [in Japanese].

[7] Fujii T. Minimum weight design of structures based on buckling strength and plastic collapse (3rd report): an improved theory on post-buckling strength of plate girders in shear. J Naval Arch Jpn 1967;122:119–128.

[8] IMO: Prevention of oil pollution in the event of collision or stranding. MARPOL 73/78, Annex I Regulation 13F; 1992.

[9] Olaru D, Fujikubo M, Yanagihara D, Yao T. Ultimate strength of girder subjected to shear/bending loads. Trans Kansai Soc Naval Arch 2001;235:133–143.

[10] Japan shipbuilding association: report of SR127 panel; 1974 [in Japanese].

[11] Olaru D. Buckling/plastic collapse behaviour of late girder members and systems. Doctoral thesis. Hiroshima University; 2002.

PROGRESSIVE COLLAPSE BEHAVIOR AND ULTIMATE STRENGTH OF HULL GIRDER OF SHIP AND SHIP-LIKE FLOATING STRUCTURES IN LONGITUDINAL BENDING

8.1 ULTIMATE LONGITUDINAL STRENGTH

When the strength of ship structures is assessed, it has been common to consider three strengths: longitudinal strength, transverse strength, and local strength. Among these, longitudinal strength, which is the hull girder strength against longitudinal bending, is the most fundamental and important strength to ensure the safety of ships.

To assess the hull girder strength, estimations of both extreme load which may act on a hull girder and the capacity of the hull girder are necessary, and many research works have been performed from this aspect. Here, attention is focused mainly on the capacity of the ship hull girder.

Before the common structural rules (CSRs) by International Association of Classification Societies (IACS) came into effect in April 2006 [1,2], structural members were fundamentally not allowed to undergo buckling in the classification society rules. In those days, the ultimate strength of a ship's hull girder as well as postbuckling capacity of structural members and systems were out of consideration since the occurrence of buckling, before the ultimate strength would be attained, was not allowed. However, since the CSRs were introduced, it has been required to evaluate the ultimate hull girder strength under longitudinal bending.

Even before the CSRs appeared, a ship may be subjected to extreme loads when it fails to escape from a storm or cargo is loaded/unloaded inadequately. Furthermore, the loads below design loads could become extreme loads if the thicknesses of plating and stiffeners are reduced due to corrosion.

When an extreme bending moment acts on a ship's hull girder, structural members located in the compression side of bending progressively undergo buckling and yielding, and finally the hull girder may attain its ultimate strength and collapse. Beyond the ultimate strength, collapsed region by buckling and yielding gradually spreads in the collapsed cross-section, and the bending capacity of the hull girder decreases. Such a collapse is called progressive collapse.

8.2 RESEARCH WORKS ON PROGRESSIVE COLLAPSE BEHAVIOR AND STRENGTH OF HULL GIRDER IN LONGITUDINAL BENDING [3,4]

8.2.1 EARLY RESEARCH WORKS ON HULL GIRDER STRENGTH

It was Thomas Young who first tried to calculate the shear force and bending moment distributions along a ship's hull caused by distributed weights of the hull girder and cargoes as well as distributed buoyancy force and wave force [5]. He assumed a wave shape to calculate the wave-induced force, and considered a ship's hull as a beam subjected to distributed loads. Young's name is well known by *Young's modulus*, but his another achievement is not so much known.

In the 1850s, Isambard Kingdom Brunel designed a huge iron ship, *Great Eastern*, of which the length was nearly twice that of many ships at that time. It is reported that he applied *Beam Theory* to calculate the bending stress in the deck and the bottom plating, and determined their thicknesses on the condition that the plate does not break under tensile load [6].

In 1874, John [7] presented a paper at the annual meeting of the Institution of Naval Architects. In his paper, John assumed a wave of which length is equal to the ship's length and calculated the bending stress applying *Beam Theory*. The calculated stress was compared with the breaking stress of the material, and the plate thickness was determined from the condition that the plate does not break.

After John, methods of analyses to calculate the loads acting on a ship's hull and to simulate the structural responses have been much improved, and the design criterion has been changed from the breaking of the material to the yielding, and then to the buckling and the ultimate strength. Recently fatigue strength has also been considered. However, the fundamental idea to assume the wave length equal to the ship's length proposed by John has not been changed until now.

8.2.2 STRENGTH TESTS ON ACTUAL SHIPS

In September 1901, *HMS Cobra* sank in a middle rough sea, breaking in a V-shape [6]. Afterwards, an investigation committee was established to find out the cause of this casualty. As one of the research activities in this committee, a series of strength test was conducted on *Cobra*'s sister ship, *Wolf*. This was the first strength test carried out on an actual ship. The measured strain/displacement was lower than the calculated one applying *Beam Theory*. The reason was later found that some structural members did not effectively carry loads due to local panel buckling and shear lag phenomena [8].

The second strength test on actual ship was conducted on two destroyers, *Preston* and *Bruce*, of the US Navy in 1930/1931. In the case of the test on *Wolf*, only the elastic response was measured. On the other hand, the *Preston* and *Bruce* were loaded until they collapsed under the sagging and the hogging conditions, respectively (see Fig. 8.1A and B). It was reported that buckling collapse of the deck or bottom structure subjected to thrust led to the overall collapse of the hull girder [9,10].

During World War II, a series of strength tests were conducted on 16 war-time standard ships in the United States to find out the cause of structural failure which occurred on many of the war-time standard ships [11]. Some tests were related to the hull girder strength, and the effectiveness of

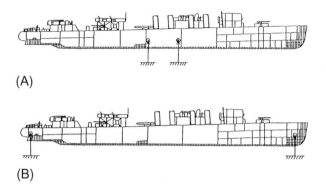

(A)

(B)

FIG. 8.1

Test setup for *Preston* and *Bruce* for hull girder collapse test. (A) *Preston* in sagging. (B) *Bruce* in hogging.

structural members such as corrugated longitudinal bulkheads and wide plating on longitudinal strength was examined.

After World War II, in 1949/1950, a collapse test was conducted on the incomplete destroyer, *Albuera*, of the Royal Navy in the UK. Fig. 8.2A shows the test setup. After applying various loads in an elastic range, the hogging load was applied until the hull girder collapsed by buckling in compression at the bottom plating and in shear at the side shell plating [12]. The collapse took place at the cross-section just after the aft-support where both bending moment and shear force are high, as indicated in Fig. 8.2B.

Since the *Albuera*'s collapse test, no full-scale strength test has been carried out on real ships to investigate into the progressive collapse behavior and the ultimate hull girder strength. However, the collapse accident of the hull girder can be regarded as a full-scale strength test when the loads acting on the hull girder at collapse were known. The MVs *Energy Concentration* [6], *Ryoyo-maru* [13], *Nakhodka* [14], and *Prestige* [15] are such cases.

8.2.3 PROGRESSIVE COLLAPSE BEHAVIOR OF HULL GIRDER UNDER LONGITUDINAL BENDING

Before explaining the calculation method of the ultimate hull girder strength, collapse behavior of the hull girder under longitudinal bending is briefly explained. According to the measured or calculated results, it is known that the structural members composing a hull girder cross-section begin to collapse one by one due to buckling or yielding with the increase in the applied bending moment, and finally the maximum capacity of the cross-section has been attained. This implies that the behavior of the structural members affects that of the cross-section.

To demonstrate this, results of calculation is introduced in Yao et al. [16]. Fig. 8.3A shows the average stress-average strain relationships of representative element on the basis of different assumptions, and Fig. 8.3B the resulting bending moment-curvature relationships calculated by Smith's method. Four fundamental cases are considered, which are

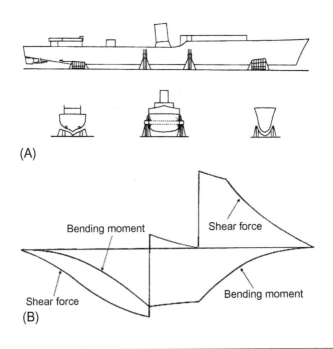

FIG. 8.2

Collapse test on *Albuera*. (A) Test setup. (B) Bending moment diagram and shear force diagram at collapse.

(a) elastic behavior without buckling;
(b) elastic-perfectly plastic behavior without buckling;
(c) buckling/yielding behavior without strength reduction beyond the ultimate strength; and
(d) buckling/yielding behavior with strength reduction beyond the ultimate strength (actual behavior).

In Case (a), bending moment-curvature relationship is also linear. In Case (b), the maximum capacity is the same under the sagging and the hogging conditions, and is equal to the fully plastic bending moment of the cross-section.

On the other hand, in Case (c), behavior is similar to that of Case (b), but the maximum capacity in sagging is different from that in hogging. This is because buckling strength is different in the deck and the bottom each other. Case (d) is the actual case, and the capacity decreases after its maximum value has been attained. Also in this case, the maximum capacity in sagging is different from that in hogging.

The difference between Cases (c) and (d) are indicated in Fig. 8.4. In Case (c), capacity of the elements do not decrease beyond their ultimate strength as is seen in Fig. 8.4A. On the other hand in Fig. 8.4B, capacity reduction is observed in the collapsed elements and the reduction is much more in the elements far from the neutral axis of the cross-section since higher strains are produced in these elements compared to those located nearer to the neutral axis.

It should be noticed that the order from (a) to (d) corresponds to the development of the calculation method.

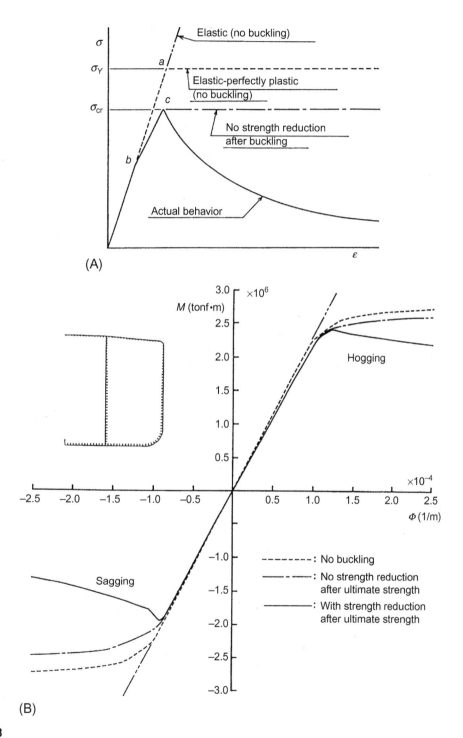

FIG. 8.3

Progressive collapse behavior of hull girder under longitudinal bending. (A) Assumed stress-strain relationship of element. (B) Bending moment-curvature relationships.

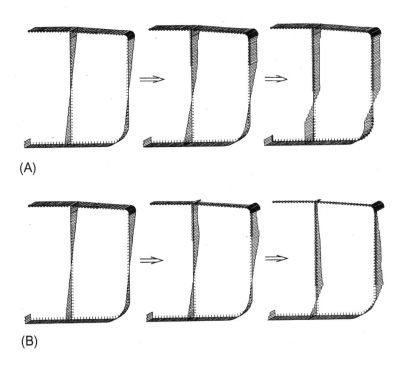

FIG. 8.4

Change in stress distribution in cross-section during progressive collapse. (A) No capacity reduction beyond the ultimate strength. (B) With capacity reduction beyond the ultimate strength.

8.2.4 CALCULATION OF ULTIMATE HULL GIRDER STRENGTH

8.2.4.1 Caldwell's method

The first attempt to calculate the ultimate hull girder strength was by Caldwell [17]. He idealized the cross-section composed of stiffened panels as that composed of panels with equivalent thickness as indicated in Fig. 8.5. When the cross-section becomes fully plastic without buckling, the stress distribution in the cross-section becomes as shown in Fig. 8.6A.

However, buckling takes place in the compression side of bending before the stress reaches yielding stress as illustrated in Fig. 8.6B. Caldwell idealized this stress distribution with buckling to that shown in Fig. 8.6C introducing strength reduction factors, and calculated the plastic bending moment of the cross-section for this stress distribution. That is, for the buckled part, the yield stress was reduced by multiplying strength reduction factors, Φ_S and Φ_D, of which magnitudes were not clearly known at that time. Caldwell's ultimate strength is evaluated as follows:

$$M_U = \sigma_Y AD \left[\Phi_D \alpha_D \gamma + 2\alpha_S \left\{ \frac{1}{2} - \gamma + \gamma^2 \frac{(1 + \Phi_S)}{2} \right\} + \alpha_S (1 - \gamma) \right] \tag{8.1}$$

where σ_Y is the yield stress of the material, and

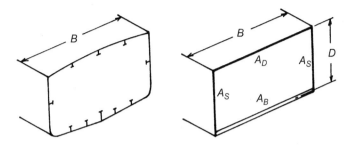

FIG. 8.5

Idealization of cross-section of ship's hull girder.

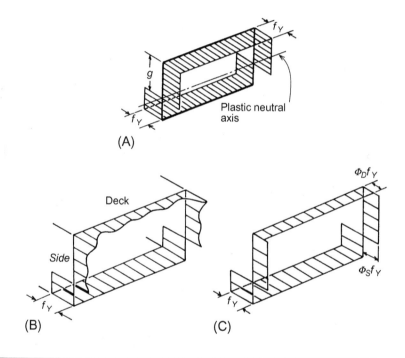

FIG. 8.6

Idealization of stress distribution in cross-section at the ultimate strength. (A) Fully plastic stress distribution. (B) Stress distribution accompanied by buckling. (C) Idealized stress distribution.

$$\gamma = \frac{g}{D} = \frac{2\alpha_S + \alpha_B - \Phi_D\alpha_D}{2\alpha_S(1 + \Phi_S)} \tag{8.2}$$

$$\alpha_D = A_D/A, \quad \alpha_B = A_B/A, \quad \alpha_S = A_S/A \tag{8.3}$$

$$A = A_D + 2A_S + A_B \tag{8.4}$$

g, D, A_D, A_S, and A_B are indicated in Fig. 8.6.

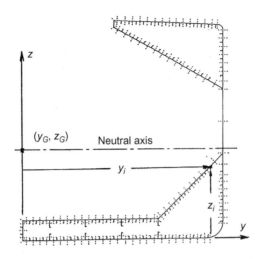

FIG. 8.7

Discretization of cross-section of ship's hull girder.

Caldwell's method corresponds to Case (c). As indicated in Fig. 8.3, the maximum capacity is overestimated in Case (c) because the strength reduction of structural members beyond their ultimate strength is assumed not to take place; see Fig. 8.3A. This also indicates that the maximum capacity is obtained assuming that all the members have attained their ultimate strength state at the same time. However, in the real structure, this is not true, since bending strain is linearly distributed toward the depth direction in the cross-section, and the time at which the ultimate strength is attained in individual element differs element by element. In addition to this, reduction in capacity beyond the ultimate strength in each element is not considered in Caldwell's method.

8.2.4.2 Smith's method

After Caldwell's proposal, more exact information was obtained regarding the strength reduction factor representing the influence of buckling. However, the problem caused by the above-mentioned time lag had not been solved until Smith [18] proposed a simplified method, which is now commonly called *Smith's method*. This method enables us to perform progressive collapse analysis on the cross-section of a hull girder subjected to longitudinal bending.

In Smith's method, a cross-section is divided into small elements composed of stiffener(s) and attached plating. Before performing progressive collapse analysis, the average stress-average strain relationships of individual elements under the axial load are derived considering the influences of yielding and buckling. Then, a progressive collapse analysis is performed assuming that a plane cross-section remains plane and each element behaves according to its average stress-average strain relationships.

Fig. 8.3A is the example of average stress-average strain relationships of elements for Smith's method, and Fig. 8.3B shows the calculated bending moment-curvature relationships of the cross-section obtained by Smith's method. This method will be explained in more detail later in Section 8.3.

After Smith, many research papers have been published, in which new methods are proposed to constitute the average stress-average strain relationship of element composed of stiffener(s) and attached plating.

8.2.4.3 Nonlinear finite element method

The nonlinear FEM can also be a powerful method to perform progressive collapse analysis on a hull girder. In 1983, ABS group [19] presented the first paper to apply the FEM to the hull girder collapse analysis. They developed special elements such as orthotropic plate element representing stiffened plate, and introduced the yielding condition in terms of sectional forces to reduce number of freedom of the calculation model. The analyses were performed on 1 + 1/2 holds model. DNV group also performed this kind of progressive collapse analysis by the nonlinear FEM employing specially developed elements [20]. The analyses were performed on 1/2 + 1/2 holds model as well as on one frame-space model.

Generally speaking, hull girder is too huge to perform progressive collapse analysis by the ordinary FEM, and some simplified methods are required. However, it has become possible to perform the FEM analysis applying explicit FEM [21], for example, using the computer code, LS-DYNA [22], although it is not common to perform such analysis in the usual design stage. As research works, for example, in Ikeda et al. [23], a series of progressive collapse analysis is performed on Handymax tanker, Aframax tanker, and VLCC applying LS-DYNA to examine the influence of thickness reduction due to corrosion on the ultimate hull girder strength. At the same time, recently, it has also become possible to perform progressive collapse analysis on a three-hold hull girder model applying implicit FEM analysis [24,25].

8.2.4.4 Idealized structural unit method

An alternative method to perform progressive collapse analysis may be the idealized structural unit method (ISUM), which was originally proposed by Ueda [26] to perform progressive collapse analysis on the transverse frame of a ship structure. Since then, new elements have been developed to perform progressive collapse analysis of a hull girder under longitudinal bending [27]. Recently, more sophisticated elements are proposed and still under development [28,29].

More recently, a total system for progressive collapse analysis of a ship's hull girder in an extreme sea has become possible with combined FEM/ISUM meshing, which will be introduced later in Section 8.7.

8.2.4.5 Simple methods

According to the ISSC report [16], the existing methods to evaluate the ultimate hull girder strength can be grouped into two: simple methods and advanced methods. Simple methods are (a) initial yielding; (b) elastic analysis; and (c) assumed stress distribution. On the other hand, advanced methods are (d) progressive collapse analysis with idealized σ-ε curves; (e) progressive collapse analysis with computed σ-ε curves; (f) ISUM; and (g) nonlinear FEM.

Method (c) is Caldwell's method, and (d) and (e) are Smith's method. As for (c), Paik and Mansour [30] assumed elastoplastic stress distribution at collapse including the influence of buckling collapse. (f) and (g) were explained in Sections 8.2.4.4 and 8.2.4.3, respectively. In the following, (a) and (b) are explained.

Initial yielding implies that the ultimate hull girder strength can be approximated by the initial yielding strength simply calculated by the following equation:

$$M_Y = Z\sigma_Y \tag{8.5}$$

where Z and σ_Y are elastic section modulus of the cross-section and the yield stress of the material, respectively.

Also in *Elastic analysis*, Eq. (8.5) is used but σ_Y is replaced by the buckling strength or ultimate strength of local panel or stiffened panel in the deck and/or bottom structure.

8.2.4.6 Assessment of available calculation methods

In the ISSC report [16], seven methods introduced in the report are assessed from the viewpoint of applicability. Each method was quantitatively graded with respect to 15 capabilities by scoring 1 through 5. It was also done qualitatively by showing the consequence of omitting capabilities by low, medium, and high. The results are shown in Table 8.1.

In this report, it is described that

Of these methods, the method based on *Initial Yielding* is an empirical one. Methods based on *Elastic Analysis* and *Assumed Stress Distribution* are direct methods whereas the remaining methods have the capability to trace out the full sequence of progressive collapse behavior of the hull girder. It is seen that the most effective among all methods is the *Progressive Collapse Analysis with Calculated σ-ε Curves*, that involves the use of numerical methods to determine the stress-strain curves of individual plate and stiffened plate elements, which are then integrated following the assumptions of simple beam theory in order to trace out the progressive collapse curve. The *ISUM* may also be an efficient method, but more rational elements have to be developed which can account the overall buckling as a stiffened panel and the tripping of stiffeners as well as the localization of yielding and deformation in the postultimate strength range of individual structural members.

8.3 SMITH'S METHOD
8.3.1 ASSUMPTIONS

The Smith's method follows the procedures described below [18].

(1) Subdivide the cross-section into elements composed of stiffener(s) and attached plating as indicated in Fig. 8.7.
(2) Derive average stress-average strain relationships of individual elements beforehand considering the influences of buckling and yielding.
(3) Give curvature incrementally on the cross-section under the condition that plane cross-section is kept plane. Then, derive tangential axial stiffness of individual elements at the present incremental step using the slope of the average stress-average strain curve at the present strain. The position of the instantaneous neutral axis during the present increment can be obtained considering the distribution of axial stiffnesses of all the elements.

Table 8.1 Assessment of Available Methods to Estimate Ultimate Hull Girder Strength in Longitudinal Bending

Capability	Simple Methods			Advanced Methods				Consequence of Omitting Capability
	Initial Yielding	Elastic Analysis	Assumed Stress Distribution	Progressive Collapse Analysis With Idealized σ-ε Curves	Progressive Collapse Analysis With Calculated σ-ε Curves	ISUM	Nonlinear FEM	
Corresponding methods in benchmark calculations in Chapter 4	–	–	Astrup Rigo (2)	Cho Rigo (1)	Dow Yao Scores	Chen Masaoka	–	–
Plate buckling	–	2	3	4	5	5	5	H
Stiffener plate buckling	–	2	3	3	5	3	5	H
Plate initial deflection	–	1	2	2	5	5	5	M
Stiffener initial deflection	–	1	3	3	5	3	5	M
Plate welding residual stress	–	2	3	3	4	4	4	H
Stiffener welding residual stress	–	1	2	3	3	4	4	M
Postbuckling behavior	–			3	5	5	5	H
Multispan model	–			2	5	5	5	M
M-Φ curve (collapse prediction)	–			3	5	5	5	H
Postultimate strength	–	–	–	3	5	4	5	M
Damage	2	2	2	3	5	3	5	–
Material modeling	1	1	1	3	5	5	5	–
Modeling/data preparation	5	4	4	3	3	3	1	
Analysis/checking results	5	3	3	3	3	3	1	–
Accuracy/reliability of results	2	1	2	3	4	3	4	–
Total (full score: 75)	15	19	30	43	67	56	63	

Notes: Score: 1, not available; 2, poor ability; 3, insufficient accuracy; 4, acceptable; 5, excellent. Consequence of omiting capability: L, low; M, medium; H, high.

(4) Evaluate the flexural stiffness of the cross-section with respect to the instantaneous neutral axis. Then, increment of bending moment is evaluated multiplying the flexural stiffness and the applied curvature increment.

(5) Calculate the strain increments in individual elements under the applied curvature increment, and their stress increments using the slope of average stress-average strain curve.

(6) Adding the obtained increments of curvature, bending moment as well as strains and stresses in the elements to their cumulative values, proceed to the next incremental step.

In Smith's method, the following two assumptions are made:

(1) Plane cross-section remains plane during progressive collapse.

(2) There exists no interaction between adjacent elements in the cross-section.

The first assumption holds until large deflection is produced in the buckled elements and large local deformation is produced. As for the second assumption, it may hold when cross-section is subjected to mainly longitudinal bending.

It should be noticed that shear force is zero at the cross-section where bending moment is the maximum. When the hull girder collapses under combined bending and shear, influence of shear force has to be considered as shall be indicated in Section 8.3.3.5.

8.3.2 STIFFNESS EQUATION

In the original Smith's method, bi-axial bending is considered when fundamental equations are derived. If an axial load is added to bi-axial bending, the stiffness equation can be written in more general form as follows:

$$
\begin{Bmatrix} \Delta P \\ \Delta M_H \\ \Delta M_V \end{Bmatrix} = \begin{bmatrix} D_{AA} & D_{AH} & D_{AV} \\ D_{HA} & D_{HH} & D_{HV} \\ D_{VA} & D_{VH} & D_{VV} \end{bmatrix} \begin{Bmatrix} \Delta u \\ \Delta \Phi_H \\ \Delta \Phi_V \end{Bmatrix} \tag{8.6}
$$

where

ΔP: increment of axial force
ΔM_H: increment of horizontal bending moment
ΔM_V: increment of vertical bending moment
Δu: increment of axial displacement
$\Delta \Phi_H$: increment of horizontal curvature
$\Delta \Phi_V$: increment of vertical curvature

and D_{AA} CD_{HH} D_{VV} are the tangential stiffness of the cross-section, and are expressed as

$$
D_{AA} = \sum_{i=1}^{n} D_i F_i / \ell_i \tag{8.7}
$$

$$
D_{AH} = D_{HA} = \sum_{i=1}^{n} D_i F_i (y_i - y_G) / \ell_i \tag{8.8}
$$

$$D_{AV} = D_{VA} = \sum_{i=1}^{n} D_i F_i (z_i - z_G)/\ell_i \tag{8.9}$$

$$D_{HH} = \sum_{i=1}^{n} D_i F_i (y_i - y_G)^2 \tag{8.10}$$

$$D_{HV} = D_{VH} = \sum_{i=1}^{n} D_i F_i (y_i - y_G)(z_i - z_G) \tag{8.11}$$

$$D_{VV} = \sum_{i=1}^{n} D_i F_i (z_i - z_G)^2 \tag{8.12}$$

Here, n is the total number of elements, and D_i, ℓ_i, and F_i represent the axial tangential stiffness, the length, and the cross-sectional area of the ith element. y_i and z_i are the coordinates of the center of geometry of the cross-section of the ith element in reference coordinate systems; see Fig. 8.7. D_i can be determined as the slope of average stress-average strain curve of the ith element at the specified strain. y_G and z_G are the coordinates of the center of geometry of the whole cross-section, which are given as

$$y_G = \left(\sum_{i=1}^{n} y_i Di F_i \right) \bigg/ \left(\sum_{i=1}^{n} D_i F_i \right) \tag{8.13}$$

$$z_G = \left(\sum_{i=1}^{n} z_i Di F_i \right) \bigg/ \left(\sum_{i=1}^{n} D_i F_i \right) \tag{8.14}$$

y_G and z_G vary at each incremental step depending on the spread of buckling and yielding.

Accuracy of the calculated results applying Smith's method largely depends on the accuracy of the average stress-average strain relationships which is used for progressive collapse analysis. In Smith's original paper [18], the average stress-average strain relationship was derived performing nonlinear FEM analysis on plates and stiffeners with attached plating. Analytical and semi-analytical methods are also proposed by many researchers. The analytical method proposed in Yao and Nikolov [31] is introduced in Section 8.8.

8.3.3 APPLICATION OF SMITH'S METHOD
8.3.3.1 Smith's works
Smith performed progressive collapse analysis on the cross-section of a destroyer [18]. This was the first proof that the cross-section of a hull girder cannot attain its fully plastic bending moment due to the occurrence of local buckling on the basis of numerical calculation. He also discussed merits and demerits in transversely stiffened system and longitudinally stiffened system from the viewpoints of ultimate hull girder strength [32]. Furthermore, he discussed the residual ultimate hull girder strength after the cross-section has been damaged due to collision, grounding, excess hydrodynamic pressure, and so on. Furthermore, he derived the ultimate hull girder strength interaction relationship under combined vertical and horizontal bending.

8.3.3.2 Application to double hull tanker
Double hull tanker for analysis and assumed initial imperfections
A series of progressive collapse analysis is performed on an existing double hull tanker (VLCC), of which a cross-section is shown in Fig. 8.8 [33]. The principal dimensions are

$$L \times B \times D = 315.0 \times 58.0 \times 30.4 \text{ m}$$

The material is HT32 except a little part of the longitudinal stiffeners. The following initial imperfections are assumed.

1. *Welding residual stress*: Rectangular distribution in Fig. 8.9 is assumed with compressive residual stress, σ_c, of 10% of the yielding stress.
2. *Initial deflection*: Buckling mode with the magnitude of 1% of the thickness is assumed for plating, while 0.1% of the span length for stiffeners, which includes vertical deflection in flexural buckling mode and horizontal deflection at the top of flange in torsional buckling mode.

Average stress-average strain relationships of structural members
The average stress-average strain relationships for typical stiffener elements are indicated in Fig. 8.9. The stiffener elements 3 and 4 are an angle-bar type, while that of other elements is a tee-bar type. In element 3, significant reduction is observed in the load carrying capacity beyond the ultimate strength. This is because coupled flexural-torsional buckling had occurred.

In the tension range, the average stress-average strain relationship is assumed to be the same with that of the material. However, the stiffness of the plate part is reduced in the elastic range to

$$\frac{d\sigma}{d\varepsilon} = \frac{\sigma_Y}{\sigma_Y + \sigma_c} E \tag{8.15}$$

where E and σ_Y are Young's modulus and the yielding stress of the material, and σ_c is the compressive residual stress in the panel. Eq. (8.15) is derived on the condition that the part of the cross-section where tensile residual stress exists can carry no more tensile load since this part has been from the beginning yielded in tension; see Fig. 8.80 in Section 8.8.

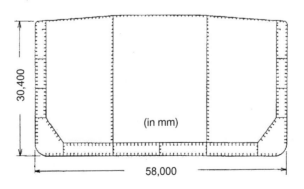

FIG. 8.8

Midship cross-section of double hull VLCC [33].

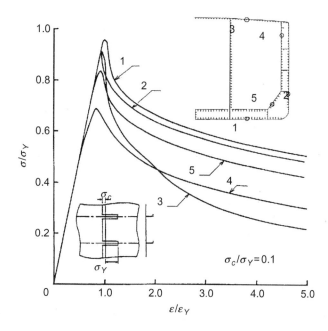

FIG. 8.9

Average stress-average strain relationships for typical stiffener elements [33].

Collapse behavior of cross-section under pure bending

The results of progressive collapse analysis on a double hull tanker are shown in Figs. 8.10–8.12. Fig. 8.10 indicates the relationships between bending moment and curvature. The dashed line is the case when buckling is excluded and the solid line is the case in which buckling is considered. In the former case, initial imperfection is not considered, and the capacity approaches to the fully plastic bending moment. The fully plastic bending moment is the same both in hogging and sagging conditions. Comparing the solid and the dashed lines, it is known that the fully plastic bending moment cannot be attained due to the occurrence of buckling. The buckling strength of deck plating is in general lower compared to that of bottom plating. This is the reason for the lower ultimate hull girder strength in sagging.

Another important issue is the flexural stiffness of the cross-section. It is known from Fig. 8.10 that the slope of the solid line is lower than that of the dashed line. This is not because of the occurrence of buckling but the lower axial stiffness of stiffener elements located in the tension side of overall bending, see Eq. (8.15).

On the other hand, Fig. 8.11A and B shows the changes in stress distribution in the cross-section during progressive collapse under the hogging and the sagging conditions together with collapsed elements at the specified points on bending moment-curvature curve in Fig. 8.10. The point where the stress is zero corresponds to the location of a neutral axis of the cross-section. This varies during the progressive collapse as indicated in Fig. 8.12.

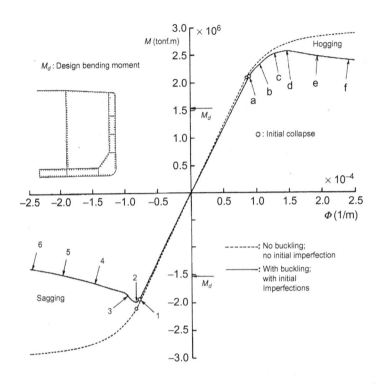

FIG. 8.10

Bending moment-curvature strain relationships of hull girder cross-section under longitudinal bending [33].

Under the sagging condition, buckling takes place at central part of the deck at point 1. Soon after this, buckled region spreads all over the deck, and the ultimate strength has been attained at point 2. Beyond the ultimate strength, the load carrying capacity rapidly decreases down to point 3, but the decreasing rate becomes lower beyond point 3. This is a reflection of the average stress-average strain relationship of the deck element indicated in Fig. 8.9. With the increase in the applied curvature, the buckled region spreads downwards in the side shell plating and the longitudinal bulkheads. Accordingly, the location of the neutral axis of the cross-section moves downwards as indicated by the solid line in Fig. 8.12. Consequently, the increasing rate of the tensile strain at the bottom part is low although the curvature is increasing, and the yielding of the bottom part in tension does not take place within the applied curvature.

Under the hogging condition, initial member collapse takes place at point a on the moment-curvature curve at the deck part by yielding in tension. This is because the elastic neutral axis locates below the mid-depth height, and the strain at the deck part is much higher than that at the bottom part. No strength reduction takes place after this collapse since no strength reduction occurs in the elements yielded in tension. Because of this, the bending moment still increases after the deck has been yielded, but the

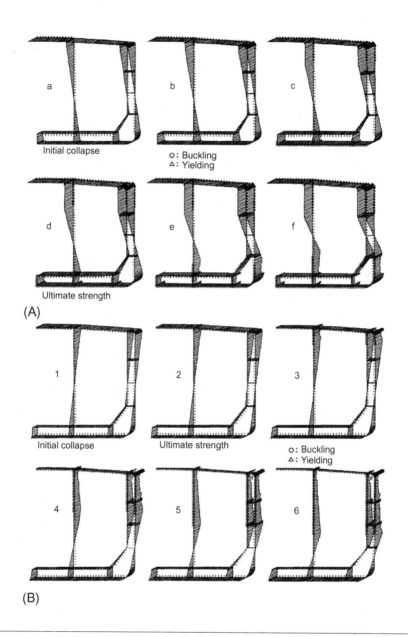

FIG. 8.11

Change in stress distribution in cross-section during progressive collapse [33]. (A) Under hogging condition. (B) Under sagging condition.

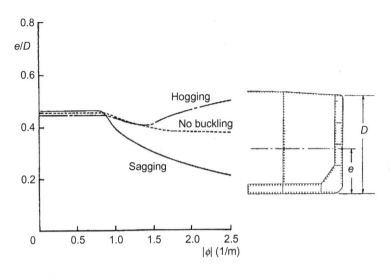

FIG. 8.12

Change in location of neutral axis in cross-section during progressive collapse [33].

flexural stiffness decreases as the yielded region spreads downwards in the side shell plating and the longitudinal bulkheads. At point c, the bottom shell undergoes buckling. At last, the inner bottom buckles at point d, and the ultimate hull girder strength is attained. The chain line in Fig. 8.12 indicates the change in location of the neutral axis under the hogging condition. After the yielding takes place at the upper part of the cross-section, the neutral axis moves downwards, since the effectiveness of the yielded part decreases. However, after the bottom part buckles, it changes to move upwards. This is because the loss of effectiveness at the buckled region in compression is larger than that at the yielded region in tension.

In general, ship structures are so designed that the bottom part is stronger than the deck part, and the neutral axis of the cross-section is located below the mid-depth height. Consequently, the deck strain is higher than the bottom strain. This indicates that the buckling at the deck under the sagging condition takes place earlier than that at the bottom under the hogging condition. So, the ultimate hull girder strength under the sagging condition is lower than that under the hogging condition.

Collapse behavior and strength under bi-axial bending

A ship's hull is in general exposed to combined vertical and horizontal bending when the ocean going ship is rolling in an oblique sea. To simulate the collapse behavior of a ship's hull girder under such bi-axial bending, a series of progressive collapse analysis is performed giving curvatures for vertical bending and horizontal bending with a constant ratio [34].

Fig. 8.13 shows the moment-curvature relationships under combined vertical and horizontal bending in sagging. The change in stress distribution during progressive collapse is shown in Fig. 8.14. It is known that the initial collapse starts at the right gunwale or the intersection part of side shell plating and

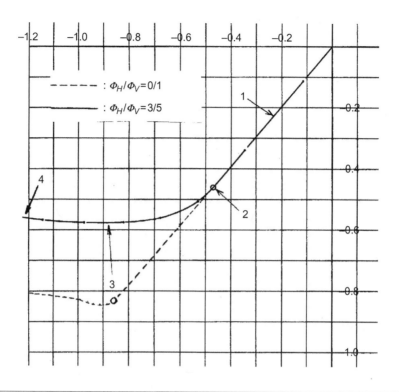

FIG. 8.13

Bending moment-curvature strain relationships of hull girder cross-section under bi-axial longitudinal bending.

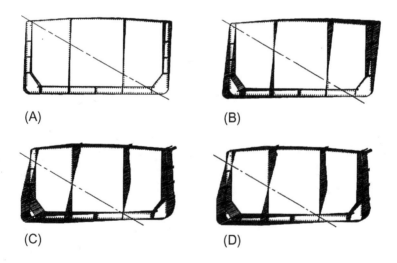

FIG. 8.14

Change in stress distribution in cross-section during progressive collapse under bi-axial longitudinal bending. (A) Point 1. (B) Point 2 (initial collapse strength). (C) Point 3 (ultimate strength). (D) Point 4.

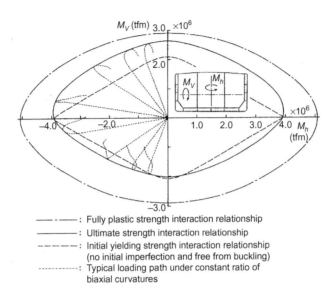

--- · --- : Fully plastic strength interaction relationship
————— : Ultimate strength interaction relationship
– – – – – : Initial yielding strength interaction relationship
(no initial imperfection and free from buckling)
·············· : Typical loading path under constant ratio of
biaxial curvatures

FIG. 8.15

Loading paths and strength interaction relationships under bi-axial longitudinal bending [34].

deck plating. Then, the collapsed parts spread in the deck, side shell plating and longitudinal bulkhead as indicated in Fig. 8.14, and the ultimate hull girder strength is attained.

The cross-section is in general almost symmetrical with respect to the center plane. So, the collapse behavior under the horizontal bending is almost the same regardless of the sign of the horizontal curvature. When only the horizontal bending moment acts on the cross-section, the initial yielding strength almost coincides with the ultimate hull girder strength. This is because the neutral axis for horizontal bending locates near the center line until buckling takes place. This implies that the magnitude of bending strain in the side shell plating is almost the same in tension and compression sides of horizontal bending. Because of this, buckling collapse in the compression side and initial yielding in the tension side take place almost simultaneously, and the ultimate hull girder strength has been attained at this instance.

The interaction curves for the initial yielding strength and the fully plastic strength are symmetrical with respect to both horizontal and vertical coordinate axes. On the other hand, that for the ultimate hull girder strength is not. This is because buckling collapse strength under vertical bending is different at the deck structure and the bottom structure. It should be noticed that the ultimate hull girder strength in sagging is lower than the initial yielding strength simply calculated by Eq. (8.2). This will be discussed later in Section 8.3.3.4

Influence of corrosion damage on the progressive collapse behavior and the ultimate strength is discussed in Yao et al. [33].

8.3.3.3 Application to bulk carrier
Bulk carrier for analysis and assumed initial imperfections

A series of progressive collapse analysis is performed on a mid-ship cross-section of the Panamax size bulk carrier shown in Fig. 8.16 [35]. HT32 steel is used for the deck plating as well as upper part of

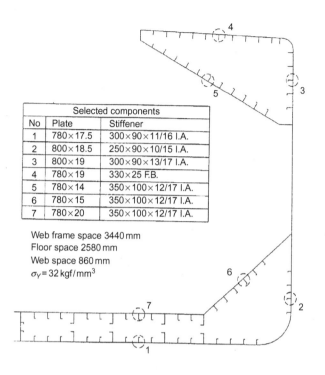

Selected components

No	Plate	Stiffener
1	780×17.5	300×90×11/16 I.A.
2	800×18.5	250×90×10/15 I.A.
3	800×19	300×90×13/17 I.A.
4	780×19	330×25 F.B.
5	780×14	350×100×12/17 I.A.
6	780×15	350×100×12/17 I.A.
7	780×20	350×100×12/17 I.A.

Web frame space 3440 mm
Floor space 2580 mm
Web space 860 mm
$\sigma_Y = 32 \, kgf/mm^3$

FIG. 8.16

Cross-section of Panamax bulk carrier [35].

side shell plating and slanted bottom plating of the top side tank, whereas mild steel is used for the remaining parts. As for the initial imperfection due to welding, the same imperfections are used as those for the double hull tanker analyzed in Section 8.3.3.2. They are

Compressive residual stress:	10% of the yielding stress, σ_Y
Initial deflection in panel:	buckling mode of which maximum value is 1% of the panel thickness
Initial deflection in stiffeners:	0.1% of the span length (in flexural buckling mode; horizontal deflection at the top)

Average stress-average strain relationships for representative elements are shown in Fig. 8.17.

Progressive collapse behavior

Using the average stress-average strain relationships shown in Fig. 8.17, progressive collapse analysis is performed on the cross-section illustrated in Fig. 8.16. The bending moment-curvature relationships are plotted by the solid line in Fig. 8.18, which is obtained with the average stress-average strain relationships in Fig. 8.17. On the other hand, the dashed line is the case when influences of initial

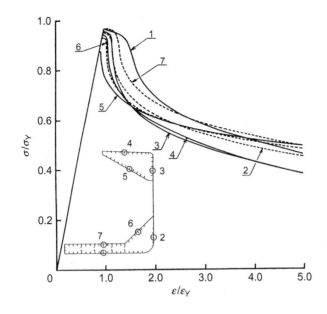

FIG. 8.17

Average stress-average stain relationships of elements of Panamax bulk carrier [35].

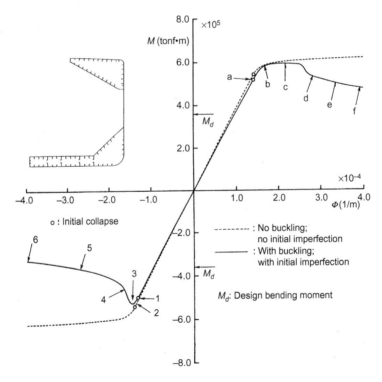

FIG. 8.18

Bending moment-curvature stain relationships of cross-section of Panamax bulk carrier [35].

imperfection as well as buckling are excluded. In this case, the maximum capacity corresponds to the fully plastic bending moment.

Here, Fig. 8.19A and B shows the stress distribution in the cross-section at the specified points on bending moment-curvature curves in Fig. 8.18. The former is under the hogging condition, and the latter under the sagging condition. Furthermore, Fig. 8.20 indicates the shift of the neutral axis of the cross-section during progressive collapse.

The dashed line in Fig. 8.18 represents the bending moment-curvature relationship when influences of buckling and initial imperfection are not considered. In this case, the capacity approaches to the fully plastic bending moment as the curvature increases. This indicates that the hull girder cannot attain the fully plastic bending moment of the cross-section because of the occurrence of buckling. The ultimate strength under the sagging condition is lower than that under the hogging condition. This is because the ultimate compressive strength of the deck part is lower than that of the bottom part, and the deck structure is weaker than bottom structure.

Under the hogging condition, deck part at the hatch end undergoes yielding initially at point a on the bending moment-curvature curve in Fig. 8.18. At point b, the bottom plating starts to buckle. At this stage, yielding spreads all over the deck plating as well as the upper part of bottom plating of the top side tank and the gunwale. At point c, the inner bottom plating undergoes buckling and the ultimate hull girder strength is attained. At this point, all the plating composing top side tank is yielded in tension. After this, between points d and e on the moment-curvature curve, buckling collapse spreads upwards in the side shell plating and the slant plate of the hopper side tank.

The chain line in Fig. 8.20 shows the shift of neutral axis during progressive collapse in hogging. Firstly, it once goes downwards after the start of yielding at deck plating. Then, it turns to go upwards after the buckling collapse takes place in the bottom part. This is because the loss of effectiveness of the buckled part in compression is more significant when it is compared to that of the yielded part in tension. Due to the rapid upwards shift of the neutral axis, the tensile strain is reduced, and the top side tank which was once yielded in tension is unloaded; see the stress distribution at point f in Fig. 8.19A.

Under the sagging condition, buckling collapse takes place at the deck adjacent to the hatch end at point 1 on the bending moment-curvature curve in Fig. 8.18. With the further increase in the applied curvature, the buckled part spread all over the deck and downwards in the side shell plating and bottom plate of the top side tank, and the ultimate hull girder strength is attained at point 3. Beyond the ultimate hull girder strength, buckling collapse further develops downwards, and the upper part of the hopper side yank also undergoes buckling collapse at point 6 on the bending moment-curvature curve.

In the stress distribution, discontinuous points are observed in a bottom plating of the top side tank. This is because the material changes from HT32 to mild steel at these points.

Here, the design bending moment, M_d, for this vessel is indicated with horizontal arrows on the vertical coordinate axis both in the hogging and the sagging ranges. From these, the safety factor from the viewpoint of the ultimate hull girder strength is derived as 1.48 and 1.67 under the sagging and the hogging conditions, respectively. For the initial member collapse strength, they are 1.39 and 1.44 for sagging and hogging, respectively.

Influence of welding residual stress on progressive collapse behavior and strength

Here, the influence of welding residual stress in the plating on progressive collapse behavior is explained. The welding residual stress of a rectangular distribution is assumed and the magnitude of compressive residual stress is varied as

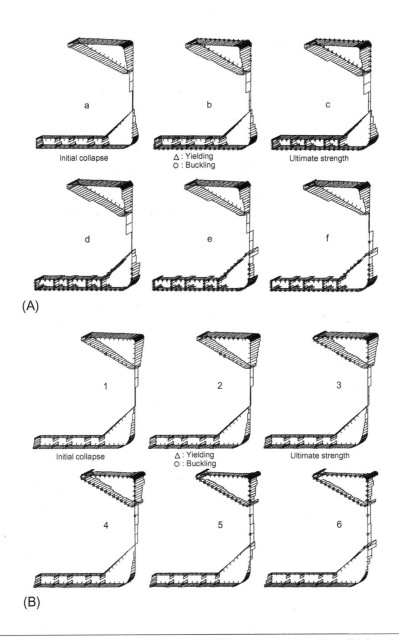

FIG. 8.19

Change in stress distribution during progressive collapse [35]. (A) Hogging condition. (B) Sagging condition.

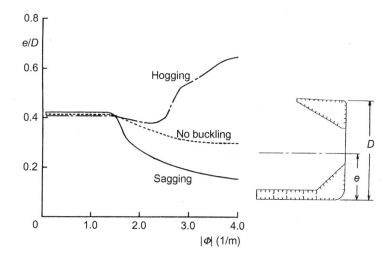

FIG. 8.20

Shift of neutral axis of the cross-section during progressive collapse [35].

$$\sigma_c/\sigma_Y = 0.0, 0.2, 0.3$$

The case of $\sigma_c/\sigma_Y = 0.1$ was already analyzed in the "Progressive collapse behavior" section.

Fig. 8.21 shows the influence of welding residual stress on the average stress-average strain relationships of the deck and the bottom elements. Firstly, in the compression range, the ultimate strength is reduced by the welding residual stress, but the strength reduction is very small within the assumed magnitude of compressive welding residual stress. On the other hand, in the tension range, in-plane stiffness is reduced because the region of the cross-section where tensile residual stress exists cannot carry further tensile load. According to Eq. (8.15), the in-plane stiffness is reduced down to 77% when $\sigma_c/\sigma_Y = 0.3$.

The bending moment-curvature relationships are plotted in Fig. 8.22 for the cases of $\sigma_c/\sigma_Y = 0.0$, 0.1, 0.2, and 0.3. It is seen that the ultimate hull girder strength is very little affected by the welding residual stress in plating, whereas the flexural stiffness decreases as the compressive residual stress increases since the area where tensile residual stress exists also increases.

Influence of initial deflection in plating on progressive collapse behavior and strength

The magnitude of initial deflection in plating has been varied as

$$A_0/t = 0.1, 0.5, 1.0$$

where t is the panel thickness. The case of $A_0/t = 0.01$ is already analyzed in the "Progressive collapse behavior" section. The initial deflection of a buckling mode is assumed.

Fig. 8.23 shows the influence of initial deflection in plating on the average stress-average strain relationships of the deck and the bottom elements. In both elements, larger reduction is observed in the ultimate compressive strength as the magnitude of initial deflection increases. The slenderness

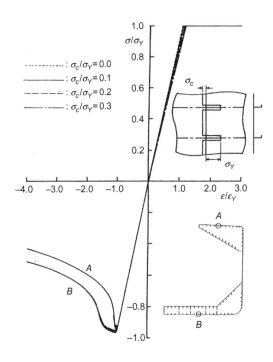

FIG. 8.21

Influence of welding residual stress on average stress-average stain relationships of elements [35].

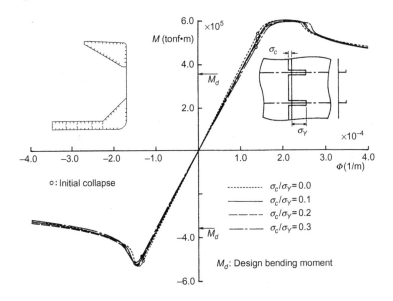

FIG. 8.22

Influence of welding residual stress on bending moment-curvature relationships [35].

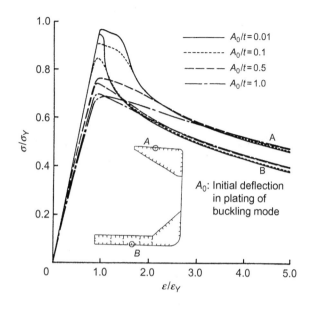

FIG. 8.23

Influence of initial deflection in plating on average stress-average stain relationships of elements [35].

ratio of the local panel, $b/t \cdot \sqrt{\sigma_Y/E}$, in deck and bottom structures is around 1.5. For this value of the slenderness ratio, ultimate compressive strength of the local panel largely decreases due to the influence of initial deflection. This could be one reason why large reduction is observed in the ultimate compressive strength due to initial deflection. However, with the increase in the compressive strain beyond the ultimate strength, each average stress-average strain curve converges to a unique curve.

The influence of initial deflection in plating on the bending moment-curvature relationship is indicated in Fig. 8.24. The shapes of individual curves are the reflection of the average stress-average strain curves of deck elements under the sagging condition, and those under the hogging condition are the reflection of average stress-average strain curves of bottom elements.

It is known that the steepness of the capacity reduction beyond the ultimate hull girder strength becomes moderate with the increase in the magnitude of initial deflection, although the ultimate hull girder strength is much reduced by the larger initial deflection in plating.

Here, as explained in Chapter 2, the deck/bottom plating is in general thick, and the maximum magnitude of initial deflection is not large. Furthermore, the buckling component which actually affect the buckling behavior is much smaller and is at most 3% of the plate thickness. So, in the actual ship structure, the ultimate hull girder strength is not so much reduced by initial deflection in plating as indicated in Fig. 8.24, since large initial deflection of a buckling mode does not exist in real local panels.

However, when a ship is constructed in a transversely stiffened system, the maximum magnitude of initial deflection directly affects the ultimate hull girder strength since the modes of initial deflection

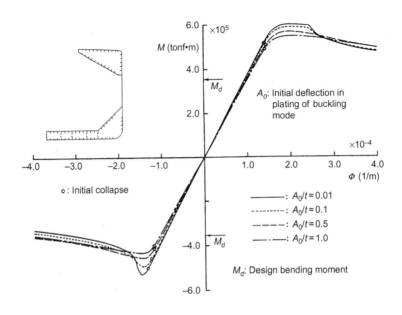

FIG. 8.24

Influence of initial deflection in plating on bending moment-curvature relationships [35].

and buckling are almost the same. In this case, large reduction will take place in the ultimate strength in accordance with the magnitude of initial deflection in plating.

Influence of initial deflection in stiffener on progressive collapse behavior and strength

Lastly, the influence of initial deflection in longitudinal stiffeners is examined varying the magnitude of initial deflection of a flexural buckling mode as follows:

$$W_0/a = 0.0002, 0.0025, 0.005$$

where a is the span length. The case of $W_0/a = 0.001$ is already analyzed in the "Progressive collapse behavior" section.

The average stress-average strain relationships of the deck and the bottom elements are derived for different W_0/a, and are plotted in Fig. 8.25. The ultimate strength is reduced according to the magnitude of initial deflection. However, decrease in the capacity beyond the ultimate compressive strength is not so much compared to the ultimate strength reduction.

Fig. 8.26 shows the bending moment-curvature relationships for different magnitude of initial deflection in longitudinal stiffeners. It is seen that the ultimate hull girder strength decreases with the increase in the magnitude of initial deflection in stiffeners. Capacity beyond the ultimate strength is also reduced by this initial deflection, but the decrease in capacity becomes smaller as the applied curvature increases. It should be noticed that flexural stiffness is also reduced by the initial deflection in longitudinal stiffeners, although it is not so much compared to reduction due to welding residual stress.

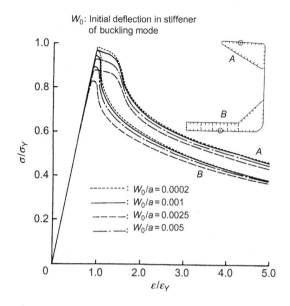

FIG. 8.25

Influence of initial deflection in stiffener on average stress-average stain relationships of elements [35].

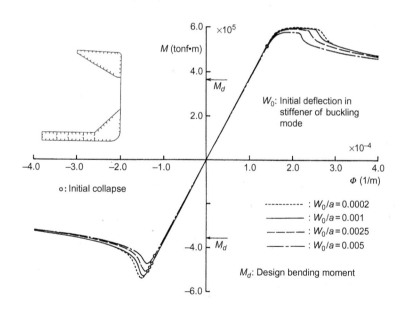

FIG. 8.26

Influence of initial deflection in stiffener on bending moment-curvature relationships [35].

8.3.3.4 Considerations on ultimate hull girder strength

Progressive collapse analysis was carried out on 10 different ships including tankers, bulk carriers, product carrier, ore carrier, pure car carrier, and container ship [36]. The ultimate hull girder strength and initial member collapse strength as well as initial yielding moment and fully plastic bending moment both in sagging and hogging are summarized in Table 8.2. The initial yielding strength is calculated by Eq. (8.5), that is

$$M_Y = Z\sigma_Y \tag{8.5}$$

where Z is elastic section modulus and σ_Y is the yield stress of the material. On the other hand, fully plastic bending moment is calculated as the bending moment corresponding to fully plastic stress distribution without considering the influence of buckling. It is expressed as

$$M_P = Z_P\sigma_Y \tag{8.16}$$

where Z_P is plastic section modulus. All the strengths except the initial yielding strength are nondimensionalized by the initial yielding strength expressed by Eq. (8.5).

M_P/M_Y represents the ratio of the plastic section modulus to the elastic section modulus, Z_P/Z, of the cross-section as is known from Eqs. (8.5), (8.16). This ratio is usually called a shape factor of the cross-section, and gives the reserve strength ratio of the ultimate hull girder strength to the initial yielding strength.

The initial yielding strength could be the most simple measure for estimation of the ultimate hull girder strength since it can be easily calculated using the elastic section modulus of the mid-ship cross-section. However, it should be noticed that the initial yielding strength is in many cases an unconservative estimate of the ultimate hull girder strength under the sagging condition. This is because the ultimate compressive strength of the deck structure as a stiffened plate is in many cases lower than its yielding strength.

Here, the initial yielding strength in Table 8.2 is calculated when the stress at the top of the deck, which is usually the center of the deck plating, reaches the yield stress. However, if the initial yielding strength is calculated with the stress at the side end of the deck, that is, if the section modulus is calculated with mold depth, D, of the cross-section, the initial yielding strength becomes very near to the ultimate hull girder strength under the sagging condition. This is shown in Fig. 8.27 [37].

8.3.3.5 Limitation and extension in application of Smith's method

Smith's method gives relatively accurate estimation of the ultimate hull girder strength under pure bending. This is the case when a ship's hull collapses at the location where working bending moment is the highest, since vertical or transverse shear force is zero at this cross-section.

On the other hand, when a ship's hull collapses under combined bending moment and shear force, the following have to be considered:

(1) influence of shear stress on yielding condition;
(2) influence of shear stress on buckling condition; and
(3) influence of warping on stress/strain distribution in a cross-section.

The first two items affect on the average stress-average strain relationships of individual elements, whereas the third item affects the fundamental assumption in Smith's method that plane cross-section remains plane during progressive collapse.

Table 8.2 Reserve Strength Factors for Ultimate Hull Girder Strength of Various Types of Ships [36]

No.	$L \times B \times D$ (in m)	M_Y (in Ton f m)	M_P/M_Y	M_{FS}/M_Y	M_{US}/M_Y	MFH/M_y	MUH/M_Y
1	$215 \times 32.2 \times 17.8$	0.5405×10^6	1.1534	0.9204	0.9813	0.9251	1.1031
2	$217 \times 32.26 \times 18.3$	0.5592×10^6	1.2316	0.8498	0.9312	0.9408	1.1296
3	$276 \times 45 \times 24.2$	0.1344×10^7	1.1584	0.8702	0.9400	0.9470	1.0803
4	$247.4 \times 36.2 \times 21.8$	0.7917×10^6	1.2260	0.9502	1.0213	0.9553	1.1698
5	$315 \times 57 \times 30.8$	0.2193×10^7	1.2681	0.9134	0.9234	1.0119	1.1172
6	$315 \times 58 \times 30.4$	0.2098×10^7	1.4147	0.9214	0.9473	1.0119	1.2312
7	$162 \times 30 \times 16.2$	0.2912×10^6	1.6326	0.8908	1.0251	1.0261	1.4214
8	$315 \times 52 \times 23.45$	0.1951×10^6	1.1886	0.9846	1.0359	0.9800	1.1220
9	$180 \times 32.26 \times 30.55$	0.4892×10^6	1.6492	0.4563	0.7482	0.9663	1.2122
10	$230 \times 32.2 \times 21.5$	0.6624×10^6	1.3439	0.9674	1.0482	0.8284	1.0652

Notes: M_Y, initial yielding strength; M_p, fully plastic bending moment; M_{FS}, initial member collapse strength under sagging condition; M_{US}, ultimate strength under sagging condition; M_{FH}, initial member collapse strength under hogging condition; M_{UH}, ultimate strength under hogging condition.

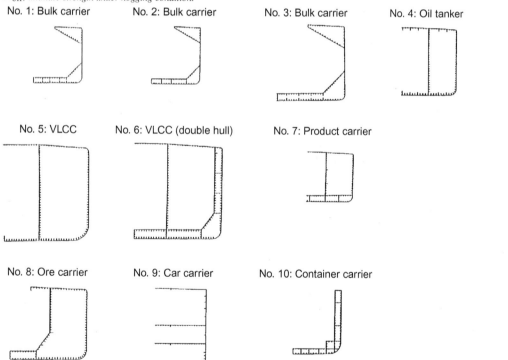

No. 1: Bulk carrier No. 2: Bulk carrier No. 3: Bulk carrier No. 4: Oil tanker

No. 5: VLCC No. 6: VLCC (double hull) No. 7: Product carrier

No. 8: Ore carrier No. 9: Car carrier No. 10: Container carrier

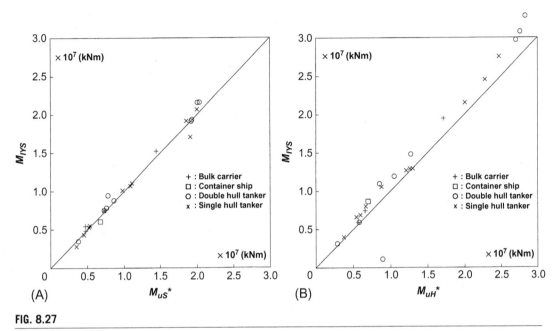

FIG. 8.27

Comparison between initial yielding strength and ultimate hull girder strength [37]. (A) In sagging.
(B) In hogging.

The first item could be solved by considering the yielding condition under combined uni-axial normal stress and shear stress as

$$\sigma^2 + 3\tau^2 = \sigma_Y^2 \tag{8.17}$$

From Eq. (8.17), the yielding stress can be modified on the assumption that the shear stress is constant [37].

$$\sigma_{Ys} = \sqrt{\sigma_Y^2 - 3\tau^2} \tag{8.18}$$

Fig. 8.28 shows the influence of shear stress on the average stress-average strain relationship of element [38].

As for the second item, buckling strength interaction relationship approximated by the following equation can be used [38]:

$$\left(\frac{\sigma}{\sigma_Y}\right)^2 + \left(\frac{\tau}{\tau_Y}\right)^2 = 1 \tag{8.19}$$

When a cross-section of the girder is subjected to shear force, warping deformation appears in the normal direction to the cross-section. This implies that the plane cross-section is no more plane after

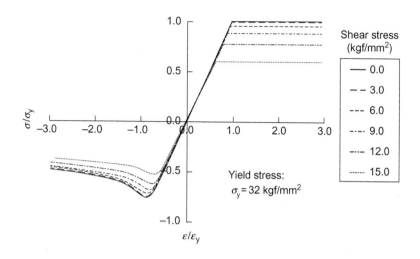

FIG. 8.28

Influence of shear stress in average stress-average strain relationship of stiffener element [38].

warping deformation takes place. The elastic warping deformation can be approximately calculated by the method proposed by Fujitani [39], which is briefly introduced later in Section 8.9.

Here, a bulk carrier under an alternate loading condition is considered. For this case, bending moment and shear force distributions are shown in Fig. 8.29, and the warping stress in the cross-sections in Fig. 8.30A and B. In Fig. 8.30, the dashed lines indicate the bending stress produced by still

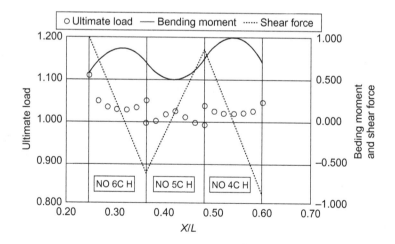

FIG. 8.29

Shear force diagram, bending moment diagram, and distribution of ultimate hull girder strength in bulk carrier under alternate loading [38].

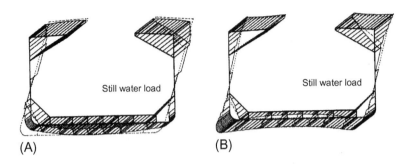

FIG. 8.30

Bending and warping stress distributions [38]. (A) No. 5 cargo hold. (B) No. 6 cargo hold.

water bending moment. It is seen that the warping stress cannot be neglected compared to the bending stress under this loading condition. The open circles in Fig. 8.29 indicate ultimate hull girder strength considering the influence of warping.

Warping strain in the cross-section can be calculated according to Fujitani's method. On the basis of the calculated results, linear strain distribution over the cross-section is modified, and Smith's method is applied with modified axial strain distribution. Fig. 8.31 shows the bending moment-curvature relationships considering the influence of shear force. Dotted lines are at the center of No. 5 and No. 6 cargo holds. Same analysis was carried out at 21 cross-sections, and the estimated ultimate hull girder strength is plotted by open circles in Fig. 8.29. As the shear force varies toward the ship's length direction, the ultimate hull girder strength also varies at cross-section by cross-section.

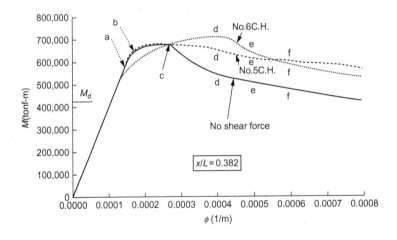

FIG. 8.31

Influence of shear force on moment-curvature relationship [38].

The influence of warping produced by torsional moment may also be accounted applying the similar method. That is, Fujitani's method to calculate warping displacement can be applied to calculate warping due to torsion.

Another problem in applying Smith's method is how to consider the characteristics of curved elements. Curved element such as bilge circle part is usually considered as a hard corner element which does not buckle. This is because the buckling strength of curved plates is very high, and the bilge circle part does not buckle when there is no thickness reduction due to corrosion.

Maeno et al. [40] performed a series of nonlinear FEM analysis on a bilge circle part indicated in Fig. 8.32, and obtained the average stress-average strain relationships shown in Fig. 8.33. In this analysis, the length, L, was taken as 4500 mm, the radius, R as 1800 mm and the thickness is varied as

$$t = 9, \ 10, \ 12, \ 14, \ 16, \ 18, \ 19.5, \ 24, \ 30 \text{ mm}$$

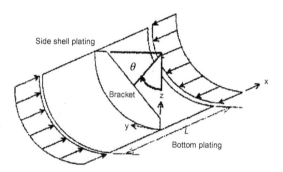

FIG. 8.32

Bilge circle structure [40].

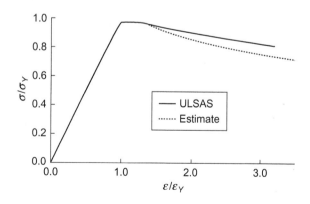

FIG. 8.33

Comparison between calculated and estimated average stress-average strain relationships [40].

With the decrease in the panel thickness, plastic buckling occurs earlier although the buckling strength is nearly equal to the yielding stress.

On the basis of the calculated results, Maeno et al. [40] developed average stress-average strain relationships for the bilge circle part as follows:

(a) The ultimate strength:

$$\frac{\sigma_u}{\sigma_Y} = \alpha \left(\frac{R}{t} - 40 \right)^2 + 0.983 \tag{8.20}$$

where

$$\alpha = -0.60 \left(\frac{\sigma_Y}{E} \right)^2 + 0.00046 \left(\frac{\sigma_Y}{E} \right) - 0.24 \times 10^{-6} \tag{8.21}$$

(b) Buckling strain:

$$\varepsilon_{cr} = 45 \cdot \frac{t}{R} + 0.803 \tag{8.22}$$

(c) Average stress-average strain relationship:

$$\varepsilon = \begin{cases} \frac{\sigma}{E} & \text{before ultimate strength} \\ \frac{\sigma}{E} + \beta \cdot \frac{R}{L} \cdot \frac{t}{R} \cdot \frac{\sqrt{12(1-v^2)}}{\pi^2} \cdot \frac{16(1-\bar{\sigma}^2)^2}{\bar{\sigma}^2(16-15\bar{\sigma}^2)} & \text{beyond ultimate strength} \end{cases} \tag{8.23}$$

where

$$\bar{\sigma} = \sigma/\sigma_Y, \quad \beta = 21\sqrt{\frac{t}{R}} - 0.615 \tag{8.24}$$

An estimated average stress-average strain relationship obtained by Eqs. (8.20) through (8.24) is compared in Fig. 8.33 with that by the nonlinear FEM. Relatively good correlation is observed.

8.4 APPLICATION OF NONLINEAR FEM

8.4.1 APPLICATION OF EXPLICIT FEM

Explicit FEM analysis is performed in general applying the computer code LS-DYNA. The fundamental idea of the explicit FEM is given in Section 8.10.

Here, three examples are introduced which simulate progressive collapse behavior of hull girders applying LS-DYNA. The first example is a tanker *NAKHODKA*, which sank in the Japan Sea in January 1997. The second is three types of tankers of different sizes. The third example is container ship models subjected to combined bending and torsion.

8.4.1.1 Analysis on casualty of NAKHODKA [14,41]
Accident happened on *NAKHODKA*

In the early morning on January 2, 1997, M.V. *NAKHODKA* had broken into two in a rough sea. After 5 h of breaking, the aft part sank, but the fore part drifted and arrived on the Japan coast.

M.V. *NAKHODKA* was built in 1970 at Polish shipyard and registered to Russian Classification Society (RS). Her principal dimensions are

$$L_{PP} \times B \times D/d = 165.98 \times 22.4 \times 12.32/9.52 \text{ m}$$

The general arrangement is shown in Fig. 8.34. Her last voyage was from Zhuzan, China to Petropavlovsk-Kamchatsky, Russia carrying 20,026.8 kL heavy oil of class C.

Loads acting on *NAKHODKA* at the time of accident

A buoy robot was set to automatically record weather and wave data at the location 103 km apart from the location of breaking. The weather and wave conditions at the moment of accident were recorded as indicated in Table 8.3.

Table 8.3 indicates that *NAKHODKA* was in a storm which is usual at Japan Sea in winter time. Therefore, *NAKHODKA* was exposed to relatively high wave bending moment. In addition to this, cargo oil was loaded partly not in accordance with the loading manual, and this produced relatively large still water bending moment.

A series of nonlinear ship motion analysis was performed applying the computer code, "SRSLAM" on the basis of the Strip Theory [14]. Head sea condition was assumed, which is approximately the same with the condition of *NAKHODKA* at her accident. A long-crested irregular wave is assumed with mean wave period of 9 s and significant wave height of 8 m. An irregular wave was produced

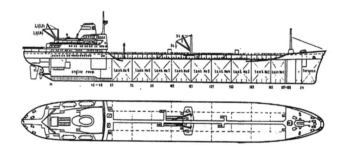

FIG. 8.34

General arrangement of M.V. *NAKHODKA* [14].

Table 8.3 Weather and Wave Conditions at Accident of M.V. *NAKHODKA* [14]	
Wind	**From WEST; 42 knots (22 m/s)**
Significant wave height	8 m
Wave period	8 s
Air temperature	6.6°C
Water temperature	15.1°C

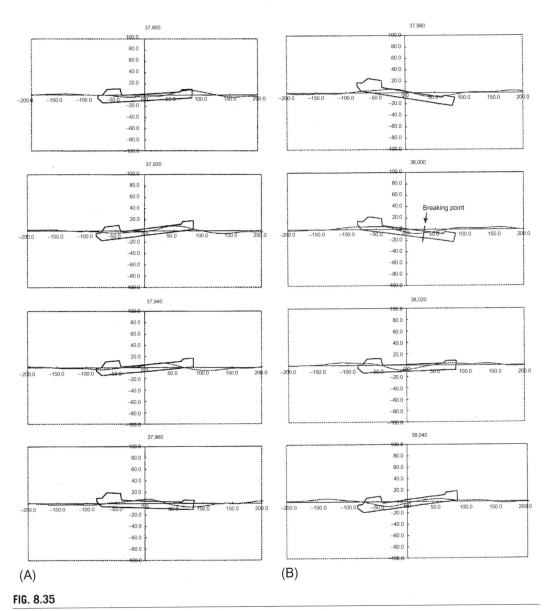

(A)

(B)

FIG. 8.35

Ship motion at braking of M.V. *NAKHODKA* [14].

by randomly combining the regular waves of which amplitudes were specified with ISSC spectrum. Several cases were simulated with different combinations of wave spectrum. The simulated hour is about 32,000 s (40,000 steps). Fig. 8.35 indicates the ship motion relative to the wave at every 20 time steps during 37,900 through 38,040 steps. At step 38,000, maximum bending moment is produced at the broken cross-section, that is Fr.137 through Fr.153.

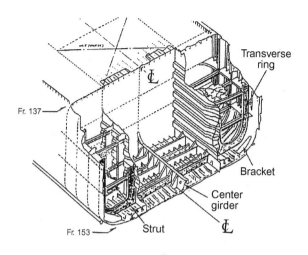

FIG. 8.36

Sketch of broken cross-section of the aft part of M.V. *NAKHODKA* [14].

Fig. 8.36 shows a sketch of the broken cross-section of the aft part on the basis of the observation by the auto-robot vehicle, DOLPHIN-3K.

Progressive collapse analysis on *NAKHODKA*

A full ship FEM model was produced as shown in Fig. 8.37. The model is composed of rigid body element except the region near the broken cross-section. The broken part was modeled by elastoplastic shell and beam-column elements as indicated in the same figure. The rigid body part is to apply distributed pressure loads which produce bending moment distribution shown in Fig. 8.37.

In accordance with the thickness measurements carried out on the fore part of the hull girder arrived at coast of Japan Sea, the thickness of structural members were reduced from the individual original values. The reduction is, for example, at deck plating 39% and at deck longitudinals 45%, respectively, in average.

Furthermore, fillet weld is assumed to break on the condition indicated in Table 8.4. These conditions are determined considering the influence of corrosion observed at the fore part of "*NAKHODKA*." On the basis of the observation in Fig. 8.36, the deck longitudinal stiffeners in the center tank are assumed to have been dropped off, that is the fillet weld is no more effective to connect the deck plating and the longitudinal stiffeners. At the same time, the bottom longitudinal stiffeners were broken at their connection. The location of breaking is where butt weld is performed to connect bottom longitudinal stiffeners. It was found that weld metal for butt welding was very few and this caused breaking of bottom longitudinal stiffeners with very low resistance at their connections. This breaking can be simulated by the condition indicated in Table 8.4.

The stress-strain relationship was determined on the basis of the measured results by tension tests on specimens cut out from the fore part of "*NAKHODKA*." LS-DYNA3D was applied to perform progressive collapse analysis considering the yielding and braking of structural members in tension

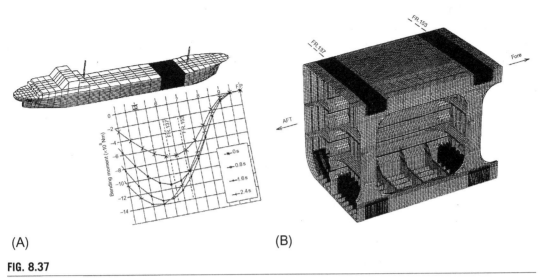

FIG. 8.37

FEM meshing of "*NAKHODKA*" (a whole ship and local) [14].

Table 8.4 Conditions for Breaking of Welded Joints [14]

	Intersection at Dkplt/Dklong	Intersection at Btplt/Btlong	ButtWld of Btlong (Near Fr.153)	Others
Leg length	3 mm fillet	3 mm fillet	Butt	3 mm
Breaking stress	196 MPa	196 MPa	107.8 MPa	377.3 MPa
Breaking strain	–	–	3 %	–

Notes: *Buttwld, butt welding; Dkplt, deck plating; Dklong, deck longitudinals; Btplt, bottom plating; Btlong, bottom longitudinals; –, given in Fig. 8.42.*

as well as buckling and yielding in compression. After trial and errors, loading rate was determined. The load was applied by distributed pressure.

The deck collapses by buckling at the time step of 0.26 and the bottom breaking follows at the time step of 0.31. The buckled deck is illustrated in Fig. 8.38 and the broken bottom in Fig. 8.39, respectively.

Fig. 8.38 indicates that buckling collapse of deck plating takes place at the fore part of Fr.153. The collapse takes place in one cross-section and not all over the deck plating. In Fig. 8.38B, buckled deck longitudinal stiffeners are seen in the center tank separated from the deck plating.

On the other hand, at the bottom structure, tensile stress in the bottom plating increases as the deck undergoes buckling collapse. Then, breaking firstly takes place at the connections of bottom longitudinal stiffeners, and then the bottom plate breaks along the intersection line with transverse bulkhead at Fr.153 as shown in Fig. 8.39.

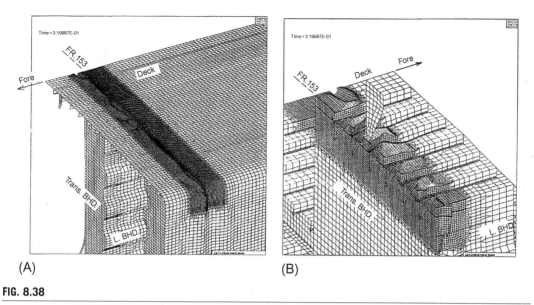

(A)   (B)

FIG. 8.38

Buckling collapse of deck part [14]. (A) Outside view. (B) Inside view.

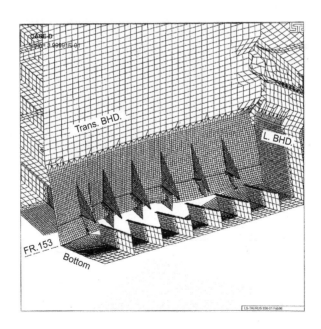

FIG. 8.39

Breaking failure of bottom plating [14].

These results of calculation are in accordance with the structural failure observed at the fore part of "*NAKHODKA*."

8.4.1.2 Analysis of three tankers of different sizes [23,43]

A series of FEM analysis was performed on three tankers shown in Table 8.5 and Fig. 8.40. To examine the influence of thickness reduction due to corrosion, three cases are considered: thickness diminution, which are intact, half diminution and full diminution. A full diminution indicates roughly 25% diminution of the original thickness at deck and bottom plating, and 30% diminution of the original for longitudinal stiffeners [43]. As for a half diminution, diminution which is a half of the extreme case is considered.

1/2 + 1/2 holds near mid-ship is modeled by shell elements and the other part by rigid body elements as in the case of "*NAKHODKA*"; see Fig. 8.37. On this model, still water loads as well as extreme wave loads were applied as distributed pressure loads on ship surface which were obtained through ship motion analysis. Mesh divisions of the shell part are shown in Fig. 8.40.

For the local panel partitioned by longitudinal stiffeners and transverse frames, initial deflection of a thin-horse mode is assumed. For the stiffener, initial deflection of a flexural buckling mode and that of torsional buckling mode are assumed.

Fig. 8.41 shows typical collapse modes for three tankers with different thickness diminutions. It is seen that the buckling/plastic collapse takes place at a certain cross-section with the concentration of the plastic deformation.

In the case of the Suez Max tanker in sagging shown in Fig. 8.41B, at the collapsed cross-section, tripping of the deck longitudinals is observed as well as local collapse of deck plating and the overall collapse as a stiffened plate by buckling. In the case of the Handy Max tanker and VLCC shown in Fig. 8.41A and C which collapsed in hogging, the buckling/plastic collapse line is observed at bottom plating. In all cases, no collapse is observed in any remaining cross-section in the ship length direction, and this results in collapse of a so-called jackknife mode.

The evaluated ultimate hull girder strength in nine cases are plotted against diminution ratio in Fig. 8.42 together with the ultimate hull girder strength calculated by applying Smith's method. A comparison of the ultimate hull girder strength proves that both results show good agreement each other by a mean difference of about 5%.

8.4.1.3 Analysis on 1/13-scale container ship model

A series of progressive collapse analysis is performed on Model 3 of the 1/13-scale container ship models in Table 8.6 explained later in Section 8.6.4 [23,44]. Fig. 8.43 shows the FEM meshing of the test model. The model is a cantilever, of which left-hand side is fixed on a rigid wall. The concentrated loads are applied at the free end of the model, the right-hand side; see Fig. 8.43.

Table 8.5 Principal Dimensions of Object Ships

Ship ID	Ship Type	Built Year	Lpp (m)	B (m)	D (m)	DWT (ton)
A	Handy Max	1982	195	27.4	16.3	40,800
B	Suez Max	1989	265	43.2	23.8	146,020
C	VLCC	1993	319	58.0	31.5	290,000

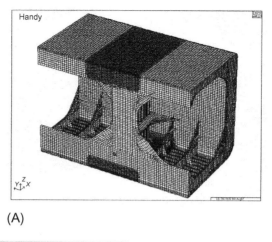

(A)

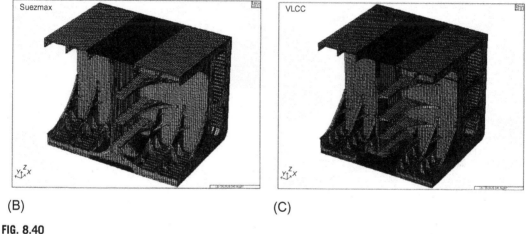

(B)   (C)

FIG. 8.40

Element representation of three tankers [44]. (A) Handy Max tanker. (B) Suez Max tanker. (C) VLCC.

Combined horizontal bending and torsion loads as well as vertical bending and torsion loads are applied changing the load ratios. For the combined vertical bending and torsion, concentrated loads, P_1 and P_2 are applied changing their ratio, whereas P_d and P_b are applied for the combined horizontal bending and torsion.

In all cases, shear buckling takes place at the side shell plating, and the compressive buckling at the bottom plating near the fixed end. Shear buckling is due to torsional moment or due to shear force depending on the ratio of torsional moment to shear force.

The magnitude of the ultimate vertical bending moment, M_U, and the ultimate torsional moment, T_U, are 6940 and 2550 kNm, respectively. It is found that the ultimate strength interaction relationships can be approximated by:

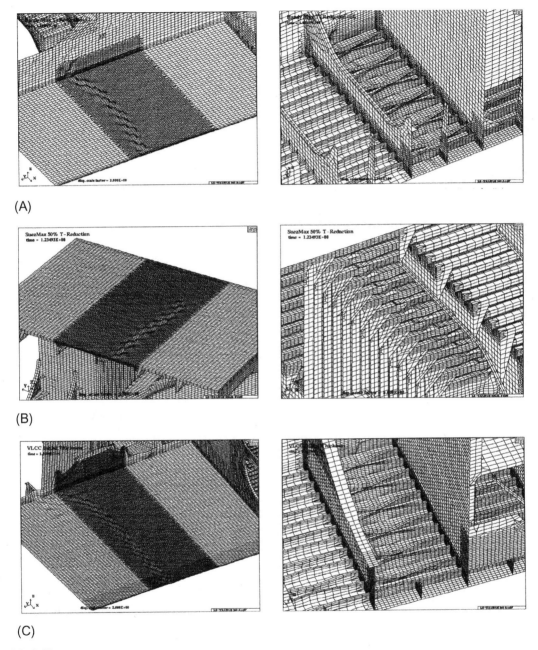

(A)

(B)

(C)

FIG. 8.41

Buckling/plastic collapse modes of different tankers with various diminutions [44]. (A) Handy Max tanker with full diminution. (B) Suez Max tanker with half diminution. (C) VLCC with intact cross-section.

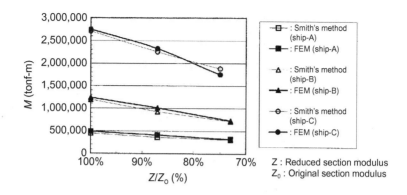

FIG. 8.42

Comparison of calculated results by FEM and Smith's method [23,44].

Table 8.6 Dimensions of Test Models

Model Type	L1	L2	L3	L4	LO	D1	D2	D3	DO	B1	B2
Model 1	1100	1100@	1050	150	7500	130	170	195	1800	195	190/180
Models 2 and 3	900	1000	650	150	6550	120	180	180	1800	180	190
Models 4 and 5	900	1000	650	150	6550	120	180	180	1800	180	190

Model Type	BO	t1	t2	t3	t4	stiff.1	stiff.2	stiff.3	stiff.4
Model 1	3200	8.93	5.61	4.44	3.01	50 × 8.65	50 × 4.29	75 × 8.65	38 × 4.26
Models 2 and 3	3200	5.94	4.35	3.13	2.28	50 × 5.97	50 × 2.92	–	50 × 2.92
Models 4 and 5	3200	5.87	4.48	3.14	2.13	50 × 5.89	50 × 2.89	–	50 × 2.89

$$\left(\frac{T}{T_U}\right)^2 + \left(\frac{M}{M_U}\right)^2 = 1 \tag{8.25}$$

The dashed circle lines in Fig. 8.44A and B are by Eq. (8.25). It can be said that Eq. (8.25) gives accurate ultimate strength interaction relationship for combined vertical bending and torsion.

On the other hand, in the case of combined horizontal bending and torsion, the ultimate bending moment, M_{HU}, and the ultimate torsional moment, T_U, are 10,700 and 2360 kNm, respectively. The ultimate torsional moment shows 7.5% difference depending on the loading method.

Fig. 8.44B indicates that Eq. (8.25) gives a little lower ultimate strength when the horizontal bending moment and the torsional moment are of the similar magnitudes.

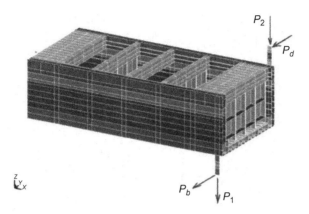

FIG. 8.43

Meshing of 1/13-scale container ship model for combined bending and torsion loading [44].

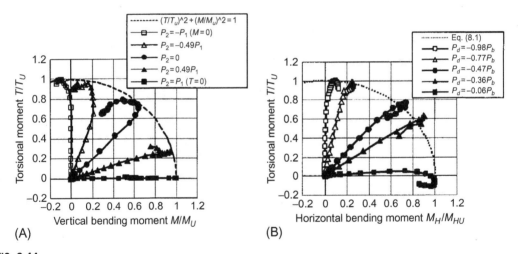

FIG. 8.44

Loading paths and ultimate strength interaction relationships [44]. (A) Vertical bending moment-torsional moment interaction. (B) Horizontal bending moment-torsional moment interaction.

8.4.2 APPLICATION OF IMPLICIT FEM

Application of implicit FEM on the hull girder collapse analysis is very few at the moment. This is because the calculation to make an inverse of the stiffness matrix takes long time in the implicit FEM compared to the case of explicit FEM. There are some results of analyses on 1/2 + 1/2 hold model of the hull girder applying pure bending. However, analysis on several hold model is very few because of the too large number of the degrees of freedom of the model.

Among the few, analysis on Cape-size bulk carrier is introduced here, which was carried out by Amlashi and Moan [24,25]. The cross-section of the subject bulk carrier is shown in Fig. 8.55. This bulk carrier was used for the benchmark calculation to evaluate the ultimate hull girder strength at the technical committee VI.2 in ISSC 2000 [16]. Principal dimensions of the vessel are $L \times B \times D = 285 \times 50 \times 26.7$ m.

Cross-section of the bulk carrier analyzed by Amlashi and Moan is indicated in Fig. 8.45. The modeling extent is 1/2 + 1 + 1/2 holds as indicated in Fig. 8.46. This model represents cargo holds No. 5 through No. 7 which are in alternative loading condition as illustrated in Fig. 8.47A. A symmetry half model is used assuming a head sea condition. For the plating and the longitudinal stiffeners in the central hold, elastoplastic elements are used which can simulate buckling/plastic collapse behavior, whereas linear elastic elements are used for the remaining region.

The number of degree of freedom in the nonlinear part is roughly 475,000 while that in linear elastic part is about 620,000.

8.4.2.1 Boundary conditions and loading conditions

Considering the profile of a cargo shape in holds with respect to the center plane, symmetry condition is imposed on two center planes of holds No. 5 and No. 7 in Fig. 8.47A. The translation of the center cross-section of holds No. 5 and No. 7 in the direction of a ship's length is uniform but free. This

No.	Dimensions	Type	σ_y (MPa)
1	390×27	Flat-bar	392
2	$333 \times 9 + 100 \times 16$	Tee-bar	352.8
3	$283 \times 9 + 100 \times 14$	Tee-bar	352.8
4	$283 \times 9 + 100 \times 18$	Tee-bar	352.8
5	$333 \times 9 + 100 \times 17$	Tee-bar	352.8
6	$283 \times 9 + 100 \times 16$	Tee-bar	352.8
7	$180 \times 32.5 \times 9.5$	Bulb-bar	235.2
8	$283 \times 9 + 100 \times 17$	Tee-bar	352.8
9	$333 \times 9 + 100 \times 18$	Tee-bar	352.8
10	$333 \times 9 + 100 \times 19$	Tee-bar	352.8
11	$383 \times 9 + 100 \times 17$	Tee-bar	352.8
12	$383 \times 10 + 100 \times 18$	Tee-bar	352.8
13	$383 \times 10 + 100 \times 21$	Tee-bar	352.8
14	300×27	Flat-bar	392

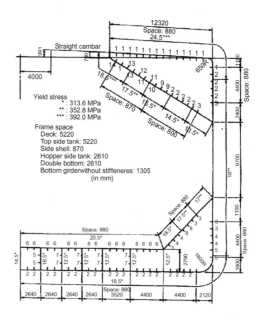

FIG. 8.45

Cross-section of cape size bulk carrier for progressive collapse analysis [24].

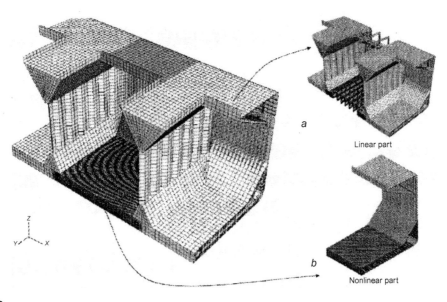

FIG. 8.46

Finite element representation of 1/2 + 1 + 1/2 holds model for progressive collapse analysis [24].

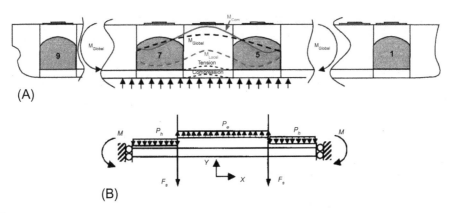

FIG. 8.47

Alternative loading condition [24]. (A) Actual. (B) Modeling.

symmetry boundary condition agrees with the fact that still water shear force is almost zero at the center of the holds located at both ends of the model under alternative loading conditions. The bending moment was applied at both ends of the model by forced rotations in addition to the distributed loads as indicated in Fig. 8.47B.

Altogether nine cases are analyzed varying initial deflection and loading conditions. Detail of the loading condition is given in Amlashi and Moan [24].

8.4.2.2 Progressive collapse behavior and strength of hull girder in combined overall and local bending

Fig. 8.48 summarizes the bending moment-end rotation relationships for seven cases together with the result applying Smith's method in Yao et al. [16] for the case of pure bending with initial deflection. The ultimate hull girder strength by Smith's method is about 10% different from the corresponding FEM result. The reason for this difference is not clear since very good correlation was observed in Fig. 8.42 between calculated results by the FEM and Smith's method.

Another point to be noticed is the capacity reduction in the postultimate strength range. When Smith's method is applied, only 1/2 + 1/2 frame space was analyzed in the ISSC calculation [16]. However, in the present calculation, 1/2 + 1 + 1/2 hold is analyzed, and elastic unloading occurs in the uncollapsed region. This is the reason of rapid reduction in capacity beyond the ultimate hull girder strength.

Here, typical collapse modes are illustrated in Fig. 8.49A and B, respectively. It is seen that buckled part is concentrated along a line in the transverse direction. This indicates that the collapse region is limited in a very narrow band spread toward the transverse direction. In the remaining region, elastic unloading took place beyond the ultimate strength due to the buckling/plastic collapse in the narrow band area. This is the cause of steep reduction in capacity beyond the ultimate strength of the hull girder.

Here, according to Fig. 8.48, drastic reduction is observed in the hull girder ultimate strength when the influence of double bottom bending is accounted for in the analysis. The ultimate hull girder strength is reduced by 3% due to initial deflection (C1 vs C3) for pure bending case, whereas it is reduced by 35% for combined loading case (bending + pressure). Pei et al. [45] also performed similar analysis on Kamsarmax bulk carrier, and reported that the reduction in the in the ultimate hull girder strength is

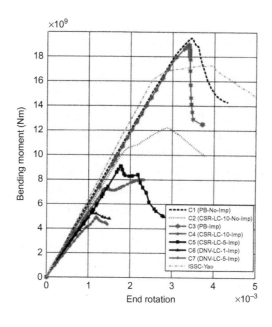

FIG. 8.48

Summary of bending moment-end rotation relationships [25].

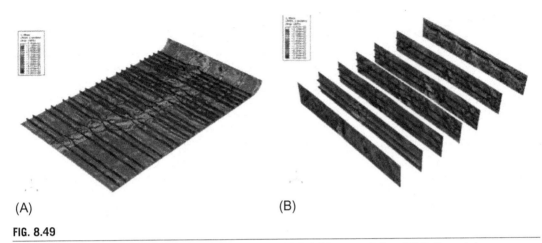

FIG. 8.49

Buckling/plastic collapse of bottom structure (Model C4) [25]. (A) Bottom plating. (B) Bottom girders.

about 20%; see Section 8.7.5. A similar analysis was carried out by Østvold et al. [46] on the Panamax bulk carrier, and the strength reduction was reported as about 14%.

8.5 APPLICATION OF THE ISUM

The ISUM could solve the problem inherent in the implicit nonlinear FEM that too many number of degrees of freedom exists by reducing the number of elements. The ISUM uses larger element compared to the ordinary FEM. There are two ways to develop the ISUM elements. One way is to use element as large as possible, for example, one stiffened panel is considered as one unit or an element. In this case, modeling of one through several holds/tanks may be possible with relatively small numbers of degrees of freedom. Such ISUM elements are summarized by Paik and Thayamballi [47,48]. Practical application can be seen, for example, in Paik et al. [49].

. The ISUM elements of this type are relatively efficient until their ultimate strength has been attained, but has difficulties to simulate the capacity reduction beyond the ultimate strength in individual elements. This in general results in higher or nonconservative ultimate strength of the whole structure.

Another way is to use smaller elements, for example, one through several elements for one local panel. Stiffeners are independent elements. Using such elements, buckling/plastic collapse behavior can be simulated exactly including the concentration of buckling/plastic deformation in the postultimate strength range of individual elements. This enables us to simulate progressive collapse behavior of a whole ship structure very accurately in combination with the exact calculation of external loads acting on a ship's hull girder.

This ISUM element was first developed by Fujikubo and Kæding [50], and is summarized by Yao [51]. The application is for example on 1/2 + 1/2 frame space of a 1/3-scale frigate model and double hull tanker [52].

Another example is the application for collapse analysis of a full ship model giving pressure load obtained by the ship motion analysis, which will be explained later in Section 8.7.

8.6 COLLAPSE TESTS ON HULL GIRDER MODELS

8.6.1 HISTORY OF COLLAPSE TESTS ON HULL GIRDER MODELS

After the collapse test on H.M.S. *Albuera*, no publication can be found regarding collapse test on existing ships. On the other hand, scale models of existing ships or simplified box girders modeling existing ships were tested. Regarding the scale model, Sugimura et al. [53] conducted collapse test in sagging on a 1/5-scale model of the *Nami* class self-defense ship. Dow [54] also conducted progressive collapse test in sagging on a 1/3-scale model of *Leander* class frigate. In both cases, the overall buckling of deck plating resulted in the collapse of a hull girder.

Although they are not complete scale models, Endo et al. [55] conducted a series of progressive collapse tests on 0.164-scale bulk carrier model, 0.129-scale ore carrier model, and 0.121-scale container ship model. They applied combined bending moment and shear force simulating a slamming load. They found that the ultimate strength had been attained after the deck buckled in compression by bending and the side shell plating yielded by shear. Mansour et al. [56] provided two box girders, one modeling a 75,600 tons oil tanker with a single hull and the other cargo ship, and applied pressure loads through air backs until they had collapsed. In both models, occurrence of flexural/torsional buckling of the longitudinal stiffeners in the compression side of an overall bending became a trigger of the overall collapse of the hull girder.

Yao et al. [57] performed progressive collapse tests on a 1/10-scale wood-chip carrier models in sagging, whereas Tanaka et al. [44] carried out a series of progressive collapse tests on five container ship models of which size is 1/13 of that of 5250 Post-Panamax class container ship.

Smaller and simpler box girder models have been also tested under pure bending, for example, by Reckling [58] and Nishihara [59]. Reckling [58] provided seven girder models changing numbers of longitudinal stiffeners, and observed some representative collapse modes. Nishihara [59] provided altogether eight test girders modeling single hull tanker, double bottom tanker, bulk carrier, and container ship; see Fig. 8.50. All models collapsed by the occurrence of overall buckling at deck or bottom plating as a stiffened panel. The test results are summarized in Table 8.7, where calculated values are by the application of Caldwell's formulas, Eqs. (8.1) through (8.4). The strength reduction factors were defined by the ultimate strength formulas for local panel and stiffened panel.

On the other hand, applying combined bending, shear, and torsion, Ostapenko [60] performed progressive collapse tests on three girder models. He found that the ultimate strength had been attained after the occurrence of overall buckling of compression flange followed by web buckling in all cases.

Sun and Guedes Soares [61] carried out pure torsion test on box girder specimens with large opening modeling a hull girder of the container ship. They compared the test results with the calculated results applying nonlinear FEM, and found that the calculated results were largely affected by the setting of boundary conditions. Gordo and Guedes Soares conducted bending collapse test on three box girder specimen made of high tensile steel [62].

In the following, the progressive collapse tests conducted by Dow [54], Yao et al. [57], and Tanaka et al. [44] are explained in more detail.

8.6.2 TESTS ON 1/3-SCALE FRIGATE MODEL [54]

Dow [54] conducted a progressive collapse test on a 1/3-scale welded steel ship modeling a *Leander* class frigate. The length, the breadth, and the depth of the model are $L \times B \times D = 18 \times 4.1 \times 2.8$ m. The

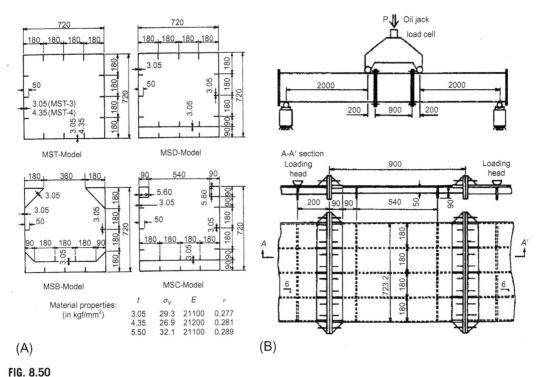

FIG. 8.50

Test specimens and test setup.

Table 8.7 Measured and Calculated Ultimate Strength

Model	Experiment M_{max} (ton m)	M_u	$M_{Y\,D,B}$	M_A	$M_{D,B}$	M_{max}/M_U	M_U/M_A	M_{max}/M_A
		\multicolumn{4}{c}{Calculation (ton·m)}			
MST-4	94.5	92.9	93.3	87.2	75.7	1.020	1.065	1.080
MST-3	57.5	59.1	70.8	57.7	43.7	0.973	1.020	0.997
MST-3	60.0	59.1	70.8	57.7	43.7	1.020	1.020	1.040
MDT-S	60.5	61.2	72.2	61.2	44.6	0.989	1.000	0.989
MDT-H	85.8	80.8	95.9	70.6	60.1	1.060	1.140	1.210
MSB-S	49.1	52.6	62.1	52.6	36.5	0.933	1.000	0.933
MSB-H	68.5	74.0	89.0	60.1	55.2	0.926	1.230	1.140
MSC-S	113.5	96.8	84.8	91.0	76.3	1.170	1.060	1.250
MSC-H	88.0	84.5	95.9	78.4	59.3	1.040	1.080	1.120

Source: Nishihara [59].

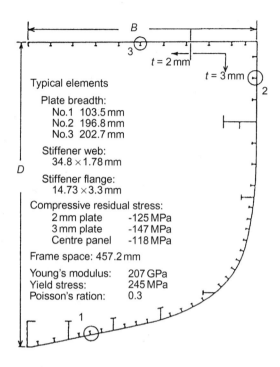

FIG. 8.51

Cross-section of 1/3-scale frigate model.

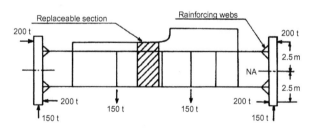

FIG. 8.52

Loading condition of combined four-point bending and pure bending.

cross-section of the model is illustrated in Fig. 8.51. The measured welding residual stress is indicated in the figure, which is very high and nearly a half of the yield stress of the material.

The load was applied by the combination of four-point bending and pure bending as indicated in Fig. 8.52. The shaded part in Fig. 8.52 is a replaceable part and the test specimen was inserted here.

Before applying sagging bending moment, both sagging and hogging moments were applied until the strain in the deck or the keel reaches 400 μ and then unloaded. By this preloading, welding residual stress is considered to be released to some extent. This shall be explained later in Section 8.11.

It is reported by Dow [54] that fairly large overall vertical deflections were observed at the main strength deck due to the overall grillage buckling, but this buckling did not show a significant influence on the hull girder moment-curvature response. He also indicated that "Approaching the ultimate load, local stiffener and plating deformations developed on the deck, superimposed on the overall deck deformation, precipitating deck collapse. The side shell clearly showed interframe buckling of the longitudinal stiffeners with associated plate buckling." The collapse mode was so-called interframe flexural buckling of deck structure.

For this model, benchmark calculation was performed at the technical committee of the ISSC 1994 [36]. Ten committee members joined the benchmark calculation. Among the 10, 6 applied Smith's method, 2 FEM, 1 ISUM, and 1 Caldwell's method. The simulation of element collapse behavior is compared in Fig. 8.53A and that of hull girder collapse behavior in Fig. 8.53B. At that time, that is in 1994, the discrepancy is as can be seen in Fig. 8.53. The estimated ultimate hull girder strength lies in between −9% and 6% of the measured result except one.

8.6.3 TESTS ON A 1/10-SCALE WOOD-CHIP CARRIER MODELS [57]

A wood-chip carrier has no top side tank different from ordinary bulk carriers. Because of this, the neutral axis of the hull girder cross-section lies at the lower location. So, high compressive stress is produced at the deck and the upper part of side shell plating under the heavy ballast condition. To avoid the occurrence of buckling at the upper part of the side shell plating, carlings are often provided at this location.

To examine the buckling/plastic collapse behavior of the upper part of the cross-section of a wood-chip carrier including the effect of carling, 1/10-scale wood-chip carrier models were prepared with and without carling [57]. General arrangements of a typical wood-chip carrier are shown in Fig. 8.54, and the modeling of cross-section in Fig. 8.55. The deck and the upper part of the side shell are the exact scale models. However, bottom structure, which comes in the tension side of the hull girder bending, is replaced by an equivalent single thick plate.

Before carrying out progressive collapse test, initial deflection and welding residual stress were measured. The measured initial deflection at deck plating was 0.3–0.5 mm in average and 2 mm at maximum. Initial deflection in side shell plating was at maximum 5 and 3 mm for the specimens without and with carling, respectively. The mode of initial deflection is in most cases in the same direction and is in a thin-horse mode.

The magnitude of initial deflection in longitudinal stiffeners is ranging between 2 and 3 mm. About a quarter is in a flexural buckling mode, whereas the rest is toward the same direction. These magnitudes are relatively large compared to those in the existing ships.

The welding residual stress was measured with stiffened plate specimens fabricated in a similar manner with the collapse test specimens. In side shell plating without carlings, measured magnitude of compressive residual stress was around 26 MPa. For the case with carlings, it was 200 MPa at maximum between the carlings. The above-mentioned compressive residual stresses may be roughly four times higher than those in actual ships.

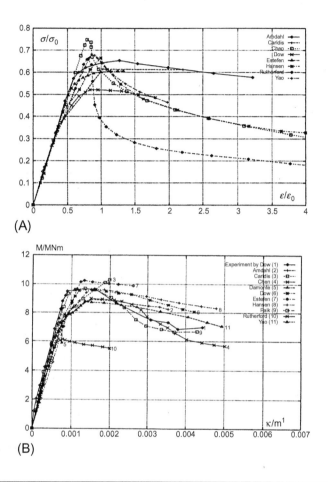

(A)

(B)

FIG. 8.53

Results of benchmark calculation at technical committee of ISSC 1994 [36]. (A) Average stress-average strain relationships. (B) Moment-curvature relationships.

Detail of the specimen is shown in Fig. 8.56. On both ends of the specimen, loading frames are fitted by bolts, and the four-point bending load was applied as illustrated in Fig. 8.57. The locations where strains and displacements were measured are also indicated in this figure.

Load-load point displacement relationships are plotted in Fig. 8.58 for two specimens. In both cases, rapid reduction is observed in the postultimate strength capacity. The specimen with carling showed lower ultimate strength.

The test specimen has three spans in the central part of the hatch opening. Among the three spans, in the case of the specimen without carlings, deck plate at the center span start to largely deflect when the load exceeds 400 kN. This is because the deck girder at hatch side underwent local collapse in this span. The deflection of the deck panel in this span is upwards, and the longitudinal stiffeners were in the compression side of overall bending.

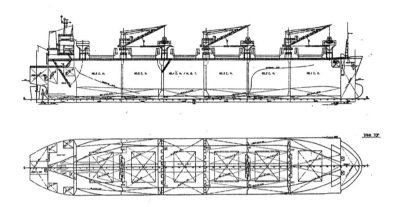

FIG. 8.54

General arrangement of typical wood-chip carrier [57].

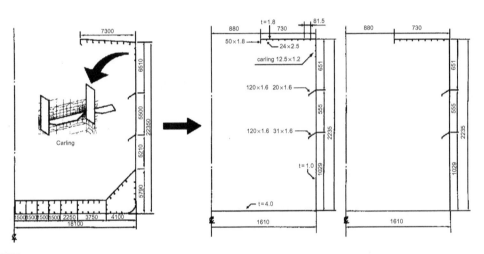

FIG. 8.55

Modeling of cross-section of wood-chip carrier [57].

After the load exceeds 500 kN, deck plates at the left-hand side and the right-hand side spans start to deflect due to the occurrence of the overall buckling. In these spans, deck deflection is downwards, and the longitudinal stiffeners were in the tension side of overall bending.

The ultimate strength was attained when the load reaches near 600 kN. Beyond the hull girder ultimate strength, deck plating of the left-hand side span deflect largely together with deck girders as shown in Figs. 8.59 and 8.61. The deflection in the remaining two spans is not so significant. However, at the longitudinal stiffeners in the center span, tripping failure is observed. Fig. 8.59 indicates that

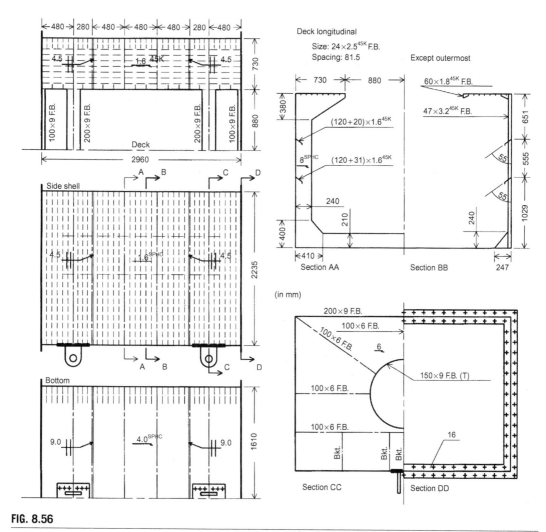

FIG. 8.56

Detail of test specimen [57].

plate induced failure occurred in the left-hand side span, whereas stiffener induced failure occurred in the center span.

Fig. 8.60B shows the plastic collapse of local panel at the upper part of the side shell plating partitioned by transverse frames and deck plating/longitudinal girders. Collapse of a typical roof mode is observed.

When carlings is provided, similar collapse behavior was observed as the specimen without carlings although the ultimate hull girder strength was lower than the specimen without carlings. This was

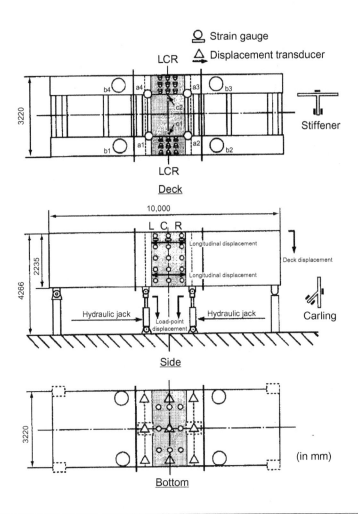

FIG. 8.57

Test apparatus and locations of strain and displacement measurement [57].

attributed to the earlier occurrence of local failure of the deck girder; see Fig. 8.61. At the last stage, carlings collapsed with side shell plating as indicated in Fig. 8.60A.

The test results were compared with the calculated results applying Smith's method in Fig. 8.62. In the analysis, four cases were considered changing the magnitude of initial deflection and welding residual stress as indicated in the figure. The fully plastic bending moment, M_P, for this cross-section is about 50 kNm. Comparing to this, the ultimate hull girder strength is roughly 50% of M_P. This is because of the low contribution of side shell plating to the ultimate hull girder strength as well as lower compressive ultimate strength of deck structure. This should be kept in mind as the characteristics of the wood-chip carrier.

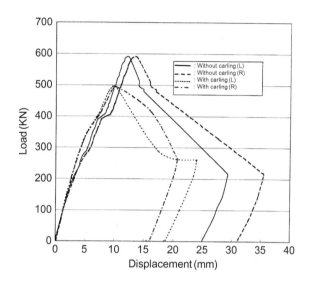

FIG. 8.58

Load-load point displacement relationships for two models [57].

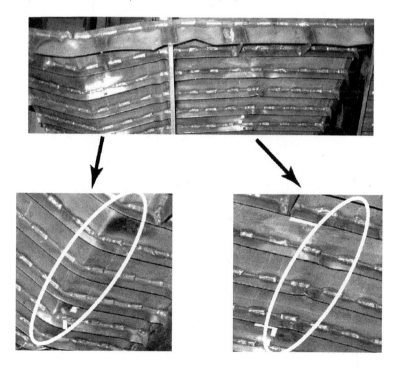

FIG. 8.59

Collapse of deck plating (model without carling) [57].

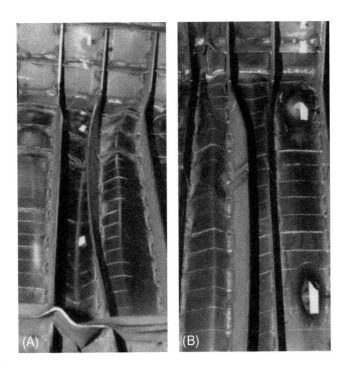

FIG. 8.60

Load-load point displacement relationships for two models [57]. (A) With carling. (B) Without carling.

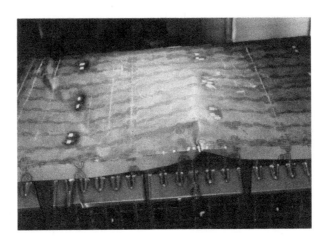

FIG. 8.61

Local collapse of deck girder at hatch side (model with carling) [57].

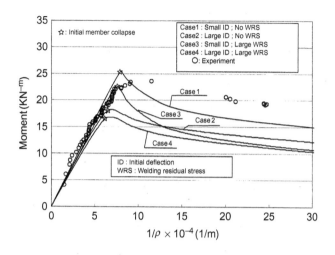

FIG. 8.62

Comparison of measured and calculated moment-curvature relationships for 1/13-scale wood-chip carrier model without carlings [57].

The calculation showed that the deck plating firstly collapses by buckling and the ultimate hull girder strength is attained almost at the same time in all cases. The ultimate hull girder strength is reduced by 11% due to initial deflection, and 28% due to welding residual stress. When both initial deflection and welding residual stress exist, the ultimate hull girder strength is reduced by 34%. However, initial deflection and welding residual stress used here are those of the specimens and are very large compared to those of the actual ships. When the actual ships are considered, influences are not so large, and the reduction may be within 10%.

8.6.4 TESTS ON 1/13-SCALE CONTAINER SHIP MODELS [44]

A series of progressive collapse tests was carried out on five 1/13-scale models of a Post-Pamamax container ship [44]. Fig. 8.63 shows the test model, and its dimensions are indicated in Table 8.7. The breadth and the depth of the cross-section are $B \times D = 3200 \times 1800$ mm. The length of model 1 is 7.5 m whereas that of the remaining models is 6.55 m. Panel thicknesses are different among model 1, 2 and 3, and 4 and 5.

The first model (model 1) was designed so that the breadth to thickness ratio, b/t_p, of the local panel and the height to thickness ratio, h/t_s, of the longitudinal stiffeners are the same with those of the existing over Panamax container ships. All the models have three cargo holds partitioned by watertight full transverse bulkheads, and partial transverse bulkhead is provided at the center of each hold. Both fore and aft holds are half-covered by cross-deck structures.

Mild steel is used instead of HT steel which is used in existing ships. Laser welding was carried out to fit longitudinal stiffeners on the panel. Models were constructed assembling the above-mentioned stiffened plates by performing ordinary CO_2 welding.

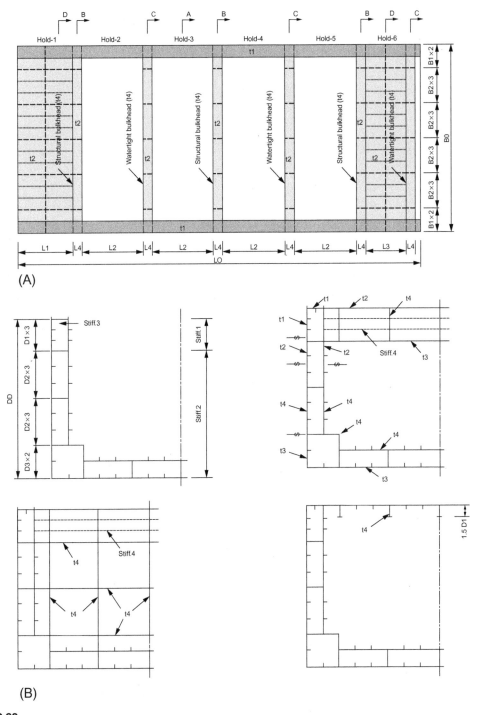

FIG. 8.63

Transverse cross-sections of test models [44].

Table 8.8 Maximum Values of Initial Deflection and Compressive Residual Stress [44]

	$W_{0\,max}/t_p$				$\sigma_{rc}/\sigma_{0.2}$			
Model	t1	t2	t3	t4	t1	t2	t3	t4
1	0.26	0.34	0.66	1.39	0.45	0.30	0.33	0.39
2	0.55	0.17	2.74	4.33	0.22	0.18	0.28	0.27
3	0.47	0.44	2.23	2.89				
4	0.24	0.55	2.35	3.13	0.57	0.30	0.24	0.25
5	0.48	0.43	2.22	3.09				

Before performing collapse tests, initial deflection and welding residual stress were measured. The measured results are summarized in Table 8.8 $\sigma_{0.2}$ stands for stress when plastic strain is 0.2%. The measured initial deflection as well as welding residual stress are large compared to those in existing ships since the welding could not be scaled down in accordance with the size of structures. Welding residual stress was measured by stress relief method on similar specimens.

A series of collapse tests was carried out fixing one end of the test model on a rigid wall and applying vertical shear forces, P_1 and P_2, on another end as indicated in Fig. 8.64. Changing the ratio of P_1 to P_2, combined bending moment and torsional moment are applied together with vertical shear force.

Table 8.9 shows the combinations of P_1 and P_2, where T and M are torsional moment and bending moment. Pure torsional moment was applied on models 1 and 4, whereas pure bending moment in hogging was applied on model 5. On models 2 and 3, combined torsional and bending moments were applied with M/T ratios of 2.7 and 0.5, respectively.

The loads were applied under the displacement control with two hydraulic jacks. Measured load-stroke relationships are shown in Fig. 8.65A–C for models 3, 4, and 5 together with the results of nonlinear FEM analyses. The measured capacities are lower than the calculated results. The cause of

FIG. 8.64

Test setup [44].

Table 8.9 Combination of Applied Torsional and Bending Moments [44]		
Model	**Loads (Downward: +)**	**Relationship Between T and M**
1	$P2 = -P1$	$M = 0$ (Torsion)
2	$P2 = 0$	$M/T = 2.7$
3	$P2 = -0.68\ P1$	$M/T = 0.5$
4	$P2 = -P1$	$M = 0$ (Torsion)
5	$P2 = P1$	$T = 0$ (Hogging moment)

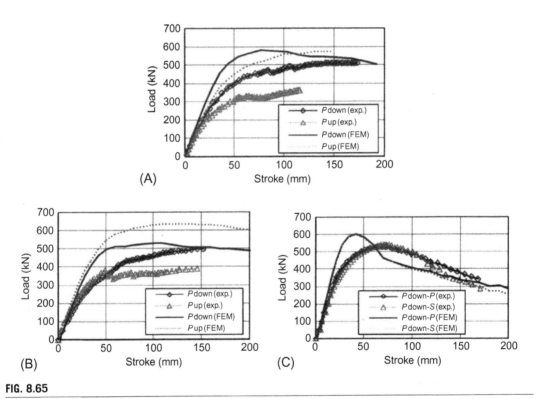

FIG. 8.65

Load-load point displacement relationships [44]. (A) Model 3 ($M/T = 0.5$). (B) Model 4 ($M = 0$).
(C) Model 5 ($T = 0$).

this discrepancy can be attributed to the fact that welding residual stress is not considered in the FEM analyses although initial deflection is considered.

In the case of model 3 subjected to combined bending and torsional moments as well as vertical shear force, shear buckling takes place due to torsional moment in the side shell plating as indicated in

Fig. 8.66A. Then, at the bilge corner part of the bottom plating adjacent to the fixed wall, compressive buckling takes place due to combined bending stress and warping stress; see Fig. 8.66B. With the extension of the buckled region, load increasing rate gradually decreases, and the maximum load is attained after a hatch corner breaks near partially closed deck; see Fig. 8.66C. Fig. 8.66D and E shows widely spread buckling deflection in bottom plating in compression and side shell plating in shear.

In the case of model 4 subjected to pure torsion, separation, that is, breaking of tagged weld joints, took place during progressive collapse at the laser weld connection between the hatch-side deck plating and the top of side shell plating as indicated in Fig. 8.66F. Because of this the maximum value of the measured load, P_2, was about 60% of the calculated value. Also in this case, compressive buckling took place at the bottom plating due to compressive warping stress, and shear buckling in the side shell plating due to torsion.

In the case of model 5 subjected to combined shear force and bending moment, shear buckling took place in the side shell plating and compressive buckling in the bottom plating in the region adjacent to the fixed wall; see Fig. 8.66G and H.

8.7 TOTAL SYSTEM FOR PROGRESSIVE COLLAPSE ANALYSIS ON SHIP'S HULL GIRDER

8.7.1 ACTUAL COLLAPSE BEHAVIOR OF SHIP'S HULL GIRDER IN EXTREME SEA

In general, progressive collapse analysis on a ship's hull girder is performed by imposing forced rotation or by applying *Arc Length Method* and derive forced rotation/bending moment equilibrium path. The ultimate strength is defined as the maximum bending moment along the equilibrium path obtained by the above-mentioned analysis.

It is very important to perform displacement-control analysis or to apply *Arc Length Method* to know he capacity of the ship's hull girder. However, even if the capacity is known, it is not possible to simulate real progressive collapse behavior of the hull girder in extreme sea. To make it possible to clarify the collapse behavior, it is necessary to perform interactive analyses between time-dependent analysis to calculate time-history of pressure distribution on outer surface of the ship's hull and progressive collapse analysis to simulate structural behavior.

In the official discussion on the report of Technical Committee III.1: Ultimate Strength of the 16th International Ship and Offshore Structures Congress (ISSC) in 2006 [63], Prof. Lehmann pointed out that "No forced displacement or arc length exists in nature. It is distributed pressure that acts on ship's hull from outside." in relation to the benchmark calculation to evaluate the ultimate hull girder strength of Passenger ship [64]. In other words, he pointed out that "Bending moment-curvature relationship obtained by force-control method or *Arc Length Method* could be far from the reality after the working bending moment has exceeded the hull girder capacity."

When working bending moment exceeds the capacity of the hull girder, one possible postultimate strength behavior is that the applied bending moment above the hull girder capacity may change to some kind of kinematic energy and large breaking motion starts to take place. Another possibility may be that large breaking deformation occurs and the distribution of buoyancy force changes, which results in the reduction of working bending moment. However, at the moment, it is not clear what shall happen.

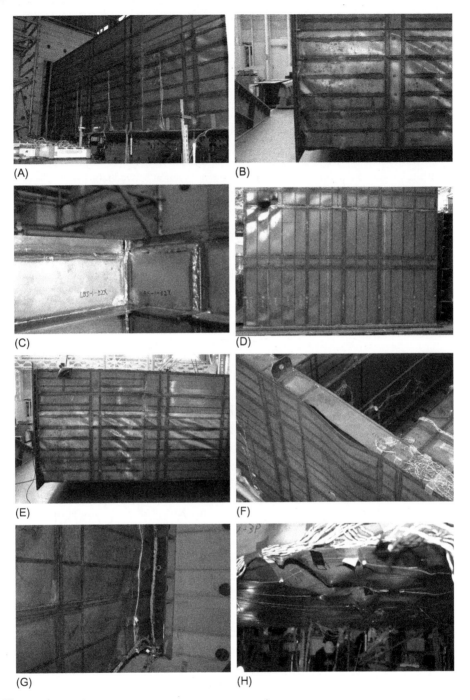

FIG. 8.66

Collapse modes observed in progressive collapse test [44]. (A) Shear buckling in side shell plating (model 3; $M/T = 0.5$). (B) Compressive buckling in bottom plating (model 3; $M/T = 0.5$). (C) Breaking of hatch coaming (model 3; $M/T = 0.5$). (D) Overall view of bottom plating after test (model 3; $M/T = 0.5$). (E) Overall view of side shell plating after test (model 3; $M/T = 0.5$). (F) Breaking of tagged weld at deck side (model 4; $M = 0$). (G) Shear buckling in side shell plating (model 5; $T = 0$). (H) Compressive buckling in bottom plating (model 5; $T = 0$).

8.7.2 START OF NEW JOINT RESEARCH PROJECT

A new research project was started to develop a total system in 2008 in Japan as a joint research project between shipbuilding company and universities. Up to now, several papers have been published [45,65–68,70] that are directly related to the new total system.

The developed total system has two phases: Phase 1 and Phase 2. The main flow of the Phase 1 system is indicated in Fig. 8.67, and can be briefly described as follows:

(1) A full ship FEM model is generated with meshing which can be used for linear FEM analysis; see Fig. 8.68A.
(2) Data representing actual mass distribution are generated considering weights of a hull girder, cargo, ballast water, etc.
(3) Time history of pressure distribution on ship hull surface is calculated under specified loading condition and wave condition applying three-dimensional *Singularity Distribution Method*. A ship's hull is assumed as a rigid body. The same meshing is used as collapse analysis for calculation of time history of the pressure distribution; see Fig. 8.68B.
(4) Finite elements is replaced by ISUM elements in one tank/hold where the hull girder may collapse; see Fig. 8.68A.

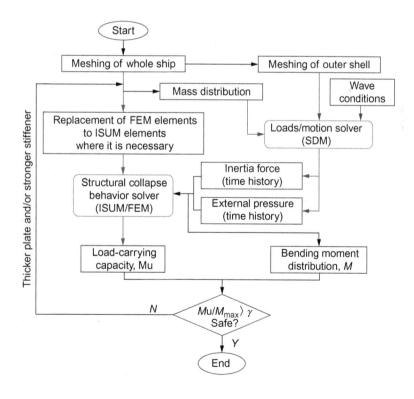

FIG. 8.67

Flow chart of the proposed total system (Phase 1) [65].

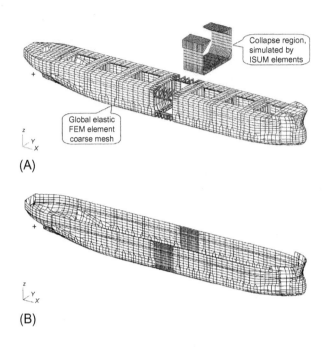

FIG. 8.68

Example of meshing of a whole ship [66]. (A) For progressive collapse analysis. (B) For pressure distribution analysis.

(5) Progressive collapse analysis is performed on the FEM/ISUM model applying time-dependent distributions of pressure and inertia forces obtained in procedures (2) and (3).

(6) It is examined if the ultimate hull girder strength has been attained within the applied loads.

The above flow is denoted as Phase 1 of the proposed system. It should be noticed that load analysis and progressive collapse analysis are performed independently in Phase 1. Such analysis may be correct until the ultimate hull girder strength is attained. On the other hand, in Phase 2, interaction between both analyses are considered. Phase 2 will be explained later in Section 8.7.6.

In the following, fundamentals are explained of the proposed system developed in this research project.

8.7.3 LOAD ANALYSIS

Assuming that the so-called memory effects are negligible and that the force is a function of displacement, velocity, acceleration, and time, the equation of motion including translational displacements and rigid body rotations can be written as

$$[M]\{\ddot{x}\} = \{F(\{x\}, \{\dot{x}\}, \{\ddot{x}\}, t)\} \tag{8.25}$$

where $[M]$ and $\{F\}$ are mass matrix and force vector, respectively. The vector $\{x\}$ represents displacement vector including rigid body motions. $\{\dot{x}\}$ and $\{\ddot{x}\}$ are velocity vector and acceleration vector, respectively.

As is known from Eq. (8.25), mass distribution will influence the floating situation and thus influence the ship motion. So, it is necessary to know the actual mass distribution. The total mass can be divided into two groups: constant mass and variable mass. The constant mass includes weights of a hull girder, paint, hull outfittings, machineries, electric wires as well as provisions. In the present system, they are automatically calculated in the preprocessing system when FEM coarse mesh is firstly generated. The variable mass includes cargo weight, fresh water weight, fuel oil weight, etc.

Eq. (8.25) can be rewritten in an incremental form as

$$[M + M_a]\{\Delta\ddot{x}\} + [B + C]\{\Delta\dot{x}\} + [K_r + K_s]\{\Delta x\}$$
$$= \{F(\{x\}, \{\dot{x}\}, \{\ddot{x}\}, t)\} - [M]\{\ddot{x}\} - [C]\{\dot{x}\} - [K_s]\{a\} \qquad (8.26)$$

where $[M_a]$, $[B]$, and $[C]$ are added mass matrix, hydrodynamic damping matrix, and structural damping matrix, respectively. $[K_r]$ represents restoring force matrix arising from hydrostatic and gravity loads, and $[K_s]$ is stiffness matrix.

Forces at the present location are schematically given as a summation of surface force and volume force as follows:

$$\{dF\} = -p(\{x\}, \{\dot{x}\}, \{\ddot{x}\}, t)dS + \rho_s\{g\}dV \qquad (8.27)$$

where $\rho_s\{g\} = \{0, 0, -\rho_s g\}^T$ is gravity loads and

p: surface loads or pressure
$\{n^t\}$: direction vector or outward normal vector
dS: area subjected to pressure
$\rho_s\{g\}$: specific gravity
dV: volume

It is noted that the forces are evaluated at the displaced position in order to consider the restoring forces and higher order nonlinear forces. Namely, the pressure p is evaluated based on the instantaneous position and similarly the normal vector is updated at each time step.

In order to evaluate the hydrodynamic pressure, a *Singularity Distribution Method* based on *Potential Theory* is employed. Usual assumptions made for *Potential Theory* are also made, that is, fluid being inviscid, incompressive, and irrotational. It is assumed that oscillatory motions with the encountering wave frequency, ω_e, are small in amplitude.

The potential satisfies Laplace's equation and boundary conditions, which are

(1) kinematic and kinetic conditions at free surface;
(2) radiation condition at infinity, sea bottom condition; and
(3) hull surface condition.

It is known that there is a difficulty to obtain a rigorous solution satisfying all boundary conditions. In the present system, a simplification is made on the condition. That is, an assumption that the encounter frequency is sufficiently high or the forward speed is sufficiently low ($\omega \gg U\delta/\delta x$) is made, which is the same assumption made in the derivation of *Strip Theory*. Then, the boundary condition on the free surface becomes identical to the free surface condition with zero forward speed and encounter frequency.

Then, Green's function with zero-forward speed can be used for the potential around the ship. Once the Green's function is obtained, the source strength distributions are determined according to the hull surface condition by solving a well-known integral equation. Thus, the potential around the ship with forward speed can be obtained. The hydrodynamic pressure is expressed on the basis of the potential expression using Bernoulli's theory.

$$\frac{p}{\rho} = \left[-\frac{\partial \Phi_{\text{tot}}}{\partial t} - \frac{1}{2}\left\{\left(\frac{\partial \Phi_{\text{tot}}}{\partial x}\right)^2 + \left(\frac{\partial \Phi_{\text{tot}}}{\partial y}\right)^2 + \left(\frac{\partial \Phi_{\text{tot}}}{\partial z}\right)^2 - U^2\right\}\right] - gz \tag{8.28}$$

where ϕ_{tot} is a total potential consists of fluctuating component $Re\lfloor \phi e^{-j\omega_e t} \rfloor$ and steady contributions. The steady contributions represent steady scattering from the ship due to forward speed. By setting up a system of equations with regards to fluid and body motions, the interaction between fluid and the motion can be solved.

The detail of the pressure calculation is given in Iijima et al. [70].

8.7.4 PROGRESSIVE COLLAPSE ANALYSIS

In the progressive collapse analysis, distributed pressure and inertia forces, which change in time, are applied on a whole ship model. Buckling/plastic collapse behavior is simulated by ISUM/FEM combined system.

The model is composed of FEM and ISUM elements. So, stiffness equation for a whole ship can be expressed as follows:

$$\left\{\begin{array}{c} \Delta F_1 \\ \Delta F_2 \\ \Delta F_3 \end{array}\right\} + \left\{\begin{array}{c} F_1 \\ F_2 \\ F_3 \end{array}\right\} = \left\{\begin{array}{c} R_1 \\ R_2 \\ R_3 \end{array}\right\} + \left[\begin{array}{ccc} K_{11} & K_{12} & 0 \\ K_{21} & K_{22} & K_{23} \\ 0 & K_{32} & K_{33} \end{array}\right]\left\{\begin{array}{c} \Delta h_1 \\ \Delta h_2 \\ \Delta h_3 \end{array}\right\} \tag{8.29}$$

Suffix 1 corresponds to the nodal points in aft and fore regions of the hull girder where FEM elements are used, whereas suffix 3 to those in the middle region where ISUM elements are used. On the other hand, suffix 2 corresponds to the nodal points on the interfaces between FEM region and ISUM region. K_{22} consists of stiffness matrices of both FEM and ISUM elements.

8.7.5 EXAMPLE: COLLAPSE BEHAVIOR OF KAMSARMAX BULK CARRIER IN ALTERNATE HEAVY LOADING CONDITION (PHASE 1 ANALYSIS)

Here, results of calculation on an existing Kamsarmax bulk carrier in alternate heavy load condition are introduced as an example [45]. Principal dimensions are $L_{pp} \times B \times D = 222.0 \times 32.28 \times 20.05$ m. The mid-ship section is shown in Fig. 8.69. Thicknesses of plating and stiffener composing a cross-section are reduced to some extent to simulate progressive collapse behavior in the specified wave height.

Deflection of the main hull under alternate heavy load condition is illustrated in Fig. 8.70. It is supposed that the double bottom of the No. 4 empty cargo hold is subjected to local bending as well as compressive load caused by hogging and may be in the most severe condition for buckling/plastic collapse. So, FEM elements in the No. 4 cargo hold is replaced by ISUM elements; see Fig. 8.68A. Performing pressure calculation, external pressure distribution is obtained as indicated in Fig. 8.71.

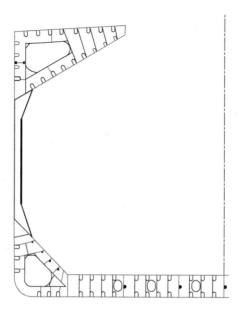

FIG. 8.69

Mid-ship cross-section [45].

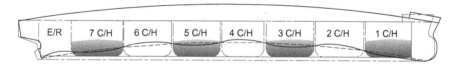

FIG. 8.70

Deflection of hull girder under alternative loading condition [45].

Progressive collapse analysis is performed with new ISUM/FEM system. Nonlinear FEM analysis is also performed applying MSC.Marc. In the MSC.Marc analysis, the ISUM elements of the model is replaced by nonlinear FEM elements with finer meshing.

In both analyses, the same external pressure and inertia forces with time history are applied. Almost the same collapse modes are obtained also by MSC.Marc analysis. The ship bottom plating locally deflects due to local bending under alternate heavy loading condition as indicated in Fig. 8.70. At the final time step, plastic deformation in the bottom concentrates along one line in the breadth direction at the center of No. 4 empty hold.

The bending moment-curvature relationships obtained by three different methods are plotted in Fig. 8.72. HULLST is based on Smith's method and gives the moment-curvature relationship under pure bending. ISUM_hogging is also under pure bending.

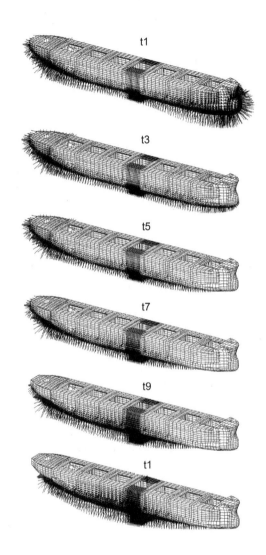

FIG. 8.71

Change in pressure distribution with time [45].

According to the results of HULLST and MSC.Marc analyses, under pure bending, the buckling takes place at the bottom plating firstly when the bending moment exceeds 5×10^9 Nm. Then, inner bottom plating undergoes buckling collapse. Almost at the same time, deck plating is yielded in tension, and the ultimate hull girder strength is attained. The ultimate hull girder strength by HULLST and ISUM/FEM is 6.15 and 6.02×10^9 Nm, respectively, under pure bending.

On the other hand, when double bottom is subjected to local bending, buckling of bottom plating in the middle hold takes place earlier since compressive stress due to local bending is superposed on the

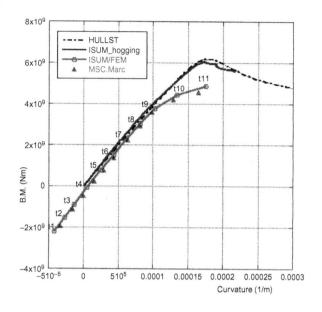

FIG. 8.72

Comparison of bending moment-curvature relationships [45].

compressive stress due to overall bending in hogging. According to ISUM/FEM and MSC.Marc results, this occurs at time step t9. The earlier buckling in bottom plating results in earlier yielding and collapse of bottom plating, and so the hull girder. The ultimate strength is considered to be attained a little after the time step t11. MSC.Marc gives a little lower ultimate strength compared to ISUM/FEM system. Calculation is by a load control method, and the postultimate strength behavior cannot be simulated.

Considering the bending moment at time step 11 as the ultimate hull girder strength, it is known that the ultimate hull girder strength is reduced by about 20% due to local bending of the double bottom structure.

The normal stress distributions in the hull girder cross-section are plotted in Fig. 8.73, where three results are shown obtained by ISUM_hogging, ISUM/FEM system, and MSC.Marc, respectively. The stress distributions at time step t11 (near ultimate strength) are illustrated.

The stress distribution obtained by ISUM_hogging is almost linear. The stress reaches yield stress (355 MPa) in tension at deck plating. The stress at bottom plating is also equal to the yield stress in compression. This is the case when the cross-section is subjected to pure bending without local bending of double bottom.

On the other hand, when double bottom is subjected to local bending, stress distribution is no more linear toward the depth direction. A knuckle point is obviously seen at the location of the top of the bilge hopper tank.

The influence of local bending of double bottom appears in the stress in inner bottom and bottom plating. In the empty hold, inner bottom plating is in a tension side of local bending whereas bottom

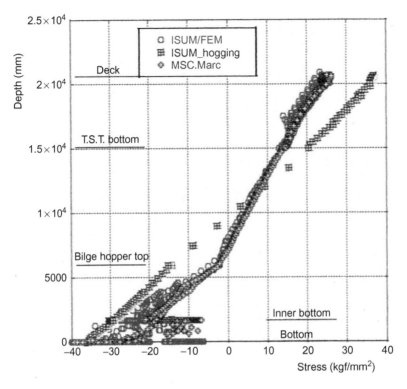

FIG. 8.73

Comparison of stress distributions at mid-ship section [45].

plating is in its compression side. Because of this, the compressive stress in inner bottom plating decreases and that in bottom plating increases as indicated by ISUM and MSC.Marc results in Fig. 8.73.

Some scatters are observed in normal stresses in the double bottom part. This could be attributed to differences in local bending curvature depending on the location in the transverse direction in the inner bottom and the bottom plating.

Fig. 8.73 also shows that the tensile stress in deck plating is lower than the yield stress when local bending acts on double bottom structure. This is because buckling/plastic collapse of bottom plating takes place earlier due to local bending of double bottom before stress in deck plating reaches the yield stress in tension.

Here, to investigate into the influence of local bending of double bottom structure on the ultimate hull girder strength as well as meaning of a knuckle point in stress distribution, calculated results are examined in more detail.

Figs. 8.74 and 8.75 indicate that local bending deformation is produced not only in double bottom but also in bilge hopper tank. This is the reason why knuckle point appears in the stress distribution at the top of bilge hopper tank under alternative heavy load condition.

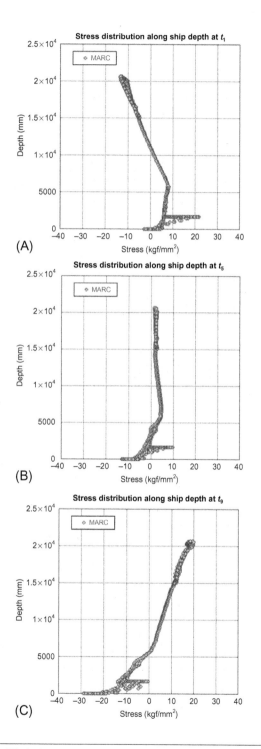

FIG. 8.74

Change in stress distribution. (A) At time step t1. (B) At time step t5. (C) At time step t9.

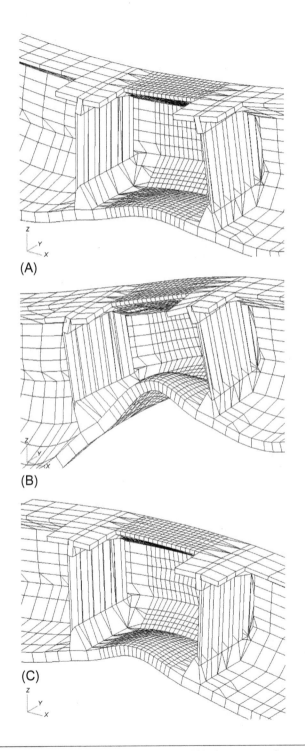

FIG. 8.75

Change in deformation. (A) At time step t1. (B) At time step t5. (C) At time step t9.

Here, stress distributions at time step t1 in Fig. 8.74A indicates that double bottom and bilge hopper tank are in hogging although overall bending is in sagging. The curvatures in regions above and below the top line of the bilge hopper tank are in opposite directions each other. This can be known also from Fig. 8.75A.

On the other hand, overall bending moment near No. 4 hold is almost zero at time step t5. Consequently, curvature above the bilge hopper tank is very small as is known from Figs. 8.74B and 8.75B.

At time step t9, both overall bending and local bending are in hogging, and compressive stresses by overall bending and local bending are superposed in bilge hopper tank and double bottom. Scatter of stresses in inner bottom and bottom plating becomes larger as the local bending moment increases.

In conclusion, the ultimate hull girder strength in hogging is reduced by about 20% due to local bending of double bottom structure when bulk carrier is in an alternate heavy load condition.

At the end, in the case of ISUM/FEM analysis, the total analysis lasted about 30 min on calculation server having 12 Cores with CPU main frequency 3.06 GHz and 32 GB memories. On the other hand, MSC.Marc analysis took about 35 h on the same server that ISUM/FEM analysis was performed.

8.7.6 FUNDAMENTAL IDEA IN PHASE 2 ANALYSIS [67,68]

As mentioned in Section 8.7.2, load analysis and progressive collapse analysis are performed independently in Phase 1 system. In the load analysis, the ship's hull is regarded as a rigid body, while in the progressive collapse analysis, global and local deformation of the ship's hull is calculated. Such calculation may be approximately correct until large deformation takes place, that is up to near the ultimate hull girder strength.

Another problem in Phase 1 is that external loads are applied by distributed forces statically, and the ultimate strength, that is, the peak pressure, cannot be obtained. The condition of the ultimate strength in Phase 1 is that the solution becomes to diverge as the pressure loads exceed a certain level. This pressure level is considered as the ultimate hull girder strength.

To simulate the actual collapse behavior including postultimate strength range, it is necessary to perform dynamic analysis considering the interaction between load analysis and progressive collapse analysis. This is Phase 2 and the fundamental idea to perform such analysis is as follows.

Firstly, total displacements, $\{x\}$, is decomposed into rigid body displacements, $\{x_R\}$, and elastoplastic displacements, $\{x_D\}$, as indicated in Fig. 8.76. Adding the damping terms, the equilibrium equation (or the equation of motion) is expressed in the following form:

$$[M](\{\ddot{x}_R\} + \{\ddot{x}_D\}) + [C_S]\{\dot{x}_D\} + [K_S]\{x_D\} = \{F(\eta, \{x_R\}, \{x_D\}, \{\dot{x}_R\}, \{\dot{x}_D\}, \{\ddot{x}_R\}, \{\ddot{x}_D\})\} \qquad (8.30)$$

where

$[M]$: mass matrix ($N \times N$ where N is degrees of freedom)
$[C_S]$: wave making dumping matrix ($N \times N$)
$[K_S]$: stiffness matrix ($N \times N$)
$\{F\}$: load vector ($N \times 1$)
$\{x_R\}$: rigid body displacement vector ($N \times 1$)
$\{x_D\}$: elastoplastic displacement vector ($N \times 1$)
η: parameter representing wave condition (wave direction, wave height, period, time, etc.)

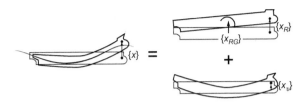

FIG. 8.76

Decomposition of displacements [67].

It is necessary to solve Eq. (8.30) with respect to $\{x_D\}$ and $\{x_R\}$. If the terms related to rigid body motion are extracted from Eq. (8.30), the following equation is obtained:

$$[M_R]\{\ddot{x}_R\} = \{F_R(\eta, \{x_R\}, \{x_D\}, \{\dot{x}_R\}, \{\dot{x}_D\}, \{\ddot{x}_R\}, \{\ddot{x}_D\})\} \tag{8.31}$$

where $[M_R]$ and $\{F_R\}$ represent mass matrix and load vector related to rigid body motion.

In Eq. (8.31), $\{F_R\}$ can be decomposed into forces related to added mass, wave making resistance, restoring force, and wave force. Among these, hydrodynamic term could be assumed to depend on rigid body displacements. On the contrary, elastoplastic displacements are considered in restoring force, since it may depends on the deformed shape. Consequently,

$$[M_R]\{\ddot{x}_R\} = \{F_{R-HD}(\eta, \{\dot{x}_R\}, \{\ddot{x}_R\})\} + \{F_{R-ST}(\{x_R\} + \{x_D\})\} \tag{8.32}$$

The above expression is derived by employing the relationship, $\{x_R\} = [T]\{x_g\}$, where $[T]$ is rigid body transformation matrix of $N \times 1$ and $\{x_g\}$ is displacement vector of 6×1 at the center of geometry of the ship's hull.

In the Phase 1 system, Shell Structure Oriented Dynamic Analysis Code (SSODAC) is used for ship motion analysis. Applying SSODAC, $\{x_R\}$, $\{\dot{x}\}$, and $\{\ddot{x}\}$ at the time $t^n = t^{n-1} + \Delta t$ are obtained. As for $\{x_D\}$, it is obtained from the progressive collapse analysis applying ISUM/FEM system. Phase 1 code has to be revised so that the influence of shape change caused by elastoplastic displacements, $\{x_D\}$, can be considered.

For the calculation of pressure distribution, the code Singularity Distribution Method (SDM) is used in the present system assuming that deformation does not affect the wave field.

For the progressive collapse analysis, Eq. (8.31) is transformed as

$$[M]\{\ddot{x}_D\} + [C_S]\{\dot{x}_D\} + [K_S]\{x_D\}$$
$$= \{F(\eta, \{x_R\}, \{x_D\}, \{\dot{x}_R\}, \{\dot{x}_D\}, \{\ddot{x}_R\}, \{\ddot{x}_D\})\} - [M]\{\ddot{x}_R\} \tag{8.33}$$

Dividing the force term into two, the following expression is obtained:

$$[M]\{\ddot{x}_D\} + [C_S]\{\dot{x}_D\} + [K_S]\{x_D\}$$
$$= \{F_1(\eta, \{x_R\}, \{x_D\}, \{\dot{x}_R\}, \{\dot{x}_D\}, \{\ddot{x}_D\})\} + \{F_2(\{\dot{x}_D\}, \{\ddot{x}_D\})\} - [M]\{\ddot{x}_R\} \tag{8.34}$$

$\{F_1\}$ is the distributed loads which are calculated by SSODAC on the basis of $\{x_R\}$, $\{\dot{x}_R\}$, and $\{\ddot{x}_R\}$ at time $t^n = t^{n-1} + \Delta t$. On the other hand, F_2 is produced by velocity and acceleration, each corresponding to the hydrostatic dumping and the added mass terms, respectively. Here, considering that rapid phenomenon is simulated, hydrostatic dumping term could be regarded as zero. Then, $F_2(\{\ddot{x}_D\})$ can be expressed as $[M_{DA}]\{\ddot{x}_D\}$, where $[M_{DA}]$ is an added mass matrix. Consequently, Eq. (8.34) reduces to:

$$[M + M_{DA}]\{\ddot{x}_D\} + [C_S]\{\dot{x}_D\} + [K_S]\{x_D\}$$
$$= \{F_1(\eta, \{x_R\}, \{x_D\}, \{\dot{x}_R\}, \{\dot{x}_D\}, \{\ddot{x}_D\})\} - [M]\{\ddot{x}_R\} \tag{8.35}$$

From Eq. (8.35), $\{x_D\}$, $\{\dot{x}_D\}$, and $\{\ddot{x}_D\}$ at time $t_n = t_{n-1} + \Delta t$ can be obtained.

Solving Eqs. (8.32), (8.35) alternatively, actual collapse behavior can be simulated approximately if the time increment is taken small enough.

8.7.7 EXAMPLE: COLLAPSE BEHAVIOR OF KAMSARMAX BULK CARRIER IN HOMOGENEOUS LOADING CONDITIONS (PHASE 2 ANALYSIS)

The same bulk carrier is analyzed as that in Section 8.7.4 but under homogeneous loading conditions [66,67]. The ship is in regular waves of which height and length are 20.9 m and the ship length, respectively. The ship speed is 4 m/s and the corresponding encounter wave period is 9.8 s. The maximum bending moment produced in this wave condition exceeds the ultimate hull girder strength at the midship cross-section. The time increment for the analysis is taken as $\Delta t = 0.08$ s.

Fig. 8.77A shows the time history of the wave elevation at mid-ship. The wave height changes with time as indicated here. The bending moment distributions at some selected time steps are shown in Fig. 8.77B. The distribution 1 represents the still water bending moment. At the midship cross-section, the maximum sagging bending moment is attained in distribution 4.

The time history of the bending curvature at midship is indicated in Fig. 8.77C. The applied bending moment exceeds the maximum capacity of the cross-section at around $t = 80$ s where the bending curvature rapidly increases at first. Further increase in the curvature is observed at around $t = 90$ s. The plastic curvature remains after the wave height is reduced.

Fig. 8.77D shows the bending moment-curvature relationship experienced at the midship cross-section. Beyond the ultimate hull girder strength, the load carrying capacity decreases with the increase in the curvature, while the applied bending moment does not decrease correspondingly. During this collapsing stage, the applied bending moment is in equilibrium not only with the reaction from the structure, that is the load carrying capacity of the hull girder, but also with some dynamic load effects including the inertia force due to the rapid increase in the rotational deformation around the collapsing cross-section. This implies that the postultimate strength behavior as shown in Fig. 8.77C and D could be obtained by conducting the time-domain simulation of both nonlinear structural behavior and the dynamic ship motion. The residual deformation is shown in Fig. 8.78. The buckling deformation is observed in the deck part, which is on the compression side of the longitudinal bending under sagging condition.

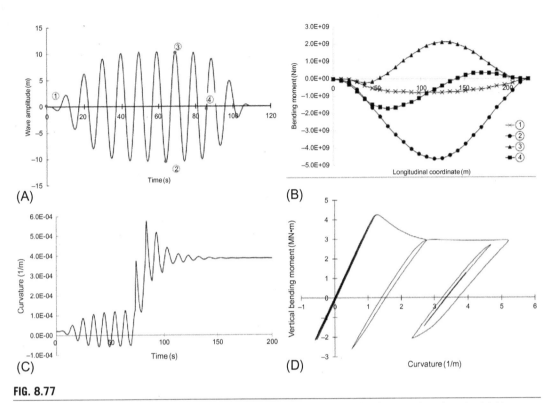

FIG. 8.77

Progressive collapse behavior under homogeneous loading condition [68]. (A) Wave elevation at midship. (B) Bending moment diagram. (C) Time history of curvature at midship. (D) Bending moment-curvature relationship.

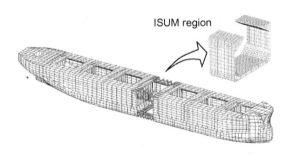

FIG. 8.78

Residual deformation after sagging collapse [68].

EXERCISES

8.1 Explain methods of initial yielding, elastic analysis, and assumed stress distribution to estimate the ultimate hull girder strength.

8.2 Explain the progressive collapse behavior of the hull girder subjected to longitudinal bending, referring to Figs. 8.3 and 8.4.

8.3 Explain the fundamental idea of Smith's method.

8.4 Explain how the neutral axis of the cross-section moved during progressive collapse under sagging and hogging condition.

8.5 Explain how the initial deflection of local panel affects the ultimate hull girder strength.

8.6 Explain how the initial deflection of longitudinal stiffener affects the ultimate hull girder strength.

8.7 Explain how the welding residual stress in the local panel affects the ultimate hull girder strength.

8.8 Explain about the ultimate hull girder strength interaction relationship under combined horizontal and vertical bending.

8.9 Explain what can be known from Table 8.2.

8.10 Explain how the local bending of double bottom affects the ultimate hull girder strength when a bulk carrier is under alternate heavy loading condition.

8.11 Calculate the initial yielding bending moment, M_Y, fully plastic bending moment, M_P, and shape factor of the cross-section, M_P/M_Y, for four cross-sections indicated in Fig. 8.50.

8.12 What is the physical meaning of shape factor?

8.13 Why does welding residual stress reduce the flexural stiffness of the hull girder cross-section?

8.14 With tensile preloading, welding residual stress is reduced. Explain why this happens.

8.8 APPENDIX: DERIVATION OF AVERAGE STRESS-AVERAGE STRAIN RELATIONSHIPS OF ELEMENTS FOR SMITH'S METHOD [29]

8.8.1 MODELING OF STIFFENED PLATING

The cross-section of a typical stiffener element is shown in Fig. 8.79A. Average stress-average strain relationship is derived when this element is subjected to axial load. In this case, deflection shown in Fig. 8.79B is produced, and three modelings can be possible, which are two single-span models, $\overline{AB}$ and $\overline{BC}$ and a double-span model, $\overline{12}$. In the following, however, a double-span model is considered, which is most realistic.

To derive the average stress-average strain relationships of stiffened plate element, the following assumptions are made:

(1) The attached plate behaves as a plate in accordance with the average stress-average strain relationship considering the influences of buckling and yielding.

(2) Plane cross-section remains as plane after deformation has taken place, and strain varies linearly in the cross-section.

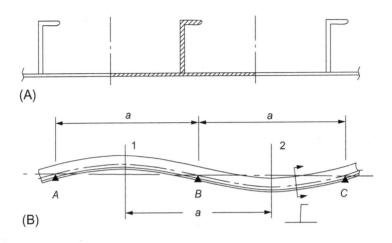

FIG. 8.79

Stiffened plate element [31]. (A) Cross-section. (B) Span.

(3) Only the axial load is considered to act on an element, and the influence of lateral pressure is not considered.

(4) Flexural, torsional, and combined flexural-torsional buckling is considered for a stiffener.

(5) Bi-linear stress-strain relationship is used as that of the material.

8.8.1.1 Average stress-deflection relationship of plating between stiffeners

The buckling/plastic collapse behavior of plating partitioned by stiffeners is simulated by a simplified method combining elastic large deflection analysis and plastic mechanism analysis. The plate is assumed to be accompanied by initial deflection and welding residual stress shown in Fig. 8.80.

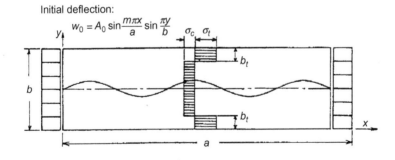

FIG. 8.80

Assumed welding residual stress and initial deflection.

In the present model, initial deflection as well as total deflection in a plate are approximated with a single deflection component, which will correspond to the collapse mode.

$$w_0 = A_0 \sin \frac{m\pi x}{a} \sin \frac{\pi y}{b} \tag{8.36}$$

$$w = A \sin \frac{m\pi x}{a} \sin \frac{\pi y}{b} \tag{8.37}$$

First, the fundamental equation for elastic large deflection for this case is expressed by Eq. (4.5) in Chapter 4 adding the term of welding residual stress as

$$\frac{m^2 \pi^2 a^2 E}{16} \left(\frac{1}{a^4} + \frac{1}{b^4} \right) (A^2 - A_0^2) A$$

$$+ \frac{\pi^2 E}{12(1 - v^2)} \left(\frac{t}{b} \right)^2 \left(\frac{\alpha}{m} + \frac{m}{\alpha} \right)^2 (A - A_0) + (\zeta - \sigma) A = 0 \tag{8.38}$$

where

$$\zeta = \frac{\sigma_t}{\pi(1 - \mu)} \sin \mu\pi \tag{8.39}$$

$\alpha = a/b$ is the aspect ratio of the plate and μ is defined as

$$\mu = 2b_t/b = \sigma_c/(\sigma_t + \sigma_c) \tag{8.40}$$

and m is defined as follows:

$$m = \begin{cases} 1 & \text{for } a/b \leq 1.3 \\ n & \text{for } n - 0.7 < a/b \leq n + 0.3 \end{cases} \tag{8.41}$$

On the other hand, assuming the plastic mechanism shown in Fig. 8.81 and applying the *Rigid Plastic Mechanism Analysis*, the average stress-average strain relationship can be derived as [71]:

- $\alpha \leq 1.0$

$$m_{45} + (1/\alpha - 1)m_{90}/2 = (2/\alpha - 1)\bar{\sigma} \cdot \bar{A} \tag{8.42}$$

- $1.0 < \alpha$

$$m_{45} + (\alpha - 1)m_0/2 = (2/\alpha - 1)\bar{\sigma} \cdot \bar{A} \tag{8.43}$$

where $\bar{\sigma} = \sigma/\sigma_Y$, $\bar{A} = A/t$ and

$$m_{90} = 1 - \bar{\sigma}^2 \tag{8.44}$$

$$m_0 = 2m_{90}/\sqrt{1 + 3m_{90}} \tag{8.45}$$

$$m_{45} = 4m_{90}/\sqrt{1 + 15m_{90}} \tag{8.46}$$

For the rectangular plate of $a \times b \times t = 800 \times 1000 \times 10$ mm, average stress-deflection relationships are plotted in Fig. 8.82. The solid line is the elastic relationship by Eq. (8.38) and the

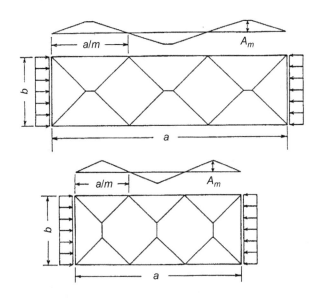

FIG. 8.81

Assumed plastic mechanisms.

chain line with a dot is the plastic relationship by Eqs. (8.42), (8.43), respectively. On the other hand, the dotted line is the elastoplastic relationship by the nonlinear FEM analysis.

In many papers, the average stress-deflection relationship is modeled by combining the elastic and the plastic relationships, and the average stress at their intersection point is regarded as the ultimate strength; see point a in Fig. 8.82. However, this figure indicates that the ultimate strength defined in such a manner is too low when it is compared with the FEM result. Furthermore, in the postultimate strength range, the plastic relationship is far below the FEM result. So, Yao and Nikolov [31] proposed to modify Eqs. (8.42), (8.43) as follows:

$$m_{45} + (1/\alpha - 1)m_{90}/2 = (2/\alpha - 1)\bar{\sigma}^2 \cdot \bar{A} \tag{8.42'}$$

$$m_{45} + (\alpha - 1)m_0/2 = (2/\alpha - 1)\bar{\sigma}^2 \cdot \bar{A} \tag{8.43'}$$

The modified plastic curve calculated by Eq. (8.42') is plotted by the chain line with two dots in Fig. 8.82, which shows very good agreement with the FEM result.

The alternative candidate for the ultimate strength is the average stress at intersection point (point b in Fig. 8.82) of elastic and modified plastic curves. However, this ultimate strength is too high compared to the FEM result. So, using the stress, σ_{iY} at initial yielding point, A, and σ_2 at point b, the ultimate strength can be approximated as follows [31].

$$\sigma_u = \sigma_{iY} + \zeta_1(\sigma_2 - \sigma_{iY}) \tag{8.47}$$

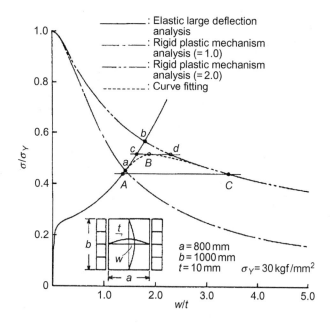

FIG. 8.82

Average stress-deflection relationships by different methods [29].

where

$$\zeta_1 = \begin{cases} \zeta_2 0.5 \left(1.0 + \cos \frac{\xi\pi}{2}\right) & (\xi \leq 4.0) \\ \zeta_2 & (4.0 < \xi) \end{cases} \qquad (8.48)$$

$$\zeta_2 = \begin{cases} 0.2 & (a/b < 0.6) \\ 0.3 - 0.1 \cos \frac{(a/b - 0.6)\pi}{0.4} & (0.6 \leq a/b \leq 1.0) \\ 0.4 & (1.0 < a/b) \end{cases} \qquad (8.49)$$

Thus, the average stress at point B, that is the ultimate strength, is defined.
The strain at the ultimate strength is defined from the condition of $\overline{cB} = \overline{cd}/3$. Then, the average stress average strain relationship along the curve $\overline{AaBC}$ is approximated as follows:

- $\overline{AaB}$

$$w = A_e + \overline{cB}\{1 - \sqrt{1 - (\sigma - \sigma_A)^2/(\sigma_B - \sigma_A)^2}\} \qquad (8.50)$$

- $\overline{BC}$

$$w = A_p - \overline{Bd}\{1 - \sqrt{1 - (\sigma - \sigma_C)^2/(\sigma_B - \sigma_C)^2}\} \qquad (8.51)$$

where A_e and A_p are the elastic and the plastic deflection when the stress is equal to σ.

In summary, the elastoplastic average stress-deflection relationship can be represented as follows:

(a) Elastic range until initial yielding takes place (up to point A): solution of elastic large deflection analysis; Eq. (8.38).
(b) Elastoplastic range; after initial yielding followed by ultimate strength and post-ultimate strength range (between points A and C): curve fitting equations; Eqs. (8.50), (8.51).
(c) Plastic range (beyond point C): modified solution of plastic mechanism analysis; Eq. (8.42′) or (8.43′).

8.8.2 AVERAGE STRESS-AVERAGE STRAIN RELATIONSHIP OF PLATING BETWEEN STIFFENERS

The average stress-average strain relationship can be expressed by Eq. (4.6) in Chapter 4, that is

$$\varepsilon = \frac{\sigma}{E} + \frac{m^2 \pi^2}{8a^2}(A^2 - A_0^2) \tag{8.52}$$

On the other hand, average stress-average strain relationship by plastic mechanism analysis can be expressed as follows:

- $\alpha \leq 1.0$

$$\varepsilon = \frac{\sigma}{E} + \frac{2m^2}{a^2}(A^2 - A_0^2) \tag{8.53}$$

- $1.0 < \alpha$

$$\varepsilon = \frac{\sigma}{E} + \frac{2m^2}{ab}(A^2 - A_0^2) \tag{8.54}$$

Average stress-deflection and average stress-average strain relationships obtained by the above procedure are compared with the nonlinear FEM results in Fig. 8.83A and B. Relatively good agreement is observed between both results.

On the other hand, when a plate is subjected to tensile load, the lateral deflection decreases with the increase in the applied tensile load, and the in-plane stiffness converges to Young's modulus, E. For this case, the in-plane stiffness is assumed to be equal to Young's modulus ignoring the influence of initial deflection.

Regarding the influence of welding residual stress when the compressive load is applied, its influence is included in ζ in Eq. (8.38). When tensile load is applied, the region where the welding residual stress in tension cannot carry any load since this region is already yielded in tension. So, the in-plane stiffness is reduced as

$$D = (1 - \mu)E \tag{8.55}$$

This was discussed in Section 4.5.2.6 in Chapter 4, and the influence of welding residual stress on average stress-average strain relationship is schematically illustrated in Fig. 8.84.

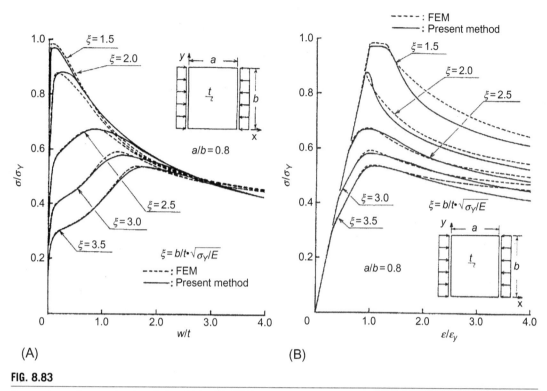

FIG. 8.83

Comparison of approximate relationships with FEM results [31]. (A) Average stress-deflection relationships. (B) Average stress-average strain relationships.

8.8.3 AVERAGE STRESS-AVERAGE STRAIN RELATIONSHIP OF STIFFENER ELEMENT WITH ATTACHED PLATING

When a stiffener element with attached plating is subjected to axial tensile load, its average stress-average strain relationship is assumed to follow the stress-strain relationship of the material. When the panel is accompanied by welding residual stress shown in Fig. 8.80, average stress-average strain relationship above the average stress, σ_Y-σ_t, can be expressed as

$$\sigma = \frac{(1 - \mu_1)F_{p1} + (1 - \mu_2)F_{p2} + F_s}{F_{p1} + F_{p2} + F_s} \cdot \varepsilon \qquad (8.56)$$

where F_{p1} and F_{p2} are the sectional area of the attached plating on the left-hand side and right-hand side, respectively, and μ_1 and μ_2 are the respective ratios of tensile residual stress to half breadth of the panel. F_s is the sectional area of the stiffener.

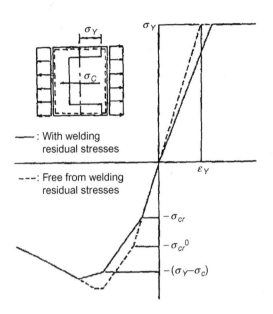

FIG. 8.84

Influence of welding residual stress on average stress-average strain relationship of plating.

Firstly, the stiffener is assumed to be accompanied by initial deflection and initial distortion (rotation) in the following forms:

$$w_0 = W_0 \sin \frac{\pi x}{a}, \quad \phi_0 = \Phi_0 \sin \frac{\pi x}{a} \tag{8.57}$$

The deflection and distortion produced by axial compressive force, P, can be decomposed into elastic and plastic components as

$$w = w^e + w^p, \quad \phi = \phi^e + \phi^p \tag{8.58}$$

Here, it is assumed that the elastic components have the same shape with initial deflection and distortion, that is

$$w^e = W^e \sin \frac{\pi x}{a}, \quad \phi^e = \Phi^e \sin \frac{\pi x}{a} \tag{8.59}$$

According to Timoshecko and Gere [72], the magnitude of the elastic components can be expressed as

$$W^e = \frac{W_0}{1 - P/P_{cre}}, \quad \Phi^e = \frac{\Phi_0}{1 - P/P_{crt}} \tag{8.60}$$

where P_{cre} and P_{crt} are flexural buckling strength and torsional buckling strength, respectively. When flexural/torsional buckling occurs, P_{cre} and P_{crt} in Eq. (8.60) are replaced by the flexural/torsional buckling strength, P_{cret}. Elastic buckling strength is given later in Section 8.12.

On the other hand, plastic components of deflection and distortion are assumed as follows:

(a) In the region of $0 \leq x \leq (a - a^p)/2$

$$w^p = \frac{2W^p x}{a}, \quad \phi^p = \frac{2\Phi^p x}{a} \tag{8.61}$$

(b) In the region of $(a - a^p)/2 < x \leq a/2$

$$\begin{cases} w^p = W^p \left[-\frac{x^2}{aa^p} + \frac{2x}{a\,p} + 1 - \left(\frac{a}{a^p} + \frac{a^p}{a} \right) \right] \\ \phi^p = \Phi^p \left[-\frac{x^2}{aa^p} + \frac{2x}{a\,p} + 1 - \left(\frac{a}{a^p} + \frac{a^p}{a} \right) \right] \end{cases} \tag{8.62}$$

Elastic and plastic components of deflection and distortion are schematically illustrated in Fig. 8.85. It is known that Eqs. (8.61), (8.62) represent the plastic deflection component which gives constant curvature at the central region.

The possible stress distributions at both ends of the double span model (at cross-sections 1 and 2 in Fig. 8.79B) are illustrated in Figs. 8.86 and 8.87 depending on the magnitude of axial strain and curvature. The curvature, K_z, in Fig. 8.87 can be expressed as

$$K_z = K_z^e + K_z^p \tag{8.63}$$

where

$$K_z^e = \frac{\pi^2}{a^2}(W^e - W_0), \quad K_z^p = \frac{4W^p}{aa^p} \tag{8.64}$$

On the other hand, curvature, K_y, can be expressed as follows:

$$K_y = \begin{cases} \frac{(h+z_0')\Phi/W}{(1+y_0')\Phi/W} \cdot K_z & \text{(angle-bar)} \\ h\left(\frac{\pi}{a}\right)^2 (\Phi^e - \Phi_0) + \frac{4h\Phi^p}{aa^p} & \text{(tee-bar)} \end{cases} \tag{8.65}$$

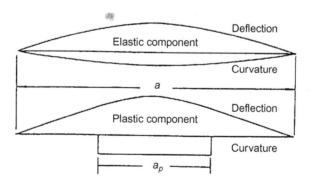

Deflection

Elastic component

Curvature

a

Deflection

Plastic component

Curvature

a_p

FIG. 8.85

Elastic and plastic components of deflection and distortion [31].

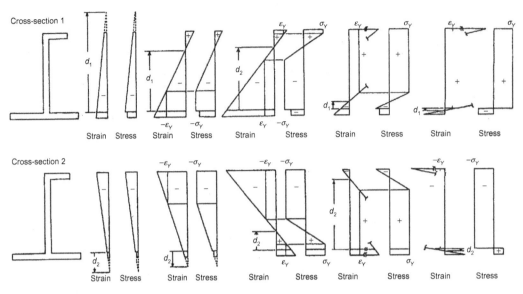

FIG. 8.86

Possible distribution of stress and strain in cross-section in web [31].

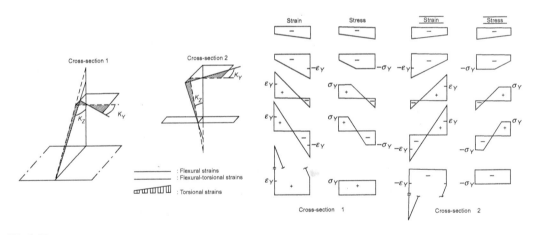

FIG. 8.87

Possible distribution of stress and strain in cross-section in flange [31].

In the double span model indicated in Fig. 8.88B, cross-sections 1 and 2 are mid-span points, where no shear force exists because they are in a symmetry plane. Therefore no reaction force is produced at the support, B. The equilibrium conditions of forces and bending moments for a double span model shown in Fig. 8.88 can be written as

$$P_1 = P_2 \qquad (8.66)$$

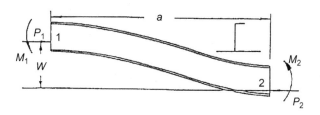

FIG. 8.88

Forces and moments acting on stiffener element [31].

$$M_1 + M_2 - W(P_1 + P_2)/2 = 0 \tag{8.67}$$

When P_1, P_2, M_1, and M_2 are calculated, the plate is assumed to behave in accordance with the average stress-average strain relationship which has been beforehand obtained. That is, average stress obtained from the average stress-average strain relationship of the plate is used as the contribution of the attached plate part.

Here, the strains of the attached plate at cross-sections, 1 and 2, are different each other and so the in-plane stiffness. Because of this, a point at which bending moment becomes zero shifts from the original supporting point. Here, denoting the distance between cross-section 1 and the point of zero bending moment as a_1, and that between cross-section 2 and the point of zero bending moment as a_2, the elastic buckling strength of each span is expressed as follows:

$$P_{cre1} = \frac{\pi^2 EI_1}{a_1^2}, \quad P_{cre2} = \frac{\pi^2 EI_2}{a_2^2} \tag{8.68}$$

where EI_1 and EI_2 are the flexural stiffnesses of the stiffener element at cross-sections 1 and 2, respectively. Considering that spans a_1 and a_2 are continuous at the point of zero bending moment, the following relationships are derived:

$$a_1 + a_2 = 2a \tag{8.69}$$

$$P_{cre1} = P_{cre2} \tag{8.70}$$

from which a_1 and a_2 are determined as follows:

$$\begin{cases} a_1 = 2a/(1+\beta) \\ a_2 = 2\beta a/(1+\beta) \end{cases} \tag{8.71}$$

where

$$\beta = \sqrt{\frac{EI_1}{EI_2}} \tag{8.72}$$

Until yielding starts, ratio of the curvature at cross-section 1 to that of cross-section 2 can be expressed as

$$\frac{K_2}{K_1} = \beta \tag{8.73}$$

At the end, the axial strain can be derived as follows:

$$\varepsilon = \varepsilon^0 + \frac{1}{2a} \sum_{i=1}^{2} \int_0^{a_i} \left[(\varepsilon_i^k - \varepsilon^0) \sin \frac{\pi x}{a_i} + \frac{1}{2} \left(\frac{dw_i}{dx} \right)^2 - \frac{1}{2} \left(\frac{dw_{0i}}{dx} \right)^2 \right] dx \tag{8.74}$$

where ε_i^k represents the axial strain at the location of the original neutral axis. ε^0 is the small displacement strain and is given by:

$$\varepsilon^0 = \frac{P}{EF} \tag{8.75}$$

The average stress-average strain relationship for the stiffener element subjected to axial thrust is derived in accordance with the procedure indicated in (Fig. 8.89).

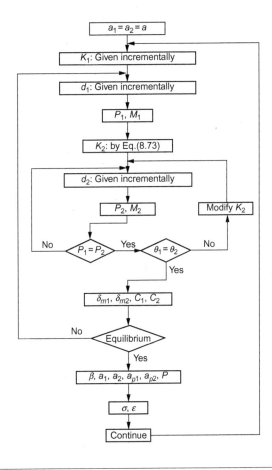

FIG. 8.89

Flow chart to derive the average stress-average strain relationship of the stiffener element.

(1) At the first step, set as $a_1 = a_2 = a$.

(2) The curvature, K_1, at cross-section 1 is increased incrementally. At each step, K_1 is fixed and the following calculation is performed.

(3) Varying the zero stress point (d_1) at cross-section 1, calculate the corresponding axial force, P_1 and bending moment, M_1, for the specified K_1 and d_1.

(4) Curvature at cross-section 2 is derived by Eq. (8.73).

(5) Fixing K_2, the location of zero stress, d_2, is searched by which axial force P_2 becomes equal to the axial force, P_1 at cross-section 1.

(6) When slope does not become continuous at the point of zero bending moment, curvature at cross-section 2 is modified so that the slope becomes continuous.

(7) Calculate deflections at cross-sections 1 and 2.

(8) It is examined if the obtained axial forces, P_1 and P_2, and the bending moments. M_1 and M_2, satisfy the equilibrium conditions, Eqs. (8.66), (8.67). If they are not satisfied, return to step (3) and change d_1. If they are satisfied, proceeds to the next step.

(9) The average strain is calculated by Eq. (8.74).

(10) Parameter, β, span lengths, a_1 and a_2, and the lengths of yielded region, d_1^p and d_2^p, are calculated. Then, next step starts returning to step (2).

The average stress-average strain relationships obtained by the above procedure are compared with the FEM results later in Fig. 9.19A through F. The open circles are the FEM results and × marks are by the present method. Relatively good agreements are observed between both results.

8.9 APPENDIX: A SIMPLE METHOD TO EVALUATE WARPING OF HULL GIRDER CROSS-SECTION [39]

8.9.1 DISPLACEMENT COMPONENTS

The coordinate system for a ship's hull girder as a thin-walled beam is taken as indicated in Fig. 8.90. Here, it is considered that a ship's hull is subjected to combined vertical bending moment, M_z, horizontal bending moment, M_y, torsional moment, T, vertical shear force, V_z, and horizontal shear

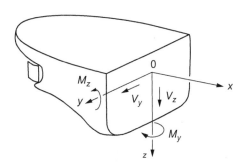

FIG. 8.90

Coordinate systems for hull girder.

force, V_y. Under the applied loads, three translations at an arbitrary point (x, y, z) in the cross-section can be expressed as

$$
\begin{cases}
U(x, y, z) = u_g + z\theta_{yg} - y\theta_{zg} + \omega(y, z)\frac{\partial \theta_{xs}}{\partial x} + U_s(x, y, z) \\
V(x, y, z) = v_s(x) - (z - z_s)\theta_{xs} \\
W(x, y, z) = w_s(x) + (y - y_s)\theta_{xs}
\end{cases}
\tag{8.76}
$$

where $v_s(x)$ and $w_s(x)$ are the displacements of the shear center in y- and z-directions, and θ_y and θ_z are rotation angles with respect to z- and y-axes, respectively. A right-handed rotation is taken as the positive rotation. $U_s(x, y, z)$ stands for the axial displacement produced by warping.

8.9.2 STRAIN AND STRESS COMPONENTS

By partially differentiating the displacement components expressed by Eq. (8.76), strain components are derived as follows:

$$
\begin{cases}
\varepsilon_x = \frac{\partial U}{\partial x} = \frac{\partial u_g}{\partial x} - z\frac{\partial^2 w_s}{\partial x^2} - y\frac{\partial^2 v_s}{\partial x^2} + \omega(y, z)\frac{\partial^2 \theta_{xs}}{\partial x^2} + \frac{\partial U_s}{\partial x} \\
\varepsilon_y = \frac{\partial V}{\partial y} = 0 \\
\varepsilon_z = \frac{\partial W}{\partial z} = 0 \\
\gamma_{yz} = \frac{\partial V}{\partial z} + \frac{\partial W}{\partial y} = 0 \\
\gamma_{zx} = \frac{\partial W}{\partial x} + \frac{\partial U}{\partial z} = \frac{\partial U_s}{\partial z} + \left\{\frac{\partial \omega}{\partial z} + (y - y_s)\right\}\frac{\partial \theta_{xs}}{\partial x} \\
\gamma_{xy} = \frac{\partial U}{\partial y} + \frac{\partial V}{\partial x} = \frac{\partial U_s}{\partial y} + \left\{\frac{\partial \omega}{\partial y} - (z - z_s)\right\}\frac{\partial \theta_{xs}}{\partial x}
\end{cases}
\tag{8.77}
$$

and the stress components as follows:

$$
\begin{cases}
\sigma_x = E\varepsilon_x \\
\sigma_y = 0 \\
\sigma_z = 0 \\
\tau_{yz} = 0 \\
\tau_{zx} = G\gamma_{zx} \\
\tau_{xy} = G\gamma_{xy}
\end{cases}
\tag{8.78}
$$

For the derivation of Eq. (8.77), the following relationships are used:

$$
\theta_y = -\frac{\partial w_s}{\partial x}, \qquad \theta_z = \frac{\partial v_s}{\partial x}
\tag{8.79}
$$

8.9.3 APPLICATION OF PRINCIPLE OF MINIMUM POTENTIAL ENERGY

The strain energy stored in the deformed beam can be expressed as

$$
\begin{aligned}
U &= \frac{1}{2}\int_V \{E(\sigma_x\varepsilon_x + \sigma_y\varepsilon_y + \sigma_z\varepsilon_z) + G(\tau_{yz}\gamma_{yz} + \tau_{zx}\gamma_{zx} + \tau_{xy}\gamma_{xy})\}dV \\
&= \frac{1}{2}\int (E\sigma_x\varepsilon_x + G\tau_{zx}\gamma_{zx} + G\tau_{xy}\gamma_{xy})dV \\
&= \frac{1}{2}\int (E\varepsilon_x^2 + G\gamma_{zx}^2 + G\gamma_{xy}^2)dV
\end{aligned}
\tag{8.80}
$$

On the other hand, denoting the distributed lateral loads in y- and z-directions as $q_y(x)$ and $q_z(x)$, respectively, the potential of the external loads can be expressed as

$$\Phi = \int_\ell (q_y v_s + q_z w_s)dx \qquad (8.81)$$

Applying the *Principle of Minimum Potential Energy*, the following relationship is derived:

$$\delta\Pi = \delta U - \delta\Phi = 0 \qquad (8.82)$$

Here, if it is assumed that displacement components, v_s, w_s, and ω are known by applying *Beam Theory*, U_s becomes the only unknown and $\delta\Phi = 0$. Then, the first derivative of the strain energy can be expressed as follows:

$$\delta U = \iiint \left[E\left(\frac{\partial u_g}{\partial x} - z\frac{\partial^2 w_s}{\partial x^2} - y\frac{\partial^2 v_s}{\partial x^2} + \omega(y,\,z)\frac{\partial^2 \theta_{xs}}{\partial x^2} + \frac{\partial U_s}{\partial x} \right) \frac{\partial \delta U_s}{\partial x} \right.$$
$$+ G\left(\frac{\partial U_s}{\partial z} + \left\{ \frac{\partial \omega}{\partial z} + (y - y_s) \right\} \frac{\partial \theta_{xs}}{\partial x} \right) \frac{\partial \delta U_s}{\partial z}$$
$$\left. + G\left(\frac{\partial U_s}{\partial y} + \left\{ \frac{\partial \omega}{\partial y} - (z - z_s) \right\} \frac{\partial \theta_{xs}}{\partial x} \right) \frac{\partial \delta U_s}{\partial y} \right] dxdydz \qquad (8.83)$$

Partially integrating the right-hand side of Eq. (8.83), the following relationship is obtained:

$$\delta U = \iint \left[E\left(z\frac{\partial^3 w_s}{\partial x^3} + y\frac{\partial^3 v_s}{\partial x^3} - \omega\frac{\partial^3 \theta_{xs}}{\partial x^3} \right) \delta U_s \right.$$
$$\left. + G\left(\frac{\partial U_s}{\partial y}\frac{\partial \delta U_s}{\partial y} + \frac{\partial U_s}{\partial z}\frac{\partial \delta U_s}{\partial z} \right) \right] dydz \qquad (8.84)$$

Substitution of Eq. (8.84) into Eq. (8.82) results in

$$G\iint \left(\frac{\partial U_s}{\partial y}\frac{\partial \delta U_s}{\partial y} + \frac{\partial U_s}{\partial z}\frac{\partial \delta U_s}{\partial z} \right) dydz = \iint \left(\frac{V_z}{I_y}z + \frac{V_y}{I_z}y - \frac{T}{I_p}\omega \right) \delta U_s dydz \qquad (8.85)$$

where V_y and V_z are shear forces in y- and z-directions, respectively, and T is torsional moment. They are related to deflection components, v_s and w_s, as follows:

$$V_z = -EI_y\frac{d^3 w_s}{dx^3}, \quad V_y = -EI_z\frac{d^3 v_s}{dx^3}, \quad T = EI_p\frac{\partial^3 \theta_{xs}}{\partial x^3} \qquad (8.86)$$

8.9.3.1 Application to thin-walled beam cross-section

In the case of a thin-walled beam cross-section, shear stress in the wall thickness direction is small and can be neglected. The shear stress in the tangential direction of the thin-wall can be expressed as

$$\tau_{sx} = \tau_{zx}\frac{dy}{ds} + \tau_{xy}\frac{dz}{ds} = G\left(\frac{\partial U_s}{\partial y}\frac{dy}{ds} + \frac{\partial U_s}{\partial z}\frac{dz}{ds} \right) = G\frac{dU_s}{ds} \qquad (8.87)$$

Here, $U_s(s)$ is a function of s along the mid-thickness line of the thin-wall. So, Eq. (8.85) reduces to:

$$G \int \frac{\partial U_s}{\partial s} \frac{\partial \delta U_s}{\partial s} t ds = \int \left(\frac{V_z}{I_y} z + \frac{V_y}{I_z} y - \frac{T}{I_p} \omega \right) \delta U_s t ds \tag{8.88}$$

It is difficult to solve Eq. (8.88) with respect to $U_s(s)$ in general. Fujitani [39] tried to solve Eq. (8.88) in the FEM-like manner.

8.9.3.2 Stiffness equation for warping

The thin-walled cross-section is divided into linear elements (segments). Here, the nodal points located at both ends of the element are denoted as i and j. If linear warping is assumed within each element, the warping displacement in the element can be expressed as follows:

$$U_s(s) = \left(1 - \frac{s}{\ell_{ij}} \right) U_i + \frac{s}{\ell_{ij}} U_j \tag{8.89}$$

where ℓ_{ij} represents the length of the element ij. The $y(s)$ and $z(s)$ coordinates can also be expressed as linear functions of s as

$$y(s) = \left(1 - \frac{s}{\ell_{ij}} \right) y_i + \frac{s}{\ell_{ij}} y_j \tag{8.90}$$

$$z(s) = \left(1 - \frac{s}{\ell_{ij}} \right) z_i + \frac{s}{\ell_{ij}} z_j \tag{8.91}$$

Substituting Eqs. (8.89), (8.90), and (8.91) into Eq. (8.88), the following stiffness equation is obtained:

$$\frac{Gt}{\ell_{ij}} \begin{bmatrix} 1 & -1 \\ -1 & 1 \end{bmatrix} \left\{ \begin{array}{c} U_i \\ U_j \end{array} \right\} = \frac{t\ell_{ij}}{6} \begin{bmatrix} 2 & 1 \\ 1 & 2 \end{bmatrix} \left(\frac{V_z}{I_y} \left\{ \begin{array}{c} z_i \\ z_j \end{array} \right\} + \frac{V_y}{I_z} \left\{ \begin{array}{c} y_i \\ y_j \end{array} \right\} - \frac{T}{I_p} \left\{ \begin{array}{c} \omega_i \\ \omega_j \end{array} \right\} \right) \tag{8.92}$$

y_i, y_j, z_i, z_j, ω_i, and ω_j of the element ij are indicated in Fig. 8.91. These are the known variables which can be calculated applying *Beam Theory*.

ω_i and ω_j in Eq. (8.92) are the Saint Venan's warping at nodal points i and j, and are evaluated by the following equation at each element:

$$\begin{bmatrix} \frac{Gt}{\ell_{ij}} & -\frac{Gt}{\ell_{ij}} \\ -\frac{Gt}{\ell_{ij}} & \frac{GT}{\ell_{ij}} \end{bmatrix} \left\{ \begin{array}{c} \omega_i \\ \omega_j \end{array} \right\} = \left\{ \begin{array}{c} \frac{Gt}{\ell_{ij}} (y_i z_j - y_j z_i) \\ -\frac{Gt}{\ell_{ij}} (y_i z_j - y_j z_i) \end{array} \right\} \tag{8.93}$$

Summing up stiffness equations for individual elements, stiffness equation for the whole cross-section is obtained with respect to n warping displacements at n nodal points. Giving the shear forces, V_y and V_z, and the torsional moment, T, at a specified cross-section, the warping displacements, U_i, are obtained. When solving the stiffness equation for a whole cross-section with respect to U_k, the following two conditions are imposed.

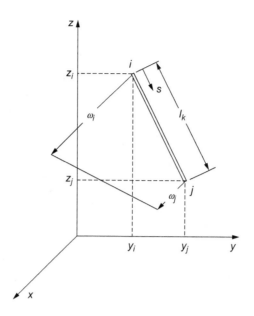

FIG. 8.91

Coordinates of nodal points of element, *ij*.

$$\sum \frac{\frac{1}{2}(U_i + U_j) - U_0 - \frac{1}{2}(z_i + z_j)\theta_0}{\Delta L} \cdot E = 0 \tag{8.94}$$

$$\sum \frac{\frac{1}{2}(U_i + U_j) - U_0 - \frac{1}{2}(z_i + z_j)\theta_0}{\Delta L} \cdot E \cdot \frac{1}{2}(Z_i + z_j) = 0 \tag{8.95}$$

where U_0 and θ_0 are the average axial displacement due to warping and the rotation of the cross-section, respectively. ΔL is a small distance between two cross-sections in which warping is calculated. The former equation represents the condition that warping displacement does not produce axial force, and the latter equation the condition of zero bending moment.

8.9.3.3 Shear stress and warping stress

Using the calculated warping displacements at nodal points, *i* and *j*, the shear stress in element, *ij*, can be derived as follows:

$$\tau_{ij} = G \times \gamma_{ij} = G\frac{dU_s}{ds} = G\frac{U_j - U_i}{\ell_{ij}} \tag{8.96}$$

On the other hand, the warping stress can be calculated as

$$\sigma_w = E \times \varepsilon_w = E\frac{dU_s}{dx} = E\frac{U_{2i} + U_{2j} - U_{1i} - U_{1j}}{2\Delta L} \tag{8.97}$$

where U_{1i} and U_{1j} are the warping displacements in nodal points, i and j, in cross-section 1, and U_{2i} and U_{2j} are those in cross-section 2 which is apart from cross-section 1 by small distance, ΔL.

Calculated warping stress was shown in Fig. 8.30 in Section 8.3.3.5

8.10 APPENDIX: FUNDAMENTAL FORMULATION IN EXPLICIT FEM [21]

Explicit FEM started as a method of analysis on weapons for military object in United States during the Cold War. To keep military secret, detail of this method was not opened at the beginning. Even at present, publications dealing with explicit FEM are not so many. Nevertheless, explicit FEM is now widely used for the collapse analysis on large scaled structures. This is because, in the process of solving a set of linear simultaneous equations, the inverse of the matrix in the stiffness equation needs not be calculated. In the following, explicit FEM is briefly explained.

The explicit method is based on equation of motion applying *Principle of d'Alembert*, and the fundamental equation is expressed as

$$[M]\{a\} + \{F_{int}\} = \{F_{ext}\} \tag{8.98}$$

where $\{a\}$, $\{F_{int}\}$, and $\{F_{ext}\}$ are acceleration vector, internal force vector, and external force vector. $[M]$ is mass matrix. Representative mass matrices are lumped mass matrix and consistent mass matrix. In the dynamic explicit FEM, lumped mass matrix is employed, and then the components of mass matrix are zero except the diagonal components. Because of this, calculation of inverse matrix of $[M]$ is quite easy.

Firstly, it is assumed that the dynamical response is known at time t^n. Then, the response after Δt, that is at time $t^{n+1} = t^n + \Delta t$, is considered. Denoting the time at the middle of time t^n and time t^{n+1} as time $t^{n+\frac{1}{2}}$, the Taylor's expansion on a general function, ϕ, gives the following expressions.

$$\phi^n = \phi^{n+\frac{1}{2}} - \frac{\partial \phi^{n+\frac{1}{2}}}{\partial t}\left(\frac{1}{2}\Delta t\right) + \frac{1}{2}\frac{\partial^2 \phi^{n+\frac{1}{2}}}{\partial t^2}\left(\frac{1}{2}\Delta t\right)^2 - \cdots \tag{8.99}$$

$$\phi^{n+1} = \phi^{n+\frac{1}{2}} + \frac{\partial \phi^{n+\frac{1}{2}}}{\partial t}\left(\frac{1}{2}\Delta t\right) + \frac{1}{2}\frac{\partial^2 \phi^{n+\frac{1}{2}}}{\partial t^2}\left(\frac{1}{2}\Delta t\right)^2 + \cdots \tag{8.100}$$

Deleting the terms of the order more than Δt^2 as small quantities, subtraction of Eq. (8.99) from Eq. (8.100) results in

$$\phi^{n+1} = \phi^n + \frac{\partial \phi^{n+\frac{1}{2}}}{\partial t}(\Delta t) \tag{8.101}$$

Method using Eq. (8.101) is called the middle finite difference method.

The location vector, $\{x\}^{n+1}$, at time t^{n+1} is represented with the velocity vector, $\{v\}^{n+\frac{1}{2}}$, at time $t^{n+\frac{1}{2}}$ as follows:

$$\{x\}^{n+1} = \{x\}^n + \{v\}^{n+\frac{1}{2}}\Delta t \tag{8.102}$$

At the same time, the velocity vector, $\{v\}^{n+\frac{1}{2}}$, at time $t^{n+\frac{1}{2}}$, can be expressed with acceleration vector, $\{a\}^n$, at time t^n as follows:

$$\{v\}^{n+\frac{1}{2}} = \{v\}^{n-\frac{1}{2}} + \{a\}^n \Delta t \tag{8.103}$$

The acceleration vector, $\{a\}^n$, at time t^n is obtained from Eq. (8.98) as

$$\{a\}^n = [M]^{-1}(\{F_{ext}^n\} - \{F_{int}^n\}) \tag{8.104}$$

At the end, merits and demerits of the dynamical explicit FEM are summarized as follows.

MERITS

(1) *Robustness*

In the dynamic explicit FEM, iterative calculation applying Newton-Raphson's method (such as arc-length method in static analysis or Newmark's β method in dynamic analysis) in implicit FEM is not necessary for convergence. Because of this, a break-down seldom takes place.

(2) *Simplicity*

Calculation is very simple. For example, in the implicit FEM, a very complex formulation is necessary to derive stiffness matrix. In the explicit FEM, a stiffness matrix is not necessary.

(3) *Saving of computer memory*

Since stiffness matrix is not necessary, required computer memory is small compared to implicit FEM.

(4) *Good compatibility with parallel processing*

It is possible to perform calculation on external and internal force in parallel.

DEMERITS

(1) *Constraint from time interval*

The time interval has to be, to a certain level, short enough to get stable solution. This results in extremely large number of calculation steps. However, recently, various factors have to be taken into consideration in the implicit FEM, and the large number of calculation steps has become a problem also in the case of implicit FEM.

Because of this, this problem is not a definite demerit.

(2) *Missing of dynamical information*

In implicit FEM, the stiffness matrix contains many dynamical information such as buckling and yielding. Such information is not provided in the explicit FEM.

8.11 APPENDIX: RELAXATION OF WELDING RESIDUAL STRESS BY PRELOADING [3]

The welding residual stress of a rectangular distribution is assumed in a rectangular plate as illustrated in Fig. 8.92. The tensile stress is equal to the yielding stress of the material, σ_Y, and the compressive residual stress is denoted as σ_c.

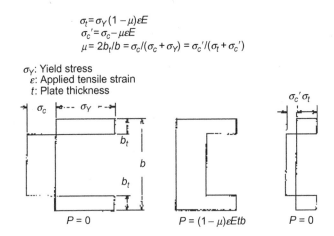

FIG. 8.92

Assumed welding residual stress [3].

Here, tensile load is applied to the plate until the tensile strain, ε is produced. The middle part of the plate is elastic and the stress of εE is produced, while in the edge part where tensile residual stress is produced, no stress is produced by the tensile strain, ε, since this part is already yielded in tension. Consequently, the average stress becomes as $(1-\mu)\varepsilon E$. At this stage, the stress in the plate is as follows:

$$\begin{aligned} \sigma_c^0 &= \varepsilon E - \sigma_c \quad \text{(In central part)} \\ \sigma_t^0 &= \sigma_Y \quad \text{(In edge part)} \end{aligned} \qquad (8.105)$$

From this state, the unloading is performed until the average stress becomes zero. During this unloading, all parts of the plate act as elastic and the stress all over the plate decreases by $(1-\mu)\varepsilon E$. Consequently, at the end of unloading, stress is released to:

$$\begin{aligned} \sigma_c' &= -\sigma_c^0 - (1-\mu)\varepsilon E = -\sigma_c + \mu\varepsilon E \\ \sigma_t &= \sigma_t^0 - (1-\mu)\varepsilon E = \sigma_Y - (1-\mu)\varepsilon E \end{aligned} \qquad (8.106)$$

Fig. 8.93 shows the influence of preloading on the average stress-average strain relationship of the element. This figure indicates that the stiffness under tension is initially equal to Young's modulus after tensile preloading. This is because the edge part is elastic after preloading. However, after the applied average stress reaches $(1-\mu)\varepsilon E$, the edge part has been yielded and the stiffness is reduced to $(1-\mu)E$ hereafter.

8.12 APPENDIX: BUCKLING STRENGTH OF STIFFENER ELEMENT WITH ATTACHED PLATING

Assuming the flexural and torsional buckling modes by Eq. (8.59), the buckling strength can be obtained by solving the eigenvalue problem of the following equations:

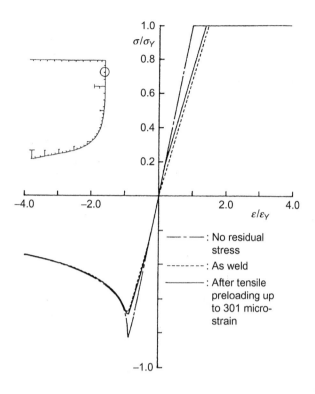

FIG. 8.93

Influence of tensile preloading on average stress-average strain relationship of element [3].

$$(P - a_1)W_e + (a_2 - y_0'\delta P)\Phi^e = 0 \tag{8.107}$$

$$(a_2 - y_0'\delta P)W^e + (\delta a_3 P - a_4)\Phi^e = 0 \tag{8.108}$$

where

$$\begin{cases} a_1 = \left(\frac{\pi}{a}\right)^2 EI_y \\ a_2 \left(\frac{\pi}{a}\right)^2 EI_{yz}'(z_0 - z_B) \\ a_3 = \frac{0'}{A'} - z_0'^2 + z_B'^2 \\ a_4 = \left(\frac{\pi}{a}\right)^2 \{C_w' + SI_z'(Z_0' - z_B')^2\} + C' + \left(\frac{\pi}{a}\right)^2 k_\phi \end{cases} \tag{8.109}$$

and

$$\delta = \frac{\text{Area of } BDEF}{\text{Area of } ABC + BDEF}, \quad I_y = \int z^2 dA$$

$$I_y' = \int z'^2 dA', \quad I_z' = \int y'^2 dA', \quad I_{yz}' = \int y'z'dA', \quad A' = \int dA' \tag{8.110}$$

$$I_0' = I_y' + I_z' + (y_0'^2 + z_0'^2)A'$$

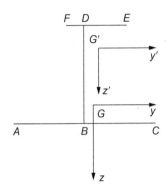

FIG. 8.94

Coordinate systems of cross-section of stiffener with attached plating.

As indicated in Fig. 8.94, the $y'z'$-coordinate system has its origin at the center of geometry of cross-section of the stiffener alone. On the other hand, yz-coordinate system has its origin at the center of geometry of cross-section of the stiffener with attached plating. In deriving Eqs. (8.107), (8.108), it is considered that the whole cross-section can deflect in z-direction, whereas only the stiffener part can deflect in y-direction and can rotate with respect to the intersection point of the stiffener and the attached plating.

k_ϕ in Eq. (8.109) stands for the spring constant representing the resistance against rotation of the stiffener from attached plating; see Fig. 8.95. For a flat-bar or tee-bar stiffener in Fig. 8.95A, the spring constant can be expressed as

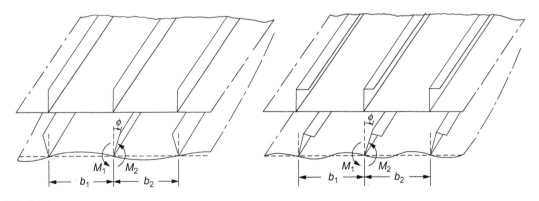

FIG. 8.95

Torsional deformation of stiffeners in stiffened plating.

$$k_\phi = \frac{E}{12}\left(\frac{t_{p1}^3}{b_1} + \frac{t_{p2}^3}{b_2}\right) \tag{8.111}$$

whereas that for an angle-bar stiffener can be expressed as follows:

$$k_\phi = \frac{E}{6}\left(\frac{t_{p1}^3}{b_1} + \frac{t_{p2}^3}{b_2}\right) \tag{8.112}$$

The buckling strength can be given as

Flat-bar and tee-bar stiffener:

$$P_{cre} = a_1, \quad P_{crt} = \frac{a_4}{\delta a_3} \tag{8.113}$$

Angle-bar stiffener:

$$P_{cret} = \frac{-b_2 + \sqrt{b_2^2 - 4b_1 b_3}}{2b_1} \tag{8.114}$$

where

$$\begin{cases} b_1 = \delta(\delta y_0'^2 - a_3) \\ b_2 = \delta(a_1 a_3 - 2y_0' a_2) + a_4 \\ b_3 = a_2^2 - a_1 a_4 \end{cases} \tag{8.115}$$

REFERENCES

[1] IACS. Common structural rules for double hull oil tankers with length 150 metres and above; 2006.

[2] IACS. Common structural rules for bilk carriers with length 90 metres and above; 2006.

[3] Yao T. Investigation into longitudinal strength of ship's hull; historical review and state of the art: focusing on ultimate longitudinal strength. Trans West Jpn Soc Naval Arch 1996;91:221–252 [in Japanese].

[4] Yao T. Ultimate longitudinal strength of ship hull girder; historical review and state of the art. Int J Offshore Polar Eng 1999;9(1):1–9.

[5] Timoshenko S. History of strength of materials. McGraw-Hill Book Co.; 1953.

[6] Rutherford S, Caldwell J. Ultimate longitudinal strength of ships: a case study. SNAME Trans 1990;98:441–471.

[7] John W. On the strength of iron ship. Trans Inst Naval Arch 1874;15:74–93.

[8] Hoffmann G. Analysis of Sir John Biles's experiment on H.M.S. *Wolf*, in the light of Piazker's theory. Trans Inst Naval Arch 1925;67:41–65.

[9] Kell C. Investigation of structural characteristics of destroyers "Preston" and "Bruce" part I—description. SNAME Trans 1931;39:35–64.

[10] Kell C. Investigation of structural characteristics of destroyers "Preston" and "Bruce" part II—analysis of data and results. SNAME Trans 1940;48:125–172.

[11] Vasta J. Lessons learnt from full-scale ship structural test. SNAME Trans 1958;66:165–243.

[12] Lang D, Warren W. Structural strength investigation of destroyer Albuera. Trans Inst Naval Arch 1952;94:243–286.

[13] Yao T. Analysis of hull girder failure. Report of DA-working group. Soc Naval Arch of Japan; 1995 [in Japanese].

[14] Japan Ministry of Transport. Report on the investigation of causes of the casualty of Nakhodka. The Committee for the Investigation on Causes of the Casualty of Nakhodka; 1997.

[15] Bahamas Maritime Authority. Report of the investigation into the loss of the Bahamian registered tanker "Prestige" off the coast of Spain on 19th November 2002; 2004.

[16] Yao T, Astrup C, Carudis P, Chen N, Cho S-R, Dow R, et al. Ultimate hull girder strength. In: Proc 14th international ship and offshore structures congress (ISSC), Nagasaki, Japan; vol. 2; 2000. p. 321–391.

[17] Caldwell J. Ultimate longitudinal strength. Trans RINA 1965;107:411–430.

[18] Smith CS. Influence of local compressive failure on ultimate longitudinal strength of a ship's hull. In: Proc int symp on practical design in shipbuilding (PRADS), Tokyo, Japan; 1977. p. 73–79.

[19] Chen K, Kutt L, Piaszczyk C, Bieniek M. Ultimate strength of ship structures. SNAME Trans 1983;91:149–168.

[20] Valsgård S, Jørgensen L, Bøe Å, Thorkildsen H. Ultimate hull girder strength margines and present class requirements. In: Proc SNAME symp on marine structural inspection, maintenance and monitoring, Arlington, USA; 1991. p. B.1–19.

[21] Benson D. Computational methods in lagrangian and eulerian hydrocodes. Comput Meth Appl Mech Eng 1992;99:235–394.

[22] Hallquist J. LS-DYNA theoretical manual; 1998.

[23] Ikeda A, Yao T, Kitamura O, Yamamoto N, Yoneda M, Ohtsubo H. Assessment of ultimate longitudinal strength of aged tankers. In: Proc 8th PRADS. Shanghai, China; 2001. p. 997–1003.

[24] Amlashi H, Moan T. Ultimate strength analysis of a bulk carrier hull girder under alternate hold loading—a case study, part 1: nonlinear finite element modelling and ultimate hull girder capacity. Mar Struct 2008;21:327–352.

[25] Amlashi H, Moan T. Ultimate strength analysis of a bulk carrier hull girder under alternate hold loading, part 2: stress distribution in the double bottom and simplified approach. Mar Struct 2009;22:522–544.

[26] Ueda Y, Rashed S. The idealized structural unit method and its application to deep girder structures. Comput Struct 1984;18(2):227–293.

[27] Paik JK, Thayamballi A, Che J. Ultimate strength of ship hulls under combined vertical bending, horizontal bending and shearing forces. SNAME Trans 1996;104:31–59.

[28] Fujikubo M, Kaeding P, Yao T. ISUM rectangular plate element with new lateral shape function (1st report): longitudinal and transverse thrust. J Soc Naval Arch Jpn 2000;187:209–219.

[29] Fujikubo M, Kaeding P. ISUM rectangular plate element with new lateral shape function (2nd report): stiffened plates under biaxial thrust. J Soc Naval Arch Jpn 2000;188:479–487.

[30] Paik JK, Mansour A. A simple formulation for predicting the ultimate strength of ships. J Mar Sci Technol 1995;1:52–62.

[31] Yao T, Nikolov P. Progressive collapse analysis of a ship's hull girder under longitudinal bending (2nd report). J Soc Naval Arch Jpn 1992;172:437–446.

[32] Smith C. Structural redundancy and damage tolerance in relation to ultimate ship hull strength. In: Proc int symp on the role of design, inspection and redundancy in marine structural reliability, Williamsburg, USA; 1983.

[33] Yao T, Fujikubo M, Kondo K, Nagahama S. Progressive collapse behaviour of double hull tanker under longitudinal bending. In: Proc 4th int offshore and polar engineering conference, Osaka, Japan; vol. IV; 1994. p. 570–577.

[34] Yao T, Nagahama S, Fujikubo M. Ultimate longitudinal strength of double hull tanker. Trans West Jpn Soc Naval Arch 1993;86:183–198 [in Japanese].

[35] Yao T, Nikolov P. Ultimate longitudinal strength of a bulk carrier. In: Proc 3rd int offshore and polar engineering conference; vol. IV; 1993. p. 497–504.

[36] Jensen J, Amdahl J, Caridis P, Chen T, Cho R, Damonte R, et al. Report of committee III.1: ductile collapse. In: Proc the 14th International Ship and Offshore Structures Congress (ISSC), St. Jones; vol. 1; 1994. p. 299–387.

[37] The Committee on Int. Common Structural Rules of Ship Structures. Part 2: ultimate hull girder strength applying JTP and JBP methods. The Japan Soc Naval Arch and Ocean Engineers; 2005.

[38] Yao T, Imayasu E, Maeno Y, Fujii Y. Influence of warping due to shear force on ultimate hull girder strength. Luebeck, Germany; 2004. p. 322–328.

[39] Fujitani Y. Analysis of beam with thin-walled cross-section. Baifu-kan; 1990 [in Japanese].

[40] Maeno Y, Yamaguchi H, Fujii Y, Yao T. Buckling/plastic collapse behaviour and strength of bilge circle and its contribution to ultimate longitudinal strength of ship's hull girder. In: Proc 14th international offshore and polar engineering conference, Toulon, France; 2004. p. 296–302.

[41] Ohtsubo H, Yao T, Sumi Y, Watanabe I, Takemoto H, Kumano A, et al. Analysis on casualty of ms nakhodka. In: Proc 18th int conf on OMAE, St. John's, Canada; 1999. p. 1–8.

[42] Japan Shipbuilding Research Association. Investigation into structural safety of aged ships. 2000 [in Japanese].

[43] Stolt-Nielsen Inc. STOLT standards for steel structure condition assessment. Tanker Structure Co-operative Forum; 1992.

[44] Tanaka Y, Ando T, Anai Y, Yao T, Fujikubo M, Iijima K. Longitudinal strength of container ships under combined torsional and bending moments. In: Proc 19th int offshore and polar engineering conf, Osaka, Japan; 2009. p. 748–755.

[45] Pei Z, Pei Z, Iijima K, Gao C, Fujikubo M, Tanaka S, et al. Collapse behaviour of ship hull girder of bulk carrier under alternate heavy loading condition. In: Proc 22nd int conf ISOPE, Rhodos, Greece; 2012. p. 839–846.

[46] Østvold T, Steen E, Holtsmark G. Non-linear analysis of a bulk carrier: a case study. In: Proc the 9th symposium on practical design of ships and other floating structures, Luebeck, Germany; 2004. p. 252–260.

[47] Paik JK, Thayamballi A. A concise introduction to the idealised structural unit method for nonlinrear analysis of large plated structures and its application. Thin-Walled Struct 2003;41:329–355.

[48] Paik JK, Thayamballi A. Ultimate limit state design of steel-plated structures. John Wiley & Sons; 2003.

[49] Paik JK, Thayamballi A, Che JS. Ultimate strength of ship hulls under combined vertical bending, horizontal bending and shearing forces. SNAME Trans 1996;104:31–59.

[50] Fujikubo M, Kæding P. New simplified approach to collapse analysis of stiffened plates. Mar Struct 2002;15(3):251–283.

[51] Yao T. Advances in analysis of ultimate limit strength of ship structures. In: Bergan P, Garcia J, Onate E, Kvamsdal T, editors. Proc int conf on computational mechanics in marine engineering, Oslo, Norway; 2005. p. 1–22.

[52] Pei Z, Fujikubo M. Application of idealized structural unit method to progressive collapse analysis of ship's hull girder under longitudinal bending. In: Proc of 15th int offshore and polar engineering conf, Seoul, Korea; 2005. p. 766–773.

[53] Sugimura T, Nozaki M, Suzuki T. Destructive experiment of ship hull model under longitudinal bending. J Soc Naval Arch Jpn 1966;119:209–220 [in Japanese].

[54] Dow R. Testing and analysis of 1/3-scale welded steel frigate model. In: Proc int conf on advances in marine structures, ARE, Dunfermline, Scotland; 1991. p. 749–773.

[55] Endo H, Tanaka Y, Aoki G, Inoue H, Yamamoto Y. Longitudinal strength of the fore body of ships suffering from slamming. J Soc Naval Arch Jpn 1988;163:322–333 [in Japanese].

[56] Mansour A, Yang J, Thayamballi A. An experimental investigation of ship hull ultimate strength. SNAME Trans 1990;98:411–439.

[57] Yao T, Fujikubo M, Yanagihara D, Fujii I. Collapse test on 1/10-scale hull girder model of chip carrier in sagging. In: Proc 13th international offshore and polar engineering conference, Honolulu, Hawaii, USA; 2003. p. 376–383.

[58] Reckling K. Behaviours of box girders under bending and shear. In: Proc ISSC'79. Paris, France. 1979. p. II.2.46–49.

[59] Nishihara S. Analysis of ultimate strength of stiffened rectangular plate (4th report): on the ultimate bending moment of ship hull girder. J Soc Naval Arch Jpn 1983;154:367–375 [in Japanese].

[60] Ostapenko A. Strength of ship hull girders under moment, shear and torque. In: Proc SSC-SNAME symposium on extreme loads response, Arlington, USA; 1981. p. 149–166.

[61] Sun H, Guedes Soares C. An experimental study of ultimate torsional strength of a ship-type hull girder with a large deck opening. Mar Struct 2003;16(1):51–67.

[62] Gold J, Guedes Soares C. Tests on ultimate strength of hull box girder made of high tensile steel. Mar Struct 2009;22(4):770–790.

[63] Yao T, Brunner E, Cho S-R, Choo S, Czujko J, Estefen S, et al. Report of committee III.1. Ultimate strength. In: Proc 16th ISSC, Southampton, UK; vol. 1; 2006. p. 356–437.

[64] Lehmann E. Discussion on report of committee III.1: ultimate strength. In: Proc 16th ISSC, Southampton, UK; vol. 3; 2006. p. 121–131.

[65] Yao T, Fujikubo M, Iijima K, Pei Z. Total system including capacity calculation applying ISUM/FEM and loads calculation for progressive collapse analysis of ship's hull girder in longitudinal bending. In: Proc 19th int conf. ISOPE conf. Osaka, Japan; 2009. p. 706–713.

[66] Pei Z, Iijima K, Gao C, Fujikubo M, Tanaka S, Okazawa S, et al. Application of new system simulating progressive collapse behaviour of ship's hull girder under extreme wave loads, Korea; 2011. p. 172–179.

[67] Goto M, Fujikubo M, Iijima K, Pei Z, Yao T. Post-ultimate strength analysis of a hull girder in waves using idealized structural unit method. In: Proc 27th TEAM on marine structures, Keelung, Taiwan; 2013. p. 593–600.

[68] Fujikubo M, Goto M, Iijima K, Pei Z, Yao T. Motion/collapse analysis of a ship's hull girder in waves using idealized structural unit method. In: Proc int conf on safety and reliability of ship, offshore and subsea structures, Glasgow, UK; 2014.

[69] Pei Z, Iijima K, Fujikubo M, Tanaka S, Okazawa S, Yao T. Simulation of progressive collapse behaviour of whole ship model under extreme waves using idealized structural unit method. Mar Struct 2015;40:104–133.

[70] Iijima K, Yao T, Moan T. Structural response of a ship in severe seas considering global hydroelastic vibrations. Mar Struct 2008;21:420–445.

[71] Okada H, Ohshima K, Fukumoto Y. Compressive strength of long rectangular plates under hydrostatic pressure. J Soc Naval Arch Jpn 1979;146:359–371 [in Japanese].

[72] Timoshenko S, Gere J. Theory of elastic stability. 2nd ed. McGraw-Hill; 1961.

THEORETICAL BACKGROUND AND ASSESSMENT OF EXISTING DESIGN FORMULAS TO EVALUATE ULTIMATE STRENGTH

9.1 RULE FORMULAS

The occurrence of buckling was not allowed in any members in ship and ship-like offshore structures in the rules specified by classification societies until the early 2000s. However, aiming at more rational design under severer wave condition, capacity in postbuckling strength range and the ultimate strength have been newly introduced as design standard in common structural rules (CSR) specified by International Association of Classification Societies (IACS), which was newly came into effect in April 2006. Firstly, two CSRs are issued: CSR-B for bulk carriers [1] and CSR-T for double hull oil tankers [2].

Later, these two CSRs are harmonized and harmonized common structural rules (H-CSR) [3] came into effect in January 2015.

9.2 ASSESSMENT OF RULE FORMULAS IN CSR-B
9.2.1 FORMULAS FOR PLATES

Rule formulas are given in Section 6.3, Chapter 6 of CSR-B [1]. Formulas are based on German DIN-Standard 18800 for design and construction of steel structures. It is on the basis of the ultimate strength considering the postbuckling capacity, but some are the buckling strength. Two formulas exist, one for plates and another for stiffeners.

The criterion to determine the ultimate strength for panels is specified as follows:

$$\left(\frac{|\sigma_x|S}{\kappa_x R_{eH}}\right)^{e_1} + \left(\frac{|\sigma_y|S}{\kappa_y R_{eH}}\right)^{e_2} - B\left(\frac{\sigma_x\sigma_y S^2}{R_{eH}^2}\right) + \left(\frac{|\tau|S\sqrt{3}}{\kappa_\tau R_{eH}}\right)^{e_3} \leq 1.0 \tag{9.1}$$

In the above equation, σ_x, σ_y, and τ are the membrane stresses and the stresses by plate bending are not considered. To evaluate the membrane stresses, the influence of thickness reduction due to corrosion has to be considered. An explanation of parameters in the above equation can be found in IACS [1].

Eq. (9.1) has the same form with yielding criterion:

$$\sigma_x^2 + \sigma_y^2 - \sigma_x\sigma_y + 3\tau_{xy}^2 \leq \sigma_Y^2 \tag{9.2}$$

Buckling and Ultimate Strength of Ship and Ship-like Floating Structures. http://dx.doi.org/10.1016/B978-0-12-803849-9.00009-5

Therefore, it may be convenient if the buckling strength assessment and the yielding strength assessment can be performed with the same equation. However, this equation has no physical meaning as buckling criterion. The coefficients in this equation also have no physical meaning. These may have empirical values adjusted to get better agreements with an exact solution.

As for σ_x and σ_y in Eq. (9.1), results of finite element method (FEM) analysis or solutions applying simple *Beam Theory* are used. The former solutions include the influence of Poisson's effect, while the latter does not. To compensate this difference, correction is made on FEM results (σ_x*, σ_y*) as follows:

$$\sigma_x = \frac{\sigma_x* - v\sigma_y*}{1 - v^2}, \quad \sigma_y = \frac{\sigma_y* - v\sigma_x*}{1 - v^2} \tag{9.3}$$

However, if a special case of uni-axial loading in x-direction is considered, FEM results become as

$$\sigma_x* = \frac{P}{A}, \quad \sigma_y* = 0 \tag{9.4}$$

and solutions by simple beam theory are also

$$\sigma_x = \frac{P}{A}, \quad \sigma_y = 0 \tag{9.5}$$

where P and A are applied uni-axial load and cross-sectional area. Here, if Poisson's correction is performed with Eq. (9.3), the following are obtained:

$$\sigma_x = \frac{P}{A(1 - v^2)}, \quad \sigma_y = -\frac{vP}{A(1 - v^2)} \tag{9.6}$$

These are different from Eq. (9.5) and are not true.

9.2.2 FORMULAS FOR STIFFENERS

Considering the influences of bi-axial and shear loads acting on stiffened plate, flexural buckling strength of a stiffener is derived. Influence of lateral pressure as well as torsional buckling are considered. However, it is not clear why simple formula of flexural buckling is not used as torsional buckling strength instead of complicated formulas. Details are given in IACS [1].

9.2.3 ASSESSMENT OF CSR-B FORMULAS ON ULTIMATE STRENGTH

The ultimate strength estimated by CSR-B formulas is compared with that by nonlinear FEM in Fig. 9.1 for continuous stiffened plates. The local panel is $a \times b = 2400 \times 800$ mm and the thickness is varied as 10, 15, and 25 mm. Four types of stiffeners are attached, which are three tee-bar stiffeners ($138 \times 9 + 90 \times 12$, $235 \times 10 + 90 \times 15$, and $387 \times 12 + 100 \times 17$ mm) and an angle-bar stiffener ($250 \times 90 \times 10/15$ mm). Young's modulus, yield stress, and Poisson's ratio of the material are 313 MPa, 206 GPa, and 0.3.

The lower strength between those of the plate and the stiffener is considered as the ultimate strength of stiffened plate. In some cases, the estimated ultimate strength is in good agreement with the FEM result, but there is no physical background for this agreement.

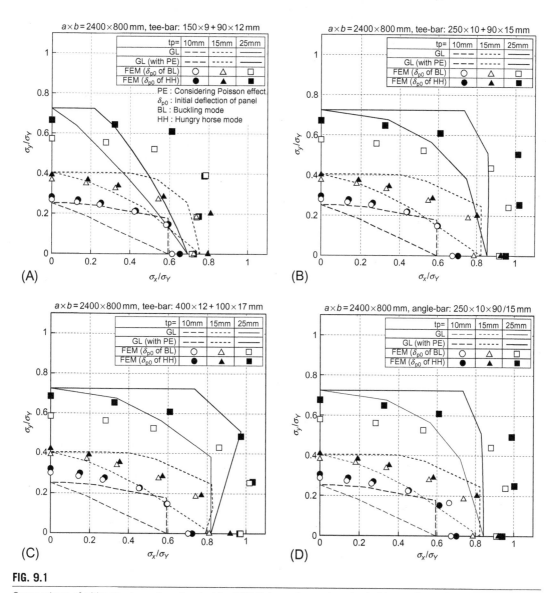

FIG. 9.1

Comparison of ultimate strength estimated by CSR-B formulas and evaluated by nonlinear FEM. (A) Tee-bar (138 × 9 + 90 × 12 mm). (B) Tee-bar (235 × 10 + 90 × 15 mm). (C) Tee-bar (387 × 12 + 100 × 17 mm). (D) Angle-bar (250 × 90 × 10/15 mm).

CSR-B formulas are also applied to estimate the ultimate strength of 720 stiffened plates in Chapter 6, and the results are summarized in Tables 9.1–9.3 in Section 9.5. CSR-B gives the same ultimate strength regardless of the number of stiffeners when the size of the local panel is the same.

9.3 ASSESSMENT OF RULE FORMULAS IN PANEL ULTIMATE LIMIT STATE (PULS)

9.3.1 THEORETICAL BACKGROUND OF PULS

PULS was developed by DnV [4] on the basis of the research works by Byklum et al.[5–9]. Elastic large deflection analysis is performed in an analytical manner. There are three models which are U3-element for local panel partitioned by stiffeners, S3-element for stiffened panel, and T1-element for complicated stiffened plate; see Fig. 9.2. The plate is assumed to be under the combined loads as indicated in Fig. 9.3.

In the stiffened plate model (S3-element), the deflection mode is decomposed into two, the local buckling mode and the overall buckling mode as indicated in Fig. 9.2A and B.

9.3.1.1 Local Bucking Model

In the local buckling mode, deflection along the stiffener lines is set as zero. Considering the interaction between local panel and stiffeners, initial and total deflections in plate are assumed as

$$w_0^L = \sum_m \sum_n A_{0mn}^L \sin\frac{m\pi x}{a} \sin\frac{n\pi y}{b} + \frac{1}{2} \sum_m \sum_n B_{0mn}^L \sin\frac{m\pi x}{a} \left(1 - \cos\frac{2n\pi y}{b}\right) \tag{9.7}$$

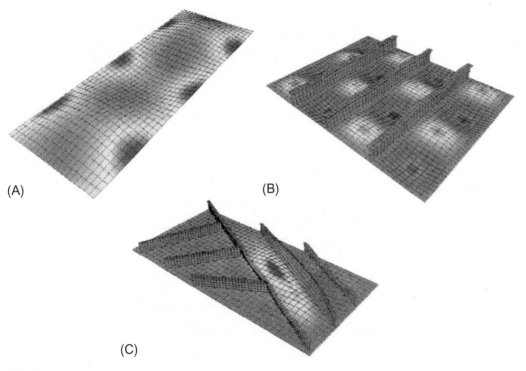

(A)

(B)

(C)

FIG. 9.2

Three models in Panel Ultimate Limit State (PULS). (A) U3-element. (B) S3-element. (C) T1-element.

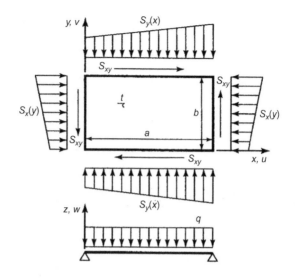

FIG. 9.3

Rectangular plate subjected to combined loads.

FIG. 9.4

Assumed deflection components of local buckling in transverse direction. (A) First term of Eq. (9.8). (B) Second term of Eq. (9.8). (C) Second term of Eq. (9.10).

$$w^L = \sum_m \sum_n A^L_{mn} \sin \frac{m\pi x}{a} \sin \frac{n\pi y}{b} + \frac{1}{2} \sum_m \sum_n B^L_{mn} \sin \frac{m\pi x}{a} \left(1 - \cos \frac{2n\pi y}{b}\right) \tag{9.8}$$

and those in stiffener web as follows:

$$v_0 = \frac{z}{h} \sum_m V_{01m} \sin \frac{m\pi x}{a} + \frac{1}{2} \sum_m V_{02m} \sin \frac{m\pi x}{a} \left(1 - \cos \frac{2\pi z}{h}\right) \tag{9.9}$$

$$v = \frac{z}{h} \sum_m V_{1m} \sin \frac{m\pi x}{a} + \frac{1}{2} \sum_m V_{2m} \sin \frac{m\pi x}{a} \left(1 - \cos \frac{2\pi z}{h}\right) \tag{9.10}$$

Fig. 9.4 shows the deflection components of the local buckling mode.

From the continuity condition of local panel and stiffener web, the following relationship is obtained:

$$\left(\frac{\partial v}{\partial z} - \frac{\partial v_0}{\partial z}\right)\bigg|_{z=0} = -\left(\frac{\partial w^L}{\partial y} - \frac{\partial w_0^L}{\partial y}\right)\bigg|_{y=y_i} \tag{9.11}$$

Shortenings of the plate and the stiffener in the longitudinal direction have to be the same since overall deflection is not allowed in the local model. Hence,

$$\bar{u}_p = \bar{u}_s \tag{9.12}$$

The applied thrust load is equal to the summation of the sectional forces in the plate, stiffener web and flange, that is

$$\int_0^b N_x^p dx + \left(\int_0^h N_x^w dz \int_0^{bf} N_x^f dy\right) = P_x \tag{9.13}$$

As for initial deflection, buckling mode is assumed. If the magnitude is not specified in the analysis, $b/200$ is automatically specified.

9.3.1.2 Overall Buckling Model
In the overall buckling model, a stiffened plate is replaced by an anisotropic plate as indicated in Fig. 9.5. The following initial and total deflections are assumed for this anisotropic plate.

$$w_0^G = \sum_m \sum_n A_{0mn}^G \sin \frac{m\pi x}{a} \sin \frac{n\pi y}{b} + \frac{1}{2}\sum_m \sum_n B_{0mn}^G \sin \frac{m\pi x}{a}\left(1 - \cos \frac{2n\pi y}{b}\right) \tag{9.14}$$

$$w^G = \sum_m \sum_n A_{mn}^G \sin \frac{m\pi x}{a} \sin \frac{n\pi y}{b} + \frac{1}{2}\sum_m \sum_n B_{mn}^G \sin \frac{m\pi x}{a}\left(1 - \cos \frac{2n\pi y}{b}\right) \tag{9.15}$$

In the cross-section of a stiffened plate, sectional forces indicated in Fig. 9.6 are produced. Also in the cross-section of the anisotropic plate model, the same sectional forces have to be produced.

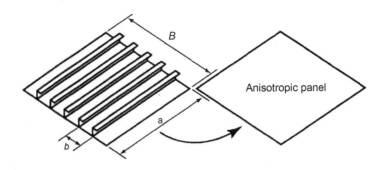

FIG. 9.5

Anisotropic plate equivalent with stiffened plate.

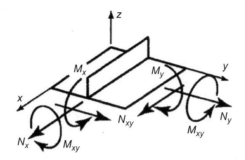

FIG. 9.6

Sectional forces/moments produced in cross-section of stiffened plate.

Here, following relationship exists between increments of sectional forces/moments and in-plane strain/curvature components.

$$
\begin{Bmatrix}
\Delta N_x \\
\Delta N_y \\
\Delta N_{xy} \\
\Delta M_x \\
\Delta M_y \\
\Delta M_{xy}
\end{Bmatrix}
=
\begin{bmatrix}
C_{11} & C_{12} & C_{13} & Q_{11} & Q_{12} & Q_{13} \\
C_{21} & C_{22} & C_{23} & Q_{21} & Q_{22} & Q_{23} \\
C_{31} & C_{32} & C_{33} & Q_{31} & Q_{32} & Q_{33} \\
Q_{11} & Q_{21} & Q_{31} & D_{11} & D_{12} & D_{13} \\
Q_{12} & Q_{22} & Q_{32} & D_{21} & D_{22} & D_{23} \\
Q_{13} & Q_{23} & Q_{33} & D_{31} & D_{32} & D_{33}
\end{bmatrix}
\begin{Bmatrix}
\Delta \varepsilon_x \\
\Delta \varepsilon_y \\
\Delta \gamma_{xy} \\
\Delta \kappa_x \\
\Delta \kappa_y \\
\Delta \kappa_{xy}
\end{Bmatrix}
\tag{9.16}
$$

where C_{ij} and D_{ij} are in-plane stiffness and flexural/torsional stiffness, whereas Q_{ij} is interactive term of in-plane and flexural/torsional stiffnesses. If the model is isotropic, $C_{13} = C_{23} = C_{31} = C_{32} = 0$, $D_{13} = D_{23} = D_{31} = D_{32} = 0$, and $Q_{ij} = 0$.

In PULS, stiffnesses are decomposed into linear and nonlinear components as follows:

$$
\begin{cases}
C_{ij} = C_{ij}^L + C_{ij}^{NL} \\
D_{ij} = D_{ij}^L + D_{ij}^{NL} \\
Q_{ij} = Q_{ij}^L + Q_{ij}^{NL}
\end{cases}
\tag{9.17}
$$

where nonzero linear terms are expressed as

$$
C_{11}^L = \frac{A_T}{B} E + \frac{v^2 t_p E}{1 - v^2}, \quad C_{12}^L = \frac{v t_p E}{1 - v^2}, \quad C_{22} = \frac{t_p E}{1 - v^2}, \quad C_{33} = \frac{t_p E}{1 + v}
\tag{9.18}
$$

$$
D_{11}^L = \frac{I_T E}{B}, \quad Q_{11} = \frac{z_{gs} A_S E}{B}
\tag{9.19}
$$

where

t_p: panel thickness
A_T: total cross-sectional area (panel + stiffener)
A_S: cross-sectional area of stiffener

z_{gs}: distance between mid-thickness plane and center of geometry of stiffener

I_T: second moment of inertia of whole cross-section with respect to mid-thickness plane

When local buckling takes place in plate and/or stiffener web, the in-plane stiffness decreases. In PULS, this effect is considered in the nonlinear terms in Eq. (9.17).

9.3.1.3 Criterion for Ultimate Strength

The ultimate strength is determined from the initial yielding condition examined at six points specified in Fig. 9.7. Each point has physical meaning as follows:

Point 1: yielding in panel after local buckling
Point 2: yielding at stiffener top in tension side of overall bending
Point 3: yielding in panel in tension after overall buckling
Point 4: yielding at stiffener top in compression side of overall bending
Point 5: yielding in panel in compression after overall bending
Point 6: yielding at the end of stiffener (deflection under lateral pressure)

As for stress components, summation of in-plane stress from local buckling model and bending stress from overall buckling model is used, that is

$$\sigma_x = \frac{1}{t_p}N_x^L + E\bar{z}\kappa_x^G, \quad \sigma_y = \frac{1}{t_p}N_y^L, \quad \tau_{xy} = \frac{1}{t_p}N_{xy}^L \tag{9.20}$$

The initial yielding condition is examined with the following equation:

$$\sqrt{\sigma_x^2 + \sigma_y^2 - \sigma_x\sigma_y + 3\tau_{xy}^2} = \sigma_Y \tag{9.21}$$

where σ_Y is the yield stress of the material.

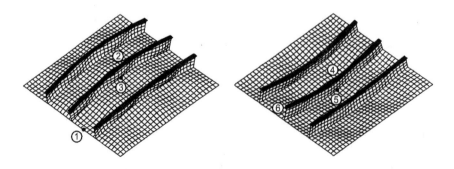

FIG. 9.7

Specified points to examine initial yielding condition.

9.3.1.4 *Procedure of Calculation Applying PULS*

Calculation applying PULS follows the procedure indicated below.

(1) With local buckling model, buckling strength and buckling mode are calculated specifying the initial deflection as zero. Initial deflection for local buckling model is then specified as that same as buckling mode.

(2) Increasing the load, load-deflection component relationship is derived for the local buckling model.

(3) In the overall buckling model, buckling strength and buckling mode is calculated specifying the initial deflection as zero. In this calculation, reduction of the in-plane stiffness in the local panel due to local buckling is considered. Increasing the load, buckling strength and buckling mode are calculated for the overall buckling model.

(4) Obtained buckling mode is given as initial deflection for the overall buckling model. Then, increasing the load, relationship between load and deflection component is calculated. In this calculation reduction of in-plane stiffness of the local panel is considered.

(5) At each incremental step, initial yielding is examined at six specified point. If the yielding condition is satisfied at one of the specified points, load at this step is considered as the ultimate strength.

9.3.2 ASSESSMENT OF PULS FORMULAS ON ULTIMATE STRENGTH

9.3.2.1 *Stiffened plates subjected to uni-axial thrust*

PULS is applied to evaluate the ultimate strength of 720 stiffened plates in Chapter 6. In Tables 9.1–9.3, the ultimate strength when numbers of stiffeners are $N = 1$, $N = 2$, $N = 8$, and $N = 100$ (nearly infinite number) is summarized. Each estimated value is compared with the corresponding FEM result in Fig. ??A through F. The PULS results for $N = 100$ is compared with the FEM assuming infinite number of stiffeners. PULS can be assessed as follows:

(1) Except for a few cases, PULS gives lower ultimate strength when slenderness ratio, $\beta = b/t \cdot \sqrt{\sigma_Y/E}$, of the local panel is below 2.5.

(2) When β is higher than 2.5, PULS gives higher ultimate strength compared to FEM.

(3) Scatter of the ultimate strength depending on the number of stiffeners becomes significant when the aspect ratio of the local panel is higher and the stiffener size is smaller.

(4) When the stiffener height becomes high, the PULS gives higher ultimate strength compared to the cases of lower stiffeners.

(5) For the stiffened plates in actual ship structures, the accuracy of the calculated ultimate strength by PULS is within 10% compared to the FEM results.

As for the item (4), the assumption of the torsional buckling mode in PULS could be the reason. According to PULS, deflection of a stiffener is as indicated in Fig. 9.9A. That is, the number of half-waves in the loading direction is the same with that of a panel local buckling. For the same stiffened plate model, stiffener deflects as indicated in Fig. 9.9B, that is in one half-wave mode between transverse members according to the results of FEM analysis. This could be the cause of higher ultimate strength in PULS when the stiffener height becomes higher.

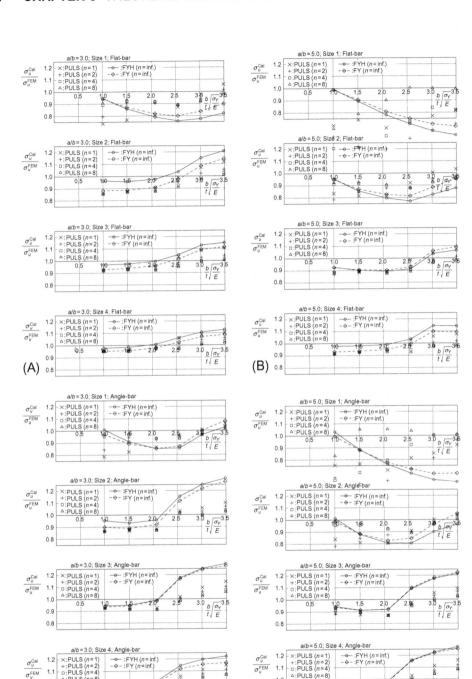

FIG. 9.8

Comparison of ultimate strength by PULS and FEM of stiffened plates. (A) With flat-bar stiffeners; $a/b = 3.0$. (B) With flat-bar stiffeners; $a/b = 5.0$. (C) With angle-bar stiffeners; $a/b = 3.0$. (D) With angle-bar stiffeners; $a/b = 5.0$.

(Continued)

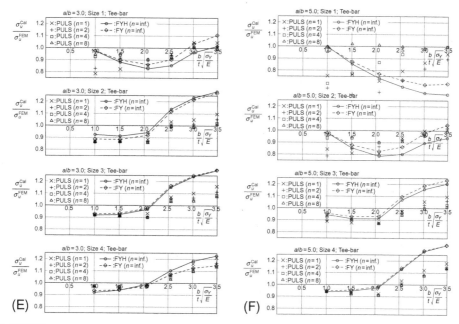

FIG. 9.8, CONT'D

(E) With tee-bar stiffeners; $a/b = 3.0$. (F) With tee-bar stiffeners; $a/b = 5.0$.

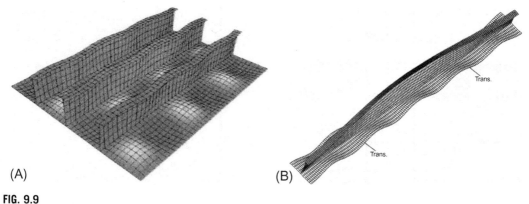

(A)　　　　　　　　　　　　　　　　　　(B)

FIG. 9.9

Comparison of collapse modes obtained by PULS and FEM analyses. (E) PULS. (F) FEM.

9.3.2.2 Stiffened plates subjected to bi-axial thrust

Stiffened plates under bi-axial thrust are considered. The size of the local panel is $a \times b = 2400 \times 800$ mm and the thickness is varied as 10, 15, and 25 mm. Five types of stiffeners are considered, which are

tee-bar ($h \times t_w + b_f \times t_f = 138 \times 9 + 90 \times 12$ mm)
tee-bar ($h \times t_w + b_f \times t_f = 235 \times 10 + 0 \times 15$ mm)
tee-bar ($h \times t_w + b_f \times t_f = 383 \times 12 + 100 \times 17$ mm)
angle-bar ($h \times t_w + b_f \times t_f = 250 \times 90 \times 10/15$ mm)
flat-bar ($H \times t_w = 250 \times 19$ mm)

In the FEM analyses, two types of initial deflection are imposed, which are of a buckling mode and of a thin-horse (hungry-horse) mode. The maximum value is both 4 mm. As for stiffeners, initial deflection of 2.4 mm is given both in flexural buckling mode and tripping mode.

Ultimate strength interaction relationships are plotted in Fig. 9.10A through E. In general, relatively good agreements are observed when the plates' thicknesses are 10 and 15 mm. On the other hand, when plate thickness increases, agreement is not so good. One of the reasons may be that influence of bending stress is not considered in PULS, which becomes higher when plate thickness increases.

In summary, it can be said that estimation by PULS is in general in good accuracy. It can be also said that it is possible in what case PULS gives higher estimation, and in what case lower estimation. This is very important when a simple method is applied.

9.4 AVERAGE STRESS-AVERAGE STRAIN RELATIONSHIP FOR APPLICATION OF SMITH'S METHOD

9.4.1 APPLICATION OF SMITH'S METHOD

In CSR, it is recommended to perform progressive collapse analysis on ship's hull girder to evaluate the ultimate hull girder strength in longitudinal bending. In the rule, some formulations are given to derive average stress-average strain relationships of stiffener elements with attached plating which compose the cross-section of a hull girder.

In this section, the accuracy of the average stress-average strain relationships of elements under axial load is examined which is specified in CSR.

9.4.2 AVERAGE STRESS-AVERAGE STRAIN RELATIONSHIPS SPECIFIED IN CSR

In H-CSRs [3], representative element composing a hull girder cross-section is a stiffener element with attached plating. For this element, four failure modes are considered, which are beam-column buckling, torsional buckling, web local buckling of stiffeners made of flanged profiles, and web local buckling of flat bar stiffeners, and plate buckling. For each failure mode, average stress-average strain relationship under axial load is specified, and that of the lowest capacity is chosen.

Here, case of beam-column failure is considered. In this case, the average stress, σ_{CR1} is expressed as a function of nondimensionalized average strain, $\varepsilon = \varepsilon_E/\varepsilon_Y$, where ε_E is the average axial strain in the element and $\varepsilon_Y = \sigma_Y/E$ is the yield strain expressed by yield stress, σ_Y, and Young's modulus, E, of the material. Here, it is assumed that yield stress of the plate is the same as that of the stiffener for simplicity. The average stress-average strain relationship is given by the following equation:

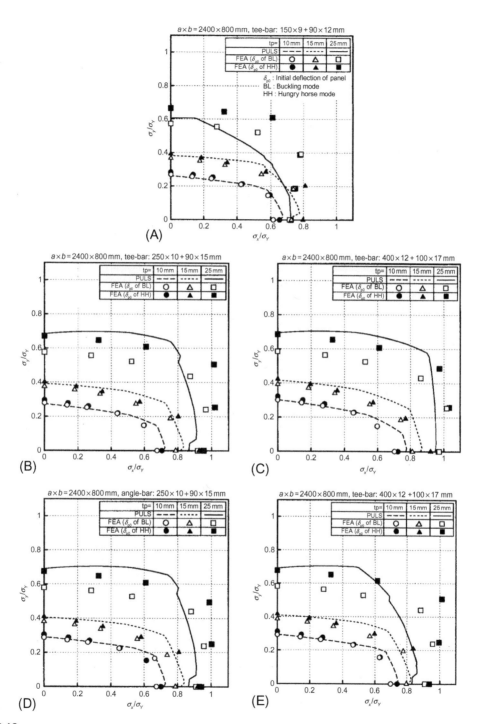

FIG. 9.10

Comparison of collapse modes obtained by PULS and FEM analyses. (A) Tee-bar stiffener ($h = 150$ mm).
(B) Tee-bar stiffener ($h = 250$ mm). (C) Tee-bar stiffener ($h = 400$ mm). (D) Angle-bar stiffener ($h = 250$ mm).
(E) Flat-bar stiffener ($h = 250$ mm).

$$\sigma_{CR1} = \Phi\sigma_{C1}\frac{A_s + b_E t_p}{a_s + b t_p} \tag{9.22}$$

ε is not explicitly seen in Eq. (9.22) but is included in Φ and σ_{C1} as follows:

$$\Phi = \begin{cases} -1 & \text{for} \quad \varepsilon < -1.0 \\ \varepsilon & \text{for} \quad -1 \le \varepsilon \le 1.0 \\ 1 & \text{for} \quad 1 < \varepsilon \end{cases} \tag{9.23}$$

$$\sigma_{C1} = \begin{cases} \frac{\sigma_{E1}}{\varepsilon} & \text{for} \quad \sigma_{E1} \le \frac{\sigma_Y}{2} \\ \sigma_Y\left(1 - \frac{\sigma_Y\varepsilon}{4\sigma_{E1}}\right) & \text{for} \quad \sigma_{E1} > \frac{\sigma_Y}{2} \end{cases} \tag{9.24}$$

$$\sigma_{E1} = \frac{\pi^2 EI}{A_E \ell^2} \tag{9.25}$$

where I is the second moment of inertial of the stiffener with attached plating of breadth, b_{E1}, and A is the net area of the stiffener with attached plating of breadth b_E. b_{E1} and b_E are specified with the slenderness ration, $\beta_E = s/t_p \cdot \sqrt{\sigma_Y/E}$, as

$$b_{E1} = \begin{cases} \frac{s}{\beta_E} & \text{for} \quad \beta_E > 1.0 \\ S & \text{for} \quad \beta_E \le 1.0 \end{cases} \tag{9.26}$$

$$b_E = \begin{cases} \left(\frac{2.25}{\beta_E} - \frac{1.25}{\beta_E^2}\right)s & \text{for} \quad \beta_E > 1.25 \\ b_E = s & \text{for} \quad \beta_E \le 1.25 \end{cases} \tag{9.27}$$

where s is spacing of adjacent longitudinal stiffeners.

For the other failure modes, similar formulas are given. However, background of these formulas is not clear.

9.4.3 ASSESSMENT OF RULE FORMULAS SPECIFYING AVERAGE STRESS-AVERAGE STRAIN RELATIONSHIPS

To assess the average stress-average strain relationships obtained by the formulas specified in H-CSR, stiffened plates analyzed in Chapter 6 are chosen for comparison. At the same time, average stress-average strain relationships obtained by the analytical method introduced in Section 8.8 are also compared. As for the FEM results, the case of $N = $ inf. is selected.

Comparison was made for all combinations of aspect ratio and the slenderness ratio of the local panel as well as the type and the size of stiffeners, which were shown in Tables 6.1 and 6.2 in Chapter 6. Here, results of comparison for six combinations are shown in Fig. 9.11A through F, in which A through C are for $a/b = 3.0$ and D through F are for $a/b = 5.0$.

The average stress-average strain relationships by H-CSR are plotted by solid lines and those by FEM and analytical method by open circles and x-marks, respectively.

In these figures, results of FEM are considered real solutions. Compared to them, CSR gives relatively good prediction in the case of $a/b = 3.0$ with Size 4 flat-bar stiffeners; see Fig. 9.11A. However, large difference is observed from FEM results when stiffener size is small and overall

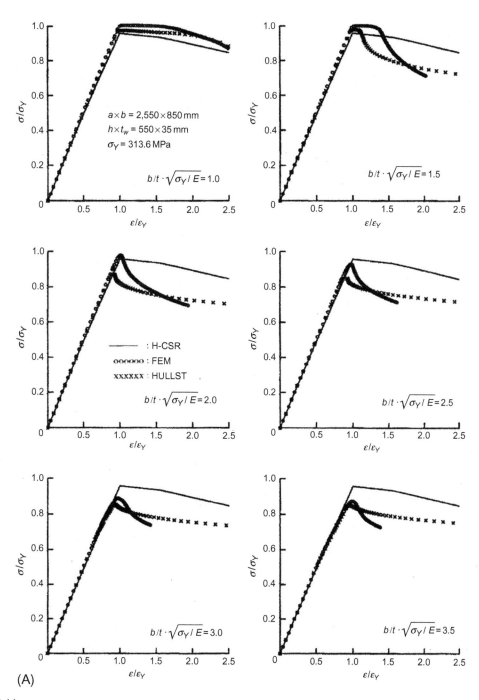

FIG. 9.11

Comparison of average stress-average strain relationships by different methods. (A) Local panel:
$a \times b = 2550 \times 850$; Size 4 flat-bar stiffeners.

(Continued)

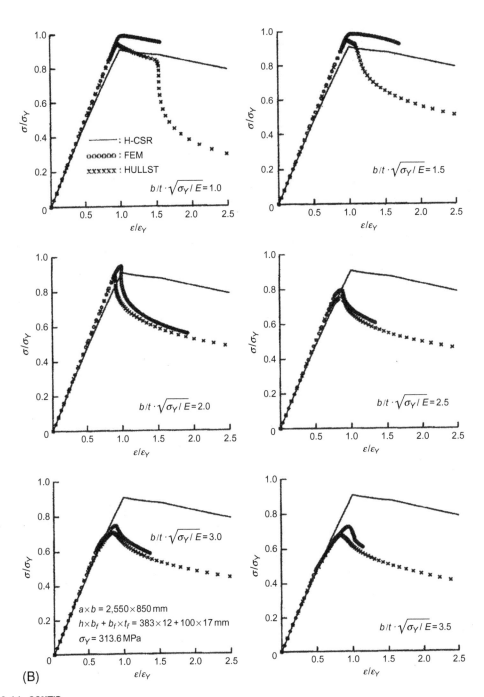

FIG. 9.11, CONT'D

(B) Local panel: $a \times b = 2550 \times 850$; Size 3 angle-bar stiffeners.

(Continued)

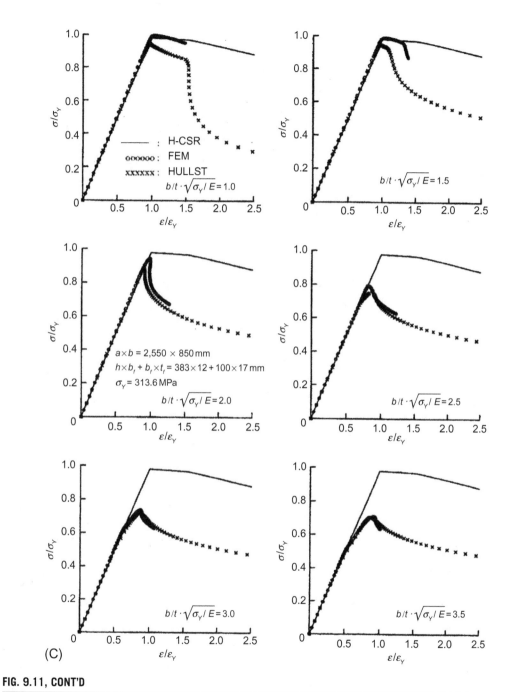

FIG. 9.11, CONT'D

(C) Local panel: $a \times b = 2550 \times 850$; Size 3 tee-bar stiffeners.

(Continued)

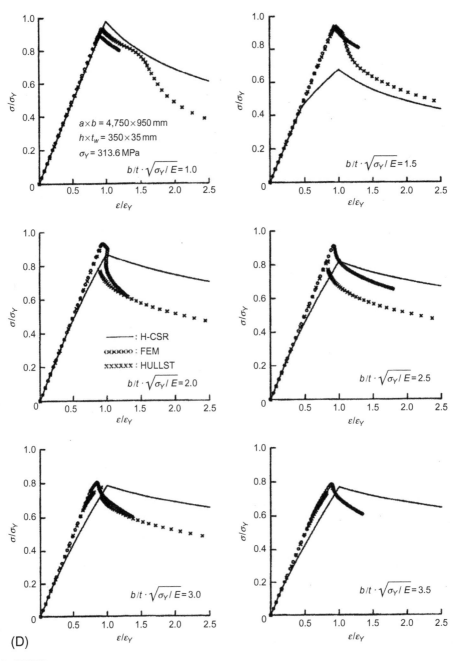

FIG. 9.11, CONT'D

(D) Local panel: $a \times b = 4750 \times 950$; Size 3 flat-bar stiffeners.

(Continued)

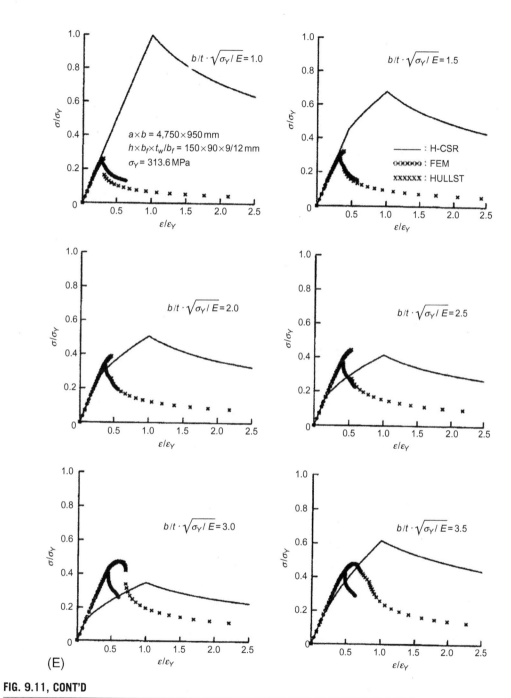

FIG. 9.11, CONT'D

(E) Local panel: $a \times b = 4750 \times 850$; Size 1 angle-bar stiffeners.

(Continued)

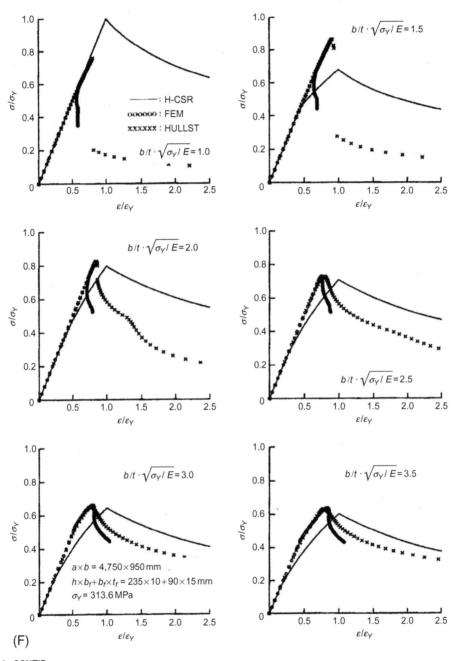

FIG. 9.11, CONT'D

(F) Local panel: $a \times b = 2550 \times 850$; Size 2 tee-bar stiffeners.

buckling takes place, see Fig. 9.11E. In some cases, the ultimate strength is well predicted. However, in general, capacity reduction beyond the ultimate strength is moderate which results in higher ultimate hull girder strength.

On the contrary, the method introduced in 8.8 in Chapter 8 gives relatively accurate average stress-average strain relationship including the ultimate strength and the capacity reduction beyond the ultimate strength.

EXERCISES

9.1 Calculate the ultimate strength of stiffened plates indicated in Table 9.1 applying FYH model explained in Section 6.3, and compare the results with those by FY method in Tables 9.1, 9.2, and 9.3.

9.2 Confirm that average stress-average strain relationship in the case of stiffened plates with $a/b = 5.0$ and Size 3 flat-bar stiffeners can be represented as indicated by solid lines in Fig. 9.10D.

9.3 Explain the theoretical background of PULS.

9.5 APPENDIX: ULTIMATE STRENGTH OF STIFFENED PLATE SUBJECTED TO UNI-AXIAL THRUST

Table 9.1 Ultimate Strength of Stiffened Plate With Flat-Bar Stiffeners

ID	FEM					CSR-B	PULS			FY
	$N = 1$	$N = 2$	$N = 4$	$N = 8$	$N = $ inf.	N	$N = 2$	$N = 8$	$N = 100$	$N = $ inf.
F3S1B10	0.9725	0.7344	0.5835	0.5495	0.5195	0.5263	0.5835	0.4974	0.4879	0.4891
F3S1B15	0.8607	0.6536	0.6112	0.5955	0.5767	0.5420	0.5931	0.5548	0.5453	0.5567
F3S1B20	0.7998	0.6777	0.6542	0.6454	0.6311	0.5772	0.6122	0.5931	0.5867	0.6213
F3S1B25	0.7238	0.6896	0.6770	0.6729	0.6619	0.5772	0.6122	0.5995	0.5931	0.6201
F3S1B30	0.6406	0.6521	0.6544	0.6549	0.6509	0.6125	0.6122	0.5995	0.5963	0.6107
F3S1B35	0.5904	0.6025	0.6114	0.6199	0.6205	0.5408	0.6122	0.6027	0.5995	0.6044
F3S2B10	0.9977	0.9810	0.9706	0.9689	0.9483	0.8044	0.8386	0.8227	0.8195	0.8376
F3S2B15	0.9823	0.9731	0.9689	0.9703	0.9562	0.8333	0.8482	0.8386	0.8355	0.8717
F3S2B20	0.9279	0.9475	0.9461	0.9486	0.9406	0.8264	0.8232	0.8259	0.8227	0.8569
F3S1B25	0.8030	0.8284	0.8521	0.8662	0.8854	0.7028	0.7908	0.7844	0.7813	0.8081
F3S2B30	0.7305	0.7613	0.7835	0.7939	0.7999	0.6125	0.7781	0.7717	0.7685	0.7631
F3S2B35	0.6931	0.7331	0.7485	0.7592	0.7632	0.5408	0.7717	0.7653	0.7621	0.7336
F3S3B10	1.0034	1.0002	0.9987	1.0017	0.9989	0.9021	0.9216	0.9152	0.9120	0.9271
F3S3B15	0.9960	0.9952	0.9940	0.9970	0.9940	0.9161	0.9247	0.9184	0.9152	0.9396
F3S3B20	0.9512	0.9713	0.9705	0.9732	0.9726	0.8264	0.8992	0.8929	0.8929	0.9205
F3S3B25	0.8430	0.8791	0.8999	0.9115	0.9223	0.7028	0.8673	0.8610	0.8610	0.8767
F3S3B30	0.7908	0.8290	0.8491	0.8557	0.8594	0.6125	0.8514	0.8482	0.8450	0.8373
F3S3B35	0.7689	0.8076	0.8280	0.8376	0.8446	0.5408	0.8450	0.8386	0.8355	0.8164

Continued

Table 9.1 Ultimate Strength of Stiffened Plate With Flat-Bar Stiffeners—cont'd

ID	FEM					CSR-B	PULS			FY
	$N=1$	$N=2$	$N=4$	$N=8$	$N=\inf.$	N	$N=2$	$N=8$	$N=100$	$N=\inf.$
F3S4B10	1.0051	1.0055	1.0046	1.0075	1.0063	0.8398	0.9566	0.9566	0.9534	0.9593
F3S4B15	0.9992	1.0014	1.0007	1.0034	1.0018	0.8069	0.9503	0.9503	0.9503	0.9476
F3S4B20	0.9572	0.9767	0.9782	0.9809	0.9780	0.7785	0.9534	0.9471	0.9471	0.9655
K3S4B25	0.8558	0.8931	0.9186	0.9281	0.9298	0.7028	0.9024	0.8992	0.8960	0.9045
F3S4B30	0.8211	0.8564	0.8772	0.8849	0.8895	0.6125	0.8897	0.8865	0.8865	0.8657
F3S4B35	0.8022	0.8331	0.8545	0.8657	0.8734	0.5408	0.8833	0.8801	0.8801	0.8474
F5S1B10	0.9971	0.8439	0.5420	0.3206	0.1799	0.2798	0.5070	0.2136	0.1977	0.1775
F5S1B15	0.9306	0.7723	0.5093	0.2954	0.2011	0.2751	0.3986	0.2328	0.2200	0.2043
F5S1B20	0.8619	0.7211	0.4625	0.2868	0.2317	0.2847	0.3667	0.2583	0.2519	0.2377
F5S1B25	0.7321	0.6209	0.4130	0.2829	0.2574	0.2961	0.3635	0.2838	0.2774	0.2661
F5S1B30	0.6403	0.5624	0.3796	0.3068	0.2822	0.3074	0.3699	0.3061	0.2997	0.2936
F5S1B35	0.5898	0.5313	0.4041	0.3244	0.3001	0.3156	0.3763	0.3221	0.3157	0.3140
F5S2B10	0.9998	0.8834	0.6649	0.6254	0.5988	0.6644	0.6920	0.5772	0.5676	0.5682
F5S2B15	0.9537	0.8227	0.7129	0.6922	0.6751	0.6983	0.6920	0.6509	0.6314	0.6433
F5S2B20	0.9069	0.7992	0.7562	0.7428	0.7291	0.7346	0.7015	0.6728	0.6633	0.6969
F5S2B25	0.7898	0.7952	0.7807	0.7717	0.7595	0.7212	0.6983	0.6792	0.6728	0.6993
F5S2B30	0.6951	0.7150	0.7226	0.7193	0.7150	0.6215	0.6792	0.6633	0.6569	0.6723
F5S2B35	0.6609	0.6817	0.6856	0.6756	0.6772	0.5577	0.6665	0.6505	0.6441	0.6422
F5S3B10	1.0015	0.9715	0.9307	0.9178	0.8780	0.8593	0.8418	0.8163	0.8068	0.8229
F5S3B15	0.9840	0.9590	0.9429	0.9411	0.9178	0.8802	0.8514	0.8355	0.8291	0.8558
F5S3B20	0.9497	0.9514	0.9458	0.9463	0.9293	0.8473	0.8386	0.8259	0.8227	0.8744
F5S3B25	0.8375	0.8673	0.8944	0.9076	0.9114	0.7212	0.8163	0.8068	0.8004	0.8331
F5S3B30	0.7665	0.8001	0.8127	0.8044	0.8033	0.6215	0.7908	0.7813	0.7781	0.7841
F5S3B35	0.7379	0.7739	0.7921	0.7926	0.7845	0.5577	0.7844	0.7749	0.7685	0.7602
F5S4B10	1.0027	0.9992	0.9933	0.9924	0.9883	0.8383	0.9120	0.9056	0.9024	0.9203
F5S4B15	0.9962	0.9950	0.9906	0.9902	0.9860	0.8059	0.9216	0.9152	0.9120	0.9264
F5S4B20	0.9661	0.9802	0.9797	0.9814	0.9778	0.7755	0.9184	0.9088	0.9088	0.9142
F5S4B25	0.8450	0.8854	0.9125	0.9279	0.9344	0.7212	0.8642	0.8578	0.8546	0.8777
F5S4B30	0.8017	0.8352	0.8537	0.8578	0.8457	0.6215	0.8418	0.8355	0.8323	0.8338
F5S4B35	0.7818	0.8208	0.8420	0.8502	0.8476	0.5577	0.8355	0.8291	0.8259	0.8092

Notes: *PID, F: flat-bar/local panel: 3: a/b = 3.0; 5: a/b = 5.0. Stiffener size: S1: Size 1; S2: Size 2; S3: Size 3; S4: Size 4. B10:*
$\beta = 1.005$; *B15:* $\beta = 1.507$; *B20:* $\beta = 2.073$; *B25:* $\beta = 2.551$; *B30:* $\beta = 3.015$; *B35:* $\beta = 3.491$ *when a/b = 3.0. B10:* $\beta = 1.002$;
B15: $\beta = 1.483$; *B20:* $\beta = 2.004$; *B25:* $\beta = 2.471$; *B30:* $\beta = 2.965$: *B35:* $\beta = 3.370$ *when a/b = 5.0.*

Table 9.2 Ultimate Strength of Stiffened Plate With Angle-Bar Stiffeners

	FEM					CSR-B	PULS			FY
ID	$N = 1$	$N = 2$	$N = 4$	$N = 8$	$N = \text{inf.}$	N	$N = 2$	$N = 8$	$N = 100$	$N = \text{inf.}$
A3S1B10	0.9836	0.8138	0.6726	0.6414	0.6156	0.6340	0.6856	0.6250	0.6186	0.5973
A3S1B15	0.9050	0.7508	0.7155	0.7033	0.6886	0.6699	0.7079	0.6856	0.6824	0.6749
A3S1B20	0.8425	0.7705	0.7555	0.7496	0.7404	0.7142	0.7111	0.7015	0.6983	0.7324
A3S1B25	0.7320	0.7288	0.7370	0.7419	0.7422	0.7028	0.6983	0.6920	0.6888	0.6951
A3S1B30	0.6432	0.6558	0.6607	0.6633	0.6621	0.6125	0.6409	0.6378	0.6346	0.6363
A3S1B35	0.5993	0.6064	0.6155	0.6209	0.6257	0.5408	0.6314	0.6250	0.6250	0.6109
A3S2B10	0.9968	0.9764	0.9612	0.9539	0.9255	0.8129	0.8418	0.8291	0.8259	0.8315
A3S2B15	0.9848	0.9687	0.9637	0.9637	0.9474	0.8493	0.8482	0.8418	0.8418	0.8733
A3S2B20	0.9200	0.9202	0.9255	0.9277	0.9263	0.8264	0.8068	0.8036	0.8036	0.8334
A3S2B25	0.7531	0.7579	0.7651	0.7690	0.7705	0.7928	0.7749	0.7717	0.7685	0.7725
A3S2B30	0.6710	0.6857	0.6999	0.7071	0.7119	0.6125	0.7047	0.7015	0.7015	0.6941
A3S2B35	0.6332	0.6512	0.6675	0.6763	0.6854	0.5408	0.6952	0.6920	0.6920	0.6610
A3S3B10	0.9995	0.9934	0.9917	0.9937	0.9896	0.9001	0.9184	0.9152	0.9152	0.9266
A3S3B15	0.9941	0.9867	0.9865	0.9879	0.9848	0.8951	0.9056	0.9024	0.9024	0.9248
A3S3B20	0.9253	0.9341	0.9389	0.9409	0.9453	0.8264	0.8546	0.8514	0.8514	0.8556
A3S3B25	0.7677	0.7766	0.7872	0.7929	0.7964	0.7028	0.7813	0.7781	0.7781	0.7641
A3S3B30	0.6969	0.7178	0.7346	0.7430	0.7485	0.6125	0.7621	0.7589	0.7589	0.6980
A3S3B35	0.6620	0.6866	0.7048	0.7149	0.7263	0.5408	0.7557	0.7526	0.7526	0.6545
A3S4B10	0.9978	0.9912	0.9906	0.9914	0.9910	0.9201	0.9566	0.9566	0.9566	0.9521
A3S4B15	0.9903	0.9804	0.9783	0.9776	0.9771	0.9171	0.9375	0.9343	0.9343	0.9182
A3S4B20	0.9349	0.9400	0.9416	0.9419	0.9441	0.8264	0.8929	0.8929	0.8929	0.8510
A3S4B25	0.7941	0.8120	0.8253	0.8324	0.8373	0.7028	0.8355	0.8355	0.8355	0.7888
A3S4B30	0.7430	0.7655	0.7822	0.7920	0.8027	0.6125	0.7259	0.8259	0.8227	0.7554
A3S4B35	0.7198	0.7441	0.7689	0.7734	0.7865	0.5408	0.8259	0.8227	0.8227	0.7358
A5S1B10	0.9981	0.8517	0.5469	0.3414	0.2426	0.3674	0.5548	0.2838	0.2679	0.2413
A5S1B15	0.9419	0.7914	0.5091	0.3097	0.2858	0.3764	0.4751	0.3253	0.3157	0.2897
A5S1B20	0.8897	0.7468	0.4889	0.3547	0.3349	0.4037	0.4656	0.3731	0.3635	0.3420
A5S1B25	0.7282	0.6532	0.4555	0.3931	0.3748	0.4290	0.4719	0.4082	0.4018	0.3844
A5S1B30	0.6473	0.5837	0.4387	0.4282	0.4109	0.4522	0.4751	0.4273	0.4241	0.4197
A5S1B35	0.5911	0.4985	0.4512	0.4430	0.4271	0.4680	0.4815	0.4464	0.4401	0.4238
A5S2B10	0.9993	0.8895	0.6332	0.5915	0.5661	0.6709	0.6952	0.5804	0.5676	0.5585
A5S2B15	0.9626	0.8286	0.6935	0.6742	0.6582	0.7197	0.7015	0.6537	0.6473	0.6489
A5S2B20	0.9183	0.8030	0.7466	0.7360	0.7242	0.7670	0.7015	0.6824	0.6760	0.7139
A5S2B25	0.7506	0.7361	0.7365	0.7372	0.7310	0.7212	0.6633	0.6537	0.6505	0.6773
A5S2B30	0.6585	0.6606	0.6594	0.6606	0.6601	0.6215	0.6473	0.6409	0.6378	0.6429
A5S2B35	0.6232	0.6212	0.6289	0.6321	0.6234	0.5577	0.6409	0.6314	0.6314	0.6268
A5S3B10	0.9985	0.9648	0.9093	0.8932	0.8789	0.8662	0.8482	0.8259	0.8227	0.8233
A5S3B15	0.9856	0.9512	0.9344	0.9319	0.9151	0.8637	0.8450	0.8355	0.8323	0.8663
A5S3B20	0.9421	0.9242	0.9291	0.9283	0.9179	0.8473	0.8099	0.8036	0.8036	0.8416
A5S3B25	0.7721	0.7796	0.7871	0.7852	0.7775	0.7212	0.7494	0.7462	0.7430	0.7684
A5S3B30	0.6917	0.7041	0.7197	0.7261	0.7285	0.6215	0.7270	0.7207	0.7207	0.7184

Continued

Table 9.2 Ultimate Strength of Stiffened Plate With Angle-Bar Stiffeners—cont'd

ID	FEM					CSR-B	PULS			FY
	N = 1	N = 2	N = 4	N = 8	N = inf.	N	N = 2	N = 8	N = 100	N = inf.
A5S3B35	0.6613	0.6764	0.6947	0.6993	0.7090	0.5577	0.7207	0.7143	0.7143	0.6884
A5S4B10	0.9984	0.9905	0.9881	0.9860	0.9817	0.8662	0.9247	0.9184	0.9152	0.9293
A5S4B15	0.9925	0.9820	0.9831	0.9807	0.9760	0.8583	0.9120	0.9056	0.9056	0.9283
A5S4B20	0.9483	0.9438	0.9473	0.9471	0.9474	0.8443	0.8673	0.8642	0.8642	0.8707
A5S4B25	0.7884	0.8006	0.8086	0.8102	0.8077	0.7212	0.8099	0.8068	0.8068	0.7884
A5S4B30	0.7159	0.7385	0.7522	0.7580	0.7649	0.6215	0.7940	0.7908	0.7908	0.7012
A5S4B35	0.6848	0.7072	0.7230	0.7303	0.7386	0.5577	0.7908	0.7876	0.7876	0.6480

Notes: ID, A: angle-bar/local panel: 3: a/b = 3.0; 5: a/b = 5.0. Stiffener size: S1: Size 1; S2: Size 2; S3: Size 3; S4: Size 4. B10: β = 1.005; B15: β = 1.507; B20: β = 2.073; B25: β = 2.551; B30: β = 3.015; B35: β = 3.491 when a/b = 3.0. B10: β = 1.002; B15: β = 1.483; B20: β = 2.004; B25: β = 2.471; B30: β = 2.965; B35: β = 3.370 when a/b = 5.0.

Table 9.3 Ultimate Strength of Stiffened Plate With Tee-Bar Stiffeners

ID	FEM					CSR-B	PULS			FY
	N = 1	N = 2	N = 4	N = 8	N = inf.	N	N = 2	N = 8	N = 100	N = inf.
F3S1B10	0.9725	0.7344	0.5835	0.5495	0.5195	0.5263	0.5835	0.4974	0.4879	0.4891
F3S1B15	0.8607	0.6536	0.6112	0.5955	0.5767	0.5420	0.5931	0.5548	0.5453	0.5567
F3S1B20	0.7998	0.6777	0.6542	0.6454	0.6311	0.5772	0.6122	0.5931	0.5867	0.6213
F3S1B25	0.7238	0.6896	0.6770	0.6729	0.6619	0.5772	0.6122	0.5995	0.5931	0.6201
F3S1B30	0.6406	0.6521	0.6544	0.6549	0.6509	0.6125	0.6122	0.5995	0.5963	0.6107
F3S1B35	0.5904	0.6025	0.6114	0.6199	0.6205	0.5408	0.6122	0.6027	0.5995	0.6044
F3S2B10	0.9977	0.9810	0.9706	0.9689	0.9483	0.8044	0.8386	0.8227	0.8195	0.8376
F3S2B15	0.9823	0.9731	0.9689	0.9703	0.9562	0.8333	0.8482	0.8386	0.8355	0.8717
F3S2B20	0.9279	0.9475	0.9461	0.9486	0.9406	0.8264	0.8232	0.8259	0.8227	0.8569
F3S1B25	0.8030	0.8284	0.8521	0.8662	0.8854	0.7028	0.7908	0.7844	0.7813	0.8081
F3S2B30	0.7305	0.7613	0.7835	0.7939	0.7999	0.6125	0.7781	0.7717	0.7685	0.7631
F3S2B35	0.6931	0.7331	0.7485	0.7592	0.7632	0.5408	0.7717	0.7653	0.7621	0.7336
F3S3B10	1.0034	1.0002	0.9987	1.0017	0.9989	0.9021	0.9216	0.9152	0.9120	0.9271
F3S3B15	0.9960	0.9952	0.9940	0.9970	0.9940	0.9161	0.9247	0.9184	0.9152	0.9396
F3S3B20	0.9512	0.9713	0.9705	0.9732	0.9726	0.8264	0.8992	0.8929	0.8929	0.9205
F3S3B25	0.8430	0.8791	0.8999	0.9115	0.9223	0.7028	0.8673	0.8610	0.8610	0.8767
F3S3B30	0.7908	0.8290	0.8491	0.8557	0.8594	0.6125	0.8514	0.8482	0.8450	0.8373
F3S3B35	0.7689	0.8076	0.8280	0.8376	0.8446	0.5408	0.8450	0.8386	0.8355	0.8164
F3S4B10	1.0051	1.0055	1.0046	1.0075	1.0063	0.8398	0.9566	0.9566	0.9534	0.9593
F3S4B15	0.9992	1.0014	1.0007	1.0034	1.0018	0.8069	0.9503	0.9503	0.9503	0.9476
F3S4B20	0.9572	0.9767	0.9782	0.9809	0.9780	0.7785	0.9534	0.9471	0.9471	0.9655
F3S4B25	0.8558	0.8931	0.9186	0.9281	0.9298	0.7028	0.9024	0.8992	0.8960	0.9045

Table 9.3 Ultimate Strength of Stiffened Plate With Tee-Bar Stiffeners

ID	FEM					CSR-B	PULS			FY
	$N=1$	$N=2$	$N=4$	$N=8$	$N=$ inf.	N	$N=2$	$N=8$	$N=100$	$N=$ inf.
F3S4B30	0.8211	0.8564	0.8772	0.8849	0.8895	0.6125	0.8897	0.8865	0.8865	0.8657
F3S4B35	0.8022	0.8331	0.8545	0.8657	0.8734	0.5408	0.8833	0.8801	0.8801	0.8474
F5S1B10	0.9971	0.8439	0.5420	0.3206	0.1799	0.2798	0.5070	0.2136	0.1977	0.1775
F5S1B15	0.9306	0.7723	0.5093	0.2954	0.2011	0.2751	0.3986	0.2328	0.2200	0.2043
F5S1B20	0.8619	0.7211	0.4625	0.2868	0.2317	0.2847	0.3667	0.2583	0.2519	0.2377
F5S1B25	0.7321	0.6209	0.4130	0.2829	0.2574	0.2961	0.3635	0.2838	0.2774	0.2661
F5S1B30	0.6403	0.5624	0.3796	0.3068	0.2822	0.3074	0.3699	0.3061	0.2997	0.2936
F5S1B35	0.5898	0.5313	0.4041	0.3244	0.3001	0.3156	0.3763	0.3221	0.3157	0.3140
F5S2B10	0.9998	0.8834	0.6649	0.6254	0.5988	0.6644	0.6920	0.5772	0.5676	0.5682
F5S2B15	0.9537	0.8227	0.7129	0.6922	0.6751	0.6983	0.6920	0.6509	0.6314	0.6433
F5S2B20	0.9069	0.7992	0.7562	0.7428	0.7291	0.7346	0.7015	0.6728	0.6633	0.6969
F5S2B25	0.7898	0.7952	0.7807	0.7717	0.7595	0.7212	0.6983	0.6792	0.6728	0.6993
F5S2B30	0.6951	0.7150	0.7226	0.7193	0.7150	0.6215	0.6792	0.6633	0.6569	0.6723
F5S2B35	0.6609	0.6817	0.6856	0.6756	0.6772	0.5577	0.6665	0.6505	0.6441	0.6422
F5S3B10	1.0015	0.9715	0.9307	0.9178	0.8780	0.8593	0.8418	0.8163	0.8068	0.8229
F5S3B15	0.9840	0.9590	0.9429	0.9411	0.9178	0.8802	0.8514	0.8355	0.8291	0.8558
F5S3B20	0.9497	0.9514	0.9458	0.9463	0.9293	0.8473	0.8386	0.8259	0.8227	0.8744
F5S3B25	0.8375	0.8673	0.8944	0.9076	0.9114	0.7212	0.8163	0.8068	0.8004	0.8331
F5S3B30	0.7665	0.8001	0.8127	0.8044	0.8033	0.6215	0.7908	0.7813	0.7781	0.7841
F5S3B35	0.7379	0.7739	0.7921	0.7926	0.7845	0.5577	0.7844	0.7749	0.7685	0.7602
F5S4B10	1.0027	0.9992	0.9933	0.9924	0.9883	0.8383	0.9120	0.9056	0.9024	0.9203
F5S4B15	0.9962	0.9950	0.9906	0.9902	0.9860	0.8059	0.9216	0.9152	0.9120	0.9264
F5S4B20	0.9661	0.9802	0.9797	0.9814	0.9778	0.7755	0.9184	0.9088	0.9088	0.9142
F5S4B25	0.8450	0.8854	0.9125	0.9279	0.9344	0.7212	0.8642	0.8578	0.8546	0.8777
F5S4B30	0.8017	0.8352	0.8537	0.8578	0.8457	0.6215	0.8418	0.8355	0.8323	0.8338
F5S4B35	0.7818	0.8208	0.8420	0.8502	0.8476	0.5577	0.8355	0.8291	0.8259	0.8092

Notes: *ID, F: flat-bar/local panel: 3: a/b = 3.0; 5: a/b = 5.0. Stiffener size: S1: Size 1; S2: Size 2; S3: Size 3; S4: Size 4. B10:* $\beta = 1.005$; *B15:* $\beta = 1.507$; *B20:* $\beta = 2.073$; *B25:* $\beta = 2.551$; *B30:* $\beta = 3.015$; *B35:* $\beta = 3.491$ *when a/b = 3.0. B10:* $\beta = 1.002$; *B15:* $\beta = 1.483$; *B20:* $\beta = 2.004$; *B25:* $\beta = 2.471$; *B30:* $\beta = 2.965$; *B35:* $\beta = 3.370$ *when a/b = 5.0.*

REFERENCES

[1] IACS. Common structural rules for double hull oil tankers with length 150 metres and above; 2006.

[2] IACS. Common structural rules for bilk carriers with length 90 metres and above; 2006.

[3] IACS. Common structural rules for bulk carriers and oil tankers; 2015.

[4] Det Norske Veritas. Nauticus hull user manual. PULS; 2005.

[5] Byklum E. Ultimate strength of stiffened steel and aluminium panels using semi-analytical methods. Phd thesis. Norwegian University of Science and Technology; 2002.

[6] Byklum E, Amdahl J. A simplified method for elastic large deflection analysis of plates and stiffened plates due to local buckling. Thin-Walled Struct 2002; 40: 925–953.

[7] Byklum E, Steen E, Amdahl J. A semi-analytical model for global buckling and postbuckling analysis of stiffened panels. Thin-Walled Struct 2004; 42: 701–717.

[8] Steen E, Byklum E, Vilming K, Østvold T. Computerized buckling models for ultimate strength assessment of stiffened ship hull panels. In: Proc 9th int symp PRADS; 2004. p. 235–242.

[9] Steen E, Byklum E. Ultimate strength and postbuckling stiffness of plate panels subjected to combined loads using semi-analytical models. In: Proc int conf on marine research and transportation (ICMRT); 2005.

BUCKLING/PLASTIC COLLAPSE BEHAVIOR OF STRUCTURAL MEMBERS AND SYSTEMS IN SHIP AND SHIP-LIKE FLOATING STRUCTURES

10.1 INTRODUCTION

In Chapters 4 through 9, buckling/plastic collapse behavior and strength of main structural members in ship and ship-like floating structures are explained. However, there are other structural members and systems which are not the main structural members but are important from the strength viewpoint.

In this chapter, the triangular corner bracket is firstly selected as a target structural member. Then, watertight transverse bulkhead, double bottom structure as well as hatch cover of bulk carriers are selected, and research works on their buckling/plastic collapse behavior and strength shall be briefly explained.

10.2 TRIANGULAR CORNER BRACKETS
10.2.1 GENERAL

Triangular corner brackets are usually fitted at the connection of vertical members such as frame or column and horizontal members such as beam or girder to smoothen the flow of force and to reduce stress concentration. In general, thickness of corner brackets are so determined that buckling does not occur. However, load above the design load could act on bracket in some cases. In these cases, it is preferable that bracket does not undergo buckling before the main strength member to which the bracket is provided collapses.

From this point of view, it is important to know the buckling/plastic collapse behavior and strength of triangular corner bracket, and many research works can be found on this subject. For example, Nishimaki et al. [1] conducted series of buckling tests on bracket specimens. Ueda et al. [2] performed eigenvalue analysis and derived elastic buckling strength of triangular corner bracket under various boundary/loading conditions. Ueda et al. [3] also carried out a series of buckling collapse tests and nonlinear finite element method (FEM) analysis were performed to clarify collapse behavior and buckling/ultimate strength of triangular corner brackets.

In this section, what were learned from Ueda et al. [3] shall be briefly introduced. Then, method to estimate the optimum thickness of a corner bracket shall be introduced [4,5].

10.2.2 BUCKLING/ULTIMATE STRENGTH OF TRIANGULAR CORNER BRACKET

A series of buckling collapse tests was carried out on various corner brackets in Ueda et al. [3]. Altogether 18 brackets are prepared including 4 types, which are isosceles right-angled triangular bracket without edge stiffener, same bracket but with 1 and/or 2 stiffeners as well as bracket with curved free edge.

The bracket was welded to the loading frame and concentrated force, P, was applied as indicated in Fig. 10.1. Initial deflection of various magnitude was given by mechanical press, and its influence was investigated how the ultimate strength was reduced.

On the other hand, buckling/plastic collapse behavior was clarified also by performing a series of nonlinear FEM analyses. Welding residual stress was measured experimentally, and also calculated applying thermal elastoplastic stress analysis.

Through experiments and nonlinear FEM analyses, it has been known that

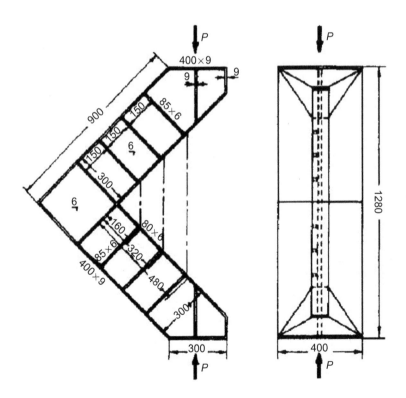

FIG. 10.1

Loading frame for buckling collapse test of corner brackets [3].

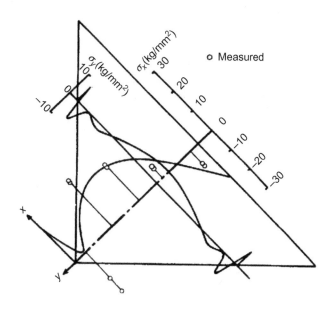

FIG. 10.2

Calculated and measured welding residual stress in triangular corner bracket [3].

(1) Welding residual stress is not self-balancing in the bracket. Self-balance condition is satisfied together with loading frame. As indicated in Fig. 10.2, residual stress in the bracket is mostly in tension.
(2) Owing to the tensile residual stress, the ultimate strength of the bracket largely increases. On the other hand, buckling strength is not so much increased comparing to the ultimate strength.
(3) Initial deflection reduces both the buckling strength and the ultimate strength.
(4) Stiffener provided at free edge increases both the buckling strength and the ultimate strength.

10.2.3 OPTIMUM THICKNESS OF CORNER BRACKET

The role of the corner bracket is to get smooth flow of forces and to reduce stress concentration. To achieve such functions, bracket has to have sufficient thickness to avoid its buckling/plastic collapse. However, bracket is the secondary strength member, and it is of no use if the bracket remains un-collapsed even after the main structural member has collapsed. The optimum situation may be that the bracket and the main strength member to which the bracket is fitted collapse simultaneously. A method is proposed to determine the thickness of a bracket on the basis of this idea [4,5].

Here, a bracket of an arbitrary shape is considered; see Fig. 10.3. This bracket is attached to both ends of a beam of which length is denoted as *L*. The optimum thickness of the bracket is determined

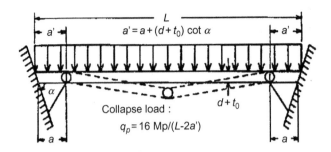

FIG. 10.3

Beam with corner brackets and its plastic collapse mechanism [5].

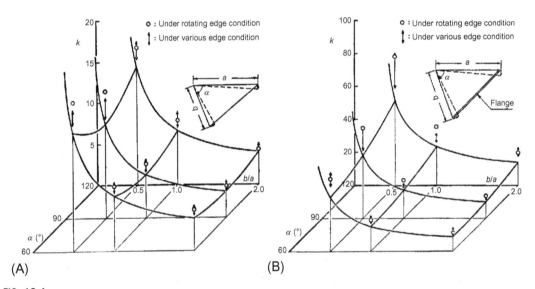

FIG. 10.4

Elastic buckling strength of triangular corner bracket [5]. (A) No edge stiffener. (B) With edge stiffener.

from the condition that the bracket collapses when the beam collapses forming three plastic hinges as indicated in Fig. 10.3.

As for the collapse of the bracket, buckling is considered beyond which in-plane stiffness decreases and loses its function (Fig. 10.4).

The elastic buckling strength of the triangular corner bracket in Fig. 10.3 is expressed as follows:

$$\sigma_{cr}^e = \frac{\pi^2 kE}{12(1 - v^2)} \left(\frac{t}{a}\right)^2 \qquad (10.1)$$

The buckling coefficient, k, in Eq. (10.1) can be expressed as [5]

$$k = \begin{cases} \left(\frac{3}{b/a - 0.2} + 1.5 \right)(1 + \cos^2 \alpha) & \text{no edge stiffener} \\ \left(\frac{10}{b/a - 0.2} + 4.5 \right)\{1.87 - \cos(\alpha - 60°)\} & \text{with edge stiffener} \end{cases} \tag{10.2}$$

k varies with respect to b/a and α as indicated in Fig. 10.4. Here, stress in Eq. (10.1) is average normal stress in the cross-section which makes 45 degrees with a beam. The application limits of Eq. (10.2) are $1/2 \leq a/b \leq 2/1$ and $60 \leq \alpha \leq 120$ degrees.

The buckling strength with edge stiffener is for the case that the height of a stiffener along the free edge is more than [5]:

$$\frac{b_f}{t} = 1.6 \frac{a}{t} \sqrt{\frac{\sigma_Y}{E}} + 4.0 \tag{10.3}$$

The formulas to obtain the optimum thickness are given later in Section 10.6.

10.3 WATERTIGHT TRANSVERSE BULKHEAD OF BULK CARRIER
10.3.1 CASUALTY OF BULK CARRIERS

Watertight transverse bulkhead is provided to keep the shape of cross-section of the hull girder and to ensure the safety of the ship in case of flooding into a hold. The watertight transverse bulkhead has to be strong enough against water pressure when flooding occurs.

In fact, serious failures occurred in plenty of bulk carriers during the late 1980s through early 1990s. Many of them sank and were lost with more than 300 fatalities. Cause of most failures was considered as the breaking of watertight transverse bulkhead due to flooded sea water into the cargo hold, and the sea water spread into other cargo holds. To prevent such casualties, there was a serious discussion in the International Maritime Organization (IMO). Finally, IMO requested International Association of Classification Societies (IACS) to provide new guidelines to strengthen the watertight transverse bulkhead to avoid the occurrence of tragic accidents.

In the present section, research works will be introduced regarding buckling/plastic collapse behavior of the watertight transverse bulkhead of bulk carriers. A simple method shall also be introduced to evaluate the collapse strength of watertight transverse bulkhead against flooded water pressure.

10.3.2 BUCKLING/PLASTIC COLLAPSE BEHAVIOR AND STRENGTH OF WATERTIGHT TRANSVERSE BULKHEAD OF BULK CARRIERS AGAINST FLOODED WATER PRESSURE

Watertight transverse bulkhead of bulk carriers is in general made of corrugated plate. The pressure from flooded water acts on individual plates of the corrugated structure in the perpendicular direction to them. Under such pressure load, the bulkhead is bent in overall mode, and local buckling may take place in the corrugate plate located in the compression side of overall bending.

In Fukuda and Yao [6], a series of nonlinear FEM analysis was performed to investigate into buckling/plastic collapse behavior and the ultimate strength of watertight corrugated bulkhead. Firstly,

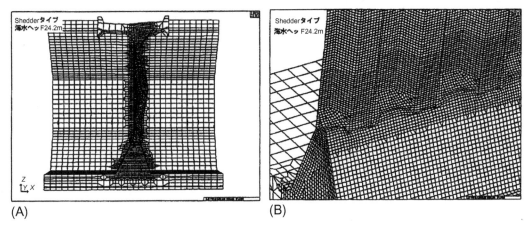

FIG. 10.5

Collapse mode at water head of 24.2 m above bottom of bulkhead (Ship R' with shedder plate) [7].
(A) Overall deflection of bulkhead. (B) Local buckling in shedder plate.

progressive collapse analysis was performed on 1/2 + 1/2 holds model performing explicit FEM analysis. Applied loads are distributed pressure on the bulkhead from the flooded aft side cargo hold, distributed pressure of cargo on inner bottom plating and sea water pressure distributed on outer skin of the hull girder. These loads were applied simulating the actual loading condition. Fig. 10.5 shows the collapsed watertight transverse bulkhead.

On the other hand, a series of alternative nonlinear implicit FEM analyses is also performed on a half-pitch model indicated in Fig. 10.6. After confirming the validity to use a half-pitch model [7], analysis was performed on 19 half-pitch models from existing bulk carriers. Results of analysis on 1/2 + 1/2 holds model and half-pitch model are compared in Fig. 10.7, where pressure-deflection relationships are plotted.

10.3.3 SIMPLE METHOD TO EVALUATE THE ULTIMATE STRENGTH OF CORRUGATED BULKHEAD AGAINST FLOODING PRESSURE

Fukuda and Yao [6] proposed a simple method to evaluate the ultimate strength of corrugated bulkhead of bulk carrier against flooding pressure load. In this method, a half pitch of the corrugate bulkhead illustrated in Fig. 10.6 is considered, and it is assumed that the lower end and the upper end are clamped and simply supported, respectively. Then, inserting two plastic hinges as indicated in Fig. 10.8, plastic collapse load is calculated. Proposed formulas were applied to 19 watertight transverse bulkheads indicated in Table 10.1. Calculated collapse loads given in the last two rows in Table 10.1, and are compared in Fig. 10.9 with the FEM results. Good correlations are observed between two results. The detail of the simple method is given later in Section 10.7.

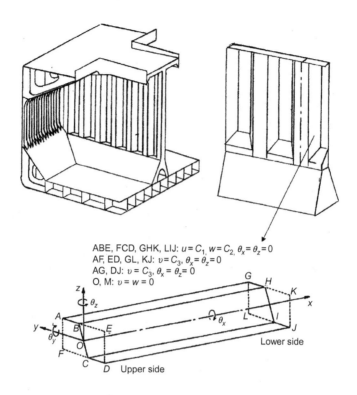

ABE, FCD, GHK, LIJ: $u = C_1$, $w = C_2$, $\theta_x = \theta_z = 0$
AF, ED, GL, KJ: $v = C_3$, $\theta_x = \theta_z = 0$
AG, DJ: $v = C_3$, $\theta_x = \theta_z = 0$
O, M: $v = w = 0$

FIG. 10.6

Half-pitch model of corrugated bulkhead [6,7].

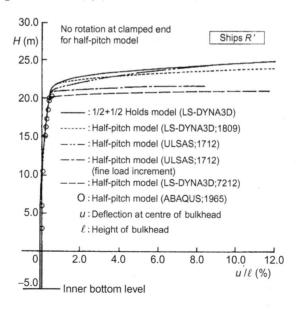

No rotation at clamped end
for half-pitch model

Ships R'

———— : 1/2+1/2 Holds model (LS-DYNA3D)
------- : Half-pitch model (LS-DYNA3D;1809)
—··— : Half-pitch model (ULSAS;1712)
——— : Half-pitch model (ULSAS;1712)
          (fine load increment)
——— : Half-pitch model (LS-DYNA3D;7212)
O : Half-pitch model (ABAQUS;1965)
u : Deflection at centre of bulkhead
ℓ : Height of bulkhead

u / ℓ (%)

Inner bottom level

FIG. 10.7

Influence of mesh size on collapse behavior (Ship R') [7].

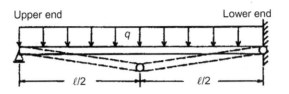

FIG. 10.8

Assumed plastic mechanism of half-pitch model.

Table 10.1 Dimensions of Corrugate Bulkheads and Collapse Loads

Ship	A	B	C	D	E	F	G	H	I	J
Size	Handy	Handy	Handy	Handy	P'max	P'max	P'max	P'max	P'max	P'max
a (m)	0.880	1.250	0.970	0.850	0.800	0.900	0.970	0.938	0.800	0.953
b (m)	0.863	1.098	1.146	0.877	0.881	0.943	1.044	1.020	0.935	1.000
d (m)	0.800	0.950	0.800	0.800	0.840	0.800	0.900	0.905	0.840	0.900
ℓ (m)	11.650	10.917	12.780	15.320	11.100	12.890	12.310	14.810	10.680	14.210
$t_{f\ell}$	12.5	20.0	16.5	22.0	14.5	12.5	15.5	15.5	14.5	16.5
$t_{w\ell}$	11.5	14.0	16.5	14.0	13.0	12.5	15.5	14.0	13.0	16.5
t_{fm}	–	16.0	–	15.0	–	–	–	12.5	–	13.5
t_{wm}	–	12.5	–	12.0	–	–	–	12.0	–	13.5
h_t	–	2.60	–	3.05	–	–	–	3.30	–	1.20
t_s	11.0	16.5	12.0	22.0	11.0	12.5	14.0	10.5	11.0	16.5
t_g	12.5	20.0	–	–	–	12.5	–	–	–	–
h_{gu}	1.40	0.45	–	–	–	1.20	–	–	–	–
σ_Y (MPa)	235	355	315	315	315	315	315	355	315	315
q_u (m) (Predected)	10.11	19.70	11.44	10.27	18.95	10.41	14.85	9.89	17.58	11.32
Q_u (m) (FEM)	10.97	18.08	12.21	10.34	19.05	9.89	15.80	9.10	18.66	10.89

Ship	K	L	M	N	O	P	Q	R	R'
Size	Cape	Cape	Cape	Cape	Cape	Cape	Cape	Cape	Cape
a (m)	1.150	1.100	1.200	0.960	1.270	1.020	1.160	1.160	1.160
b (m)	1.217	1.077	1.208	0.937	1.350	1.315	1.286	1.159	1.159
d (m)	1.100	1.000	1.100	0.850	1.170	1.100	1.055	1.000	1.000
ℓ (m)	15.000	15.000	13.200	16.303	16.120	18.810	15.310	14.542	14.542
$t_{f\ell}$	19.5	17.5	18.5	22.5	20.0	25.0	21.0	19.0	19.0
$t_{w\ell}$	17.0	17.5	18.5	15.5	20.0	18.5	21.0	17.0	17.0
t_{fm}	17.0	–	15.0	–	17.0	17.5	–	16.5	16.5
t_{wm}	15.5	–	15.0	–	17.0	14.0	–	16.0	16.0
h_t	3.50	–	6.50	–	5.95	2.00	–	3.50	3.50
t_s	12.0	14.5	16.0	15.0	13.0	16.0	10.0	12.5	12.5
t_g	19.5	–	–	–	20.0	25.0	–	19.0	–
h_{gu}	0.40	–	–	–	0.30	0.50	–	0.40	–
σ_Y (MPa)	355	315	315	315	315	315	235	355	315
q_u (m) (Predected)	14.91	13.73	16.74	13.22	12.09	9.30	11.15	12.90	14.00
Q_u (m) (FEM)	14.50	11.30	16.60	13.65	11.54	9.40	11.15	12.90	14.00

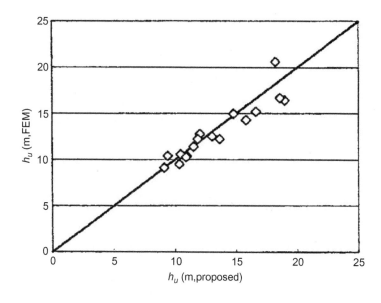

FIG. 10.9

Comparison between calculated and predicted collapse loads [2].

10.4 DOUBLE BOTTOM OF BULK CARRIER
10.4.1 DOUBLE BOTTOM STRUCTURE IN BULK CARRIER

The double bottom structure consists of bottom plating, inner bottom plating, girders and floors as well as longitudinal stiffeners on bottom and inner bottom plating. Girders and floors have many opening for piping and passing. On the double bottom structure, direct pressure loads act such as cargo weights and sea water pressure. In addition to the pressure loads, tensile or compressive loads act produced by longitudinal bending. In the case of ore carriers, large shear force also acts under alternate heavy loading condition. Furthermore, in the case of large vessel, transverse compressive load acts due to sea water pressure on side shell plating.

Consequently, structural members in double bottom is subjected to complicated combined loads. Because of this, collapse behavior of double bottom structure is also complicated. The followings have to be considered as for the buckling/plastic collapse behavior of double bottom structures [8–11].

(1) Buckling/plastic collapse behavior of local panel in bottom and inner bottom structures subjected to combined bi-axial thrust and lateral pressure.
(2) Buckling/plastic collapse behavior of stiffened panel in bottom and inner bottom structures subjected to combined bi-axial thrust and lateral pressure.
(3) Buckling/plastic collapse behavior of girders and floors subjected to combined shear and bending as well as in-plane compression/tension.

(4) Influence of opening on buckling/plastic collapse behavior of girders and floors subjected to combined shear and bending as well as in-plane compression/tension.
(5) Influence of vertical and horizontal stiffeners on buckling/plastic collapse behavior of girders and floors subjected to combined shear and bending as well as in-plane compression/tension.
(6) Buckling/plastic collapse behavior of double bottom structure subjected to lateral pressure load.
(7) Buckling/plastic collapse behavior of double bottom structure subjected to combined uni-axial thrust and lateral pressure load.
(8) Buckling/plastic collapse behavior of double bottom structure subjected to combined bi-axial thrust and lateral pressure load.

In the above items, (1) through (5) are already explained in Chapters 4 through 8 [8–10].

10.4.2 BUCKLING/PLASTIC COLLAPSE BEHAVIOR AND STRENGTH OF DOUBLE BOTTOM STRUCTURE

The behavior of double bottom structure can be simulated by analytically modeling the double bottom as an orthotropic plate [8]. However, it is only the elastic behavior that an orthotropic plate model can simulate. Contrary to this, it is possible to obtain plastic collapse load by applying *Rigid Plastic Mechanism Analysis*. This method can give the collapse load, but cannot simulate progressive collapse behavior including local buckling of structural members.

On the other hand, applying FEM or idealized structural unit method (ISUM), progressive collapse behavior can be simulated and the ultimate strength can be evaluated. ISUM is in the framework of FEM, but shorter computing time is attained by using larger but sophisticated elements compared to ordinary FEM. Therefore, new elements have to be developed to analyze new problem. Ishibashi et al. [9] and Ishibashi [10] developed a new ISUM element representing girders or floors in double bottom structures.

Here, the results of an analysis applying nonlinear FEM and ISUM are introduced on double bottom structures of Panamax size as well as Cape size bulk carriers [9–11]. In existing ships, a center girder is provided. However, to reduce the computation time introducing symmetry condition, double bottom structure without a center girder is considered in the analysis. Fig. 10.10A and B shows FEM quarter models of double bottom of Panamax size and Cape size bulk carriers, respectively. No opening is provided in this model in girders and floors.

The model is, as mentioned above and indicated in Fig. 10.10A and B, a quarter of the whole double bottom in one cargo hold. Symmetry condition is imposed along two center planes. On the other hand, the end plane connecting to lower stool is considered to be simply supported. On the end plane connecting to bilge hopper tank, springs are inserted to constrain the rotation of this end plane assuming simply; see Fig. 10.11. Spring constant can be determined considering the size of bilge hopper tank. These two end planes are supported at their mid-depth lines and can move freely in the horizontal direction perpendicular to the end planes with uniform displacement.

Here, collapse behavior of Panamax size bulk carrier is briefly explained. Three conditions are considered in the ISUM analysis as follows:

Case 1: All the panel can buckle.
Case 2: All the panel cannot buckle.

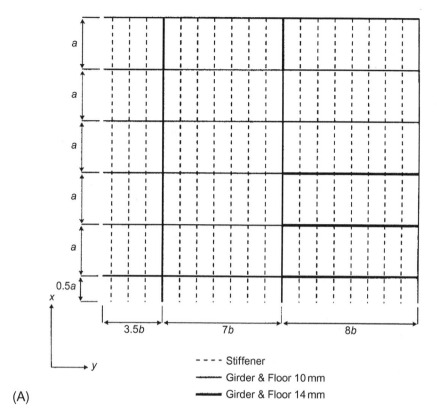

(A)

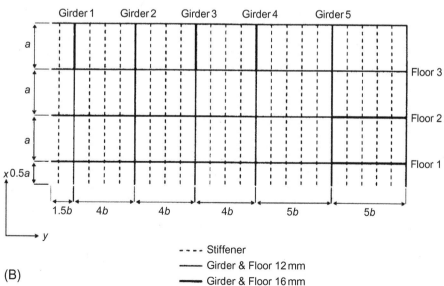

(B)

FIG. 10.10

Arrangement of girders and floors of double bottom structure [10]. (A) Panamax size bulk carrier.
(B) Cape size bulk carrier.

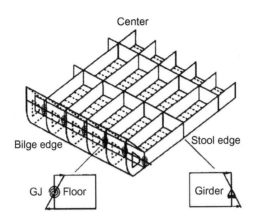

FIG. 10.11

Boundary conditions imposed on quarter model of double bottom [10].

Case 3: Bottom and inner bottom plating can buckle but webs of girders and floors cannot buckle.

Both nonlinear FEM and ISUM analyses were performed in load control scheme, and the Newton-Raphson method was applied for the convergence of solution. In the FEM analysis, only Case 2 was analyzed. As this was a load control analysis, analysis was up to the maximum pressure, at which the stiffness becomes low enough. Fig. 10.12 shows calculated pressure-central deflection relationships.

As for the results of ISUM analyses, Case 2 shows different result from Case 1. This is because buckling of bottom plating, which is simulated by Case 1 analysis at water head of 5 m, is not simulated in Case 2 analysis. On the other hand, the result of Case 3 starts to differ from that of Case 1 when water head exceeds 13 m. At this water head, webs of girders and floors undergo buckling in Case 1 analysis, whereas it is not simulated by Case 3 analysis. The result of Case 1 analysis shows good correlation with FEM result in which all the structural members can buckle.

Although the buckling behavior is different in three cases, the ultimate strength is almost the same. This is because, the ultimate strength of the double bottom structure is attained when the end cross-sections of girders and floors collapse by shear yielding.

Fig. 10.13A and B shows how the average shear stress varies with the increase of the applied pressure load in Case 2. Both results of nonlinear FEM and ISUM analyses are indicated, and good correlations are observed between the two results. At water head of 12–12.5 m, average shear stress at the stool end cross-section of Girder 1 reaches nearly to the shear yield stress, and becomes fully plastic. Above this water head, end cross-section of Girder 1 sustains no more shear force. Because of this, shear force which Girder 2 has to sustain increases hereafter as indicated in Fig. 10.13A. The end cross-section of Girder 2 also becomes fully plastic when water head reaches 15.5 m.

On the other hand, floors also show the similar behavior. Above the water head of 15.5 m that all the girders have collapsed, end cross-sections of Floors 3 and 5 sustain more shear force. At the water head of 18 m, end cross-sections of all the girders and floors become fully plastic by shear and the double bottoms structure can sustain no more pressure load.

Fig. 10.13 indicates that double bottom structure collapses when end cross-sections of all the girders and floors have fully yielded by shear.

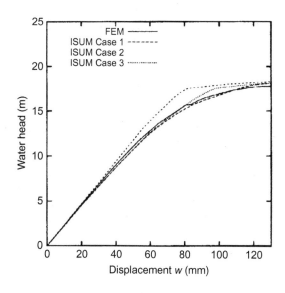

FIG. 10.12

Pressure-lateral deflection relationships (Model P1) [10].

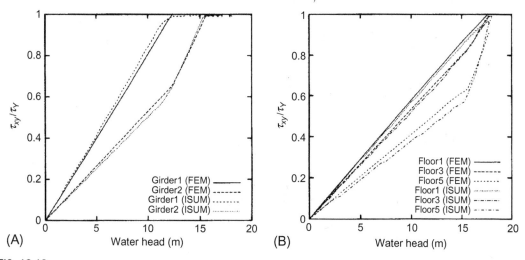

FIG. 10.13

Shear stress-pressure relationships in girder and floor end cross-sections (Model P1) [10]. (A) Stress in girder end cross-sections. (B) Stresses in floor end cross-sections.

10.4.3 SUMMARY OF FINDINGS REGARDING BUCKLING/PLASTIC COLLAPSE OF DOUBLE BOTTOM STRUCTURES

The findings regarding buckling/plastic collapse of double bottom structures are as follows:

(1) In the case of Cape size bulk carrier, local buckling takes place in the bottom plating, which is located in the compression side of overall bending, by transverse thrust, and buckling deflection develops.

(2) With further increase of water pressure, end cross-section of the center girder collapses by shear yielding, and then the other girders also collapses by shear yielding.

(3) After all the girders attained fully plastic condition by shear in their end cross-sections, floors sustain the further pressure load, and their end cross-sections also become fully plastic one by one.

(4) After all the end cross-sections of all the girders and floors become fully plastic by shear, the double bottom attain the ultimate strength.

(5) If the web thickness of girders and floors are doubled, end cross-sections of floors collapse by buckling in bending after the end cross-sections of all the girders have collapsed by shear yielding.

(6) In the case of Panamax bulk carrier, floors are the main strength members of the double bottom structure. On the other hand, in Cape size bulk carrier, girders are the main strength members. However, in both cases, the double bottom shows fundamentally the same collapse behavior.

(7) If the occurrence of local buckling in web of girders and floors are neglected, collapse load of the double bottom structure may be overestimated.

(8) Although the occurrence of local buckling in a clamped mode at bottom plating is not considered in the ISUM used here, results of analysis showed good correlations with the FEM results including collapse load.

(9) In the ISUM used here, shear buckling of girders and floors as well as compressive buckling of floors are not considered. When these buckling cannot be ignored, this ISUM may overestimate the collapse load.

10.5 HATCH COVER OF BULK CARRIERS
10.5.1 REGULATION ON HATCH COVER OF BULK CARRIERS

The British Cape size bulk carrier, *Derbyshire*, sank and lost in Typhoon near Okinawa in August 1880, and 44 crew died. After the casualty, Investigation Committee was launched in UK and the cause of casualty was inquired. In 1994, the committee succeeded in finding the fore part of *Derbyshire* laying on a sea bottom 4200 m below sea level, and photographs were taken. Based on the observation, the committee concluded that the hatch cover of No. 1 cargo hold was damaged by the blue water on bow and sea water submerged into the cargo hold and the *Derbyshire* sank and was lost [12].

The hatch cover had been designed in accordance with the rule specified in International Convention on Load Lines (ICLL) [13], at that time. However, on the basis of conclusion of the Investigation Committee, IACS discussed on reinforcement of hatch cover of bulk carrier, and new Unified Rules, UR-S21, came into effect in 1997 to strengthen the hatch cover of bulk carriers [14].

Research works to inquire the cause of the *Derbyshier*'s casualties had been continued after that, and in 2000, a tank test was conducted to estimate the blue water pressure load on a ship's bow using scale model of bulk carriers in UK [15].

In this section, results of nonlinear FEM analysis are introduced to investigate into the buckling/ plastic collapse behavior and the ultimate strength of hatch cover of bulk carriers under lateral pressure loads. Then, the simple method is also introduced to evaluate the collapse load of hatch cover against blue water.

10.5.2 BUCKLING/PLASTIC COLLAPSE BEHAVIOR OF HATCH COVER

In Refs. [16–18], two types of hatch covers were considered: folding type and side sliding type. As for the folding type, that of Handy size bulk carrier was considered, and as for side sliding type, those of Panamax size and Cape size were considered. For each, two hatch covers were considered which are designed by ICLL rule [13] and UR-S21 [14], respectively. In addition to this, two cases are analyzed in each hatch cover, that is, with and without corrosion margin. Thus, a nonlinear FEM analysis was performed on altogether 12 hatch covers.

The hatch cover of a folding type consists of four box girders as indicated in Fig. 10.14A. Covers A and B as well as B and C are connected by hinges. Covers A and D are supported along three sides and the remaining side is free. On the other hand, B and C are supported along two shorter sides. Because of the difference in supporting condition, deflection of A and B as well as C and D along their junction lines are different. According to the results of analysis modeling four box girders together, high strain was produced in the hinges connecting box girders [19]. However, it was also reported that the collapse behavior is dominated by the collapse of box girders, B and C. So, in the analysis of hatch cover of a folding type, only B or C was analyzed.

On the other hand, in the case of hatch cover of a side sliding type, the hatch cover consists of two covers as indicated in Fig. 10.14B. In this case, each cover is considered to be supported along three sides and the remaining side is free. In all cases, pressure load acting on the top plate is always perpendicular to the plate surface.

Fig. 10.15 shows the collapse mode of hatch cover of Handy size bulk carrier without corrosion margin. This hatch cover is designed in accordance with UR-S21. It is observed that overall buckling as a stiffened pate occurred in the top plate which is in the compression side of overall bending.

The findings obtained through nonlinear FEM analysis on 12 hatch covers can be summarized as follows:

(1) In the top panel in the compression side of overall bending, local panel buckling or overall buckling as a stiffened plate takes place. Because of this, so-called plastic collapse load obtained by *Rigid Plastic Mechanism Analysis* is not attained.

(2) Except the hatch covers of Handy size and Cape size bulk carriers designed by UR-S21 specified by IACS, overall buckling as a stiffened plate takes place in the top plate after local panel buckling has occurred. The overall buckling becomes a trigger of the collapse of a whole hatch cover.

(3) In the case of the hatch covers of Handy size and Cape size bulk carriers designed by UR-S21 specified by IACS, plastic mechanism is formed in the middle of the top plate. This becomes a trigger of the collapse of a whole hatch cover.

(4) The hatch covers of Panamax size and Cape size bulk carriers designed by UR-S21 specified by IACS have enough strength above design load even in the condition that corrosion margin is zero.

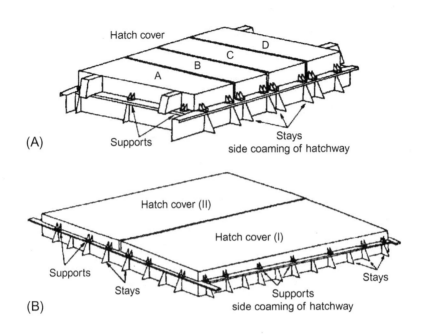

FIG. 10.14

Representative hatch covers [16,17]. (A) Folding type. (B) Side-sliding type.

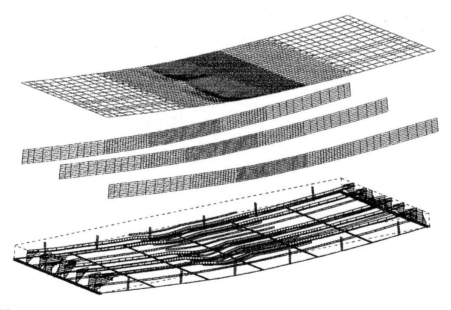

FIG. 10.15

Collapse mode of hatch cover (Handy size; IACS; with corrosion margin) [17].

10.5.3 **SIMPLE METHOD TO EVALUATE COLLAPSE STRENGTH OF HATCH COVER**

A simple method was proposed in Refs. [16–18] to evaluate the collapse strength of hatch cover of bulk carriers on the basis of the results of nonlinear FEM analyses. In this method, the hatch cover of a folding type is modeled as a both end simply supported beam. As for the hatch cover of a side sliding type, the hatch cover is modeled as an inisotropic orthogonal plate. Elastic pressure-average stress relationships are derived for the beam and the orthogonal inisotropic plate models, where average stress is the stress in top plate of the hatch cover.

On the other hand, plastic pressure-average stress relationship is derived as the ultimate strength interaction relationship for the top plate of a hatch cover modeled as a stiffened plate. Collapse load is obtained as the pressure at the intersection point of elastic and plastic curves. The influence of local panel buckling is considered introducing effective width in the postbuckling range. Detail of the formulas are given in Yao et al. [18].

The collapse loads obtained by the simple method and nonlinear FEM analysis are compared in Fig. 10.16. It is known that the proposed simple method gives relatively accurate collapse loads.

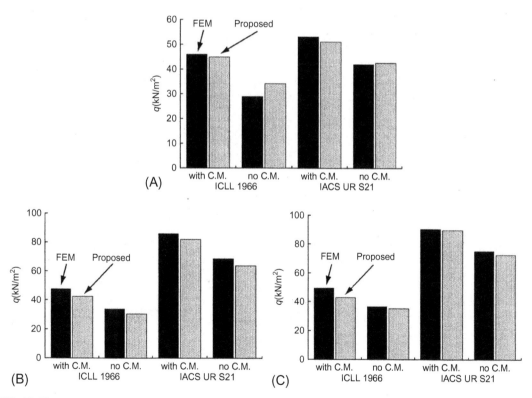

FIG. 10.16

Comparison of calculated and predicted collapse loads [18]. (A) Handy size. (B) Panamax. (C) Cape size.

EXERCISES

10.1   Describe the fundamental idea to determine the optimum thickness of a corner bracket.

10.2   Evaluate the collapse pressure load of watertight corrugated transverse bulkhead of bulk carrier, A, B, C, D, and E in Table 10.1 applying the simple method explained in Section 10.7, and compared the calculated results with those in Table 10.1.

10.3   Explain the collapse behavior of double bottom structure.

10.4   Explain the fundamental idea to determine the collapse pressure of hatch cover of a folding type.

10.5   Explain the fundamental idea to determine the collapse pressure of hatch cover of a side sliding type.

10.6 APPENDIX: OPTIMUM THICKNESS OF TRIANGULAR CORNER BRACKET

10.6.1 FUNDAMENTAL IDEA TO DETERMINE OPTIMUM THICKNESS OF CORNER BRACKET

The optimum thickness of the corner bracket is determined from the condition that the bracket and the main strength member to which the bracket is fitted collapse simultaneously. The plastic collapse load of the beam with corner brackets indicated in Fig. 10.3 and the shear force at the bracket end are expressed as

$$q_p = \frac{16M_P}{(L - 2a')^2} \tag{10.4}$$

$$V_P = \frac{8M_P}{(L - 2a')} \tag{10.5}$$

The normal and shear forces distributions are assumed as

$$N_y = \lambda_1 \cos\frac{\pi x}{h_2} + \lambda_2 \tag{10.6}$$

$$N_{xy} = \lambda_3 \sin\frac{\pi x}{h_2} - \lambda_2 \frac{1}{h_2}(x - h_2) \tag{10.7}$$

The distributions expressed by Eqs. (10.6), (10.7) are indicated in Fig. 10.17. The average normal stress in the bracket is expressed as

$$\sigma_m = \frac{1}{t(h - h_1)} \int_{h_1}^{h} N_y dx = \frac{\beta}{t} \tag{10.8}$$

where

$$N_y = \frac{h_2}{h_1 - h}\left(\sin\frac{\pi h}{h_2} - \sin\frac{\pi h_1}{h_2}\right)\lambda_1 + \lambda_2 \tag{10.9}$$

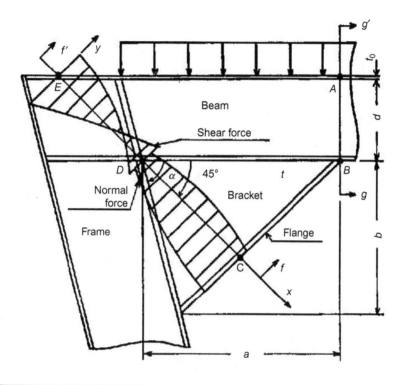

FIG. 10.17

Assumed distributions of normal and shear forces at cross-section of beam and bracket [5].

Considering the equilibrium conditions of forces and moments indicated in Fig. 10.18, following equations are derived:

$$\begin{cases} a_{11}\lambda_1 + a_{12}\lambda_2 + a_{13}\lambda_3 = B_1 \\ a_{11}\lambda_1 + a_{12}\lambda_2 + a_{13}\lambda_3 = B_1 \\ a_{11}\lambda_1 + a_{12}\lambda_2 + a_{13}\lambda_3 = B_1 \end{cases} \tag{10.10}$$

where

$$\begin{cases} a_{11} = m_1 + m_4 + \dfrac{1}{\sqrt{2}}(m_7 + m_9) \\ a_{12} = m_2 + m_5 + \dfrac{1}{\sqrt{2}}(m_8 + m_{10} - m_{11}) \\ a_{13} = m_3 + m_6 - \dfrac{1}{\sqrt{2}}m_{12} \\ a_{21} = \dfrac{1}{\sqrt{2}}(m_7 + m_9) \\ a_{22} = \dfrac{1}{\sqrt{2}}(m_8 + m_{10} + m_{11}) \\ a_{23} = \dfrac{1}{\sqrt{2}}m_{12} \end{cases}$$

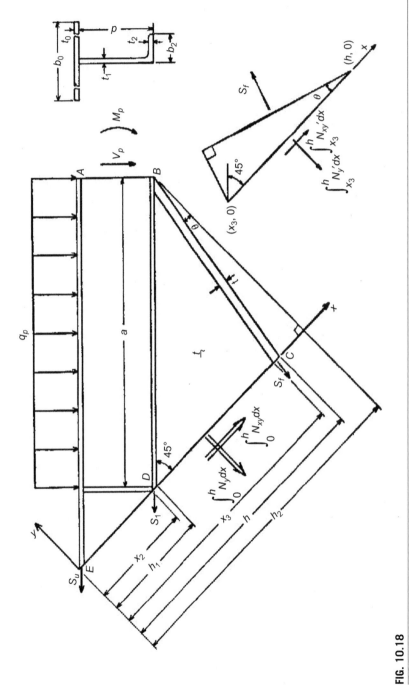

FIG. 10.18

Assumed distributions of normal and shear forces at cross-section of beam and bracket [5].

$$\begin{cases} a_{31} = \frac{1}{2}t_0 m_1 + \left(\frac{h_1}{\sqrt{2}} - \frac{t_2}{2}\right) m_4 + m_7 d_0 + m_{13} \\ a_{32} = \frac{1}{2}t_0 m_2 + \left(\frac{h_1}{\sqrt{2}} - \frac{t_2}{2}\right) m_5 + m_8 d_0 + m_{14} \\ a_{33} = \frac{1}{2}t_0 m_3 + \left(\frac{h_1}{\sqrt{2}} - \frac{t_2}{2}\right) m_6 \end{cases}$$

$$\begin{cases} b_1 = 0 \\ b_2 = -a' q_p - V_P - v_2 M_P \\ b_3 = -\left(t_0 + d + a - \frac{1}{2}a'\right) a' q_p - (t_0 + d + a)V_P - M_P - v_3 M_P \end{cases}$$

$$\begin{cases} m_1 = \frac{b_0 h_2}{\sqrt{2}\pi} \sin \frac{\sqrt{2}\pi t_0}{h_2} \\[2mm] m_2 = \frac{b_0 t_0^2}{\sqrt{2} h_2} \\[2mm] m_3 = -\frac{b_0 h_2}{\sqrt{2}\pi}\left(1 - \cos \frac{\sqrt{2}\pi t_0}{h_2}\right) \\[2mm] m_4 = \frac{b_2 h_2}{\sqrt{2}\pi}\left(\sin \frac{\pi h_1}{h_2} - \sin \frac{\pi x_2}{h_2}\right) \\[2mm] m_5 = \frac{b_2}{2\sqrt{2} h_2}(h_1^2 - x_2^2) \\[2mm] m_6 = -\frac{b_2 h_2}{\sqrt{2}\pi}\left(\cos \frac{\pi x_2}{h_2} - \cos \frac{\pi h_1}{h_2}\right) \\[2mm] m_7 = \frac{b_f h_2}{\pi}\left(\sin \frac{\pi h}{h_2} - \sin \frac{\pi x_3}{h_2}\right) \\[2mm] m_8 = b_f(h - x_3) \\[2mm] m_9 = \frac{h_2 t_3}{\pi} \sin \frac{\pi h_1}{h_2} + \frac{h_2 t}{\pi}\left(\sin \frac{\pi h}{h_2} - \sin \frac{\pi h_1}{h_2}\right) \\[2mm] m_{10} = t_3 h_1 + t(h - h_1) \\[2mm] m_{11} = -\frac{t_3}{h_2}\left(\frac{1}{2}h_1^2 - h_1 h_2\right) - \frac{t}{h_2}\left\{\frac{1}{2}(h^2 - h_1^2) - h_2(h - h_1)\right\} \\[2mm] m_{12} = -\frac{t_3 h_2}{\pi}\left(1 - \cos \frac{\pi h_1}{h_2}\right) + \frac{t h_2}{\pi}\left(\cos \frac{\pi h_1}{h_2} - \cos \frac{\pi h}{h_2}\right) \\[2mm] m_{13} = -\frac{t_3 h_2}{\pi}\left\{h_1 \sin \frac{\pi h_1}{h_2} - \frac{h_2}{\pi}\left(1 - \cos \frac{\pi h_1}{h_2}\right)\right\} \\[2mm] \qquad + \frac{t h_2}{\pi}\left\{\left(h \sin \frac{\pi h}{h_2} - h_1 \sin \frac{\pi h_1}{h_2}\right) - \frac{h_2}{\pi}\left(\cos \frac{\pi h_1}{h_2} - \cos \frac{\pi h}{h_2}\right)\right\} \\[2mm] m_{14} = \frac{1}{2}t_3 h_1^2 + \frac{1}{2}t(h^2 - h_1^2) \end{cases}$$

Solving Eq. (10.10), the following are obtained:

$$\lambda_1 = \frac{c_{22}d_1 - c_{12}d_2}{c_{11}c_{22} - c_{12}c_{21}}, \quad \lambda_2 = \frac{-c_{21}d_1 + c_{11}d_2}{c_{11}c_{22} - c_{12}c_{21}}, \quad \lambda_3 = -\frac{a_{11}}{a_{13}}\lambda_1 - \frac{c_{12}}{c_{13}}\lambda_2 \qquad (10.11)$$

where

$$\begin{cases} c_{11} = a_{21}a_{13} - a_{11}a_{23} \\ c_{12} = a_{22}a_{13} - a_{12}a_{23} \\ c_{21} = a_{31}a_{13} - a_{11}a_{33} \\ c_{22} = a_{32}a_{13} - a_{12}a_{33} \end{cases}$$

$$\begin{cases} d_1 = -a_{23}b_1 + a_{13}b_2 \\ d_2 = -a_{33}b_1 + a_{13}b_3 \end{cases}$$

The optimum thickness is obtained equating σ_m and the buckling strength, and is expressed as follows [5]:

$$\sigma_m = \sigma_{cr}^e \quad \left(\sigma_{cr}^e < \frac{1}{2}\sigma_Y\right) \tag{10.12}$$

$$\sigma_m = \sigma_u \quad \left(\frac{1}{2}\sigma_Y \leq \sigma_{cr}^e\right) \tag{10.13}$$

σ_{cr}^e is given by Eq. (10.1) and σ_u as follows:

$$\sigma_u = \sigma_Y(1 - (a/\pi t))\sqrt{3(1 - v^2)\sigma_Y/2kE} \tag{10.14}$$

10.7 SIMPLE METHOD TO EVALUATE COLLAPSE LOAD OF CORRUGATED BULKHEAD SUBJECTED TO LATERAL PRESSURE

10.7.1 RIGID PLASTIC MECHANISM ANALYSIS

A half-pitch model of a corrugated bulkhead is modeled as a beam of which one end is clamped and the other end simply supported. Parameters of the model are given in Fig. 10.19. Uniformly distributed load is considered instead of triangular or trapezoidal distribution for simplicity. Under this loading and boundary conditions, plastic hinges are formed at the clamped end and the point which is 0.586ℓ apart from the clamped end, where ℓ is a span length. However, also for simplicity, plastic hinges are assumed to be formed at the clamped end and the mid-span point as indicated in Fig. 10.8 in Section 10.3. This change causes only 1.2% difference in the collapse load.

For the model shown in Fig. 10.8, the beam collapses when the lateral pressure per unit area reaches

$$q_u = \frac{8}{C \cdot s_\ell \cdot \ell^2}(0.5M_{\ell e} + M_n) \tag{10.15}$$

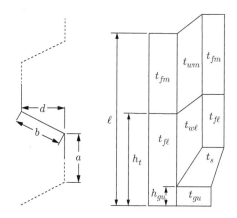

FIG. 10.19

Parameters representing geometry of corrugated bulkhead.

where $M_{\ell e}$ and M_m are fully plastic bending moments at the clamped and the mid-height point, respectively, and are expressed as

$$M_{\ell e} = Z_{\ell e} \cdot \sigma_{Yf}, \quad M_m = Z_m \cdot \sigma_{Yf} \tag{10.16}$$

In the above expression, σ_{Yf} is the yield stress of the flange plate. $Z_{\ell e}$ and Z_m are the plastic section moduli at the clamped end and the mid-height point.

C in Eq. (10.15) is 1.0 for the exact solution of the *Rigid Plastic Mechanism Analysis*, but is taken as 1.1 to get better agreement with the collapse load obtained by the FEM analyses.

Here, the influential factors which may affect the plastic collapse load are considered as follows:

(1) local buckling in compression flange;
(2) effectiveness of web plating at the clamped end;
(3) shedder plate and gusset plate; and
(4) shear force at the clamped end.

In the following, it is described how these influences are taken into account in the plastic section modulus.

10.7.2 INFLUENCE OF LOCAL BUCKLING IN COMPRESSION FLANGE

To account for the influence of local buckling in the compression flange, the full width of the compression flange, a, is replaced by the effective width, a_{ef}, which is given by

$$a_{ef} = C_e \cdot a \tag{10.17}$$

where

$$C_e = \begin{cases} 2.25/\beta - 1.25/\beta^2 & (1.25 < \beta) \\ 1.0 & (\beta \le 1.25) \end{cases} \tag{10.18}$$

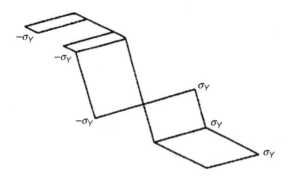

FIG. 10.20

Fully plastic stress distribution considering local buckling.

$\beta = a/t_f \cdot \sqrt{\sigma_{Yf}/E}$ is the slenderness ratio of the flange plating. The fully plastic stress distribution with buckled flange is illustrated in Fig. 10.20. The plastic section modulus for this stress distribution can be expressed as follows:

$$Z_P = \frac{d}{2b}(2e_p^2 - 2e_p b + b^2)t_w \gamma + \frac{1}{2}\{b - (1 - C)e_p\}at_f \tag{10.19}$$

where

$$e_p = \frac{1}{2}b + \frac{1}{4}(1 - C_e)a\frac{t_f}{t_w}\frac{1}{\gamma} \tag{10.20}$$

and $\gamma = \sigma_{Yw}/\sigma_{Yf}$ is the ratio of yielding stress of the web plate to that of the flange plate. The first and the second terms in the right-hand side of Eq. (10.19) are the contributions of the web and the flange plates, respectively.

As for the section modulus, Z_m, at mid-height, Z_P expressed by Eq. (10.19) can be used.

10.7.3 EFFECTIVENESS OF WEB PLATING AT CLAMPED END

The flange plate of the corrugated bulkhead is welded to the lower stool, and the axial force in the flange is fully transmitted to the sloped wall of the lower stool. On the other hand, the web plate is welded to the top plate of the lower stool. In general, however, there exists no continued structural member on the backside of the top plate of the lower stool to which the axial force in the web plate can be transmitted.

Because of this, axial stress in the web plate is much reduced at the lower end. Judging from the results of the FEM analyses, the effectiveness of the web plate at the clamped end is regarded as 30%. On this basis, the thickness of the web plate is reduced to 30% of its original thickness when the plastic section modulus at the lower end of the model is calculated. Thus, modifying the first term of Eq. (10.19), the reduced section modulus at the lower end can be expressed as

$$Z'_P = \frac{3d}{20b}(2e_p^2 - 2e_p b + b^2)t_w \gamma + \frac{1}{2}\{b - (1 - C)e_p\}at_f \tag{10.21}$$

where

$$e'_p = \frac{1}{2}b + \frac{5}{6}(1 - C_e)a\frac{t_f}{t_w}\frac{1}{\gamma}$$ (10.22)

10.7.4 INFLUENCES OF SHEDDER PLATE AND GUSSET PLATE

Shedder plate and gusset plate are often provided at the lower end of the corrugated bulkhead. To examine the influences of shedder and gusset plates, a series of FEM analyses has been performed on bulkheads of Ships *J*, *K*, and *M* changing the height of the gusset plate. It was found that the collapse load increases owing to the shedder plate. This may be because some part of the axial force in the web plate is transmitted into the shedder plate.

When a gusset plate is provided, the collapse load firstly decreases a little with an increase in the height of the gusset plate. However, it turns to increase as the height of the gusset plate further increases. Then, the collapse load saturates to a certain level when the gusset height reaches to 1.0 through 1.5 times the flange width. This may be because the transmitted force from the web into the shedder plate and the gusset plate saturates as the gusset height reaches a certain value.

To examine the effects of gusset and shedder plates, the average axial stresses in the flange and the gusset plates at the lower ends are compared in Fig. 10.21. The ordinates represent the ratio of the average stress in gusset plate to that in flange plate. The solid line in Fig. 10.21 is the approximation of

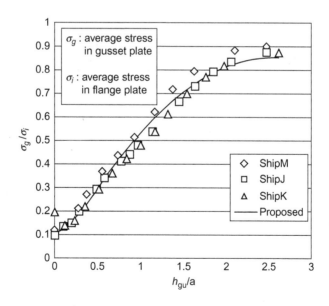

FIG. 10.21

Change in stress ratio in gusset and flange plates [2].

the stress ratio, which is expressed as follows:

$$
f = \begin{cases} \frac{77}{192}\left(\frac{h_{gu}}{a}\right)^2 + 0.13 & (h_{gu}/a < 0.4) \\ -\frac{7}{48}\left(\frac{h_{gu}}{a} - 2.6\right)^2 + 0.9 & (0.4 \le h_{gu}/a < 2.6) \\ 0.9 & (2.6 \le h_{gu}/a) \end{cases}
\tag{10.23}
$$

Using the stress ratio expressed by Eq. (10.23), the plastic section modulus when shedder and gusset plates are provided can be expressed as

$$
Z_P'' = \frac{3d}{20b}(2e_p^2 - 2e_p b + b^2)t_w\gamma + \frac{1}{2}\{b - (1 - C)e_p\}at_f + \frac{1}{2}fbC_e a' t_{g1} d
\tag{10.24}
$$

where $a' = a + 2\sqrt{b^2 - d^2}$ is the width of the gusset plate, and t_{g1} is defined as follows:

$$
t_{g1} = \begin{cases} t_g & \text{(with shedder and gusset plates)} \\ \sqrt{2}t_s & \text{(with shedder plate only)} \end{cases}
\tag{10.25}
$$

10.7.5 INFLUENCE OF SHEAR FORCE ON FULLY PLASTIC STRENGTH AT CLAMPED END

If no shear force acts at the lower end of the bulkhead, Z_P'' calculated by Eq. (10.24) can be used as the plastic section modulus at the lower end. However, shear force is not zero at the lower end, and the influence of shear force on the plastic section modulus has to be considered for the accurate estimation of the plastic collapse load applying the half-pitch model.

Here, it is assumed that the fully plastic strength interaction relationship between shear force and bending moment can be approximated as

$$
\left(\frac{M}{M_P}\right)^2 + \left(\frac{V}{V_P}\right)^2 = 1
\tag{10.26}
$$

On the other hand, when a beam of which one end is clamped and another end is simply supported is subjected to uniformly distributed lateral load, shear force and bending moment at the clamped end can be expressed as follows:

$$
V = \frac{5}{8}q\ell, \quad M = \frac{1}{8}q\ell^2
\tag{10.27}
$$

where ℓ and q are the length of a beam and the magnitude of distributed lateral load. From Eq. (10.27), the following relationship is obtained:

$$
V = \frac{5}{\ell}M
\tag{10.28}
$$

Here, the fully plastic shear force and the fully plastic bending moment at the clamped end are expressed as

$$
V_p = A_w \cdot \sigma_{Yf}/\sqrt{3}, \quad M_P = Z_P'' \cdot \sigma_{Yf}
\tag{10.29}
$$

where $A_w = bt_w$ is the web area. Using V_P and M_P expressed by Eq. (10.29), Eq. (10.28) becomes as

$$\frac{V}{V_P} = \frac{5}{\eta} \frac{M}{M_P} \tag{10.30}$$

where

$$\eta = \ell V_P / M_P \tag{10.31}$$

Substituting Eq. (10.30) into Eq. (10.26), bending moment at the intersecting point of Eqs. (10.26), (10.30) is obtained in the following form:

$$\frac{M}{M_P} = \sqrt{\frac{1}{1 + 25/\eta^2}} = \alpha \tag{10.32}$$

Hence, the plastic section modulus at the clamped end becomes as

$$Z''' = \alpha Z_P'' \tag{10.33}$$

Consequently, setting $Z_{\ell e}$ and Z_m in Eq. (10.16) as

$$Z_{\ell e} = Z_P''', \quad Z_m = Z_P \tag{10.34}$$

and substituting Eq. (10.16) into Eq. (10.15), the collapse load is calculated as follows:

$$q_u = \frac{8}{C \cdot s_\ell \cdot \ell^2} (0.5 Z_P''' + Z_P) \sigma_{Yf} \tag{10.35}$$

REFERENCES

[1] Nishimaki K, Urata A, Hara Y. Collapsing strength of ordinary bracket. Hitachi Zosen Tech Rev 1964;25(1):1–6 [in Japanese].
[2] Ueda Y, Matsuishi M. Elastic-plastic analysis of steel structures using finite element method (3rd report)—buckling of triangular plate under various loading conditions. J Kansai Soc Naval Arch 1968;131:27–32 [in Japanese].
[3] Ueda Y, Kuramoto Y, Yao T. Effects of initial imperfection due to welding on rigidity and strength of triangular corner bracket. Trans JWRI 1977;6(1):39–45.
[4] Ueda Y, Yao T. A method to determine the necessary thickness of a corner bracket. J Soc Naval Arch Jpn 1982;152:286–296 [in Japanese].
[5] Ueda Y, Yao T. A method to determine the necessary thickness of a corner bracket (2nd report). J Soc Naval Arch Jpn 1983;154:356–366 [in Japanese].
[6] Fukuda N, Yao T. A simple method to estimate collapse strength of transverse bulkhead of bulk carrier subjected to flooding load. In: Proc. 17th technical exchange and advisory meeting on marine structure, Tainan, Taiwan; 2003. p. 545–554.
[7] Konishi H, Yao T, Shigemi T, Kitamura O, Fujikubo M. Design of corrugated bulkhead of bulk carrier against accidental flooding load. In: Proc. 7th int symp on PRADS practical design of ships and mobile units, The Hague, The Netherlands; 1998. p. 157–163.
[8] Terazawa K, editor. Strength of ship structures. Tokyo: Kaibundo; 1971. p. 243–255 [in Japanese].

[9] Ishibashi K, Fujikubo M, Yao T. Development of new ISUM element to simulate buckling/plastic collapse behaviour of ship's double bottom structures. Jpn Soc Naval Arch Ocean Eng 2007;5:217–225 [in Japanese].

[10] Ishibashi K. Study on collapse analysis of double bottom structure applying idealised structural unit method. Doctoral thesis, Osaka University, 2007 (Chapter 5, in Japanese).

[11] Ishibashi K, Fujikubo M, Yao T. Collapse analysis of ship's double bottom structures with ISUM. In: Proc 2nd int conf computational methods in marine engineering, Barcelona, Spain; 2007.

[12] Department of Environment, Transport and Regions. MV Derbyshire Surveys, UK/EC Assessors' Report; March 1998.

[13] IMO. International convention of load lines, annex I: determinations for load lines. Regulation 15 in Chapter 11; 1996.

[14] IACS. Evaluation of scantlings of hatch covers of bulk carrier cargo holds. IACS Unified Requirement 1997; vol. 1. 1997. p. S21.

[15] Vassalos D, Guarin L, Milne S. Results of the UK bulk carrier sea keeping model testing programme; April 2000.

[16] Yao T, Koiwa T, Hayashi S, Sato S. Plastic collapse strength of hatch cover of bulk carrier against green sea load (1st report)—hatch cover of folding type. J Soc Naval Arch Jpn 2000;187:517–524 [in Japanese].

[17] Yao T, Koiwa T, Hayashi S, Sato S. Plastic collapse strength of hatch cover of bulk carrier against green sea load (2nd report)—hatch cover of side sliding type. J Soc Naval Arch Jpn 2001;190:747–754 [in Japanese].

[18] Yao T, Magaino A, Koiwa T, Sato S. Collapse strength of hatch cover of bulk carrier subjected to lateral pressure load. Mar Struct 2003;16:687–709.

[19] Abdul Rahim M, Pereira M. Collapse analysis of bulk carrier structural parts. In: Proc 12th technical exchange and advisory meeting on marine structure (TEAM'98). Kanazawa, Japan; 1998. p. 11–20.

CHRONOLOGICAL TABLE OF STUDY ON BUCKLING/ULTIMATE STRENGTH

A

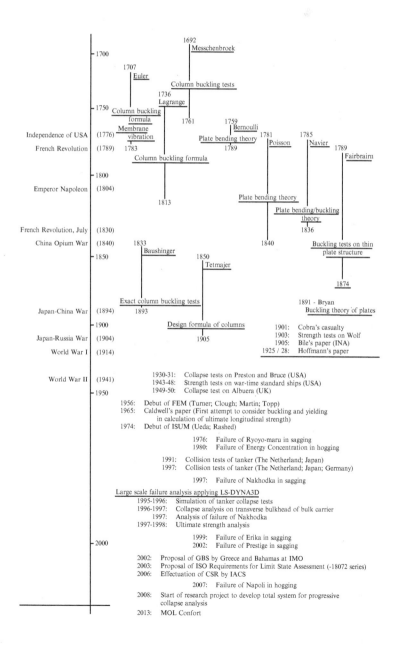

FUNDAMENTALS IN IDEALIZED STRUCTURAL UNIT METHOD (ISUM)

B.1 SHORT HISTORY OF ISUM DEVELOPMENT

The idealized structural unit method (ISUM) of the first generation was proposed by Ueda et al. [1] to analyze progressive collapse behavior of a transverse ring of a tanker. The first ISUM element was as a beam-column element with two nodal points representing a girder, and the influence of local buckling in web was considered. No lateral deflection was considered explicitly, and the reduction of the in-plane rigidity by buckling was considered by introducing effective width concept. The buckling and the ultimate strength interaction relationships were used in terms of sectional forces (axial force, shear force, and bending moment) and the ultimate strength interaction relationship was considered as a plastic potential when the stiffness matrix beyond the ultimate strength was derived. The detail of this ISUM element is explained also in Ref. [2].

On the basis of a similar concept, a rectangular plate element with four nodal points at its corners was developed by Ueda et al. [3,4] and Paik [5]. A stiffened panel as well as a rectangular panel partitioned by stiffeners were considered as an element. They are classified by Ueda as the ISUM of the first generation. Collapse behavior under in-plane thrust load was simulated with this element. It should be noticed that special intuition and engineering sense were required to formulate such elements, and the formulation is rather complex and difficult to understand.

The ISUM of the second generation employed eigen-functions to represent the deflection in an element [6]. The formulation had become more mathematical and easier to understand. Here, if the panel is initially completely flat or accompanied by initial deflection of a buckling mode, deflection of the buckling mode develops and fundamentally similar periodical deflection develops even beyond the ultimate strength. Such behavior can be simulated by the assumed eigen-function.

However, panels in a ship structure have initial deflection of a complex mode. In this case, localization of plastic deformation take place beyond the ultimate strength, and the deflection is no more periodical [7]. It should be noticed that such collapse behavior cannot be simulated by the ISUM elements of the first and the second generations.

The ISUM of the third generation can simulate the localization of buckling/yielding deformation [8–10]. Prior to the formulation of this element, a series of the finite element method (FEM) analyses was performed, and the collapse behavior of rectangular plates under longitudinal and transverse thrust was studied. On the basis of the simulated collapse behavior, new lateral shape function is assumed and implemented in the ISUM element. In this case, a long plate is modeled by several elements. This increases the number of nodal points compared to the ISUM elements of the first and the second

481

generations, but enables to simulate the localization of buckling/yielding deformation which takes place in actual structures. At the same time, this modeling enables the use of this plate element for the analysis of overall collapse behavior of a stiffened plate. Even if the number of nodal points increases, it is still very few when compared to the ordinary FEM modeling.

In the following, fundamental features of the new ISUM plate element of the third generation are introduced, and the results of some example calculation are briefly introduced.

B.2 FORMULATION OF NEW ISUM ELEMENT

B.2.1 COLLAPSE BEHAVIOR OF RECTANGULAR PLATE UNDER THRUST

When a thin rectangular plate is subjected to uni-axial thrust in the direction of its longer side, buckling takes place of which mode is m half-waves in the loading direction. Here, the plate is assumed to be simply supported along its four sides, and attention is focused on one half-wave. The elasoplastic large deflection behavior of this one half-wave is investigated on the basis of the results of FEM analysis. Fig. B.1 shows average stress-average strain and average stress-central deflection relationships during progressive collapse.

The breadth, b, of the plate is taken as 1000 mm, and the length, a, is changed as 600, 800, and 1000 mm. The slenderness ratio of the plate, $b/t \cdot \sqrt{\sigma_Y/E}$, is changed as 1.5, 1.95, 2.6, 3.0, and 3.5. Young's modulus, E, is set as 205.8 GPa, and the yield stress, σ_Y, as 313.6 MPa.

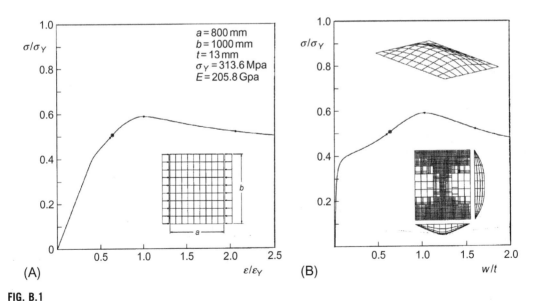

FIG. B.1

Collapse behavior of rectangular plate under thrust [5]. (A) Average stress-average strain relationship.
(B) Average stress-central deflection relationship.

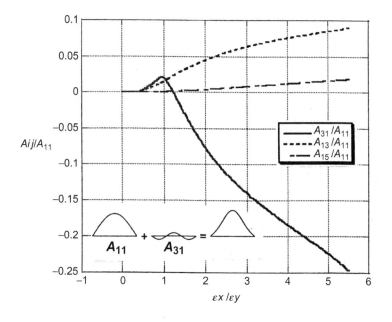

FIG. B.2

Average strain-deflection component relationships [5].

Fig. B.1A and B shows the average stress-average strain relationship and the average stress-central deflection relationship, respectively, for the plate of $a/b = 0.8$ and $b/t \cdot \sqrt{\sigma_Y/E} = 3.0$. The buckling mode for this plate is one half-wave of a sinusoidal mode.

In the case of a simply supported plate, the deflection under thrust can be expressed as

$$w = \sum_i \sum_j A_{ij} \sin \frac{\pi i x}{a} \sin \frac{\pi j y}{b} \tag{B.1}$$

The obtained relationships between the applied average strain and some deflection components are illustrated in Fig. B.2 for the case of $a/b = 0.8$ and $b/t \cdot \sqrt{\sigma_Y/E} = 3.0$. The average strain is divided by the yield strain and the deflection components by one half-wave component, A_{11}. All other components are very small and are not shown in this figure. Above the buckling load, deflection component of the buckling mode, A_{11}, increases, and so A_{13} and A_{31} do too, although the latter two components are small compared to A_{11}. Beyond the ultimate strength, A_{31} and A_{13} begin to increase more, which causes the flattening of deflection in breadth direction resulting in a roof mode deflection; see Fig. B.1B. The change in deflection from a sinusoidal mode to a roof mode is a characteristic phenomenon for the plate of $1.5 \le b/t \cdot \sqrt{\sigma_Y/E} \le 3.5$.

B.2.2 ASSUMED DISPLACEMENT FIELD IN THE ISUM ELEMENT

The ISUM rectangular plate element has four nodal points at its corners as indicated in Fig. B.3. Displacements in the element are expressed by linear interpolation of the nodal displacements as

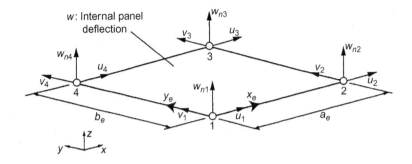

FIG. B.3

Nodal displacements of ISUM element [5].

$$\begin{cases} u = \left(1 - \frac{x}{a_e}\right)\left(1 - \frac{y}{b_e}\right)u_1 + \frac{x}{a_e}\left(1 - \frac{y}{b_e}\right)u_2 + \frac{x}{a_e}\frac{y}{b_e}u_3 + \left(1 - \frac{x}{a_e}\right)\frac{y}{b_e}u_4 \\[2mm] v = \left(1 - \frac{x_e}{a_e}\right)\left(1 - \frac{y_e}{b_e}\right)v_1 + \frac{x_e}{a_e}\left(1 - \frac{y_e}{b_e}\right)v_2 + \frac{x_e}{a_e}\frac{y_e}{b_e}v_3 + \left(1 - \frac{x_e}{a_e}\right)\frac{y_e}{b_e}v_4 \\[2mm] w_n = \left(1 - \frac{x_e}{a_e}\right)\left(1 - \frac{y_e}{b_e}\right)w_{n1} + \frac{x_e}{a_e}\left(1 - \frac{y_e}{b_e}\right)w_{n2} + \frac{x_e}{a_e}\frac{y_e}{b_e}w_{n3} + \left(1 - \frac{x_e}{a_e}\right)\frac{y_e}{b_e}w_{n4} \end{cases} \tag{B.2}$$

where a_e and b_e are the length and the breadth of the element, respectively. w_n is introduced to use this plate element as plating of stiffened plate which undergoes overall buckling.

Regarding lateral deflection, w, inside the element, the following mode is assumed:

$$w = w_n + w_\ell + w_t + w_p \tag{B.3}$$

where

w_n: linearly interpolated deflection of nodal point deflections (z-direction)
w_ℓ: deflection produced by longitudinal thrust (x-direction)
w_t: deflection produced by transverse thrust (y-direction)
w_p: deflection produced by lateral pressure (z-direction)

For longitudinal thrust, w_ℓ is expressed as follows [8–11]:

$$w_\ell = A_\ell \left\{ \left(\sin \frac{\pi x}{a} + f_{31} \cdot \sin \frac{3\pi x}{a} \right) \sin \frac{\pi y}{b} + f_{13} \sin \frac{\pi x}{a} \sin \frac{3\pi y}{b} \right\} \tag{B.4}$$

In the original new ISUM element, f_{13} was not included, whereas in the modified element [11] it is included. In Eq. (B.4), a and b are the length and the breadth of the panel, and f_{31} and f_{13} are given as follows:

$$f_{31} = \begin{cases} 0 & (\bar{\varepsilon}_x/\varepsilon_Y < 1 - p) \\ -m \cdot \ln(\bar{\varepsilon}_x/\varepsilon_Y + p) & (\bar{\varepsilon}_x/\varepsilon_Y \geq 1 - p) \end{cases} \tag{B.5}$$

$$f_{13} = p_1 \cdot ATAN\{p_2\beta \cdot (\bar{\varepsilon}_x/\varepsilon_Y) + p_3\} + p_4 \tag{B.6}$$

where

$$p = \begin{cases} 0.2(\beta - 1.0) - 0.2 & (\beta < 3.0) \\ 0.2 & (\beta \geq 3.0) \end{cases} \tag{B.7}$$

$$m = \begin{cases} 0 & (\beta < 1.0) \\ (0.5\beta - 0.5)(0.35\alpha + 0.02) & (1.0 \leq \beta < 1.5) \\ 0.25(0.35\alpha + 0.02) & (1.5 \leq \beta < 2.0) \\ (0.1\beta + 0.05)(0.35\alpha + 0.02) & (2.0 \leq \beta < 3.5) \\ 0.35(0.35\alpha + 0.02) & (3.0 \leq \beta) \end{cases} \tag{B.8}$$

$$\alpha = a/b, \quad \beta = b/t \cdot \sqrt{\sigma_Y / E} \tag{B.9}$$

and

(1) $\beta < 2.0$:

$$\begin{cases} p_1 = 0.07(1/\alpha) \\ p_2 = 0.06\beta + 0.095 \\ p_3 = 0.3\beta - 0.9 \\ p_4 = -0.01 \end{cases} \tag{B.10}$$

(2) $2.0 \leq \beta < 2.5$:

$$\begin{cases} p_1 = 0.014(1/\alpha)(\beta + 3.0) \\ p_2 = 0.02\beta + 0.175 \\ p_3 = 0.3\beta - 0.9 \\ p_4 = -0.03\beta + 0.05 \end{cases} \tag{B.11}$$

(3) $2.5 \leq \beta < 3.0$:

$$\begin{cases} p_1 = 0.014(1/\alpha)(\beta + 3.0) \\ p_2 = 0.12\beta - 0.075 \\ p_3 = -0.15 \\ p_3 = 0.3\beta - 0.9 \\ p_4 = -0.03\beta + 0.05 \end{cases} \tag{B.12}$$

(4) $\beta < 2.0$:

$$\begin{cases} p_1 = 0.084(1/\alpha) \\ p_2 = 0.285 \\ p_3 = -0.15 \\ p_4 = -0.04 \end{cases} \tag{B.13}$$

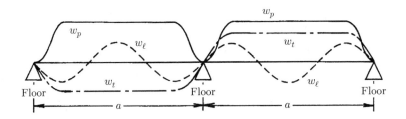

FIG. B.4

Shape functions of ISUM element [5].

For transverse thrust, w_t is expressed as

$$w_t = g \cdot w_{\text{coll}} + (1 - g) \cdot A_t \sin \frac{\pi x}{a} \sin \frac{\pi y}{b} \tag{B.14}$$

where

$$w_{\text{coll}} = \begin{cases} A_t \sin \frac{\pi x}{a_0} \sin \frac{\pi y}{b} & (0 \leq x \leq a_0) \\ A_t \sin \frac{\pi y}{b} & (a_0 \leq x \leq a - a_0) \\ A_t \sin \frac{\pi (x-a+a_0)}{a_0} \sin \frac{\pi y}{b} & (a - a_0 \leq x \leq a) \end{cases} \tag{B.15}$$

$$g = \begin{cases} 0 & (\bar{\varepsilon}_y < \varepsilon_{ycr}) \\ 1 - \exp\left(-c \cdot \frac{\bar{\varepsilon}_y - \varepsilon_{ycr}}{\varepsilon_Y}\right) & (\bar{\varepsilon}_y \geq \varepsilon_{ycr}) \end{cases} \tag{B.16}$$

Deflection w_p produced by lateral pressure is assumed as

$$w_p = \begin{cases} \frac{1}{4} A_p \left(1 - \cos \frac{2\pi x}{b}\right)\left(1 - \cos \frac{2\pi y}{b}\right) & (0 \leq x \leq \frac{b}{2}) \\ \frac{1}{2} A_p \left(1 - \cos \frac{2\pi y}{b}\right) & (\frac{b}{2} \leq x \leq a - \frac{b}{2}) \\ \frac{1}{4} A_p \left(1 - \cos \frac{2\pi (x-a+b/2)}{b}\right)\left(1 - \cos \frac{2\pi y}{b}\right) & (a - \frac{b}{2} \leq x \leq a) \end{cases} \tag{B.17}$$

Deflection components expressed by Eqs. (B.9), (B.19), and (B.22) are shown in Fig. B.4. The coefficients of deflection components, A_ℓ, A_t, and A_p are treated as an additional degrees of freedom. So, this element has 15 degrees of freedom, which are u_1, u_2, u_3, u_4, v_1, v_2, v_3, v_4, w_{n1}, w_{n2}, w_{n3}, w_{n4}, A_ℓ, A_t, and A_p.

B.2.3 GENERALIZED STRAIN AND STRESS COMPONENTS

The generalized strain vector is defined as follows:

$$\{\varepsilon\} = \{\varepsilon_x \quad \varepsilon_y \quad \gamma_{xy} \quad \kappa_x \quad \kappa_y \quad \kappa_{xy}\}^T \tag{B.18}$$

Introducing the initial deflection, w_0, which has the same components with w, the in-plane strain components are represented as

$$
\begin{cases}
\varepsilon_x = \frac{\partial u}{\partial x} + \frac{1}{2}\left(\frac{\partial w}{\partial x}\right)^2 - \frac{1}{2}\left(\frac{\partial w_0}{\partial x}\right)^2 \\
\varepsilon_y = \frac{\partial v}{\partial y} + \frac{1}{2}\left(\frac{\partial w}{\partial y}\right)^2 - \frac{1}{2}\left(\frac{\partial w_0}{\partial y}\right)^2 \\
\gamma_{xy} = \frac{\partial u}{\partial y} + \frac{\partial v}{\partial x} + \frac{\partial w}{\partial x}\frac{\partial w}{\partial y} - \frac{\partial w_0}{\partial x}\frac{\partial w_0}{\partial y}
\end{cases}
\tag{B.19}
$$

and the bending components as

$$
\begin{cases}
\kappa_x = -\left(\frac{\partial^2 w}{\partial x^2} - \frac{\partial^2 w_0}{\partial x^2}\right) \\
\kappa_x = -\left(\frac{\partial^2 w}{\partial y^2} - \frac{\partial^2 w_0}{\partial y^2}\right) \\
\kappa_{xy} = -2\left(\frac{\partial^2 w}{\partial x \partial y} - \frac{\partial^2 w_0}{\partial x \partial y}\right)
\end{cases}
\tag{B.20}
$$

On the other hand, the generalized stress vector corresponding to generalized strain vector is defined as follows:

$$
\{N\} = \{N_x \quad N_y \quad N_{xy} \quad M_x \quad M_y \quad M_{xy}\}^T
\tag{B.21}
$$

where the in-plane components are

$$
\begin{cases}
N_x = \int_{-t/2}^{t/2} \sigma_x dz \\
N_y = \int_{-t/2}^{t/2} \sigma_y dz \\
N_{xy} = \int_{-t/2}^{t/2} \tau_{xy} dz
\end{cases}
\tag{B.22}
$$

and the bending components are

$$
\begin{cases}
M_x = \int_{-t/2}^{t/2} z\sigma_x dz \\
M_y = \int_{-t/2}^{t/2} z\sigma_y dz \\
M_{xy} = \int_{-t/2}^{t/2} z\tau_{xy} dz
\end{cases}
\tag{B.23}
$$

The above generalized stress components are defined to avoid integration towards the thickness direction when stiffness matrices of elements are calculated.

B.2.4 RELATIONSHIP BETWEEN STRAINS AND NODAL DISPLACEMENTS

Substituting Eqs. (B.2), (B.4), (B.14), (B.17) into Eqs. (B.19), (B.20), strain components are expressed as follows.

$$
\{\varepsilon\} = \left([B_0] + \frac{1}{2}[C][B_2] + \frac{1}{2}[C_n][B_3]\right)\{h\}
\tag{B.24}
$$

where

$$\{h\} = \{u_1 \quad u_2 \quad u_3 \quad u_4 \quad v_1 \quad v_2 \quad v_3 \quad v_4 \quad w_{n1} \quad w_{n2} \quad w_{n3} \quad w_{n4} \quad A_\ell \quad A_t \quad A_p\}^T \tag{B.25}$$

$$[C] = \begin{bmatrix} \frac{\partial \bar{w}}{\partial x} & 0 & \frac{\partial \bar{w}}{\partial y} & 0 & 0 & 0 \\ 0 & \frac{\partial \bar{w}}{\partial y} & \frac{\partial \bar{w}}{\partial x} & 0 & 0 & 0 \end{bmatrix}^T \tag{B.26}$$

$$[C_n] = \begin{bmatrix} \frac{\partial w_n}{\partial x} & 0 & \frac{\partial w_n}{\partial y} & 0 & 0 & 0 \\ 0 & \frac{\partial w_n}{\partial y} & \frac{\partial w_n}{\partial x} & 0 & 0 & 0 \end{bmatrix}^T \tag{B.27}$$

$$[B_2] = \begin{bmatrix} 0 & 0 & 0 & 0 & 0 & 0 & 0 & 0 & 0 & 0 & 0 & 0 & G_{1\ell} & G_{1t} & G_{1p} \\ 0 & 0 & 0 & 0 & 0 & 0 & 0 & 0 & 0 & 0 & 0 & 0 & G_{2\ell} & G_{2t} & G_{2p} \end{bmatrix} \tag{B.28}$$

$$[B_3] = \begin{bmatrix} 0 & 0 & 0 & 0 & 0 & 0 & 0 & 0 & -\frac{1}{a_e}\left(1 - \frac{y}{b_e}\right) & \frac{1}{a_e}\left(1 - \frac{y}{b_e}\right) & \frac{1}{a_e}\frac{y}{b_e} & -\frac{1}{a_e}\frac{y}{b_e} & 0 & 0 & 0 \\ 0 & 0 & 0 & 0 & 0 & 0 & 0 & 0 & -\frac{1}{b_e}\left(1 - \frac{x}{a_e}\right) & -\frac{1}{b_e}\frac{x}{a_e} & \frac{1}{b_e}\frac{x}{a_e} & \frac{1}{b_e}\left(1 - \frac{x}{a_e}\right) & 0 & 0 & 0 \end{bmatrix} \tag{B.29}$$

and

$$\bar{w} = w_\ell + w_t + w_p \tag{B.30}$$

$$\begin{cases} \frac{\partial \bar{w}}{\partial x} = G_{1\ell}A_\ell + G_{1t}A_t + G_{1p}A_p \\ \frac{\partial \bar{w}}{\partial y} = G_{2\ell}A_\ell + G_{2t}A_t + G_{2p}A_p \end{cases} \tag{B.31}$$

Here, it is considered that the element is in equilibrium under the action of nodal forces. With the increase in the nodal force, the strain expressed by Eq. (B.24) becomes

$$\{\varepsilon + \Delta\varepsilon\} = \left([B_0] + \frac{1}{2}[C + \Delta C][B_2] + \frac{1}{2}[C_n + \Delta C_n][B_3]\right)\{h + \Delta h\} \tag{B.32}$$

The strain increment is defined as the difference between Eqs. (B.32) and (B.24), and is expressed as

$$\{\Delta\varepsilon\} = \left([B_0] + [C][B_2] + [C_n][B_3] + \frac{1}{2}[\Delta C][B_2] + \frac{1}{2}[\Delta C_n][B_3]\right)\{\Delta h\} \tag{B.33}$$

Here, replacing $[B_0] + [C][B_2] + [C_n][B_3]$ by $[B_1]$, Eq. (B.33) can be rewritten as follows:

$$\{\Delta\varepsilon\} = \left([B_1] + \frac{1}{2}[\Delta C][B_2] + \frac{1}{2}[\Delta C_n][B_3]\right)\{\Delta h\} \tag{B.34}$$

where

$$
[B_1] =
\begin{bmatrix}
-\frac{1}{a_e}\left(1-\frac{y}{b_e}\right) & \frac{1}{a_e}\left(1-\frac{y}{b_e}\right) & \frac{1}{a_e}\frac{y}{b_e} & -\frac{1}{a_e}\frac{y}{b_e} & 0 & 0 \\[2mm]
0 & 0 & 0 & 0 & -\frac{1}{b_e}\left(1-\frac{x}{a_e}\right) & -\frac{1}{b_e}\frac{x}{a_e} \\[2mm]
-\frac{1}{b_e}\left(1-\frac{x}{a_e}\right) & -\frac{1}{b_e}\frac{x}{a_e} & \frac{1}{b_e}\frac{x}{a_e} & \frac{1}{b_e}\left(1-\frac{x}{a_e}\right) & -\frac{1}{a_e}\left(1-\frac{y}{b_e}\right) & \frac{1}{a_e}\left(1-\frac{y}{b_e}\right) \\[2mm]
0 & 0 & 0 & 0 & 0 & 0 \\[2mm]
0 & 0 & 0 & 0 & 0 & 0 \\[2mm]
0 & 0 & 0 & 0 & 0 & 0
\end{bmatrix}
$$

$$
\begin{bmatrix}
0 & 0 & p_{11} & p_{12} & p_{13} & p_{14} & r_{11} & r_{12} & r_{13} \\[2mm]
\frac{1}{b_e}\frac{x}{a_e} & \frac{1}{b_e}\left(1-\frac{x}{a_e}\right) & p_{21} & p_{22} & p_{23} & p_{24} & r_{21} & r_{22} & r_{23} \\[2mm]
\frac{1}{a_e}\frac{y}{b_e} & -\frac{1}{a_e}\frac{y}{b_e} & p_{31} & p_{32} & p_{33} & p_{34} & r_{31} & r_{32} & r_{33} \\[2mm]
0 & 0 & 0 & 0 & 0 & 0 & C_{1\ell} & C_{1t} & C_{1p} \\[2mm]
0 & 0 & 0 & 0 & 0 & 0 & C_{2\ell} & C_{2t} & C_{2p} \\[2mm]
0 & 0 & \frac{1}{a_e b_e} & -\frac{1}{a_e b_e} & \frac{1}{a_e b_e} & -\frac{1}{a_e b_e} & C_{3\ell} & C_{3t} & C_{3p}
\end{bmatrix}
\tag{B.35}
$$

Here, p_{ij} is the function of w_{n1}, w_{n2}, w_{n3}, and w_{n4}, and r_{ij} is the function of A_ℓ, A_t, and A_p. The virtual incremental strain is obtained from Eq. (B.34) as follows:

$$
\{\delta\Delta\varepsilon\} = ([B_1] + [\Delta C][B_2] + [\Delta C_n][B_3])\{\delta\Delta h\}
\tag{B.36}
$$

B.2.5 NONLINEAR CONTRIBUTION OF SHAPE FUNCTION TO IN-PLANE STRAIN COMPONENTS

Here, r_{ij}, $C_{i\ell}$, C_{it}, and C_{ip} in Eq. (B.35) are nonlinear contributions of the assumed lateral shape function in the element to the strain components. They are derived through elastic large deflection analysis in an analytical manner.

That is, the compatibility equation of the panel with large deflection can be written as

$$
\frac{\partial^4 \Phi}{\partial x^4} + 2\frac{\partial^4 \Phi}{\partial x^2 \partial y^2} + \frac{\partial^4 \Phi}{\partial y^4} = E\left[\left(\frac{\partial^2 \overline{w}}{\partial x \partial y}\right)^2 - \left(\frac{\partial^2 \overline{w}_0}{\partial x \partial y}\right)^2 - \left(\frac{\partial^2 \overline{w}}{\partial x^2}\right)\left(\frac{\partial^2 \overline{w}}{\partial y^2}\right) + \left(\frac{\partial^2 \overline{w}_0}{\partial x^2}\right)\left(\frac{\partial^2 \overline{w}_0}{\partial y^2}\right)\right]
\tag{B.37}
$$

where Φ is Airy's stress function, and the in-plane stress components are defined as follows:

$$
\sigma_x = \frac{\partial^2 \Phi}{\partial y^2}, \quad \sigma_y = \frac{\partial^2 \Phi}{\partial x^2}, \quad \tau_{xy} = -\frac{\partial^2 \Phi}{\partial x \partial y}
\tag{B.38}
$$

Substituting Eq. (B.30) into Eq. (B.37), Airy's stress function, Φ is obtained, and then the stress components by Eq. (B.38). Assuming the plane stress condition, the strain components are given as

$$
\left\{ \begin{array}{c} \varepsilon_x \\ \varepsilon_y \\ \gamma_{xy} \end{array} \right\} = \frac{1}{E} \left[\begin{array}{ccc} 1 & -v & 0 \\ -v & 1 & 0 \\ 0 & 0 & 2(1+v) \end{array} \right] \left\{ \begin{array}{c} \sigma_x \\ \sigma_y \\ \tau_{xy} \end{array} \right\}
\tag{B.39}
$$

The strain components thus derived correspond to the nonlinear terms in Eq. (B.19). It should be noticed that nonlinear strain components produced by large deflection is derived through elastic compatibility equation. The influence of yielding on large deflection is considered through coefficients, f_{31} and f_{13} in Eq. (B.4) as well as g in Eq. (B.14).

In the above derivation, interaction between $\overline{w}$ and w_n is not considered.

B.2.6 RELATIONSHIP BETWEEN GENERALIZED STRESS AND GENERALIZED STRAIN

The relationship between generalized stress and generalized strain in the elastic range is expressed as

$$
\left\{ \begin{array}{c} N_x \\ N_y \\ N_{xy} \\ M_x \\ M_y \\ M_{xy} \end{array} \right\} = \left[\begin{array}{cccccc} D_1 & vD_1 & 0 & 0 & 0 & 0 \\ vD_1 & D_1 & 0 & 0 & 0 & 0 \\ 0 & 0 & (1-v)D_1/2 & 0 & 0 & 0 \\ 0 & 0 & 0 & D_2 & vD_2 & 0 \\ 0 & 0 & 0 & vD_2 & D_2 & 0 \\ 0 & 0 & 0 & 0 & 0 & (1-v)D_2/2 \end{array} \right] \left\{ \begin{array}{c} \varepsilon_x \\ \varepsilon_y \\ \gamma_{xy} \\ \kappa_x \\ \kappa_y \\ \kappa_{xy} \end{array} \right\}
\tag{B.40}
$$

or

$$
\{N\} = [D^e]\{\varepsilon\}
\tag{B.40'}
$$

where

$$
D_1 = \frac{tE}{1-v^2}, \quad D_2 = \frac{t^3 E}{12(1-v^2)}
\tag{B.41}
$$

The yield function is defined as

$$
f = Y - N_p
\tag{B.42}
$$

where $N_p = t\sigma_Y$ is the fully plastic sectional force and Y is the Egger's yield function [12] defined with equivalent plastic strain $\overline{\varepsilon}_p$ as

$$
Y^2 = \frac{Y_1^2(\overline{\varepsilon}_p)^A + Y_0^2 \cdot B}{(\overline{\varepsilon}_p)^A + B}
\tag{B.43}
$$

where

$$
Y_0^2 = N_P^2 \cdot (Q_n + 3|Q_{nm}| + 2.25Q_m)
\tag{B.44}
$$

$$
Y_1^2 = N_P^2 \cdot (Q_n + 0.5Q_m + \sqrt{0.25Q_m^2 + Q_{nm}^2})
\tag{B.45}
$$

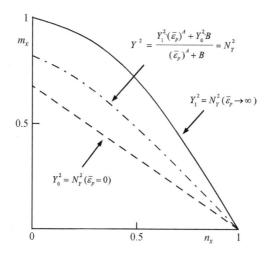

FIG. B.5

Egger's yield function [11].

and

$$Q_n = n_x^2 - n_x n_y + n_y^2 + 3n_{xy}^2$$

$$Q_m = m_x^2 - m_x m_y + m_y^2 + 3m_{xy}^2$$

$$Q_{nm} = n_x m_x + n_y m_y - 0.5 \times (n_x m_y + n_y m_x) + 3n_{xy} m_{xy}$$

$$n_x = N_x/N_P, \quad n_y = N_y/N_P, \quad n_{xy} = N_{xy}/N_P$$

$$m_x = M_x/M_P, \quad m_y = M_y/M_P, \quad m_{xy} = M_{xy}/M_P$$

$$M_P = \frac{1}{4}t^2\sigma_Y, \quad A = 0.81, \quad B = 0.0003$$

The Egger's yield function is shown in Fig. B.5.

The stress-strain relationship in the plastic range is derived considering the yield function as a plastic potential, and is given as follows:

$$\{d\sigma\} = [D^p]\{d\varepsilon\} \tag{B.46}$$

where

$$[D^p] = [D^e] - \frac{[D^e]\left\{\frac{\partial f}{\partial N}\right\}\left\{\frac{\partial f}{\partial N}\right\}^T[D^e]}{H_{SEC} + \left\{\frac{\partial f}{\partial N}\right\}^T[D^e]\left\{\frac{\partial f}{\partial N}\right\} - \frac{\partial f}{\partial \bar{\varepsilon}_p}} \tag{B.47}$$

B.2.7 **DERIVATION OF STIFFNESS EQUATION**

Applying the *Principle of Virtual Work* in an incremental form, the stiffness equation is obtained in the following form:

$$\{F\} - \{R\} + \{\Delta F\} = ([K_w] + [K_{wn}] + [K_1])\{\Delta h\} \tag{B.48}$$

where

$$\{R\} = \int_A [B_1]^T \{N\} dA \tag{B.49}$$

$$[K_w] = \int_A [B_2]^T [S][B_2] dA \tag{B.50}$$

$$[K_{wn}] = \int_A [B_3]^T [S][B_3] dA \tag{B.51}$$

$$[K_1] = \int_A [B_1]^T [D][B_1] dA \tag{B.52}$$

$$[S] = \begin{bmatrix} N_x & N_{xy} \\ N_{xy} & N_y \end{bmatrix} \tag{B.53}$$

B.2.8 **EXTENT OF THE ELEMENT**

In the case of the new ISUM element explained above, the extent of the element could be freely chosen. For example, the deflected plate can be modeled by the ISUM element as indicated in Fig. B.6. With the same shape function, the extent of the element is different. This characteristic widens the applicability of this ISUM element.

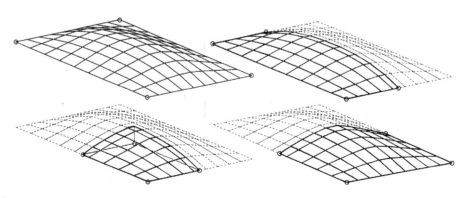

FIG. B.6

Extent of ISUM element in compressed plate [5].

B.3 ACCURACY OF THE PROPOSED SHAPE FUNCTIONS

Taking longitudinal thrust as an example, the results of calculation on uni-axially compressed rectangular plate are shown in Figs. B.7–B.9. For the analysis here, a modified new ISUM element [11] is used. In each figure, (A) shows the average stress-average stain relationships and (B) the average stress-deflection relationships. It is seen that the accuracy decreases as the aspect ratio of the plate increases and the plate becomes thinner.

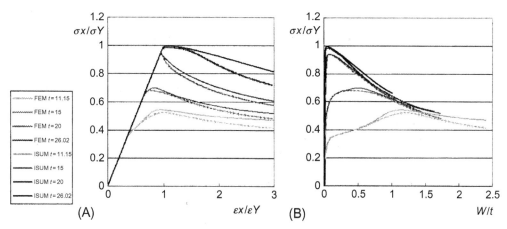

FIG. B.7

Comparison of calculated results by FEM and ISUM ($a/b = 0.6$) [5]. (A) Average stress-average strain. (B) Average stress-deflection.

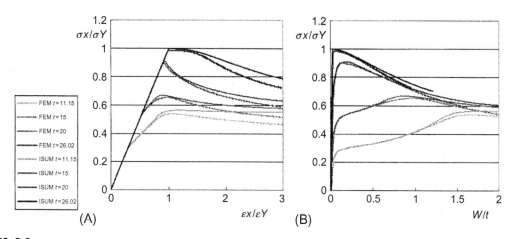

FIG. B.8

Comparison of calculated results by FEM and ISUM ($a/b = 0.8$) [5]. (A) Average stress-average strain. (B) Average stress-deflection.

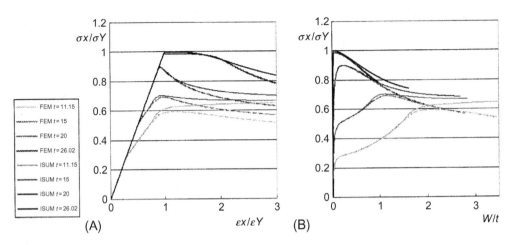

FIG. B.9

Comparison of calculated results by FEM and ISUM ($a/b = 1.0$) [5]. (A) Average stress-average strain.
(B) Average stress-deflection.

REFERENCES

[1] Ueda Y, Rashed S. An ultimate transverse strength analysis of ship structure. J Soc Naval Arch Jpn 1974;136:309–324.

[2] Ueda Y, Rashed S. The idealized structural unit method and its application on deep girder structure. Comput Struct 1984;18:277–293.

[3] Ueda Y, Reshed S, Paik JK. Plate and stiffened plate units of the idealized structural unit method. Under in-plane loading. J Soc Naval Arch Jpn 1984;156:366–376.

[4] Ueda Y, Rashed S, Adbel-Nasser Y, An improved ISUM rectangular plate element taking account of postultimate strength behavior. Mar Struct 1993;6:139–172.

[5] Paik JK. Advanced idealized structural unit considering excessive tension-deformation effects. J Hydrospace Tech 1995;1:125–145.

[6] Ueda Y, Masaoka K. Ultimate strength analysis of thin plated structures using Eigen-function (2nd report). Rectangular plate element with initial imperfection. J Soc Naval Arch Jpn 1995;178:463–471.

[7] Yao T, Fujikubo M, Yanagihara D, Murase T. Postultimate strength behavior of rectangular plate subjected to uni-axial thrust. In: Proc 11th int offshore and polar eng conf. Stavanger, Norway; 2001. p. 390–397.

[8] Fujikubo M, Kæding P, Yao T. ISUM rectangular plate element with new lateral shape function (1st report). Longitudinal and transverse thrust. J Soc Naval Arch Jpn 2000;187:209–219.

[9] Kæding P. Development of new idealized structural unit method for the collapse analysis of stiffened plate structure [Doctoral thesis]. Hiroshima University; 2001.

[10] Fujikubo M, Kæding P. New simplified approach to collapse analysis of stiffened plates. Mar Struct 2002;15(3):251–283.

[11] Nagashio Y. Consideration on shape function of new ISUM element [Master thesis]. Osaka University; 2005.

[12] Egger H, Kroplin B. Yielding of plates with hardening and large deformations. Int J Num Mech Eng 1978;12:737–750.

STRUCTURAL CHARACTERISTICS OF REPRESENTATIVE SHIP AND SHIP-LIKE FLOATING STRUCTURES

C.1 BULK CARRIERS

C.1.1 STRUCTURAL CHARACTERISTICS

A bulk carrier has a top side tank and a hopper side tank which are connected by side frames as illustrated in Fig. C.1. The double bottom structure consists of bottom plating, inner bottom plating, girders, and floors. The number of holds is usually 5–9, and each hold has a large opening in the deck part for loading and unloading cargoes.

Dry bulk of cargoes are mainly loaded, ranging from light to heavy. The former cargoes are grains and coals, and the latter ores. Light cargo is loaded homogeneously, as indicated in Fig. C.2, while heavy cargo is loaded alternately, as indicated in Fig. C.3. For cargo with intermediate weight, block loading is performed; see Fig. C.4. In the light condition without loading cargoes, it is usual to fill the ballast water in the cargo hold located at mid-ship. In a rough sea, other cargo holds are also filled by sea water to ensure necessary draft.

The top plate of a hopper side tank is inclined about 45–50 degrees from the horizontal level considering the convenience in grain cargo unloading. On the other hand, the bottom plate of a top side tank is inclined around 30 degrees, which is a rest angle of the piled-up cargo, to prevent the cargo from breaking loose. In both tanks, transverse webs are provided at every several side frames to support longitudinal stiffeners in the tanks. The transverse bulkhead is made of corrugated plate, and stools are provided at its lower and upper edges in many cases.

C.1.2 ATTENTION FROM A STRUCTURAL STRENGTH VIEWPOINT

Relatively high stresses are produced at the following locations.

C.1.2.1 Double bottom structure

Double bottom is a grillage structure composed of girders and floors, and the stress in the double bottom is determined by the cargo/ballast water pressure on inner bottom plating as well as water pressure on bottom plating. Under the homogeneous loading condition, cargo pressure and bottom water pressure cancel each other, and the bending stress is small. On the other hand, under the alternate loading condition, the double bottom is nearly simply supported along the transverse bulkheads, and large bending deformation, and so large bi-axial bending stress, are produced in the bottom and inner bottom plating at the center of the hold.

495

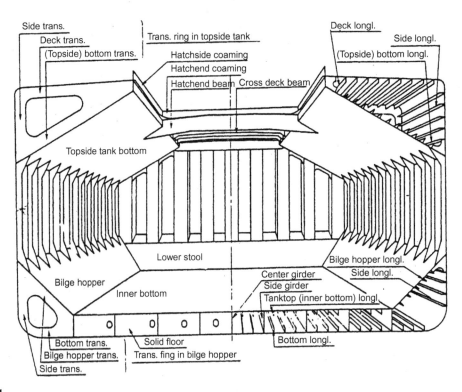

FIG. C.1

Structure of bulk carrier.

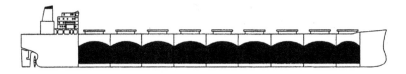

FIG. C.2

Homogeneous loading.

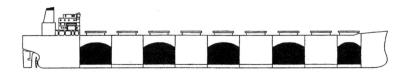

FIG. C.3

Alternate loading.

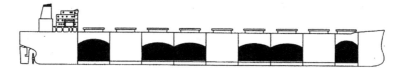

FIG. C.4

Block loading.

The longitudinal edges of the double bottom are connected to hopper side tanks. Because of this, torsional stiffness of the hopper side tank as well as in-plane stiffness of transverse webs in the hopper side tank largely affect the boundary condition against edge rotation of the connecting plane. If the stiffness is high, bending stress in the central part becomes low.

The distributed loads on double bottom structure are transmitted to hopper tanks or transverse bulkheads at the ends of girders and floors as shear forces. Therefore, it is important to assess the strength of girders and floors at their ends near hopper side tanks and transverse bulkheads.

On the hopper side tank, shear force and bending moment act from the double bottom. At the same time, cargo pressure acts on its slanted top plate and water pressures on bottom, bilge, and side shell plating. Transverse webs in the hopper side tanks are placed connecting to the floors and the web, and are in a complex stress state. The stress is high especially at the corner of the transverse web, and the strength against buckling and fatigue cracking has to be carefully examined.

In addition to the above, a knuckle line is formed at the connection of the inner bottom plate and the slanted top plate of a hopper side tank. Stress concentration also takes place at the knuckle line.

C.1.2.2 Side frame

Due to the lateral pressure on side shell plating, high bending stress is produced at a mid-span part as well as at both ends connected to a top side tank and a hopper side tank. This has to be carefully examined in the design stage.

C.1.2.3 Stool

It is usual to provide stools at the upper and the lower ends of corrugated transverse bulkheads. Especially when one hold is empty and the adjacent hold is fully loaded, high stress is produced at the connections between stool and corrugated plate or the connection between lower stool and inner bottom plating. This has to be examined carefully.

C.1.2.4 Others

When a block loading in Fig. C.4 is carried out, large tensile load acts on the transverse bulkhead located between two loaded cargo holds due to the doubled cargo weights. In this case, high stress is produced at the connection between corrugated bulkhead and cross deck.

When the ballast/cargo hold is filled with ballast water, higher pressure acts on bulkheads and side shell plating compared to cargo loading, and higher stress is produced at connections.

C.2 SINGLE HULL TANKER

After severe pollution of sea environment caused by the grounding accident of *Exxon Valdes*, the Marine Environmental Protection Committee of the International Maritime Organization (IMO) adopted new regulations to strengthen the tanker structure. According to these regulations, it became an obligation to make a tanker structure as double hull structure. So, no single hull tanker exists as a newly built tanker. However, single hull tankers will still exist for a while, although they are going to fade out.

C.2.1 STRUCTURAL CHARACTERISTICS

According to the international regulation to protect the sea from spilled oil pollution, tank capacity is limited to a certain volume. To keep this regulation, tanker ship is partitioned into many tanks by longitudinal and transverse bulkheads. In the case of tankers with relatively long tanks, a transverse members-main structural system is adopted. In this system, several transverse rings are provided between transverse bulkheads as main transverse strength members. A typical transverse ring is illustrated in Fig. C.5, where transverse (or horizontal) girders are provided at the deck and bottom in the center tank. In the wing tank, vertical girders are provided along longitudinal bulkhead and side

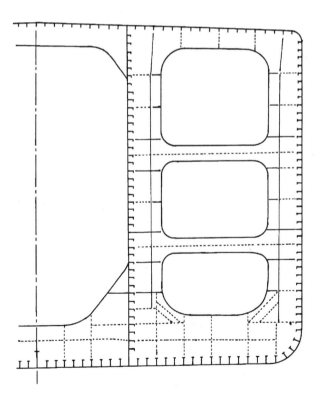

FIG. C.5

Structure of a single hull tanker (transverse members-main structural system).

shell plating in addition to transverse girders at the deck and bottom. Cross-ties are also placed between two vertical girders as indicated in Fig. C.5. The transverse ring supports outer shell plating subjected to lateral pressure and has a function to keep the shape of the cross-section.

On the other hand, long tanks cannot be adopted in the case of a large tanker because of the limitation on tank capacity. Because of this, horizontal members-main structural system is adopted in the case of a large tanker, in which horizontal girders in the longitudinal direction are the main girders, and the transverse rings are supported by the horizontal girders; see Fig. C.6. No cross-tie is provided, but horizontal girders are at the height of cross-ties in the case of transverse members-main structural system.

C.3 ATTENTION FROM A STRUCTURAL STRENGTH VIEWPOINT

In both structural systems, the central tanks are fully loaded and side tanks are empty in full load condition, while the central tanks are empty and the side tanks are fully ballasted in ballast condition. In both cases, large shear deformation appears in the transverse rings in side tanks, and high shear stress is produced. At the same time, high normal stress is produced at corners of transverse rings. So, buckling strength and fatigue strength have to be carefully examined at the corners.

In addition to this, in case of a horizontal members-main structural system, high stress is produced at the corners of horizontal girders indicated in Fig. C.6. The horizontal girder is subjected to also longitudinal bending, and normal stress due to longitudinal bending is produced in the horizontal girders. Therefore, influence of longitudinal bending loads has to be considered when several tanks are analyzed.

C.4 DOUBLE HULL TANKER

Newly built tankers after 1995, the year of adoption of international regulations at the IMO, are double hull tankers. Structural characteristics of the double hull tanker are known from Fig. C.7. It is seen that side and bottom structures are double structures of which width is 3–5 m. All the double hull parts are used as ballast tanks.

A large hopper is provided at the bilge part to reduce stress concentration when transverse strength is examined. The shape of the transverse ring is similar to that of a single hull tanker, although the side and bottom parts are the doubled structures.

One or two cross-ties (strut) are placed in the center tank or side tank as a transverse strength member. In the double side structure, 3 or 4 side stringers (horizontal girders) are provided for inspection in the case shown in Fig. C.7A. At the same level, horizontal girders are provided on transverse bulkhead.

Also in the case shown in Fig. C.7B, 3 or 4 side stringers (horizontal girders) are provided on the side part of the transverse ring in the double side structure.

Recently, new notations by classification societies often specify to determine the dimensions of structural members on the basis of the results of direct calculation. From the viewpoint of structural strength, the followings have to be examined, which are similar to those for a single hull tanker.

1. Strength of connection between horizontal girder on transverse bulkhead and side stringer in double side and its stiffening.
2. Strength of connection between R-part of the bilge hopper and side girder/stringer and its stiffness.

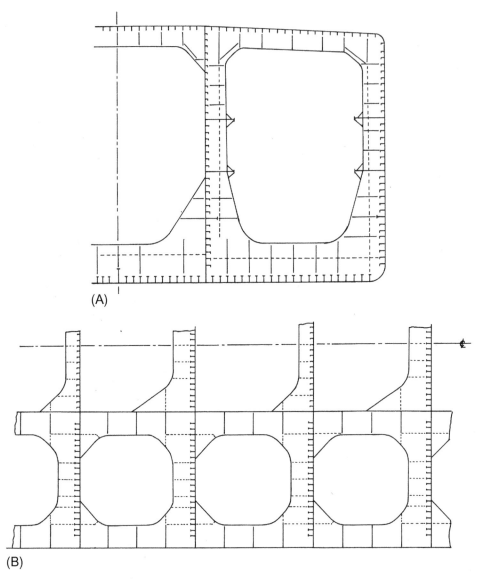

(A)

(B)

FIG. C.6

Structure of a single hull tanker (horizontal members-main structural system). (A) Mid-ship section.
(B) Horizontal girder.

3. Strength of connection between longitudinal stiffeners on side shell plating and transverse bulkhead.
4. Structural details of the end of cross-tie (strut).

The hull weight of a double hull tanker increases by 25% compared to a single hull tanker. The fuel tanks may also become doubled structure in future in addition to cargo tanks.

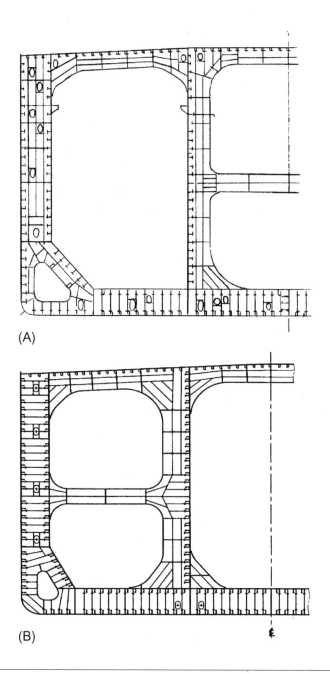

(A)

(B)

FIG. C.7

Mid-ship section of a double hull tanker. (A) Cross-tie in center tank. (B) Cross-tie in wing tank.

C.5 CONTAINER SHIP

C.5.1 STRUCTURAL CHARACTERISTICS

In a container ship, the volume and the shape of the hold are so designed that the maximum number of containers can be loaded. Consequently, a container ship has a large opening in the deck and the width of the double side structure is the minimum. Because of this, torsional stiffness of the hull girder is low.

In case of the old type container ship, one or two longitudinal girders are placed in the upper part of the cargo hold, on which hatch covers are mounted; see Fig. C.8B. However, it became possible to remove longitudinal girders and make smaller space between containers since the gusset-less hatch cover was approved; see Fig. C.8A.

Another change is the increase in the number of containers to be carried. This increases the ship size and therefore the thickness of deck plating. Recently, high tensile steel of $\sigma_Y = 40$ kgf/mm^2 has been used, of which the thickness is 60–80 mm.

The container ship has a thin hull form to attain a higher speed. At the same time, the container ship has a large flare to increase number of containers and to fit mooring equipments by increasing deck area. Furthermore, special apparatus is included, such as a cell-guide and rushing bridge.

C.5.2 ATTENTION FROM A STRUCTURAL STRENGTH VIEWPOINT

The hull of container ship is of an open cross-section due to large deck openings. Because of this, torsional stiffness is low, and large torsional deformation is produced with high shear and warping stresses. Therefore, it is important to assess the torsional strength in wave. Furthermore, torsional deformation is constrained at the corner of the opening and stress concentration takes place here. Because of this, special attention has to be paid to the shape and local thickness of the corner parts.

Recently, guidance is given by the classification societies on the finite element method (FEM) analysis for a whole ship, and the strength assessment can be performed including fatigue strength for the corner of the deck opening on the basis of the calculated results. Another issue is the large

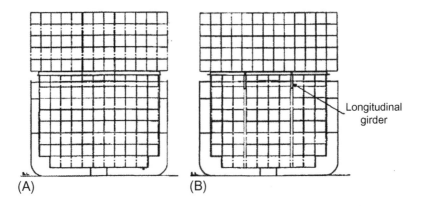

(A)   (B)

FIG. C.8

Hold system of a container ship. (A) One-line loading system. (B) Three-line loading system.

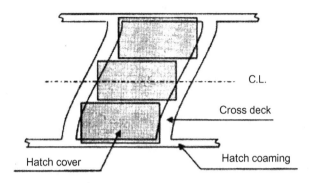

FIG. C.9

Deformation of deck openings due to torsion.

deformation of deck openings. This causes aberration between hatch opening and hatch covers as indicated in Fig. C.9. So, deformation of deck opening also has been carefully examined.

In the case of post-Panamax and over-Panamax size container ships, deflection of cross-deck due to forces in the longitudinal direction come from containers is not negligible, and has to be carefully examined.

It becomes important to assess the structural strength of plating and stiffeners in bow structure against bow flare slamming, since a container ship has a large flare and navigates with high speed. Bow flare slamming produces also additional bending moment in sagging. For this additional bending moment, buckling strength of deck structures has to be examined.

The container is not so heavy and the draft of the container ship is small. Consequently, stresses at the double bottom and transverse bulkhead are not so high. However, local stiffening is necessary in some locations. So, the necessary thickness has to be examined from the viewpoints of buckling and fatigue.

C.6 PURE CAR CARRIER

C.6.1 STRUCTURAL CHARACTERISTICS

A pure car carrier (PCC) is a ship that carries only cars, and for this, many decks are provided to park cars. Drivers drive in cars onto the car decks passing on ramp ways between car decks as well as between a quay and the ship. Since the cars move on the deck, it is required to make the numbers of pillars and bulkheads as small as possible. In the PCC, water-tight bulkheads are provided only under the strength deck. In old car carriers, partial transverse bulkheads were provided above the strength deck to prevent racking deformation of the transverse cross-section as indicated in Fig. C.10. Recently, however, deep web frames of a cantilever type have been provided instead of partial bulkheads, as indicated in Fig. C.11.

To increase the number of cars as much as possible, distances between car decks as well as height of deck beams are reduced as much as possible and small brackets are used at the ends of the deck beams.

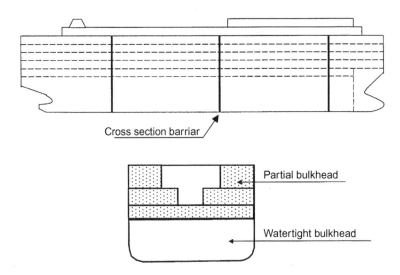

FIG. C.10

Pure car carrier in old times.

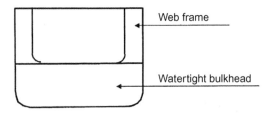

FIG. C.11

Pure car carrier of web frame system.

C.6.2 ATTENTION FROM A STRUCTURAL STRENGTH VIEWPOINT

Because of the small numbers of transverse bulkheads, large racking deformation is produced in the transverse cross-section as illustrated in Fig. C.12. Due to the racking deformation, high stress is produced at the lower end of the web frame, and the fatigue strength at this location has to be carefully examined at the design stage.

From a transverse strength viewpoint, strength of the deck beams has to be carefully examined. At the same time, their vertical deflection produced by static and dynamic car weight should also be examined considering the influence of initial imperfection due to welding. This is to ensure clearance between the bottom of the deck beam and the car roof. In such manner, there exist more restrictions in structural arrangements in the case of a PCC compared to bulk carriers and tankers.

In addition to the above, attention should be paid to the problem of bow flare slamming, since PCCs also have large bow flare structure as container ships, although their ship speed is not so high. In this regard, the thickness of plating and dimensions of stiffeners in bow structures should be examined carefully.

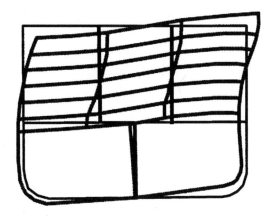

FIG. C.12

Racking deformation of cross-section of pure car carrier.

C.7 LNG CARRIER (MOSS-TYPE SPHERE TANK SYSTEM)
C.7.1 STRUCTURAL CHARACTERISTICS

Independent spherical tank system was developed by Moss-Rosenberg Co. Ltd, in Norway, which is usually called "Moss-type."

As indicated in Fig. C.13, the hull girder is of a double hull structure, and a spherical tank is connected along its equator to a circular skirt which is placed on the double bottom. The spherical tank and a part of the skirt are covered by a heat insulator. In such a manner, the deformation of the hull girder does not easily transmit to the sphere tank.

A tank itself has no stiffener and curved thick plates are welded to make the spherical tank. The material of the tank is aluminum alloy or 9% nickel steel. A dome is provided at the tank top to introduce pipes into the tank. The pipes and pumps are placed and protected in the pipe tower, which is installed at the center of the spherical tank.

C.8 ATTENTION FROM A STRUCTURAL STRENGTH VIEWPOINT

The Moss-type tank system is a simple axi-symmetric structure as indicated in Fig. C.13, and it is possible to perform accurate analysis considering axi-symmetry condition.

In the spherical structure, buckling strength has to be examined when the working compressive membrane stress is high. In general, liquid cargo collides to the tank wall with relatively high speed when the tank is half-filled by liquid cargo and the natural frequency of the liquid cargo coincides with the rolling frequency of the ship's hull. This is called sloshing. In the Moss-type system, however, liquid smoothly moves along the tank surface since the surface is sphere shaped and no stiffener is fitted. Therefore, the influence of sloshing is considered small in the Moss-type system.

A circular skirt supports the weight of the tank and liquefied natural gas. So, the skirt has to have sufficient strength so that the skirt does not buckling. Another problem in the skirt is thermal

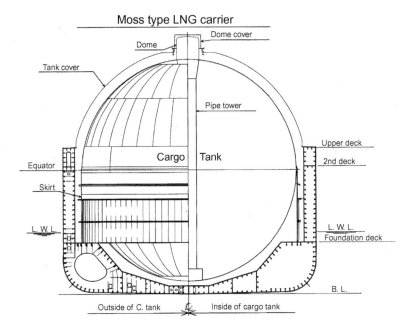

FIG. C.13

Moss-type spherical tank.

deformation. When liquefied natural gas is loaded in the tank, the temperature at top of the skirt is very low even if the sphere is covered by heat insulator. At the same time, the bottom of the skirt is connected to the double bottom, and is a room temperature. Because of this, large heat shrinkage is produced at the top of the skirt, while it is almost zero at its bottom. In other words, skirt is so designed that the heat shrinkage is absorbed by the skirt in a Moss-type system.

C.9 LNG CARRIER (MEMBRANE TANK SYSTEM)
C.9.1 STRUCTURAL CHARACTERISTICS

Membrane-type system stands for the structural system illustrated in Fig. C.14. In this system, the inner shell is covered by heat insulator which has strength against pressure loads. On the insulator, thin metal membrane made of stainless steel or nickel alloy is stuck so that the liquid-tightness is ensured. The insulator is 200–400 mm thickness with double layers structure, between which secondary protecting wall is placed. The heat insulator and the secondary protecting wall play a role to protect the ship's hull from extremely low temperature of the liquefied natural gas; see Figs. C.14 and C.15. The design concept of the membrane-type system is to absorb thermal shrinkage by the shape of membrane and the characteristics of heat insulator and to support the weight of tank and liquid cargo by the inner hull structure.

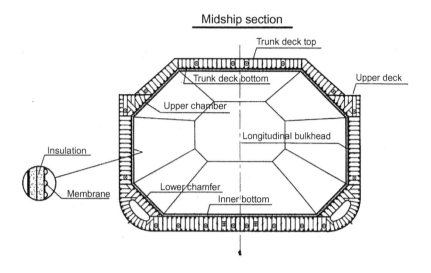

FIG. C.14

Mid-ship section of membrane-type LNG ship.

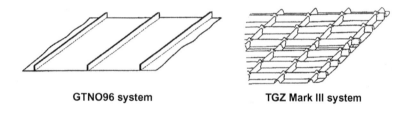

FIG. C.15

Shape of membrane structures.

In case of liquefied membrane-type system, it is necessary to flatten all the inner surface of the tank to fit a heat insulator, and the hull is of a double hull structure covering the tank.

At the beginning when the Moss-type system has developed in the 1960s, various types were proposed. Now, two types are the main of which licenses are owned by Gas Transport & Technigas Co. Ltd. One is the GT N096 system and the other is the TGZ Mark III system. Both systems are compared in Table C.1.

The characteristics of the membrane-type system which are common in both systems are as follows:

(1) Not strength but liquid-tightness is required to the membrane sheet. Therefore, use of material could be the minimum.

(2) As the membrane is thin, no temperature gradient occurs in the thickness direction of the membrane. This results in smooth shrinkage or elongation of the membrane in its in-plane direction.

Table C.1 Kinds of membrane type

	GT NO96 System (Gas Transport System)	TGZ Mark III System (Technigas System)
Membrane material	Imper steel (36% nickel steel)	Stainless steel (SUS304L steel)
Secondary protection wall	Imper steel membrane	Triplex
Heat insulation	Plywood made box filled with silicon treated perlite	Reinforced polyurethane form and plywood
Mechanism to absorb heat shrinkage	Low expansion rate of the imper steel	Corrugation in membrane

(3) No stiffening is necessary for the tank itself since the weight of the tank and the liquid cargo is supported by the inner hull structure through heat insulator.

(4) The inside of a hull is efficiently used as a tank space. Therefore, the ship size could be compact. It is also an advantage that the visibility from the bridge is not hindered.

C.9.2 ATTENTION FROM A STRUCTURAL STRENGTH VIEWPOINT

As mentioned above, the weight of the tank and the liquid cargo is supported by the inner hull structure through heat insulator. Therefore, strength design has to be performed considering hull structure, heat insulator, and membrane systematically. Another issue is thermal stress which is produced by the temperature difference by 150°C between tank inner wall and inner hull structure. This has to be examined.

Here, large sloshing loads could act on the inner wall of a tank since there exists no structural members to prevent motion of liquefied natural gas. So, some restriction has to be specified on the level of the liquid cargo to prevent sloshing in a certain case.

In the construction stage, accuracy of the thickness of heat insulator should be carefully kept and the work to place them on the inner surface of the tank should be carefully controlled to keep flatness of the surface of placed insulator. Otherwise, unexpected stress may be produced in the membrane member.

Regarding the ship hull structure, connection at inner bottom plating and slanted hopper plate, which is specially called "Lower Chamfer," as well as Lower Chamfer and longitudinal bulkhead should be carefully designed and constructed to avoid the initiation of fatigue cracking.

C.10 ORE CARRIER
C.10.1 STRUCTURAL CHARACTERISTICS

The side tank structure of an ore carrier has almost the same structure as a single hull tanker of a transverse member-main system, although the center hold has a double bottom. The lower part of the longitudinal bulkhead is in many cases inclined toward the center for the convenience of unloading of ore cargoes as indicated in Fig. C.16.

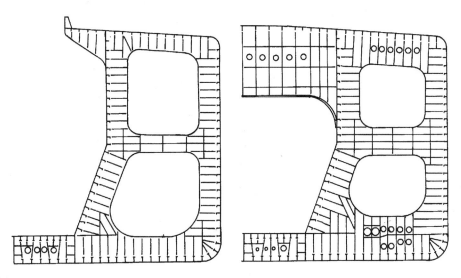

FIG. C.16

Cross-section of a ore carrier.

In general, the number of transverse bulkheads in the center hold is fewer than those in side tanks. Corrugated plate is usually used for transverse bulkheads in the center hold. The breadth of the center hold is narrow and its height is large compared to the holds of ordinary bulk carriers since heavy ore cargo is loaded. Because of the same reason, number of floors in the center hold is double of that of the transverse rings in the side tank.

C.10.2 ATTENTION FROM A STRUCTURAL STRENGTH VIEWPOINT

As in the case of tanker, side tanks are empty under full load condition and the center hold is empty under the ballast condition. Because of this, large shear deformation takes place in the transverse rings in side tanks, and high stresses are produced at the root of the cross-tie and the ends of deck beams in the transverse rings in side tanks. The shear deformation makes relative displacement between the longitudinal bulkhead and the ship side, which is enlarged by the knuckle in the longitudinal bulkhead.

Furthermore, half of the floors in the double bottom of the center hold are not connected to the transverse rings in the side tanks, which makes transmittal mechanism of forces complex and high stress is generally produced at the end of floors which are not connected to transverse rings in the side tanks.

Under the full load condition, the strut in the side tank is subjected to axial thrust due to water pressure on side shell plating and ore pressure on longitudinal bulkhead, and the occurrence of buckling of the strut has to be examined. The buckling of the cross deck has to also be examined due to the action of water pressure on side shell plating.

C.11 FLOATING PRODUCTION, STORAGE, AND OFFLOADING SYSTEMS
C.11.1 STRUCTURAL CHARACTERISTICS

Floating production, storage, and offloading systems (FPSOs) have become the primary method for many offshore oil and gas producing regions around the world. FPSO for liquefied natural gas and the related floating facilities are called FLNG. They are mostly ship-shaped, and either converted from existing tankers or purpose-built. The hull structural scanting design for tankers may be applicable FPSOs. However, FPSOs have their own unique characteristics from various operational requirements. For instance, stools that support topside plants and mooring systems (external turrets, spread moorings, etc.) are installed on the upper deck. The helicopter deck is also installed and the accommodation space is increased with the increase of the crews (Fig. C.17).

C.11.2 ATTENTION FROM A STRUCTURAL STRENGTH VIEWPOINT

One of the most important aspects of FPSO structural design and assessment is the hull girder ultimate strength under extreme load conditions. The weight of the topside plants is included in the working loads. For the case of turret mooring, the head sea condition is the primary wave condition due to a weather vane. The sea state at the installation area is used as the design condition.

Because of a relatively long-term service period out of the dock compared to normal ships, for example, 20 years for the long case, the condition assessment of corrosions and other deteriorations are particularly important. The structural safety assessment against fire and explosions and collisions with shuttle tankers are also the major strength design aspect.

FIG. C.17

FPSO.

Source: Courtesy of MODEC.

Index

Note: Page numbers followed by f indicate figures and t indicate tables.